Classic American Tractors

THE ALLIS-CHALMERS STORY

C.H. WENDEL

Published by

Krause Publications, a division of F+W Media, Inc.
700 East State Street • Iola, WI 54990-0001
715-445-2214 • 888-457-2873
www.krausebooks.com

To order books or other products call toll-free 1-800-258-0929
or visit us online at www.krausebooks.com

ISBN-13: 978-0-87349-927-9
ISBN-10: 0-87349-927-1

Printed in the United States of America

FOREWORD

Probably one of the greatest worries for the writer, especially the historical writer, is that upon completion of a project, it will be discovered that an important feature was omitted. Perish the thought! In the case of this project however, such a possibility does indeed exist. Early in our research it became evident that the awesome and prodigious line of products leaving the Allis-Chalmers factories presented a formidable challenge.

The challenge soon took on a plural connotation — challenges abounded! Sheer numbers of machines were of course one challenge — just the crawler tractor and wheel tractor lines with their countless variations number into the hundreds. Beyond the challenge of sheer numbers within a product line however, came the challenge of diverse lines — products totally unrelated to each other, and in fact, belonging to totally independent industries! While one specific section might cover balers — something with which everyone in agriculture is familiar — an alphabetical listing of A-C products might well follow with the company's Beach Sanitizer, a unique tractor-drawn machine illustrated in this volume. From there however, the research goes to the Betatron, the first X-ray machine used for the treatment of cancer, and a machine developed by Allis-Chalmers. Thus, the challenge of compiling an in-depth history of Allis-Chalmers as viewed through their products required an immense amount of research, often into areas with which the agricultural historian is unfamiliar.

Gaining a reasonable adroitness and proficiency with these many specialized areas required uncounted hours of research not usually associated with the history of a company. So extensive and so enormous was the Allis-Chalmers line at its zenith that our research efforts occasionally strayed into British, French, and German publications that were able to provide specific engineering or historical data. In this connection, a most ironic and unexpected phenomenon appears — European engineers or engineering journalists seem to have, for the most part, almost totally ignored the prestigious and outstanding contributions of A-C men like Edwin Reynolds to the whole field of mechanical engineering. After an extensive study of many different articles and commentaries, the Author is content in believing that the Reynolds philosophy of building an engine or machine virtually from a pencil sketch on the back of an envelope was just too extreme and too radical a method of machine design for most of his contemporaries to follow. In England and on the Continent, good engineering required stacks of drawings and empirical formulas as a prerequisite. The fact was, that while the engineering academics, whether residing in America or elsewhere, were poring over their designs and drawings, Reynolds and his team were getting the machines built and worrying about the fine points as an afterthought!

Given the wide range of company activities, precise details of many company projects are very limited, and in a few cases, may not have been presented herein. Occasionally, the matter of whether or not to include certain material or photographs came down to subjective judgement on the part of the Author. It is hoped however, that no area of major importance in the daily comings-and-goings of Allis-Chalmers has been overlooked.

In some instances, information regarding a certain machine or product line may have been minimal or completely lacking. In other cases, the available information may have even been contradictory, and as previously indicated, the simple lack of space may have prevented the inclusion of other portions of the A-C product line. Should any readers have additional information or literature which might be helpful to further editions of this book, their assistance will be greatly appreciated. The chance for errors, typographical or otherwise, always exists, and wherever these may have occured, kindly inform the Author so that proper corrections can be made. Write: C. H. Wendel, R.R. 1, Box 28-A, Atkins, Iowa 52206.

Introduction

Perhaps to its own harm, Allis-Chalmers has at times represented the ultimate in diversification. Even in the early days when Decker & Seville was scarcely established, the firm built stone buhr flour mills, along with manufacturing gray iron castings, and numerous other activities. Before long, sawmills were added, then steam engines, then rock crushers, stamp mills, and all manner of mining equipment. The 1901 merger that resulted in Allis-Chalmers Company gave substance to the new trademark which advertised A-C the "Company of the Four Powers — Steam, Gas, Water, and Electricity."

In pursuance of "the Four Powers," Allis-Chalmers was, from its beginnings, heavily involved in purchase of various companies which helped to perpetuate this avowed prominence in virtually every field of heavy industry. Charting the company's progress in this regard is nearly impossible to achieve through the usual methods of body copy, captions, and photographs. In desperation, a flow chart of Allis-Chalmers activities has been included on Page Five of this volume to better delineate the many and varied activities of Allis-Chalmers, its immediate predecessors, and its acquired manufacturing operations. Hopefully, this chart will help to simplify the confusion which now exists regarding who-did-what and "whodunit." The flowchart does not include companies acquired by Allis-Chalmers International after its formation in 1957. Some of the companies acquired by A-C International include: Thomas C. Pollard Pty. Ltd., Newcastle, Australia; Industrial Dufermex, S.A., Mexico City, Mexico; Vendor; S.p.A., Cusano, Italy; Establissements de Constructions Mecaniques de Vendeuvre S.A., Paris, France; Allis-Chalmers-Rumely Ltd., Guelph, Ontario, Canada.

Although Allis-Chalmers entered some very interesting and challenging areas of engineering, this particular research project was concerned with Allis-Chalmers tractors and farm equipment as the first priority. A study of company activities reveals however, that engineering achievements in other areas of the company often resulted in major improvements within existing farm equipment lines. A primary example is Allis-Chalmers' pioneering use of drop forgings and high-strength steel sections wherever possible in its farm machinery. These improvements, plus the use of high tensile strength and light weight castings had seen frequent application in other areas of the company, particularly involving steam turbines and other machines. The expertise of A-C engineers in these specialized areas of metallurgy proved to be very beneficial as the company entered the 1930s with an entirely new line of tractors and implements.

Many stories could have been included in this book — stories by old employees about certain phases of activity. One such story has it that the forerunner to the famous WC tractor was built up of an old car engine and miscellaneous other parts that could be scrounged together for the effort. Reports have it that lots of conversations passed before the Board of Directors sanctioned further development. As things turned out, the WC put Allis-Chalmers firmly into the farm tractor business. Sadly, no photographs of these early attempts have been located. The very few photos illustrating early or experimental machines have for the most part been totally devoid of any information that might connect them with later developments.

Finally, there is bound to be an occasional reader who will not find a particular A-C product within. From the sheer standpoint of numbers, this is to be expected — just the company index of farm equipment items requires over 30 pages!

For the convenience of the reader, special sections have been included on some acquired companies such as Advance-Rumely, Monarch, and others. Also included at the back of this volume is an extensive serial number listing for Allis-Chalmers tractors and certain other products. Included too are listings for much of the Advance-Rumely line. These lists have been compiled with extreme care, but neither the Author nor the Publisher can guarantee their reliability. Every possibility exists that the original lists may contain heretofore undetected errors, and despite our own efforts, additional errors may have been unintentionally added. Therefore, where absolute accuracy is imperative, the reader is referred to the original manufacturer or the successor.

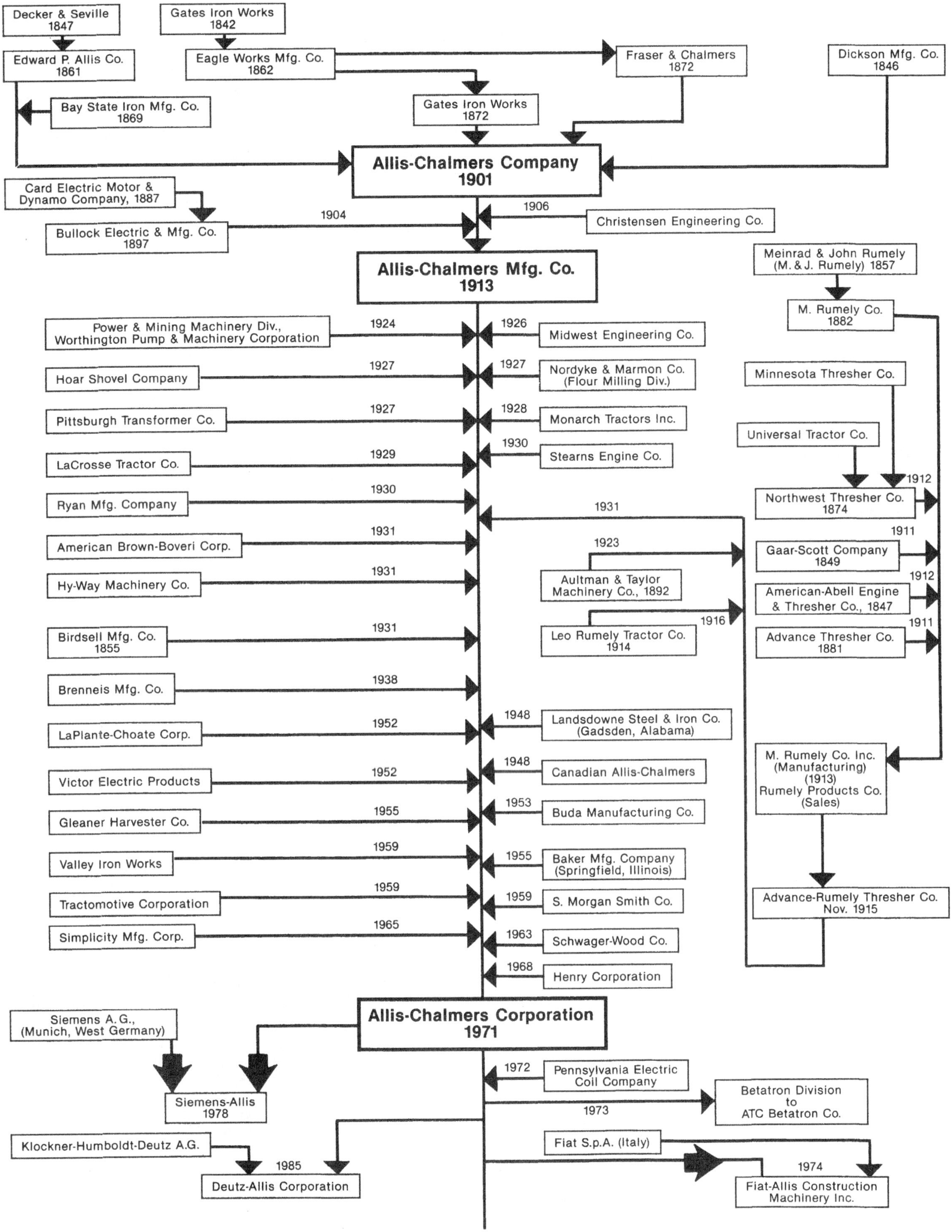
Decker & Seville 1847
Gates Iron Works 1842
Edward P. Allis Co. 1861
Eagle Works Mfg. Co. 1862
Fraser & Chalmers 1872
Dickson Mfg. Co. 1846
Bay State Iron Mfg. Co. 1869
Gates Iron Works 1872
Allis-Chalmers Company 1901
Card Electric Motor & Dynamo Company, 1887
1904
Bullock Electric & Mfg. Co. 1897
1906
Christensen Engineering Co.
Allis-Chalmers Mfg. Co. 1913
Meinrad & John Rumely (M. & J. Rumely) 1857
M. Rumely Co. 1882
1924
Power & Mining Machinery Div., Worthington Pump & Machinery Corporation
1926
Midwest Engineering Co.
1927
Hoar Shovel Company
1927
Nordyke & Marmon Co. (Flour Milling Div.)
Minnesota Thresher Co.
1927
Pittsburgh Transformer Co.
1928
Monarch Tractors Inc.
Universal Tractor Co.
1929
LaCrosse Tractor Co.
1930
Stearns Engine Co.
1912
1930
Ryan Mfg. Company
1931
Northwest Thresher Co. 1874
1931
American Brown-Boveri Corp.
1923
1911
Gaar-Scott Company 1849
1931
Hy-Way Machinery Co.
Aultman & Taylor Machinery Co., 1892
1912
American-Abell Engine & Thresher Co., 1847
1916
1931
Birdsell Mfg. Co. 1855
Leo Rumely Tractor Co. 1914
1911
Advance Thresher Co. 1881
1938
Brenneis Mfg. Co.
1952
LaPlante-Choate Corp.
1948
Landsdowne Steel & Iron Co. (Gadsden, Alabama)
1952
Victor Electric Products
1948
Canadian Allis-Chalmers
M. Rumely Co. Inc. (Manufacturing) (1913) Rumely Products Co. (Sales)
1955
Gleaner Harvester Co.
1953
Buda Manufacturing Co.
1959
Valley Iron Works
1955
Baker Mfg. Company (Springfield, Illinois)
1959
Tractomotive Corporation
1959
S. Morgan Smith Co.
Advance-Rumely Thresher Co. Nov. 1915
1965
Simplicity Mfg. Corp.
1963
Schwager-Wood Co.
1968
Henry Corporation
Siemens A.G., (Munich, West Germany)
Allis-Chalmers Corporation 1971
1972
Pennsylvania Electric Coil Company
Siemens-Allis 1978
Betatron Division to ATC Betatron Co.
1973
Klockner-Humboldt-Deutz A.G.
Fiat S.p.A. (Italy)
1985
1974
Deutz-Allis Corporation
Fiat-Allis Construction Machinery Inc.

Formative Years

When this photograph was taken in 1877, Edward P. Allis was already 53 years old, but time hadn't removed the glitter from his eyes. Despite financial problems at various times, the techniques of business management developed by Allis were worked to their ultimate maximum. This consisted of finding and hiring the best available engineers for each specific endeavor, and leaving them to do their work while he concentrated his efforts on financial affairs. A New York native, Allis came to Milwaukee in the 1840's and joined a partner in a leather shop. Eventually he left this business, and seemed to play the real estate market for a time, finally purchasing the Reliance Works of Decker & Seville. This opened the final act for his life's drama, since from these roots the huge factory arose that would become world famous for its steam engines, flour mills and sawmills. Edward P. Allis died in 1889.

Edward P. Allis

The year is 1846. The bustling city of Milwaukee had that year exported over 213,000 bushels of wheat. Over 1¾ million pounds of lead had been exported — this being the product of numerous lead mines in the region. Plank roads were under construction, and the city was anxious to gain railroad service. During the spring of 1846, there arrived in Milwaukee a young man by the name of Edward P. Allis. Neither he nor anyone else could have imagined the immense fame that would eventually come to himself or his chosen city of residence.

Edward Phelps Allis was born at Cazevonia, N.Y. on May 12, 1824. His parents, Jere and Mary Allis educated their son at the nearby Cazevonia Academy, then at Geneva College, and finally in 1845 Edward P. Allis received a Bachelor of Arts Degree.

It was no chance meeting that put Allis and his friend William Allen into a business partnership. The two had been college classmates. For their part, the Allen family had formerly resided in New York, but moved their leather business to Milwaukee after the supply of tan bark was exhausted in their native state. Thus, it came as a natural succession of events that Allis made the move West to Wisconsin. Little time was lost in establishing the partnership, since records indicate that their "Empire Leather Store" opened for business on May 2, 1846.

Allis still had allegiances to the people back in New York. After getting the leather store well established, he went back to Geneva, N.Y., and on September 12, 1848 he married Mary Marie Watson. Early company records indicate that a few months later he withdrew nearly $3,000 from the company accounts to purchase and furnish a home. William, the first of their 12 children, arrived into the new home on November 14, 1849.

For reasons unknown, Edward Allis sold his interest in the Wisconsin Leather Company during 1856. Possibly he anticipated the Financial Panic of 1857, or perhaps he had tired of the leather business and was making plans to move into another area of business. For whatever reason, the sale came at the perfect time for Allis. By waiting out the financial storm, Allis was in a position to move into an entirely different business venture, and at a bargain price!

The grand opportunity came with the crumbling of Decker & Seville, a firm established back in 1847. Known as the Reliance Works, the company found itself

Almost nothing remains of the Decker & Seville days, but this heading from an old invoice gives an engraver's perspective of the Reliance Works. Dated 1858, it indicates that Reliance Works manufactured "French Burr Mill Stones, Castings, Screws, Machinery, and Portable Grist Mills." Like most shops of the time, blacksmith work was also a part of the business; this probably included new items built to order, as well as repair work. Founded by Charles Decker and James Seville in 1847, the Reliance Works flourished until the financial panic of 1857. Four years later, Reliance Works was sold at a sheriff's sale, and in May of 1861 Edward P. Allis became the owner. Despite numerous buyouts of other firms, the firm of Decker & Seville is the main limb from which Edward P. Allis & Company was to succeed as a machine builder.

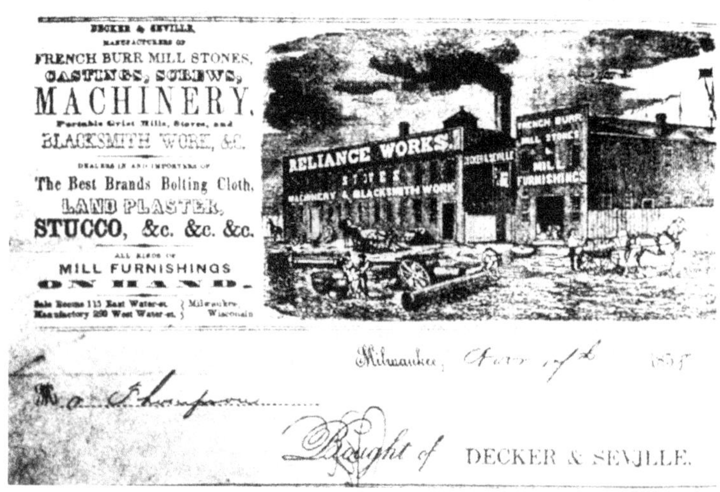

deep into trouble by 1857 as a result of the Financial Panic, and on February 7, 1861, the property was sold at a sheriff's sale to satisfy a judgement. Various historians have attempted to recreate the events surrounding Allis' purchase of Decker & Seville, and from all accounts it would appear that Allis was not the final bidder at the sheriff's sale. Rather, it seems a reasonable conjecture that another party did the actual bidding, probably at Allis' behest and under his instructions. Regardless of the circumstances, the Milwaukee Sentinel of May 27, 1861, carried an advertisement which read in part: *"Reliance Works of Edward P. Allis & Co. (late Decker & Seville)".*

Decker & Seville owed it origins to James Seville, the operator of the firm, and his financial partner, Charles Decker of Dayton, Ohio. Reliance Works early gained a role of leadership in the production of sawmills, flour milling equipment, and castings made-to-order. Before its decline in 1857, Reliance Works was one of Milwaukee's largest employers with a payroll of about 75 men.

During the early 1860's Allis employed about 40 men — their working hours were from 7 AM to 6 PM Monday through Friday, and 7 AM to 5 PM on Saturday. One hour was given off for lunch. By the close of the Civil War, Reliance Works was operating at full capacity, with its principal products being imported French buhr stones and various mill machinery.

Edward P. Allis enjoyed a superb relationship with his employees. Allis seemed to sense a closeness to his employees that was totally uncharacteristic of most labor-management relationships. Perhaps for this reason, Allis was unable to understand the militance of labor, not only at Reliance Works, but throughout the country. Terrence V. Powderly, Grand Master Workman of the Knights of Labor called for a national strike on May 1, 1886, to force management into an eight-hour workday. By the time May 1 rolled around, the Allis employees were asking for ten-hours pay to be given for eight hours of work. Hope for a settlement vanished, and on May 3, 1886 an attack on the works was held back by police and fire hoses. Governor Jeremiah Rusk provided police protection from the Watertown Rifles, the Beloit City Guards and the Darlington Rifles. Under their protection Allis reopened Reliance Works on May 5. The Haymarket riot in Chicago of May 4, plus another disturbance in Milwaukee on May 5, all but ended the strike. Within a week, nearly all the employees had returned to work. This photograph of May 5, 1886, shows government troops assembled at Reliance Works with fixed bayonets.

Since the original site of Reliance Works offered little room for expansion, a new site was developed in 1866 and occupied in 1867. This 1871 photograph of Reliance Works at its new location at Clinton and Florida Streets illustrates a main building, together with a series of additions, probably constructed in accordance with short-term need, and with little scientific planning. The cupola at the roof peak provided ventilation in summer, and probably offered extra light throughout the year. In our age, it seems impossible that for the 1871 factory, coal oil lamps were probably the best that could be expected except for natural light. The great number of windows in the Reliance Works was typical construction for the period. As if factories weren't already cold enough in winter, the huge windows probably made the entire plant little more than an above average windbreak. Even with these conditions, the workday ran from 7 a.m. to 6 p.m., Monday through Friday. Saturday was an exception — the final whistle blew at 5 p.m. The only noticeable benefits consisted of getting paid once a week.

Since the original Reliance Works site did not permit further expansion, a 20 acre site at Florida and Clinton Streets was chosen already in 1856. Ten years later, construction of an entire new plant and foundry commenced. From this new factory would come an immense variety of innovative machines, including the first Allis steam engine. The first whistle blew on New Year's Eve, 1868. Until 1874, Allis catalogs illustrated a simple and reliable steam engine as a part of the total offering.

During 1869 Allis bought out his largest rival, Bay State Iron Mfg. Co. Even the combined production of the two plants failed to bring Allis to the point of prominence that he probably deserved. All this changed with his successful bid on a major portion of the cast iron water main for Milwaukee's water project.

The first step was in hiring two experts, John Pennycook, formerly of Scotland and William Wall, an expert foundryman. The second step was construction of a $100,000 pipe foundry. In all, the Allis factories secured initial bids on 2,600 tons of pipe. Pumps were required to feed the new water system, and Allis successfully received that contract also, even though his bid of $168,000 was the highest of the bidders. For

Formative Years

this project, the renowned mechanical engineer, Robert W. Hamilton, was commissioned by Allis to draw up the plans. Each pump was to have a capacity of 8 million gallons of water each 24 hours against a head of 150 feet. Although the two engines were to be completed in September, 1873, it would be another year before the waterworks engines finally went into motion. There is no doubt that this project finally propelled the Allis enterprise into the industrial role it would assume for decades to come.

Having reached the age of 50, Allis took his three eldest sons into the organization. William, born in 1849; Edward P. Jr., born in 1851; and Charles, born in 1853. Louis, born in 1866 joined the business for a time and eventually set up his own organization, the Louis Allis Company. The remaining Allis children never assumed an active role in company affairs.

Despite the successful installation of Milwaukee's huge pumping engines, the Financial Panic of 1873 was taking its toll on the Allis organization. Even though the patient looked healthy, the simple fact was that from a financial standpoint, the illness was terminal. In other words, Edward P. Allis was virtually bankrupt. In late 1876 a meeting of creditors established this fact without a doubt. Since Allis had established an excellent reputation in the business community, little difficulty was experienced in reorganizing the company and establishing new lines of credit. In the interim, much of the debt load was secured by company property, and Allis himself signed notes promising to repay the debts. To the chagrin of some, the notes were retired in a short time.

When George Madison Hinkley joined the Allis organization in the Fall of 1873, circular sawmills of the pattern shown here marked the zenith of sawmill design. In fact, circular sawmills had no lengthy history at this time, but they represented a radical departure from the hand methods used since the beginning of time. Hinkley was a considered expert in sawmill design, and with his entrance to the Allis offices, he brought along a number of patents covering various sawmill devices. Although this engraving illustrates the head saw and top saw, nothing is shown of the carriage feed mechanism or other important saw details. The fact is that a great deal of engineering was involved to build a successful sawmill, and through Hinkley's efforts the circular mill was brought to a state of perfection unchanged except for the advent of modern hydraulic and electrical systems. Allis continued building circular sawmills into the early 1900s.

Going back to the college days, the Wisconsin Leather Company, and the eventual acquisition of Decker & Seville, Edward P. Allis was developing business management techniques that would ultimately lead to financial success. The financial problems of the 1870's were no deterrent, but rather served to further hone the ideas developed over many years. A basic tenet of the Allis technique was to hire the best possible experts in a given field. Since Allis himself was trained in financial management, and probably found himself most at home in this endeavor, the continued success of the company depended mightily on individuals with inordinate skill. Thus came the now-famous triumvirate of William Dixon Gray, George M. Hinkley, and Edwin Reynolds. From these individuals, the Allis operations gained worldwide recognition for flour mill machinery, sawmills, and steam engines.

By 1889 Allis employed almost 1,500 people. Annual production came to some $3 million, and more than $700,000 of this was paid out in wages. Edward P. Allis was in his mid-sixties. Remember, this was a time when 70 was a ripe old age for the majority of Americans. On Friday, March 29, 1889 Edwin Reynolds and Edward P. Allis walked through the shops for the last time. On Monday, April 1, Edward P. Allis died. So ended an era that saw Milwaukee grow from a bustling town to a full-fledged city, during which Allis had been a major force in its growth. By April 1, 1889, it could truly be said in Milwaukee that Edward P. Allis Company was a household term!

George M. Hinkley

Although Edward P. Allis came into financial embarrassment in 1876, the problem was soon confined and eventually resolved. Various historians have concluded that problems with the pipe foundry and the erection of the huge Milwaukee pumping engines were partly to blame; in other words, the Allis organization simply overextended itself. Despite these problems however, Allis was not to be deterred from his goal of establishing a unique business management method for the time. It consisted of hiring the best available specialists in each field, leaving them to the work of design, engineering, and sales, while Allis concentrated his efforts on his chosen area of financial management. Innovative for its time, the Allis management techniques soon were adopted by virtually every industry. Although the hiring of Pennycook and Hamilton were among the first such applications of the Allis technique, and ultimately helped move the company to bankruptcy, the hiring of George M. Hinkley to manage the sawmill department proved conclusively that the idea would work.

In October, 1873, George Madison Hinkley came to

work for Edward P. Allis Company. Until this time the sawmill operation was bringing in something on the order of $1,000 per year. Proof of Hinkley's genius both as an engineer and as a salesman is seen in an annual sales figure of $400,000 by 1889. Hinkley was born at Seneca, N.Y. on May 24, 1830. After becoming a master carpenter, Hinkley went on to become a millwright and worked in various sawmill operations in and around Michigan. During the Civil War, Hinkley enlisted in Company I, Sixth Michigan Cavalry. Before the War ended, Hinkley saw the infamous Andersonville prison from the inside, being paroled from that place in 1864.

By 1870, George M. Hinkley had gained a wide reputation in the sawmill business, having already secured several patents for improvements to sawmill machinery. In joining the Allis organization, Hinkley contributed his patents and went on inventing all the while. At his death in 1905, over 35 patents had been issued to Hinkley for a wide range of sawmill machinery and accessories.

In December, 1885 Hinkley announced through Edward P. Allis & Co. that "We have just completed our new band saw mill." Although the original machine was a 9 ft. mill built to carry a ten-inch saw, an 1886 company catalog illustrates this mill in five models. The smallest used wheels 8 feet in diameter and could make a cut 44 inches deep. It weighed 22,000 pounds. In contrast, the huge No. 9 mill used ten-foot wheels, carried a saw 14 inches wide, and could make a cut 90 inches deep! This mill weighed 38,000 pounds. Although the band mill had been entertained prior to Hinkley, it was his genius that reduced it to a practical machine, the final test of all invention. Historians of mechanical technology generally credit George M. Hinkley for the development and perfection of the band saw mill. In addition to being an extraordinary inventor and machinist, Hinkley was also an excellent businessman. The Allis sawmill department was entirely in his charge, and by 1889 it noted sales of $400,000.

Even though the sawmill department remained one of the smaller parts of the company, its strong showing during the 1870s and after helped immensely to bring the company out of the doldrums following the 1876 bankruptcy. Circular sawmills were the rule of the day — bandsaws had been invented, but it was George Hinkley who reduced the sawmill to a practical machine. With the advent of the bandsaw, the entire logging industry was revolutionized. In late 1885 Hinkley announced that the new Allis band mill would be set up at the factory, and after a short exhibition time would be set up at the Jump River Lumber Company at Dorchester, Wis. Thus began a machine that eventually would be sold in all parts of the world.

There can be no doubt that despite Hinkley's many inventions covering sawmill equipment, perfection of the band mill was his crowning achievement, and remains a monument to his inventive genius. At his death in 1905 the *American Lumberman Magazine* wrote: "Mr. Hinkley was as great a man in his line of business as Carnegie in his..... His improvement of sawmill machinery was almost akin to that of Edison to electrical development or of Ericsson to naval construction....."

William Dixon Gray

After hiring George Hinkley, the next member of the famous engineering triumvirate to grace the Allis organization was William Dixon Gray. Like Hinkley, Gray was a talented inventor, and also displayed many of Hinkley's talents in business management.

William Gray was born in Lauder, Scotland, on July 22, 1843. In 1854 he and his family came to a farm in Canada, and in 1861 he became a millwright's apprentice. This was followed by a move to Minneapolis in 1865 to become a full-time millwright. At Minneapolis he went into the employ of Pray & Webster, a firm then erecting a large flour mill for Cadwallader C. Washburn. The intervening years provided the necessary experience for Gray to join E. P. Allis & Company in January, 1877, as superintendent of the milling machinery department.

When Gray went to work for Allis, stone buhr mills were almost universally accepted as the only means of reducing wheat and other grains to flour. Imported French buhr stones were considered to be the superior material, so the Allis factories were always busy shaping millstones. In addition, the company sold some 20,000 square yards of silk bolting cloth to sift the flour during 1876.

Two years after coming to Allis, Gray contracted for the erection of a thirty-run mill at Niagara Falls — this after constructing some seventy runs during 1877, placing them in operation all over the United States, Canada, and several foreign countries.

Stone buhr mills, despite differences in design, all used the same basic principle of reducing whole grain to flour. One stone remained stationary, and the runner stone turned about its axis. Grain entering the center

Formative Years

hole or eye of the stationary stone was thus reduced to flour by the time it reached the outer periphery. A series of furrows or lands cut into the two stones caused a gradual reduction of the grain, and many times the process became one of crushing the grain rather than obtaining the shearing and rolling action considered optimal. Methods of dressing millstones varied greatly, and this is well witnessed by a search of the Patent Office records which lists hundreds, yes even thousands of patents covering a specific layout for the dress. A certain dress suitable for a specific grain might work poorly or not at all with another, or perhaps regional differences in a grain such as wheat might bring about the need for a somewhat different dress of the stones. Proper management called for only enough pressure between the stones to secure the shearing and rolling action of the grain that would reduce it to flour. Unfortunately for the stone milling process, the hulls, kernel, and germ all ended up in the same hopper, and the result was that the germ made its way into the middlings rather than into the finished product. For these reasons, and the high maintenance costs of the stone buhr mill, William D. Gray cast about for a better way to make flour.

Crude roller mills were available in Europe during the early 1870s. The Wegmann mill of Wegmann, Bodmer & Co. of Zurich, Switzerland, was used in Budapest, Hungary during 1874. Gray was convinced that this process, with some improvements, was the ideal method of milling flour. Therefore E. P. Allis & Company acquired import and sales rights in 1878.

Gray was dissatisfied with the Wegmann mill — its lightweight frame and gear-driven rolls were the primary problem from his viewpoint. This led to development of the Gray roller mill, the first of which was installed in June, 1878, at the Eagle Mill in Milwaukee. Resistance to change is an innate part of human nature, and human nature certainly was alive and well when it came to converting the milling trade from the old way

William Dixon Gray joined Edward P. Allis & Company in June, 1876. At this point in time, virtually all flour milling was achieved by the use of mill stones, and Allis was a leader in manufacturing and dressing the imported French buhr stones which the majority of millers preferred over granite. Roller mills were already known in Europe, at least to a limited extent by 1873. During 1877, Edward P. Allis & Co. became importers of the Wegmann machine built in Switzerland. Of gear-driven design as shown in this photograph, the Wegmann mill was soon redesigned by Gray to feature a belt drive. Gray was convinced that the roller mill offered a much better alternative to flour milling than the stone buhr mill. Despite convincing evidence, most millers were reluctant to make the change, and in fact, many small mills refused to use a roller mill. Theoretically, a properly dressed millstone is supposed to reduce wheat to flour with a rolling and shearing motion between the dressed stones. Practically, the pressure between the stationary and runner stones often crushed the grain into an inseparable mass containing the outer husk, the starchy kernel, and the oily germ. The gradual reduction process of the roller mill offered a means of separating these three elements effectively. Although it has been impossible to positively identify this mill, it appears to be of the original Wegmann design.

Gray's first major installation of the roller mill was in C. C. Washburn's "B Mill" at Minneapolis. This was done in such secrecy that by the spring of 1879 it was rumored that the entire venture was a failure. Soon the success of the Gray roller mill was evident — Washburn converted both his "A" and "B" mills to rollers. Shortly thereafter, the brothers Pillsbury at Minneapolis installed rollers in their Excelsior mill. This photograph shows one of the units destined for the Excelsior installation. There is no doubt that Gray revolutionized the milling industry. Other writers have reported that while Wegmann was primarily concerned with reducing middlings to flour, Gray saw the roller mill differently — in Gray's eyes the roller mill "was a means to mass production." In short, Gray had Americanized a foreign product, but in so doing he had given milling companies far greater capacity. Gray's eminently successful roller mill saw more units sold by 1884 than did the total output of all other manufacturers combined. By 1894, over 85% of all flour milled in the Minneapolis, Duluth, and Milwaukee mills was reduced with Gray's Patent Noiseless Roller Mill.

to the new. Most of the old-time millers were fully convinced the idea would never work, no matter what. Others felt that the roller mill might have a place in the process, but that stone mills were still necessary to handle fine middlings. Under these circumstances, Gray found some comfort in his dealings with C. C. Washburn at Minneapolis. In early 1878 Washburn was building a new mill, with Allis supplying the equipment. When installation was nearly completed, it was discovered that extra space was available for additional mills. Gray convinced Washburn to install a small experimental roller mill. The secrecy of the experiments led to wide conjecture about total failure of the process, when in fact, just the reverse was true. Roller mills were so successful that eventually both the Washburn A and B mills were converted to rolls.

Success with Washburn convinced Gray to approach Charles A. Pillsbury of Minneapolis. After convincing a somewhat reluctant Pillsbury to convert one of his Excelsior mill to rolls, Gray lost no time in removing the old equipment. After following the flour from mill to bakery, and seeing complete success with the product, the Pillsbury's too began phasing out the stone mill in favor of rolls.

Gray's early success with the roller mill led to the purchase of an old mill operation in Milwaukee. This was subsequently remodeled to include roller mills and the full line of Allis flour machinery. Known as the Daisy Mill, it was an ideal sales tool, as well as providing a point of interest for visiting sightseers.

An almost immediate success with roller mills prompted the development of various other mill machinery. Included were automatic flour packers, centrifugal reels, the flour purifier, and numerous other machines that played the dual role of making the flour mill more efficient and provided a better product for the consumer.

Edwin Reynolds

As previously indicated, Edward P. Allis was highly skilled at business management. The Allis plan therefore called for the very best engineers in each field of endeavor. Although the great steam pumping engines furnished for Milwaukee were designed by Hamilton, there is a strong inference in various company records and other histories that Allis was not altogether satisfied with Hamilton's engineering methods. Thus, in early 1877 Allis placed an advertisement in the relatively young journal known as *American Machinist*. The advertisement stated that a Superintendent for the Reliance Works was required. Edwin Reynolds replied to the ad and assumed the position on July 1, 1877. With Reynolds on board the great engineering triumvirate which also included Hinkley and Gray was now complete. Their individual and collective engineering skills were to dominate, or perhaps dictate would be a better word, the Allis line for several decades. The simple fact is that their designs became a legacy which followed long after them. Hinkley's sawmills, Gray's roller mills, and Reynolds' steam engines were so well built, and so eminently successful that once the market was saturated, sales declined to the point that the only major aftermaket was in repair parts. Of the three men, Hinkley, Gray, and Reynolds, the latter was the best known, and was perhaps, one of the best engineers of his time.

Born March 23, 1831, at Mansfield, Conn., Edwin Reynolds eventually apprenticed as a machinist. The following years saw a succession of machinist jobs, and during the Civil War he worked for a time on the machinery installed in Ericsson's famous *Monitor*. Finally in 1871, Reynolds entered the employ of the famous Corliss Steam Engine Co. of Providence, Rhode Island. George H. Corliss (1817-1888) was one of the great engine builders of the Nineteenth Century. Edwin Reynolds, as Superintendent of the Works, was intimately involved with the construction of the mammoth Centennial Engine installed in Philadelphia's Exposition Hall for America's Centennial Celebration in 1876. A walking beam condensing-type engine, it used cylinders of 40 inch bore and 120 inch stroke. This 1,400 horsepower engine was placed in the center of Machinery Hall and operated all the exhibits in the building.

Despite a reduction in salary, Reynolds accepted the position with Allis primarily because he was given the liberty to create new and more efficient engines. Despite his years of service with Corliss, Reynolds never gained the rite of passage so far as his own engineering acumen was concerned — the Corliss Co. insisted on following its own designs. Given this opportunity at Allis, an entirely new engine, the first of the Reynolds-Corliss designs was announced in October, 1877.

Initialy, it was announced that the Reynolds-Corliss was heavier and stronger than its competitors, and in addition it was capable of much higher speed. The Reynolds girder frame design gave much better rigidity, and the improved valve gear mechanism offered much higher speed than was formerly possible. The Reynolds-Corliss engines were from the beginning designed so that either right-hand or left-hand engines could be purchased with a bare minimum of extra patterns, castings, or machine work at the factory. This substantially lowered the manufacturing cost. To help gain a foothold in the engine market, E.P. Allis & Co. also offered to retrofit existing engines with the Reynolds-Corliss cylinders and valve gear. Within a short time the company dropped the offer to retrofit existing engines and concentrated their sales effort on the "Improved Corliss".

The Reynolds sales attack went after the common slide-valve engine full bore. One such situation often related by historians promised the installation of a 150 horsepower Reynolds-Corliss for $3,000. In addition, Reynolds guaranteed a 50% reduction in fuel costs, going even further by stating that if this claim didn't measure up, then the engine would not cost a cent! The

Formative Years

engine was installed and it measured up to expectations, and even more. Ten years later the owner calculated a savings of $20,000 in fuel.

By the late 1870s Allis was finally able to supply a complete sawmill or a complete flour mill, including the engine. The ability to ship and erect a package unit provided a substantial competitive advantage and served to boost sales even further. Thus, the combination of manufacturing volume, innovative engineering, and the ability to provide a complete service from drawing board to erection on the customer's site brought the Allis business technique to its epitome.

Engine production for 1878 totaled about 20 units, with this figure rising over four times by 1882. Reynolds challenged each new demand with an engine designed specifically for the application. An example is found in two large blowing engines built for Joliet Steel Works in 1880. Intended to provide draft for the huge cupolas, these engines were the first of the direct-acting metallic valve design ever built. With these huge engines, Edwin Reynolds revolutionized the industry, putting to rest the leather valve design that hearkened back to the Middle Ages. Some years later the American Society of Mechanical Engineers in its *Transactions* noted that "the essential features of this design have not been improved upon."

The Reynolds blowing engine design captivated the industry and nearly annihilated the competition, at least for a time. Both Reynolds and Allis gained immense prestige through this great success, and equally important, the bank account grew by $30,000 each time one was sold. The mind boggling dimensions of these engines included a total height of 40 feet. Joliet Steel ordered another pair of Allis blowing engines in 1881, and in 1886 the Carnegie interests went with the Reynolds design. By 1900 Carnegie Steel alone was operating over thirty Allis blowing engines.

In summation, the Reynolds-Corliss design proved eminently successful from the very first engine. The demand for large power fed upon itself, with Reynolds and Allis at the ready to design, manufacture, and install units to fit virtually any application. From ordinary belted engines to blowing engines and large pumping engines, the Reynolds design found itself thoroughly adaptable. With few changes, the Reynolds-Corliss was built until the advent of cheap and reliable electric power made these engines obsolete.

The fertile and inventive mind of Edwin Reynolds was not limited to steam engine design. During the early 1880's the hometown of Milwaukee was faced with a growing problem of sewage disposal. Reynolds met the challenge with a huge centrifugal sewage pump capable of handling 70 million gallons a day, making it by far the largest pump of its time. This huge pump was powered by a cross-compound Reynolds-Corliss engine,

Earlier in the Nineteenth Century, George H. Corliss had revolutionized steam engine design with the engine that still bears his name. During 1871 Edwin Reynolds assumed the role of General Superintendent at the Corliss works. As America readied itself for the Centennial Celebration of 1876, Corliss built the great Centennial Engine of 1,400 horsepower. In July, 1877 Edwin Reynolds became superintendent of the Reliance Works of Edward P. Allis & Co. Reynolds soon proved himself to be one of the most talented designers of his time, and also was fortunate to be an excellent business manager. After the death of Edward P. Allis in 1889, Reynolds in fact, if not in title, was the practical head of the company. When the merger that resulted in Allis-Chalmers came about in 1901, it was Reynolds who functioned as the catalyst in bringing about an agreement. After Allis-Chalmers Co. was formed, Revnolds remained as its Chief Engineer until his retirement.

Almost immediately after joining the Allis organization in 1877, a new "Improved Corliss Engine" was announced. Its girder frame design and a greatly improved valve mechanism soon brought the Reynolds-Corliss to the forefront as a leading contender. The Reynolds valve gear permitted somewhat higher crankshaft speeds than had previously been possible with the original Corliss designs. The girder frame provided exceptional rigidity, and simplified the manufacturing process. Included in the Reynolds design was a means of making the engines interchangeable from right-hand to left-hand with no additional foundry patterns being required. In so doing, great expense was saved in the manufacturing and machining processes, and inventories could be much smaller to achieve the same goal. By early 1878 the company was deluged with orders, working at full capacity, and holding a sizeable backlog of business.

Formative Years

using the huge twelve-foot impeller as a flywheel. Thus began another important part of the Allis operation, the building of pumps to fit every requirement and application. Sometime after, Reynolds met another challenge of Milwaukee. The slow-moving water in the city made it imperative that some means be devised of hastening this material along to its ultimate destination. Several years earlier, Reynolds had conceived the idea of a screw pump, but had never had the opportunity to build one. Now came the chance to build one of immense proportions — it had a capacity of 40,000 cubic feet per minute for a total of over 400 million gallons per day.

Granted, the Reynolds pumps were larger than anything heretofore built by anyone, but even so, these pumps were far from the optimum in design. Primarily for this reason, Allis was unable to top the competition in the pump business. All this changed in 1884 when Irving H. Reynolds, a nephew of Edwin joined the firm. Irving Reynolds is said to have left school at age 16 to join a whaling expedition. When he came to Milwaukee at age 22, he had no formal engineering education, noting that this way he had nothing to "unlearn." Although his Uncle Edwin eventually held some fifty different patents, Irving himself received a half dozen, the most important of which was the patent covering the triple-expansion pump. This particular development was of significant value to the pump business.

The stage was now set. When Edward P. Allis died in 1889, his plan for successful business management was in full effect. With the help of Hinkley, Gray, and Reynolds the Allis legacy would continue for decades to come. Shortly after the dawn of the Twentieth Century, Allis-Chalmers would be born — a merger that would dominate American industry for many years.

Allis-Chalmers Company

In the decade following the death of Edward P. Allis the company continued with great achievements, especially in respect to steam engines and pumping machinery. Some of the world's largest steam engines and pumps were being shipped from the Allis factory, with Edwin Reynolds apparently being the chief architect of the company's direction. Various writers have indicated that the Allis children had nothing of the enthusiasm for the business that their father had experienced, although some of the boys maintained a part of company management.

As the Twentieth Century appeared, Edward P. Allis & Company was the largest steam engine builder in the United States. The Reliance Works was at its capacity, and had no room for further expansion. Thus in 1900, a 100 acre site was purchased in what was to become West Allis. This plant opened in September, 1902.

A chance meeting in a Chicago hotel lobby between Edwin Reynolds and William J. Chalmers would soon change several small companies into a corporate giant. In his book, *Industrial Heritage*, Dr. Walter F. Peterson comments that "the Allis boys had regarded Reliance largely as a source of income. Chalmers eagerness to promote this merger arose from the fact that his firm was headed for serious financial difficulties ..." Reynolds as Chief Engineer envisioned a merger whereby three immense factories across the United States could virtually monopolize the engine trade. Under these circumstances then, Reynolds and Chalmers became engaged in a conversation which ultimately mentioned the possibility of a merger. Chalmers was President of the Chicago firm of Fraser & Chalmers, well known for their mining machinery, stamp mills, and ore concentrating machinery. From this meeting in late 1900, serious negotiations soon developed between the Fraser & Chalmers and the Allis interests. At a point which remains unclear to this writer, Gates Iron Works of Chicago, and Dickson Mfg. Co. of Scranton, Penn. became involved. Negotiations continued, and on May 8, 1901, articles of incorporation were filed in New Jersey for Allis-Chalmers Company. The next day, May 9th, Charles Allis was elected President; Will Allis was named as Chairman of the Board; and Edwin Reynolds was appointed as Chief Engineer for the company.

Gates Iron Works of Chicago was a notable builder of rock and ore crushing machinery, and had recently introduced the Gates Gyratory Rock & Ore Crusher.

Dickson Mfg. Co. of Scranton was known for its blowing engines and air compressors among other items.

The steam engines, saw mills, and flour mills of Allis when combined with the mining machinery of Fraser & Chalmers and the output of the other two partners gave the new firm of Allis-Chalmers Co. a very extensive line that covered many phases of American industry, and for that matter, world trade.

At this point in time Allis-Chalmers was entering the relatively new field of steam and hydraulic turbines, but found itself without a means of building its own electrical generating equipment. During 1904, arrangements were made to lease the Bullock Electric & Mfg. Co. of Norwood, Ohio, a suburb of Cincinnati. During that same year Allis-Chalmers acquired Bullock outright, calling it the Norwood Works. This early acquisition might have suggested that Allis-Chalmers would continue a pattern of acquisition, but in fact, no major purchases of outside companies were made for nearly twenty years thereafter.

Bullock Electric traced its roots back to 1887 with the formation of the George F. Card Mfg. Co. of Cincinnati. This company built small electric motors. In 1890 the company name was changed to Card Electric Motor & Dynamo Co., and in 1897 a reorganization brought the title of Bullock Electric Mfg. Co. Bullock moved to Norwood in 1899. A year earlier they had built their first a-c generators. During 1900, Bullock acquired American rights to the a-c equipment being built by Oerlikon Co. of Switzerland. Although Bullock carried little of the grandeur commonly placed on Westinghouse and General Electric, the company was very agressive, and obviously on the move upward.

Formative Years

For Allis-Chalmers, the purchase of Bullock now rounded out its line, and with this came a new approach to the public. Now the company could use in its trademark: "Ours the Four Powers: Steam, Gas, Water, Electricity." No other company could build under a single management a prime mover using any one of the four powers at this time.

During 1906 Allis-Chalmers made a brief hiatus into the air-brake business through the purchase of Christensen Engineering Co. of Milwaukee. From all appearances, this venture evaporated by 1910. About the only thing salvageable was the experience gained in building air compressors, and this was used to advantage for manufacturing a host of sizes and styles during the period up to about 1930.

Except for the acquisition of Bullock Electric and Christensen Engineering, Allis-Chalmers Company seemed more content to expand within itself rather than adopt a specific plan of acquisitions. In February, 1904, Allis-Chalmers announced its entry into the field of hydraulic turbines with Clemens Herschel as the chief consultant. The first unit was built at the Scranton Works (formerly Dickson Mfg. Company). Of the Francis type, it was rated at 1,250 horsepower. This unit remained in operation until 1952.

Already in 1902 the newly organized Board of Directors was exploring the possibility of entering the steam turbine business. By 1904 Allis-Chalmers had joined the Steam Turbine Advisory Syndicate in England, for which Allis-Chalmers was to receive full manufacturing rights in the United States, Canada, and Mexico. The following year, Allis-Chalmers gained a direct arrangement with C. A. Parsons & Co. of England to acquired American rights to the Parsons patents.

The first Allis-Chalmers steam turbine was built at West Allis. Its was installed in the Williamsburg Power Plant of the Brooklyn Rapid Transit Company during 1905. Of the Parsons horizontal type, it was rated at 5,500kw, operating at 750 rpm. Even before this unit was installed it was obvious to Edwin Reynolds that the days of the Corliss engine were numbered, and they were.

During 1903 Allis-Chalmers also explored the possibility of building gas engines as another source of power, and subsequently acquired American rights for the designs of the Nurnberg engine built in Germany. The following year Allis-Chalmers Americanized the design, offering it under their own name.

Despite the many announcements of new prime movers, and even though sales were relatively good, Allis-Chalmers found itself in trouble almost constantly during its first decade. The officers of the company did not get along well together, with each seeming to pull in his own direction. A sagging national economy held sales down — Americans were still hearing the echoes of William Jennings Bryan's "Cross of Gold" speech.

By New Year's Day of 1912 it was obvious that the company was on the brink of bankruptcy. On April 8 of that year it was announced that Delmar W. Call and Otto H. Falk had been appointed the receivers of the company. Thus it was that on this date Otto H. Falk would begin his long odyssey at the helm of Allis-Chalmers.

Allis-Chalmers Mfg. Co. was incorporated on April 16, 1913, with General Falk as its president. He remained in this position until 1932. General Falk obtained his military rank with the Wisconsin National Guard partially because of his outstanding service during the Milwaukee and Chicago labor riots of 1886. He retired from the Guard in 1911 with the rank of Brigadier General. The Falk Co. of Milwaukee was a large machinery builder, and it was General Falk's brother Herman who chaired this organization.

Prior to the reorganization, a strong-minded Board of Directors had consistently chosen presidents who would do their bidding. With General Falk in command, Allis-Chalmers gained a professional executive who believed that a Board of Directors should not dictate policy, but advise the president in this regard. Falk believed that the president "should thoroughly control the business," and he did during his entire career as president.

It was almost a quarter century ago that Edward P. Allis had died, and his grand vision, while not altogether dead, was at the very least seriously ill. Falk saw about him a series of plants scattered around the country. The old Reliance Works was still in use, even though the West Allis Works had been in operation for some time. True enough, the old regime had made a small move toward consolidation by closing the Gates Iron Works, also known as No. 1 Chicago Works. But this move was too little, and too late to do any good. Likewise, Scranton works was closed in 1911, but it was not until 1914 that the real estate was placed on the market.

Falk began by phasing out No. 2 Chicago Works, earlier the site of Fraser & Chalmers. The mining, cement machinery, and crusher lines being built at No. 2 Chicago Works would be moved to West Allis. Reliance Works was closed in September, 1921. An analysis of company records indicates that by 1923 General Falk had reduced six plants to only two, and these were producing more machines than the combined capacity of the six original factories.

To restore confidence among company personnel, from Directors to floor sweepers, Falk instituted a plan of common sense and ordinary fairness seldom found in any social rank, wealthy or poor. Although his ideas of interpersonal relationships are extensive, they could well be summed up by his statement, "Treat others the way you like to be treated."

General Falk and his family were an important part of the German-American community of the Milwaukee area. However, when Allis-Chalmers received a subcontract to build shell casings for use in World War One, the German-American Alliance was highly critical of this move. Falk was not about to be pressured into a reversal — his primary goal was the ultimate success of Allis-Chalmers. For several years following this 1915 war contract, General Falk was not well received by Americans of German ancestry, and German loyalty.

By the 1920s General Falk felt that the company was capable of handling additional operations. The mining, crushing, cement, and creosoting machinery lines of Worthington Pump & Machinery Corp. were purchased in 1924. Two years later the Hoar Shovel Co. was acquired — they were building a small shovel designed specifically for use within mines. Midwest Engineering Co. also came into the fold in 1926, as did the flour milling business of Nordyke & Marmon Co., while 1927 saw the purchase of Pittsburgh Transformer Co. In 1928, Monarch Tractor Co. of Springfield, Ill. came under the protective wing of Allis-Chalmers. The following year saw the addition of LaCrosse Plow Co., and in 1930 Stearns Motor Co. joined the list.

The pattern of acquisition continued into 1931 with the purchase of Advance-Rumely Thresher Co. and Birdsell Co. A final purchase of that year, that of Highway Machinery Co. broadened the company's line of roadbuilding equipment.

With the retirement of General Falk in 1932, Max W. Babb took over at the helm of Allis-Chalmers. During the Falk years, Allis-Chalmers had regained a sound footing, and in addition to its extensive product lines, it had also become a farm tractor builder.

Farm tractors were the ultimate rage of the 1920s, but Allis-Chalmers seemed to gain only a small portion of the market. While International Harvester, Deere, and J. I. Case already had established dealer organizations, Allis-Chalmers remained at the garden gate trying to reap the reward, but at arm's length. The acquisition of LaCrosse Plow Co. served to broaden the agricultural line, but the 1931 purchase of Advance-Rumely Thresher Co. provided the elusive goal of acquiring a dealer organization. The Rumely line was internationally known, and was highly respected by the vast majority of farmers and machine men. The Rumely people had long before established their own dealer organization, and this sales base was the ticket that finally gained entry for Allis-Chalmers as a farm equipment builder of significant importance.

During the 1930s Allis-Chalmers pioneered many "firsts" in the farm equipment industry, including the first rubber-tired tractor in 1932. Also of major significance during the 1930s was the introduction of the "All-Crop" harvester — a little machine that effectively spelled the end for the grain binder and threshing machine.

Although the early years of tractor sales seemed to always lose money, Falk pressed forward with this pet project, and in 1926 he hired Harry Merritt as manager of the tractor department. Within a short time the company's tractor experiments started to pay off. It was Harry Merritt who, after a 1929 trip to California, came up with the Persian Orange color for Allis-Chalmers equipment. Merritt had seen acres of brilliant orange poppies in bloom, and decided that was the color he wanted for the A-C tractors. Compared to the dull green finish of earlier years, the bright, majestic orange was an instant attention getter, especially in a time when farm tractors were usually gray or green, and almost always dull. It would appear that the Persian Orange affair was one of many sales gimmicks devised by Merritt, but it worked, and soon there were bright orange machines operating all over the United States.

Walter Geist assumed the presidency of Allis-Chalmers in 1942, holding that office until 1951. Geist made his climb up the executive ladder primarily through his earlier development of the multiple V-belt drive which Allis-Chalmers marketed as the Texrope drive. Because the Geist patents were very broad in nature, and because of their great significance to all phases of industry, Allis-Chalmers enjoyed great success from this invention through its own sales of the Texrope drive and from royalties paid to the company from other manufacturers. Walter Geist died in office on January 29, 1951, at the age of 56 years.

W. A. (Bill) Roberts was President from 1951 until his sudden death in 1955. During his short tenure, Roberts moved to reconsolidate the organizational structure of the company. In a manner somewhat different from his predecessors, Roberts looked to the future of the company, and it was through his efforts that the Hawley plant in West Allis became involved in nuclear power. The Hawley plant had been built during World War Two specifically for work on the Manhattan Project which later resulted in the atomic bomb. Roberts was convinced that nuclear power should be explored for peacetime use, so the Hawley plant gained new life.

Ascending to the presidency after Roberts' death was Robert S. Stevenson who held this office from 1955 to 1965. Stevenson moved quickly to revise company operations. The tractor and general machinery divisions were split into five operating divisions, each with its own profit and loss statement. Under the Stevenson plan, farm equipment, construction machinery, industrial equipment, power equipment, and general products were separate divisions. The Buda division at Harvey, Ill, remained a separate entity.

Many lines were completely phased out. Corliss engines had not been built for some time, the sawmill business had evaporated, and about the only demand in the flour mill division was for repair parts. Most of these lines were completely abandoned, including parts service. In December, 1962, the company ended its career in the steam turbine business.

Already in September, 1957, Stevenson announced the formation of Allis-Chalmers International as a distinct operating division. It was to be responsible for all manufacturing and sales outside of the United States and Canada. Within a few years, Allis-Chalmers was operating factories in Mexico, Australia, England, France, and other countries.

During the Stevenson years the Nuclear Energy Division was closed, but in an exciting new development, Allis-Chalmers engineers were performing pioneering work with fuel cells.

After leaving the presidency, Stevenson was followed for three years from 1965 to 1968 by Willis G. Scholl. He continued the same general approach to company management that had been espoused by Stevenson.

During the 1960s Allis-Chalmers became a prime target for forced takeover. A-C was a blue chip company with an excellent reputation for quality, and prospects looked good for continued growth. During 1967 Scholl was faced with a tender offer from LTV (Ling-Temco-Vought) at a price of $24 per share for A-C common stock. Simultaneously came an offer by General Dynamics, but by late 1967 the Board of Directors unanimously turned down the LTV offer. Other merger talk followed, and in early 1968 Gulf & Western proposed a tender offer for up to 3 million shares of A-C stock. Against this background Scholl left the President's chair at Allis-Chalmers, and was succeeded by David G. Scott, an appointee said to have been hand-picked by Stevenson.

In October, 1968, came an announcement that White Consolidated Industries was about to take over Allis-Chalmers. By December, Allis-Chalmers had filed suit in U. S. District Court resisting the takeover attempt. After numerous courtroom battles, proxy fights, and rulings from the Federal Trade Commission, White Consolidated disposed of its A-C stock holdings, and so the crisis passed.

Scott was however faced with huge losses the company was experiencing, and this at a time when new capital was needed for plant modernization. Many of the A-C plants were obsolete and inefficient. New money was needed for research and development. His solution was to immediately stop the torrential losses that were plaguing the company. Improved plant efficiency was a major item, and further investment in areas of known performance and certain opportunity was another. Then in 1971 the company received a new name, Allis-Chalmers Corporation — symbolizing a new image, new ideas, and new products.

Standard Steel Corp. was acquired in 1969. Renamed as Stansteel Corp., this firm expanded existing markets for asphalt plants, small kilns, and similar machinery. The same year, A-C entered into a licensing agreement with Voest Co. of Austria to build and sell Voest's continuous casting machinery in the United States.

Stephenson-Adamson, an old-line company that was building bulk material handling equipment came into the fold in 1974. However, in June of that year, A-C sold its paper manufacturing equipment division to J. M. Voith GmbH of West Germany.

January, 1974, saw the completion of negotiations with Fiat S.p.A. of Italy to form a joint venture. Both firms contributed their resources for manufacturing and marketing construction machinery. Since Fiat held 65% equity in the joint venture, with A-C holding 35%, this agreement essentially ended A-C involvement in the construction machinery business.

By 1977, Allis-Chalmers consisted of several separate groups. The Process Equipment Group manufactured equipment for processing high-volume solids and fluids. To elaborate, solids processing equipment included crushers, grinding mills, screens, feeders, conveyors, and many other items. Fluids processing machinery included hydraulic turbines, pump-turbines, water and industrial pumps, valves, and an extensive line of compressors.

The Agricultural Equipment Group built a rather complete line of farm machinery, plus diesel engines, engine-generator sets, and light industrial tractors.

The Electrical Group built and sold steam turbine generators, plus an extensive line of electrical and electronic equipment, with this line becoming a joint venture known as Siemens-Allis in 1978.

The Material Handling Group included an extensive line of lift trucks, while Outdoor Power Equipment Group offered an extensive line of lawn and garden tractors, snow removal equipment, and other related items.

During the 1970s corporate decisions were made which instead of resulting in corporate profits, led to huge operating losses. Great expectations by the electric power industry of doubling, and quadrupling the use of electricity proved to be wrong, and in this case, Allis-Chalmers was only one of many big losers. These and other losses were covered in part by the sale of various portions of the company to other interests. Unfortunately, with their sale also went a certain portion of income.

Even under these circumstances however, Allis-Chalmers might have continued as a major farm equipment builder, but the boom of the 1970s, its huge tractors, even bigger machinery, and apparently no end in sight, began to collapse by 1980. Memos circulating within the company indicate that at the highest echelons, A-C was looking for a merger partner, but despite continuing rumors, nothing became evident until early 1985.

Evidence was mounting however, that changes were imminent. A letter of October 29, 1984, announced that tractor and combine production had been recently shut down, and would remain down for at least three months. Salaried employees associated directly with production activities were put on layoff status, and other salaried employees were placed on a 4-day work week, with salary adjusted to a 4-day schedule.

Finally on March 28, 1985, a news release was issued, reading in part as follows:

"Allis-Chalmers Corporation said today that it had reached a preliminary agreement with a subsidiary of Klockner-Humboldt-Deutz AG of West Germany to purchase the Allis-Chalmers agricultural equipment business, including the Allis-Chalmers Credit Corporation.

Finalization of the transactions would include a cash payment of approximately $100 million, and Allis-Chalmers will have an opportunity to participate in future profits of the new business. This transaction involves businesses carried on the Allis-Chalmers books at an investment of $260 million."

Advance-Rumely Thresher Co.

Through the purchase of Advance-Rumely Thresher Co. in 1931, Allis-Chalmers gained the rank and reputation of a major farm equipment builder. Rumely's OilPull tractor had earned its laurels as one of the best tractors of its time, not from the viewpoint of price alone, nor of exterior appearance. Rather, the OilPull built its reputation on economy and reliability. With the purchase of Advance-Rumely, Allis-Chalmers gained an already established sales and collection organization. Given the economic conditions in the decade of the 1930s, had A-C not purchased Rumely or a similar builder of like reputation, chances are that the company might never have appropriated the immense amount of capital required to build a first class sales and collection organization.

There is no doubt that General Falk had taken the Allis-Chalmers farm tractor under his wing, and had nursed it as with special attention. The 1920s however, were not conducive to tractor sales by anyone, especially a firm with no real ties to the farmer. Selling sawmills, flour mills, and steam turbines was one thing, but these A-C endeavors were almost unknown to the average farmer.

Another major factor was an ever increasing farmer preference for tractors built by established and reliable companies. For a few years following World War One, tractor companies appeared overnight, and disappeared just as quickly. A good many farmers were taken in by big promises to the point that a great deal of distrust existed between farmers and the people trying to sell tractors. For the most part, buying from J. I. Case, Deere, International Harvester, or Rumely virtually assured reliability and satisfaction. Not to be neglected, Henry Ford made his mark in the 1920s tractor market with the Fordson. Ford was a household word — after all, it was Henry Ford who put rural America on wheels with his famous Model T. Ford already had the name recognition, and already had an established sales organization for his cars. Thus, farmers bought Fordsons by the thousands, all in an attempt to mechanize their farm at the least possible cost. Allis-Chalmers desperately needed to acquire an established farm equipment company if there was to be any hope of bucking a very competitive market, and if there was to be any chance of success in the tractor business. The chance came with the purchase of Rumely. Now, Allis-Chalmers could point with pride to a history of farm machinery development reaching back to 1853.

The history of Advance-Rumely begins with the birth of Meinrad Rumely in 1823. In 1848, Rumely left his native Germany for America. Two brothers, Jacob and John had earlier made the same voyage.

Meinrad Rumely had learned the millwright trade as a young man. After arriving in America, he worked with brother Jacob at Canton, Ohio, for a short time, making wooden pumps. During 1850 Meinrad took employment with Russell & Company of Massillon, Ohio, and a year later he was in Chicago working as a machinist. Within a short time, wanderlust reappeared, this time taking Meinrad to Clinton, Iowa and from there to St. Paul, Minnesota. In a month he was back in Chicago, but heard that the railroad was coming through LaPorte, Indiana.

The railroad from Toledo became a reality, and at the urging of friends in LaPorte, he set up a blacksmith and machine shop. His brother John now came to LaPorte, and together they operated a foundry and machine shop. With the railroad shops coming to LaPorte, M. & J. Rumely kept busy with custom work. In addition, the brothers Rumely developed and built cane mills to press the juice from sorghum cane.

During 1854, M. & J. Rumely began to build grain threshers. Some years later, Meinrad's grandson Edward asked his grandfather why he began building threshers. Grandfather and grandson were on a drive in the country, and Meinrad pulled up to some grain stacks and told young Edward to "Get me that flegel." (the German word for flail). "Now," he said, "take that flegel and pound all the grain from this bundle." By the time he was done, young Edward was completely worn out. Then Meinrad told him in effect, "Now you see why I wanted to build a machine that would eliminate this job."

In 1859 the Rumely separator won First Prize at the U. S. Fair in Chicago. By 1862 a small corn sheller and a hay press were added to the line.

In a curious parallel to some of the Allis-Chalmers ancestry, the Rumely firm began building a steam-powered ore stamp in 1867. Several carloads of gold-

Meinrad and Therese Rumely were married in 1855. Of their nine children, two sons, William N. and Joseph J. appear to have been the most active in the family business. Meinrad was born in Germany in 1823, coming to America in 1848, and settling at LaPorte, Indiana in 1852. Meinrad and his brother John operated under the title of M. & J. Rumely until 1882 when Meinrad bought out his brother's interest. William N., the eldest son, received a number of patents, primarily on improvements to steam engines and thrashers. Several of these were acquired jointly with his father. Beginning in the 1870s the Rumely firm assumed a management technique that was both agressive and progressive. Numerous patents indicate a rapid evolvement of Rumely engines and threshers during the last quarter of the nineteenth century. Despite the active participation of William N. and Joseph H. Rumely in the family enterprise, Meinrad remained as President of the company until his death in 1904.

Advance-Rumely Thresher Co.

bearing ore were shipped to LaPorte from Colorado, and after successful testing of the prototype, a limited number were built, with Rumely advertising the Forbes steam stamp as late as 1871.

Despite the brief excursion into stamp mills, the Rumely brothers maintained an intense effort to improve their threshers. Probably in response to the introduction of the portable steam engine by J. I. Case in 1869, Rumely followed suit with a similar unit in 1872.

Another curious journey to the unknown was Rumely's introduction of an ice elevator in 1879. Ice houses were the rule of the time, so a number of these units were sold locally. In addition, a substantial number were sold to breweries and other firms requiring a year-around supply of ice. The annual supply was harvested each winter from area rivers and lakes, and was stored for use throughout the rest of the year.

The New Rumely thresher was introduced in 1881. With little change, it was built until the Rumely Ideal took its initial bow in 1904. During 1882, Meinrad bought out the business interest of his brother John, reorganizing the firm as M. Rumely Co.

During 1886 Rumely announced a new straw-burning steam engine — this was probably a traction engine, but the actual introduction date of the Rumely traction engine remains unclear. In late 1891 Rumely announced a new self-feeder for their threshers, eliminating the tiring and dangerous job of feeding the grain into the separator by hand.

M. Rumely Co. continued a pattern of steady growth during the 1890s. By 1896 the firm offered an extensive line of portable engines, traction engines, threshers, and sawmills. During 1901, clover hullers were added to the line.

Meinrad Rumely died in March, 1904. When he bought out his brother's share of the business in 1882 and reorganized the company, Meinrad named himself as president, with son Joseph J. as secretary-treasurer and son William N. as vice-president and superintendent. Although Meinrad and Therese Rumely had nine children, these two seem to have been most active in the family business.

To Joseph J. and Margaret E. Rumely was born their eldest son, Edward A. Rumely in 1882. As a youngster, Edward took a great interest in the family business. Upon reaching college age, Edward enrolled at Notre Dame University, but after a year, he went to England, there enrolling at Oxford. The following year Edward was a student at Heidelberg University in Germany. During this period he became acquainted with Rudolf Diesel of engineering fame, and also became a lifelong friend of Prof. Hugo Junkers. Finally, Edward entered Freiburg University in southern Germany to study medicine, graduating magna cum laude. He also earned a dual degree in economics. Thus armed, Edward returned to LaPorte in 1907, and soon took an active role in operation of the family business.

Almost immediately after assuming a position with the company, Edward set about applying many of the economic and business ideals he had absorbed in Europe. Beyond this, he saw the need for a reliable farm tractor using an internal combustion engine. Numerous writings of Rumely in later years seem to indicate that he had been strongly influenced by Rudolf Diesel in this regard.

John A. Secor of New York City had been experimenting with internal combustion engines since 1885. Ultimately this led to the volume governing system still in use today. Secor also subjected the internal combustion engine to extensive scientific tests in order to determine how to extract the maximum amount of power with the least amount of fuel. Dr. Rumely learned of Secor's work and persuaded him to move to LaPorte. Arriving there in 1908, Secor, Dr. Rumely, and Wm. H. Higgins, the Rumely factory superintendent set about building a tractor. A careful analysis of newspaper accounts, Dr. Rumely's correspondence, and other existing data seems to indicate that one or more prototypes were built in 1908. Not until March, 1909 was word leaked by the press that Rumely was

The earliest extant illustration of the Rumely engine and thresher is this engraving from the early 1870s. The heading notes that this is "M. & J. Rumely's Portable Engine & Improved Separator at Work." Although the engraver waxed idyllic, as was the rule rather than the exception in these times, this illustration portrays the Rumely thresher with an integral straw stacker. Presumably the back half of the stacker could be folded over the machine for a transport position. Despite numerous searches, virtually nothing can be located concerning the sizes or styles available for this thresher. Quite possibly the early Rumely engines owed at least part of their design to I. W. Forbes of LaPorte. Rumely built the Forbes Quartz Battery during the early 1870s ..... this ore stamp, although apparently successful, was Rumely's only venture into ore and mining machinery.

Advance-Rumely Thresher Co.

contemplating a tractor factory, and not until October of that year was it announced that testing of the new tractor had been successfully concluded.

The brand-new OilPull tractor plant opened on February 21, 1910. By November 1 of that year, the first 100 OilPull tractors had been built — not a bad output in a time when the total accumulated output of many tractor builders was somewhat less than 100.

It appears that the 1910 model acquired the nickname of "Kerosene Annie," an obvious reference to its use of kerosene fuel. A study of newspaper accounts and in fact, Dr. Rumely's own writings indicates that the OilPull received cosmetic changes from the hand of J. C. Junkin. To quote Dr. Rumely in this regard: "We took him (Jack Junkin) when we bought the Advance Thresher Co. in 1911. He experted the OilPull and helped finish off the design." J. C. Junkin had been with Advance Thresher Co. for several years, and appears to have done design work for various competing companies on a concurrent basis. It was Junkin who designed the Twin City tractor offered by Minneapolis Steel & Machinery Co., as well as the short-lived Advance tractor. This probably explains why the Advance and Twin City tractors bear an uncanny resemblance.

In 1911 Rumely acquired Advance Thresher Co. of Battle Creek, Michigan. This was followed in October of that year by the purchase of the Gaar-Scott Co. of Richmond, Indiana. The following year, 1912, saw the purchase of American-Abell Engine & Thresher Co. of Toronto, Ontario, for $2.5 million. Northwest Thresher Co. of Stillwater, Minnesota, came into the fold during October, 1912.

M. Rumely Company reorganized in April, 1913, as two separate and distinct entities. The original title of M. Rumely Co. now applied only to engineering and manufacturing operations, while the new title of Rumely Products Co. applied to sales and collections. This change was made primarily as a tax advantage.

During 1914 Rumely Products Co. acquired certain sales rights to the Falk gas engines built by the Falk Co. at Milwaukee. Whether Rumely purchased Falk's gas engine operation outright, or made a selling arrangement with this firm remains unclear. What is clear however, is that the Rumely-Falk engines carried the famous Secor-Higgins carburetor as used on the OilPull tractors. Indications are that Rumely Products Co. also consummated a stock deal with the Olds gas engine people at Lansing, Michigan. In return, Rumely marketed the so-called Rumely-Olds engines, cream separators, and other items.

Already in 1913, rumors were circulating of financial problems for the Rumely organization. These persisted through the following year, and finally on January 21, 1915, the house of Rumely caved in. By September, announcement was made of a reorganized firm of Advance-Rumely Thresher Co. After nearly a year of

The New Rumely separator shown here carried a "Patents Applied For" notice, indicating that this machine or parts of it were soon to be under patent protection. Already in 1879, John Rumely was awarded Patent No. 214,056 under the title of "Thrashing Machine" but the patent itself covered a unique form of tapered socket for the cylinder teeth. John Rumely also acquired Patent No. 251,642 in 1881. It gave specific protection for the tailings conveyor used on the New Rumely thresher. Close examination of this engraving shows a slatted conveyor just above the front axle and inclined upward toward the threshing cylinder. Apparently the cut grain was fed downward onto the traveling conveyor and then taken into the cylinder for threshing. Threshers of the 1870s were considerably improved over those of 20 years earlier, but it would be yet another two decades before thresher design reached the degree of perfection that was acceptable to builders and users alike. It seems likely that Meinrad Rumely would have been familiar with threshing machines prior to leaving his native Germany, especially with his background as a millwright. The six-volume *Encyclopaedie der gesammten Landwirtschaft* published in 1850 describes and illustrates a variety of threshing machines, some of which were powered by portable steam engines.

The LaPorte newspapers of December 10, 1891, announced that M. Rumely Co. would be building a band cutter and self-feeder attachments for their threshers beginning with the 1892 season. The self-feeder was a great labor saving device, since it eliminated the need for two men to stand on a platform in front of the cylinder, cutting bands and hand-feeding the grain into the machine. Now the bundle pitchers could stay right on their rack and pitch into the feeder. Note the divide board in the center of the feeder. The divide board kept the grain bundles straight and provided even feeding, even with men pitching bundles into the machine from both sides. As the bundles approached the head of the feeder the band knives cut the twine around each bundle of grain. This illustration from an 1898 M. Rumely Co. catalog illustrates the tremendous design changes that had taken place. The machine shown here was a 36 x 56, which translates into a threshing width of 36 inches and a separating or cleaning area that was 56 inches wide.

legalities, M. Rumely Co. and Rumely Products Co. was sold at a forclosure sale for $4 million. Five days after the sale, on December 14, 1915, Advance-Rumely Thresher Co. was formally incorporated. Canadian Rumely Co. of Toronto was part of the sale, with Advance-Rumely owning the entire stock of this firm.

Several factors were involved in the demise of M. Rumely Co. Sales figures had been excellent — too good perhaps. The majority were made on time, but poor crop prices in some areas and crop failures in others left with what is now called "a negative cash flow." To cover the temporary shortfall of income, Dr. Rumely attempted to get additional time from the bankers — they refused.

Another important, and often overlooked factor was the fact that the Rumely organization had simply grown too fast. The acquisition of four major companies in a two-year period strained the Rumely finances to a point where virtually no cushion was left for "contingencies," nor could any additional cushion be secured so that the company could weather the storm.

Speculation is easy, but an analysis of the net profit figures in the years following reorganization shows Advance-Rumely common stock for 1918 ranging from 11 to 26⅛ on the NYSE. This stock earned $3.19 per share. Thus, had Rumely been able to get through the disastrous 1913-15 period, the company under Dr. Rumely's leadership might have taken many a different turn in the years to follow. For 1919, Advance-Rumely stock sold from 21 to 54 per share and yielded $12.03 on common. After this point Advance-Rumely fell from grace, with common hitting a low of 6½ in 1923. Earnings dropped to nil and remained there for some time.

During 1923 Advance-Rumely introduced a new all-steel thresher and offered a silo filler. On December 28 of that year Aultman-Taylor Machinery Co. of Mansfield, Ohio, ceased to exist and came under the wing of Advance-Rumely.

A series of new lightweight OilPull tractors was unveiled in October, 1924. The lightweights retained many of the basic principles set forth in the original OilPull design, but did so with higher engine speeds and a lot less iron than before. Farmers welcomed the new design and Advance-Rumely gained a temporary reprieve.

The announcement of the Advance-Rumely combine gave the company additional income that helped offset sluggish tractor sales. The No. 1 combine of 1925 was offered in Prairie and Hillside styles.

Staying in the tractor market required something beside the OilPull. Even the lightweight designs were far too heavy for row-crop work. Thus, on April 13, 1928, the Rumely DoAll was announced. Only a few of these combination machines were built the first year, with several hundred being assembled in 1929.

By mid-1928, rumors circulated that a merger of Advance-Rumely with another firm was imminent. Finley P. Mount, president of Advance-Rumely denied the rumors, but by October, 1929, Mount was in contact with Otto H. Falk, president of Allis-Chalmers.

At this early stage of negotiations with A-C, enters once again Dr. Edward Rumely. After M. Rumely Co. went into receivership in 1915, Dr. Rumely went to New York and acquired control of the New York Evening Mail. All went well until Dr. Rumely fell victim to the anti-German sentiment so pervasive during World War One. An over-zealous government official had Dr. Rumely arrested on charges of conspiracy, with the government accusing him of financing the Evening Mail through German financial sources. Despite his

From existing serial number lists and other data, it would appear that the Rumely Ideal thresher was first marketed in 1904, with production continuing until 1924. During that period, some 20,000 units were produced. The total range of sizes was apparently not built at the same time — the old standby 28 x 48 and 32 x 52 sizes seem to have been the most popular, although a substantial number of 36 x 60 threshers were built, and Rumely sold some huge 40 x 64 threshers for at least a few years. At the small end of the scale were the Rumely Ideal threshers in 22 x 36 & 44 inch sizes. These units were primarily intended for the small farmer or for those with limited tractor power. Rumely catalogs indicate that the 22 x 36 was ideally matched with the 12-20 OilPull tractor, while the 28 x 44 machine was suitably used with the 16-30 OilPull.

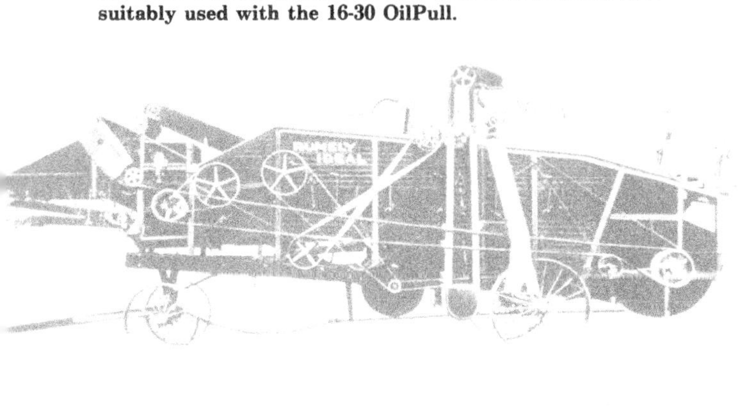

A 1911 file photograph illustrates the 20-man crew of the thresher shop posing with another machine loaded on a flatcar and ready to begin its duties. The Rumely Ideal wood separators required an immense amount of precision carpentry. Beginning with the superstructure and continuing to the interior straw racks, all of this material was fashioned from wood. The process began with the careful selection of the wood itself for soundness and strength. A major selling point for the Rumely Ideal was its strength of construction. Following Meinrad Rumely's Old World tradition of only the very best workmanship, each machine was carefully built to exacting specifications, then carefully tested before leaving the factory. The threshing cylinder was constructed of twelve double-thickness steel bars set upon three grey iron cylinder heads. The bars were held in place by the cylinder teeth, and in addition, three iron bands were shrunk over the cylinder for additional strength. Each cylinder after being built was carefully balanced so as to secure a smooth running machine.

Advance-Rumely Thresher Co.

contention to the contrary, and in spite of virtually nothing but circumstantial evidence, Dr. Rumely was convicted. After a series of appeals he was sentenced to prison, but was released after one month. Finally, President Coolidge granted a full pardon for Dr. Rumely in 1925.

Although Dr. Rumely had not shared an obvious role in the Advance-Rumely organization, he may have served in an advisory capacity to Mr. Mount. Dr. Rumely's subsequent activities as a mediator between Advance-Rumely and Allis-Chalmers surely could not have been possible without the consent and cooperation of Mount or his board of directors.

In January, 1930, Dr. Rumely met with General Falk at Milwaukee regarding the merger. Falk was presented with a balance sheet showing Advance-Rumely assets of nearly $33 million. Of this amount, nearly $8 million was in customer notes, with another $13 million ascribed to "trade names, trade marks, patents, designs, good will, and other intangible values."

Dr. Rumely pointed to the advantages of a merger. These included: (a) Liquidation of farmer notes would create funds which Allis-Chalmers could than use to carry its own tractor paper; (b) Allis-Chalmers would gain an experienced collection organization with average losses of less than 1%; (c) the merger would automatically give Allis-Chalmers excellent position in the farm equipment business in terms of total size and in terms of an established dealer network; (d) Allis-Chalmers would gain an extensive branch house organization.

Negotiations continued off and on until a final settlement was reached on June 1, 1931. Dr. Rumely remained active in promoting the merger, and continued stressing the points made in his initial contact with General Falk. It would appear that a sizeable number of A-C directors were opposed to the purchase, primarily because of Rumely's enormous accounts receivable in the form of customer notes. On the other hand, Advance-Rumely stockholders hoped for a lot more money than they finally settled for.

With the purchase of Advance-Rumely, A-C became the nation's fourth largest farm equipment builder. All Rumely tractor models were phased out, including the OilPull, DoAll, and Rumely Six. Although Allis-Chalmers continued to build the Advance-Rumely steel threshers until 1935, all other equipment lines were terminated. Under the title of LaPorte Works, Allis-Chalmers had now solidified its position in the farm equipment business.

As a footnote to Dr. Rumely, he continued his efforts toward adaptation of diesel power to farm uses; this despite losing control of M. Rumely Company in 1915. In 1922, Dr. Rumely was instrumental in organizing Super Diesel Tractor Corp. at LaPorte. Phillip L. Scott was the consulting engineer for the company. The salient feature of Super Diesel design was a complete and progressive air-fuel mixing process derived from a small auxiliary cylinder and piston in the cylinder head. Throughout the 1930s the experiments continued, but Super Diesel never seemed to get off the ground to any extent, and certainly never got into tractor production.

In later life, Dr. Rumely recalled meeting Charles F. Kettering at Dayton, Ohio, in 1908. Kettering was experimenting with a gasoline engine. Under load the engine knocked slightly. Kettering took a cloth, poured a few drops of iodine on it, and held it a few inches away from the air intake. The knocking ceased! Dr. Rumely explained to Kettering that his approach to the problem of knocking was by adding water to the air-fuel mixture. A few years later, Kettering developed the use of tetraethyl lead as an anti-knock additive to motor fuels.

Dr. Rumely died in 1964. Throughout his life he

A study of early parts books indicates that the Rumely Ideal Jr. thresher was built in an 18 x 40 inch size between 1909 and 1917. For 1918 only, the Ideal Jr. was built in a 20 x 36 inch size. It used a hand-feed, together with a slat-stacker. Probably intended for areas where stack threshing was still in vogue, this machine permitted the continuance of earlier threshing methods but did so with a greatly improved design. Certain areas continued the practice of threshing from a barn, and a small machine of this type was quite adaptable to this use. Like the larger Ideal separators, the Ideal Jr. used wood throughout its construction. According to Rumely catalogs, the Ideal Jr. models were also intended for the requirements of the small farmer and those having small tractors with limited power. Production of each individual style is not possible to determine — threshers were consecutively numbered, regardless of size.

For a short time following the 1911 purchase of Gaar, Scott & Co., Rumely continued to build the Gaar-Scott line, renaming it the "Tiger Line" so-called because of the famous trademark of the Gaar-Scott people. The Tigerbilt threshing machinery was marketed long enough to dispose of remaining stocks, and after that Rumely continued to offer repair parts for the Gaar-Scott line. Rumely continued parts service for the TigerPull tractor until 1928 and finally discontinued parts service on the remainder of the Gaar-Scott line in 1931. Abram Gaar had founded the company in 1849, with the company being incorporated under the Gaar-Scott name in 1870. Although the thresher shown here was somewhat different in exterior appearance than the Rumely Ideal, it was nevertheless an excellent competitor. Rumely's purchase of Gaar-Scott effectively removed a major contender from the business.

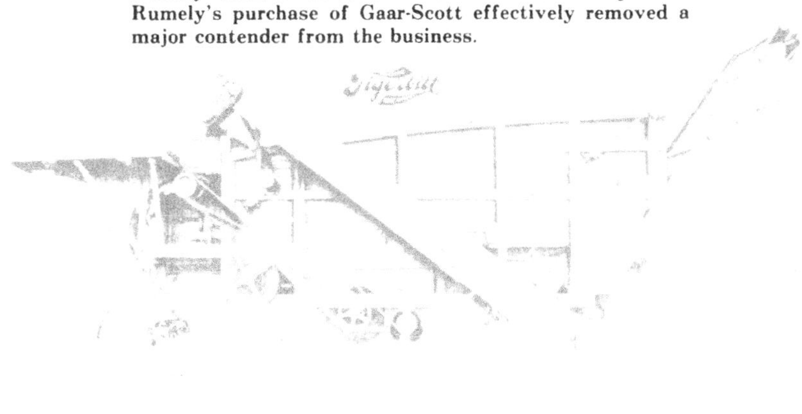

Advance-Rumely Thresher Co.

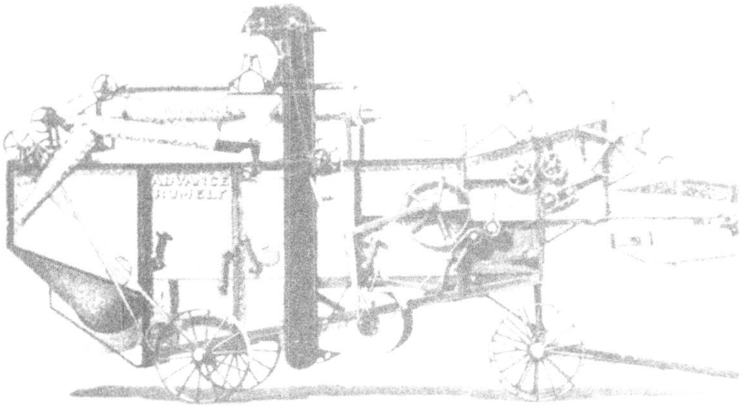

In addition to its larger all-steel separators, Rumely offered this machine in 22 x 36 and 24 x 38 inch sizes. The latter size was introduced in 1924 and remained on the market for a time, but the 22 x 36 model, first built in 1918 remained a part of the line until 1933, some two years after Allis-Chalmers purchased Advance-Rumely. Designed specifically for the small farmer, the 22 x 36 model listed at $1,028 F.O.B. Saskatoon, Saskatchewan in 1928. This price included the windstacker, self-feeder, and weigher — in other words, a complete, ready-to-go machine. After purchasing Advance-Rumely, Allis-Chalmers apparently changed this machine slightly and marketed it as a 22 x 38 thresher during the 1932-36 period. This model and the larger 28 x 46 model were the last Allis-Chalmers-Rumely threshers built. By 1936, Allis-Chalmers was well into the combine business, as threshing machines were fast falling from favor.

Advance Thresher Co. of Battle Creek, Michigan, was incorporated in 1881. Over the next 20 years it grew to a factory of immense proportions. In fact, the 1901 year book states that theirs was "the most complete thresher factory in the world." Not only did Advance build an excellent thresher — the company also pioneered the steam traction engine, offering ten different styles for 1901. These ranged from a small 6 horsepower simple engine up to a big 26 horsepower compound. Advance engines used the Marsh reverse gear, together with a double-ported valve. The latter feature gained twice the effective valve opening during a specified travel, thus giving the cylinder the greatest amount of steam in the shortest possible time. Advance separators for 1902 were offered in seven sizes ranging from a small 24 x 40 inch model up to a big 36 x 52 inch thresher. Also a part of the Advance line were the "Pony" sawmills available in several sizes.

appeared to be at least a full generation ahead of his time, often expounding ideas that seemed radical at the time, but are generally accepted today. Already in 1908 he established Interlaken School, a boys school that stressed the idea of "learning by doing." His professional training had convinced him of the need for vitamins as an essential part of the daily diet for man and animals alike. In 1925 he established Vitamin Food Co. to promote this idea.

Already in the 1920s Dr. Rumely advocated the use of nitrogen fertilizers in crop production; he also promoted the use of urea as a supplement to animal feeds.

During the great depression, Dr. Rumely organized the Committee for the Nation to promote monetary reform; when Franklin D. Roosevelt attempted to pack the Supreme Court, Rumely and others responded with the Committee for Constitutional Government. Their mass mailings effectively turned public sentiment against the idea, with Harold Ickes later denouncing the committee as "mail order government."

Some 50 years ago production of the Rumely OilPull tractor ceased. Yet, the OilPull still maintains a special place in farm tractor development. Various marketing surveys indicate that despite the many years since it has been built, the OilPull is still of third rank in popularity by vintage tractor enthusiasts. Perhaps these remaining examples will serve to perpetuate the Rumely mystique.

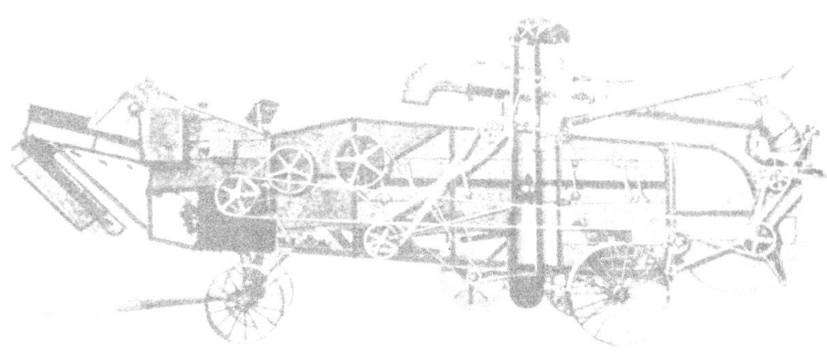

Like most other thresher manufacturers, Rumely developed an all-steel separator. J. I. Case Co. had taken the lead some years earlier, but the Ideal steel separator made its debut in 1916. From that time through 1920, Ideal threshers were built at the former Battle Creek, Mich., factory of Advance Thresher Co. Beginning in 1921 all threshers were built at LaPorte. The Rumely steel thresher in its larger sizes was offered in 28, 30, 32, and 36 inch models. Company records indicate that the 28-inch size was built at various times with separator sizes of 44, 46, and 48 inches. The 30 x 48 inch model was apparently built only in 1924, while the 32 x 52 was produced from 1916 to 1928. The big 36 x 60 machine was introduced in 1916, with intermittent production continuing through 1930. A 1926 price book lists the 32 x 52 Ideal Steel Thresher at $1,350. Beyond this figure was $300 for the windstacker, $325 for the feeder, and $145 for the grain weigher and its attachments. The total figure of $2,120 still did not include a drive belt should a new one be required. A 125-foot 7-inch wide belt added yet another $76 to the total cost.

Advance-Rumely

Advance Thresher Company was organized at Battle Creek, Mich., in 1881. By the turn of the century, Advance was a major builder of steam traction engines and threshers, this despite formidable competition. The Advance trademark illustrated here was as unique as it was colorful. The "Advance" flagbearer rides a chariot pulled by three racing horses — the one to the left representing ingenuity; the one to the right, industry; and the large black horse in the center representing invention. M. Rumely Company purchased Advance Thresher Co. in 1911. Twenty years later, Allis-Chalmers bought out Rumely.

Advance began building steam traction engines about 1888, and by the time the company went into the Rumely fold in 1911, some 23,000 engines had been built. This 16 horsepower model of 1901 more or less typified the line at that point. This particular engine used an 8½ x 10 inch bore and stroke. All simple engines of this series were designed for a boiler pressure of 120 psi, but the 15, 18, and 26 hp compound engines carried a pressure of 150 psi.

The 1901 Advance separator line included seven sizes having cylinder lengths from 24 to 36 inches, and corresponding separating widths of 40 to 52 inches. All machines were regularly equipped for steam power, but the five smallest sizes could also be furnished with the special bevel gears for operation with a horsepower. For 1901 at least, Advance offered the Dingee-Woodbury horse-power, probably the best known of these machines. It could be furnished for use with as many as 18 horses if desired.

Almost immediately after taking over Advance Thresher Co., Rumely made their Battle Creek plant another site of operations. Thresher production seems to have been centered at Rumely's LaPorte plant, although after the merger, M. Rumely Company seems to have taken the best features of their Rumely Ideal and combined them with salient features of the Advance design shown here. Likewise, the Rumely steam engine was blended with the Advance design to give what many historians now deem as one of the finest traction engines built. Rumely continued to provide parts service for the Advance line until being purchased by Allis-Chalmers in 1931, at which time parts service was suspended.

By the time Rumely took over Advance Thresher Company their steam engine design had been refined to this general appearance. Advance engines were highly respected for their quality workmanship, along with an excellent boiler design. The engine shown here is a 21 horsepower tandem compound which briefly explained means that two steam cylinders were used on a common piston rod. Steam from the boiler was first admitted to the smaller high pressure cylinder. After expanding somewhat in this cylinder it moved into the so-called low pressure cylinder. This cylinder was somewhat larger in area. The idea of increasing the size was to permit approximately the same mean effective pressure on both pistons. Despite the attempts at promoting fuel economy with a compound engine, the fuel savings was largely consumed on the substantially higher cost compared to a simple engine.

Advance-Rumely Thresher Co.

Production of Rumely Rice Separators is intermingled with production figures on standard threshers, therefore no specific production dates can be determined. The Rice Separator was simply a modified version of the standard machine — modifications presumably included some changes at the cylinder, and possibly in the straw racks. Over the years, Rumely Ideal rice threshers were built in 28 x 48, 32 x 52, and 36 x 60 inch sizes. The all-steel machines were initially listed as "grain and rice threshers," indicating that the machine was substantially the same for both applications. One other significant difference is apparent — Rumely catalogs noted that the Ideal wood-frame machine was constructed of special material to withstand the climatic conditions of rice territory. Presumably this meant the use of different kinds of wood than were found in the standard machine.

When Abram Gaar purchased the Spring Foundry in 1849 he renamed it A. Gaar & Co. This Richmond, Ind., firm also included a partner, William G. Scott. In later years, Abram Garr's son John M. assumed an active role in the business. During 1870 the company was incorporated under the title of Gaar, Scott & Co. An exact timeframe for this company's entry into the engine and thresher business remains unclear, but indications are that Abram Gaar's skills as a cabinet-maker were used virtually from the beginning of the company to build threshers. An early engraving illustrates the Spring Foundry as a typical two-story factory of the 1850s. By 1911 when Gaar-Scott was absorbed by M. Rumely Co., the firm had grown to a considerable size. Rumely continued to build Gaar-Scott threshers for a short time at least, calling this the "Tigerbilt" line. The Gaar-Scott tractor, developed by 1911, was absorbed into the Rumely operation with production being swiftly brought to an end.

Probably in response to the needs of specific crop applications, Rumely offered their Bean and Pea Huller during the 1920s. Total production was apparently quite small, and production time seems to have paralleled the output. Edible beans, soybeans, and various types of peas were the obvious crops for this machine. Built only in a single model, it used two cylinders. The front cylinder ran at 225 rpm, just fast enough to shatter the pods. A second cylinder running twice as fast was placed some distance back, and this cylinder apparently did the cleanup work of threshing out the remaining seed pods. With the coming of the combine, harvesting methods changed dramatically, and so the need for the bean and pea huller evaporated. This machine was of substantial size — nearly 8 feet wide and 25 feet long. For 1926 the Advance-Rumely pea and bean huller listed at $1,275.

With farmers making increasing use of clover as a forage crop, an efficient method of producing clover seed became essential. Bailey's *Cyclopedia of American Agriculture* notes that in 1900, nearly 85% of total U. S. clover seed production was concentrated in the states of Indiana, Ohio, Illinois, Wisconsin, and Michigan. Thus it seems logical that Rumely developed a clover huller to capture a portion of this market. Over 1.3 million bushels of clover seen were harvested in 1900, so getting even a small portion of this market meant that a good many hullers could be sold. The Rumely huller of 1902 was rather typical of its contemporaries — like other machines, it used a separate recleaner after the seed left the huller. Clover hulling consisted essentially of rubbing the seed from the heads. With the Rumely huller the clover first entered a stemming cylinder that separated the seed heads from the stems. As the stems made their way out of the huller they passed over slatted racks which permitted any remaining seed to drop through. With this design the hulling cylinder was not overloaded with the stems, giving it greater capacity to do its intended job, that of rubbing the tiny seeds from the heads.

Advance-Rumely Thresher Co.

Gaar, Scott & Co. of Richmond, Ind., has a history going all the way back to 1836. Abram Gaar started out as a cabinet maker and millwright, finally ending up in the foundry business by 1849. His son, John Milton Gaar took a special interest in threshing machines, and since these were built primarily of wood, the skills of both a cabinet maker and a millwright were required. Along the way, William G. Scott became a partner, with Gaar, Scott & Co. being incorporated in 1870. By the turn of the century Gaar-Scott had adopted the tiger as its mascot, calling their products a part of the "Tiger Line." Shown here is a 1910 model Gaar-Scott thresher, complete with gearless stacker, self-feeder and a grain weigher.

For 1910, Gaar-Scott offered their single cylinder "coal and wood" engine in 13, 16, and 18 hp sizes. The regular-geared engine with the Universal boiler was offered in 16 and 18 hp sizes. Following this were the regular-geared double cylinder engines in 18 and 22 hp sizes. For plowing and heavy traction work, heavy-gear engines were built in single cylinder sizes of 22, 25, and 32 hp. A 39 hp tandem compound was also available, and topping the line were 22 and 25 hp double cylinder models. King of the line was this huge 40 hp model featuring double tandem-compound cylinders. Advance-Rumely continued parts service for these engines until 1931.

Although some thresher builders attempted to convert ordinary threshing machines into clover threshers when needed, this effort was usually somewhat less than satisfactory. Some companies preferred to build a special clover huller, tailored to the particular needs of this valuable seed crop. Included among these was the Gaar-Scott clover huller, with a 1910 model being illustrated here. Like most clover huller designs, the Gaar-Scott used a threshing cylinder to remove the majority of leaves and stems, with the final threshing taking place in the hulling cylinder. This particular machine had a threshing width of 35 inches and a 42 inch separator width. The finished seed then passed into a recleaning device shown at the center of the machine, and from there to waiting grain sacks for the trip to the granary.

Gaar-Scott steam traction engines were well known, with the Tiger Line being quite popular in some areas of the country, with at least a sprinkling of these engines going across the entire United States. The "Universal" boiler was so-called for its adaptability to wood or coal, and for certain areas of the country, to the use of straw as fuel. On the open prairies, coal was expensive, wood was virtually non-existent, while straw was plentiful and best of all, it was free. By using special grates, a firebrick arch in the firebox, and a special feeding chute, these engines could be easily converted to burning straw. When necessity demanded, it could then be changed back to conventional fuels.

A 1910 Gaar-Scott catalog illustrates the link reverse as used on these engines. The link reverse was one of the earliest styles, and was widely used on locomotives. Although slightly slower in its action than a radial gear such as the Woolf, the link had the distinct advantage of providing fairly uniform steam distribution even though it might be worn to some extent. Radial gears, including the Woolf gear, were not so tolerant of worn parts or misadjustment. Gaar-Scott added a special feature to the link reverse on their engines. When the engine was to run only in one direction for some time, as in sawmill work, one eccentric could be unhooked, and by adding a special cam rod, only one eccentric was required, leaving the link mechanism at rest, and preventing unnecessary wear.

Advance-Rumely Thresher Co.

M. Rumely Co. apparently introduced this clover huller in 1904, concurrently with the Rumely Ideal grain threshers. The design apparently remained virtually without change at least into the 1920's, but no specific production data has been located. This machine was offered in two sizes — the smaller Standard used a 34-inch stemming cylinder and a 40-inch hulling cylinder. The somewhat larger Special carried a 40-inch stemming cylinder, along with a 52-inch hulling cylinder. Advance-Rumely catalogs of the 1920s list this machine as a Clover and Alfalfa Huller, reflecting its additional application to a relatively new forage crop. For 1926, Advance-Rumely listed the "Standard" at $1,150, and the "Special" at $1,350. Additional required equipment not included in the base price was a windstacker for $210 and a self-feeder for $250.

Although production of Rumely steam traction engines seems to have begun in the late 1870s, no specific data can be located concerning the first attempts for a prime mover. Already in 1876, Meinrad and John Rumely applied for Letters Patent covering a friction clutch — it is of the same general design as used on the 1895 model shown here. The friction clutch gained further refinement and subsequent patent protection through William N. Rumely's Patent No. 428,634 of 1890. Rumely's link reverse valve gear of these early engines resulted from Patent No. 190,083 granted to Theo. Scheffler of Patterson, N. J. Granted in 1877, this link design was apparently used for many years. The New Rumely engine shown in this photograph was of the rear-geared design, or in steam engine parlance, rear-mounted. Standard equipment included a link reverse, although the New Rumely could be equipped with the Arnold valve gear if so requested. Four sizes were available for 1895 — 8, 10, 12 and 15 horsepower. An Excelsior governor was standard equipment. The New Rumely carried some unique features that included elliptical springs to carry the front end. Another unique feature was the final drive design which utilized internal tooth gears.

M. & J. Rumely were probably experimenting with steam power during the 1860s, and by the early 1870s they offered this little portable to operate their threshers. The *Patent Office Gazette* abounds with developments of the Rumely people — Meinrad received several patents during the 1870s, as did his brother John. Meinrad's son, William N. Rumely seems to have been especially interested in steam power since numerous patents concerning steam engine components were issued to him during the 1880s. The so-called wet-bottom boiler is obvious in this photograph — it represented the best design of the time, although engineers in general could never agree as to whether this style or the dry-bottom was the best. (With a wet-bottom boiler, water completely encircles the firebox, but with the dry-bottom, water extends down all sides of the firebox, but not across the bottom). Although barely visible in this photograph, the feedwater heater just below the connecting rod carries the words, "M. & J. Rumely, LaPorte, Ind."

To meet the specific needs of wheat producing areas, Rumely offered this model built specifically to burn straw as fuel. Large areas of the Plains States had little or no available wood, and transporting coal to the threshing site was also a burden. Straw burning traction engines were equipped with a special straw chute. In order to keep cold air away from the fire it was necessary to keep the chute full at all times, pushing straw in gradually, and in small forkfuls. Feeding the straw too fast would check the flame, and feeding too slowly would not maintain the heat necessary to keep up steam. Through 1901 Rumely used a return flue boiler on their straw burning engines, but for 1902 a direct-flue, locomotive-type boiler was used — it was equipped with special deflector and grates. The same boiler could also be retrofitted to burning coal or wood as desired.

Advance-Rumely

About 1895 Rumely introduced a compound traction engine using the Woolf valve gear. This engine was built in 13, 16, and 20 horsepower sizes. As is evident from this illustration, the tandem compound design included a forward high pressure cylinder, with the low pressure cylinder mounted directly to the engine frame. In practice, the exhaust from the high pressure cylinder, passed into the low pressure cylinder for further expansion. Because of the expansion that had already occured in the high pressure cylinder, the low pressure cylinder was of substantially larger area. The vast majority of traction engine builders offered either a tandem compound or a cross compound design at some time during their career. Of these, the vast majority discontinued their production after a few years. Compounding was widely practiced with large stationary engines, particularly those of the Corliss design, and with excellent results. For farm engines the advantage was much smaller. Farm engines were relatively inexpensive to operate, and the accrued savings from compounding was hardly justified by the additional investment. A major reason for the failure of compound traction engines generally, at least in the 1895 era was the relatively low boiler pressures then in use. Using a boiler pressure of say, 100 pounds gives a saturated steam temperature of 328 degrees F. Without the benefit of an indoor environment and excellent insulation from heat loss, the expansive process was relatively small after leaving the high pressure cylinder of a compound. In later years, successful compound traction engines of other builders utilized boiler pressures of 150 to 200 psi.

Rumely's double cylinder traction engine for 1902 was offered in 15, 20, and 25 horsepower sizes. Instead of the Arnold reverse, the double cylinder was available only with the link reverse. Its action was somewhat slower than the Woolf and other radial valve gears, but Rumely compensated for this by using a double ported valve. In effect, a valve opening of one-eighth inch actually gave an effective opening of twice that amount. Rumely first offered the double cylinder engine in 1901. The double cylinder design was favored by many engineers simply because with the double cylinder engine there were no dead centers to contend with. Single cylinder engines if stopped on dead center must be pushed away from the center by hand, but the double cylinder design placed the cranks at a 90-degree angle. With this arrangement, dead centers were virtually impossible.

In 1902 the Rumely traction engine took on a new appearance. Although this year's production still featured the unique internal tooth bull gears, the engine now had a classy look, especially with the addition of a three-quarter length canopy. Rumely catalogs continued to stress the importance of their rear-mounted design which they claimed was far superior to the side-mounted method used by some other builders. (Side-mounted engines carried the drive wheels on heavy brackets riveted to the side of the firebox). Rumely single cylinder engines carried either the Arnold or the Rumely gear reverse. A study of early Rumely parts catalogs leads to the conclusion that both styles were used, either at the wish of the buyer or at the whim of the builder. All 1902 Rumely engines with the exception of the 25 horsepower model used a $10\frac{1}{4}$ inch stroke, varying the bore to suit the specific power requirements. The 25 horsepower model carried a $10\frac{5}{8}$ x 11 inch bore and stroke.

"No piece of modern machinery receives harder usage than the Traction Engine." So stated the introduction to Rumely's 1908 catalog. This initial statement shows that Rumely recognized the use and abuse to which traction engines were subjected, so every precaution was taken to build each part heavy enough to stand the most severe strain. A 1908 catalog shows the single cylinder series to consist of 12, 16, 20, and 25 horsepower sizes. Features included the Arnold reverse, a drop forged connecting rod, integral crosshead pump for automatic boiler feed, and a special anti-friction metal in all bearing parts. Rumely often pointed to this latter feature, claiming their anti-friction metal to be different from all others. Patent research discloses that William N. Rumely was indeed granted Patent No. 550,293 in 1895 for Antifriction Metal. This material was an alloy of lead, tin, antimony, copper, and phosphorous.

Advance-Rumely Thresher Co.

Rumely's double cylinder engine for 1908 was offered in 16, 20, and 25 horsepower sizes. The General Purpose engine used gearing sufficiently heavy for use, but a special Plowing Engine was built in 25 and 30 horsepower sizes. The plowing engines used extra heavy gearing as a major difference from the standard models. Shafting and boxings were much heavier, and the coal and water bunkers offered much greater capacity. Plowing engines used a slightly larger boiler — its design included additional flues for extra heating surface. A heavier flywheel was featured, and the drivewheel width was increased for greater traction.

The Rumely double cylinder steam plowing engines carried a higher boiler pressure than that used on the standard threshing engine. During the 1905-1910 period many steam engine builders opted for the largest possible design, with Rumely culminating their effort in a huge 40 horsepower model. Far more popular however, were the 25, 30, and 36 horsepower engines. An engine typical of this series is illustrated from a 1908 catalog. It featured a jacketed boiler to reduce heat loss and conserve both fuel and water. The drivewheels were built of steel to withstand the strains encountered in plowing. The 36 horsepower model could pull fourteen plows with ease in most soils. Those persons unfamiliar with steam engine ratings would likely think that 36 horsepower was indeed a very small amount of output for such a gargantuan. The entire matter of horsepower ratings has long been a matter of discussion, and it appears that by some means, many years ago, the decision was made that this figure represents the pull that would be exerted by 36 horses — horsepower in the purest sense of the term. In fact, a 36 horsepower steam engine would likely have a brake horsepower rating of approximately 130.

When M. Rumely Co. bought out Gaar-Scott in 1911, they also bought all rights to their traction engines. The Gaar-Scott engine shown here used a side-mounted design — the very design that seemed so abhorrent to the Rumely people. Despite this complaint, the Gaar-Scott engine was in fact very simple, and very well built. Its simplicity included a link reverse, along with other then-current features. Gaar-Scott apparently began building traction engines sometime prior to 1885, and by the time the company was absorbed into Rumely of LaPorte, over 11,000 traction engines had been built. Gaar-Scott production records indicate that after Rumely acquired the company, another 1,316 engines were built, probably to use the remaining inventory of castings. After 1914 the Gaar-Scott engine was no more.

Like Gaar-Scott, Advance Thresher Co. of Battle Creek came into the Rumely organization in 1911. Advance began building traction engines in 1885. The late-model Advance shown here was offered in 10, 12, 16, 20, 22, and 30 horsepower sizes. From all appearances, Advance-Rumely Thresher Co. eventually incorporated some of the Advance features into the Advance-Rumely Universal engine. By 1911, Advance had built over 12,000 traction engines. Limited production continued at Battle Creek after the Rumely takeover, with the last Advance engine, No. 14638, being built in 1917. Advance engines in general used the Marsh reverse gear. Although these engines were of side-mount design, each rear stub axle was supported within a framework resting on very heavy springs. Additionally, the Advance stub axle was of hollow design to permit easy and effective lubrication. Straw burning engines were available, and Advance also built a compound traction engine in virtually the same time frame as Rumely.

Advance Rumely

The Advance-Rumely Universal engine first appeared in 1915. From all appearances, this was little more than the same engine that had formerly been built by M. Rumely Co. especially since Advance-Rumely was still building the Advance engines at the Battle Creek Works. Sometime during 1916 it would appear that the Universal first took its place as a distinctively new design. The single cylinder engines were built in 16, 18, 20, 22, and 25 horsepower sizes. During 1917 a double cylinder compound engine was offered, followed in 1918 by an 18 horsepower compound. Not until 1920 did Advance-Rumely again offer a compound; this time a 25 horsepower model was built. It used a 12-inch stroke with a 9-inch diameter on the high pressure cylinder and a 13-inch diameter on low pressure cylinder. The last Advance-Rumely Universal engine, a 20 horsepower model with a 9 x 11 inch bore and stroke, was built in 1924.

John Alstyne Secor was born in New York City on June 28, 1847. Mechanical engineering was well established in the family ancestry, with John following the same vocation. Secor began experimenting with internal combustion engines in 1885, giving special consideration to the use of liquid fuel. By 1897 Secor began building small engines using various low-grade liquid fuels. With his experiments Secor established the volume governing system used to this day on virtually all non-diesel engines. Secor likewise recognized the imperative need for substantially higher cylinder jacket temperatures in order to successfully burn low-grade fuels. Some of Secor's early engines used no water jacket of any kind, although it was soon discovered that the increased heat thus established carried the point too far for successful lubrication. Dr. Rumely persuaded Secor to move to LaPorte in 1908. In two years time, Dr. Rumely, John Secor, and William H. Higgins, the Rumely factory superintendent had perfected the Rumely OilPull tractor. John Secor died in 1933 at the age of 88 years.

A 1911 magazine photograph of Dr. Edward A. Rumely was taken when the M. Rumely Co. was at its zenith. Dr. Rumely was in fact a medical doctor, having graduated magna cum laude from Freiburg University in Germany. Dr. Rumely also acquired a degree in economics at the same time. During his years in Germany, Dr. Rumely became acquainted with men like Rudolph Diesel and Hugo Junkers. Their revolutionary developments in the world of internal combustion engines profoundly influenced Dr. Rumely's thought concerning the requirements for a really successful farm tractor. After returning to America in 1907, Dr. Rumely set about gathering the people and information to achieve this goal, with the result being the world-famous Rumely OilPull. After losing control of the M. Rumely Co., Dr. Rumely distinguished himself with other achievements. In 1925 he established the Vitamin Food Co., a firm whose ultimate goal was to educate Americans to the value of a good diet and the proper vitamins. With this venture, Dr. Rumely was fully two generations ahead of his time. Also during the 1920s, Dr. Rumely helped organize the Super Diesel Tractor Co., using as a basis some of the Hugo Junkers patents. Regrettably, the time was not yet ripe for a diesel tractor, with developments apparently getting no further than several years worth of drawings, countless letters, and an immense amount of work. Dr. Rumely never lost interest in good health, and published a title, "Smoking and Health" in 1960. Dr. Rumely died in 1965 at the age of 82.

This previously unpublished photograph illustrates what was probably the first of the Rumely prototypes preceding the OilPull tractor. Included at this point was the exhaust-induced draft for cooling the radiator. Likewise, this model used a very large double-weight governor system which can be observed directly behind the flywheel. Although the main frame is totally uncharacteristic of the final OilPull design, the inclined cylinders remained a keystone in the production model. At this point the tractor was not yet called "OilPull," it merely carried a stencil noting, "I Burn Kerosene Too." The original photograph carried no information whatever regarding this tractor, but from the general design and the lack of a "finished" appearance, it is most likely that this was indeed a prototype, and probably one of the very first, since it uses design features not found in any other experimental or production models.

"The Starved Rooster's Lament" became an Aultman-Taylor trademark at least by the 1890's. This reproduction illustrates a shamefully undernourished rooster atop an Aultman-Taylor strawstack. The message was obvious, "Use an Aultman-Taylor thresher — it will put all the grain in the sack, and none in the stack!" Aultman-Taylor had roots going back to 1867, and despite reasonably good sales of its threshers, engines, and tractors, the company seems to have had more lean years than good ones. From the historical perspective, the Advance-Rumely purchase of Aultman-Taylor in 1924 is baffling at the very least. Aultman-Taylor had little to offer Rumely except for its threshers and a series of gargantuan tractors — tractors which were on the way out in favor of lightweight, row-crop designs. Even more baffling is WHY Rumely even considered buying Aultman-Taylor, since Advance-Rumely also had not made its shareholders any substantial profits for several years.

An 1887 Aultman-Taylor catalog notes that their first traction engine was built in 1885. The company's description notes that "two horses are used in these engines, not to pull it, but to guide it on the road." Obviously, a steering mechanism had not yet been developed. Two sizes were built — the Standard Traction was a 10 hp model using a 7 x 10¼ inch bore and stroke, while the Eureka Traction was a 12 hp unit using a 7½ x 10¼ inch bore and stroke. These engines were ordinarily equipped with the Straub reverse gear, but for an additional $100 a standard link reverse could be furnished.

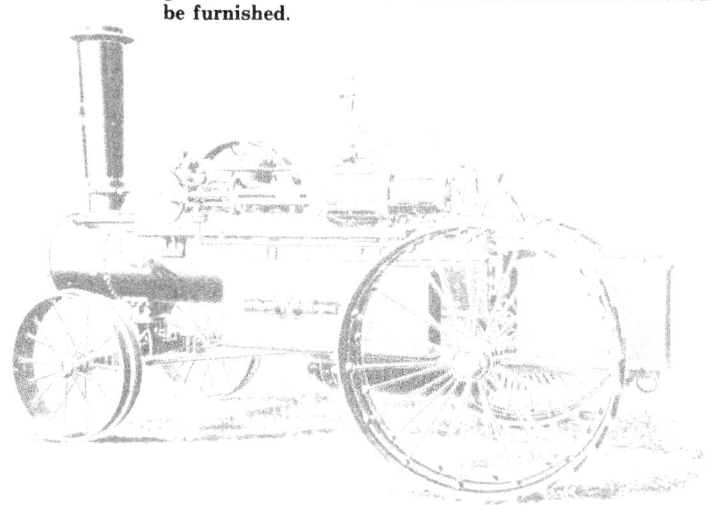

By the late 1890s Aultman-Taylor, like most other engine builders, was offering a compound engine. For 1903 Aultman-Taylor offered compound engines in 10, 15, and 20 hp sizes, using either the Hercules or Eureka designations. This cut illustrates the tandem compound arrangement whereby steam from the boiler entered the smaller, or high pressure cylinder. Expansion was partially completed in this cylinder and then traveled to the larger, or low pressure cylinder. The plan was to be more economical in the use of steam with a consequent savings in fuel and water. Under average conditions however, the savings was minimal at best, and probably did little to pay off the additional investment.

Another Aultman-Taylor catalog of the early 1900s illustrates the unique bevel gear drive mechanism. Much of the speed reduction was accomplished by the first set of gears which in turn transmitted power through the inclined drive shaft directly to the "compensating gear" or differential. This cut from a 1901 catalog also has the prices penned in — they indicate that the simple (not compounded) engines listed at $1,225, $1,435, and $1,650 for the respective sizes of 8, 12, and 16 hp. The compounds in 10, 15, and 20 hp sizes were priced respectively at $1,330, $1,600, and $1,860.

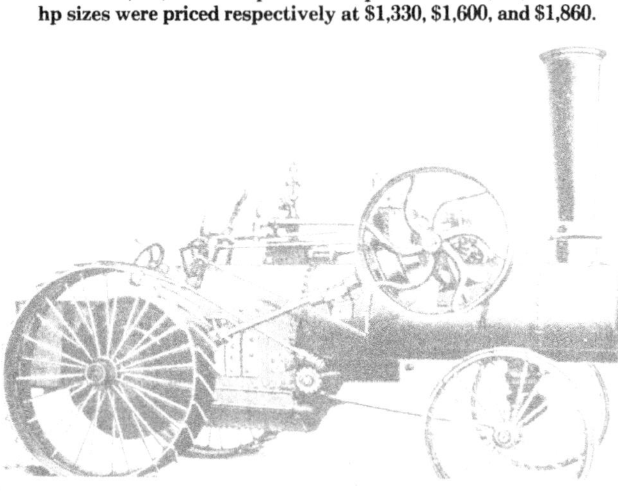

The 1901 Aultman-Taylor catalog makes some strong selling points for the compound engines. For Aultman-Taylor, as for several other engine builders, the compounds were built on the patented Woolf designs. The Eureka Jr. Compound used cylinders of 5⅝ & 8 x 9 inches for 10 horsepower. Slightly larger, the Eureka Compound derived its 15 hp from 6 & 8½ x 10¼ inch cylinders. The largest size, the 20 hp Hercules was equipped with 6½ & 9½ x 10¼ inch cylinders.

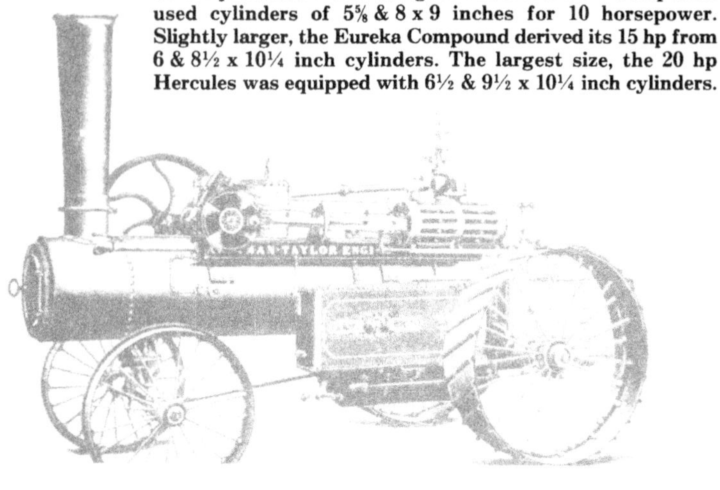

During the early 1920s Aultman-Taylor ceased production of steam traction engines entirely, preferring instead to concentrate on production of their large and rather successful gasoline tractors. However, the 1913 catalog still shows a rather extensive steam offering ranging from 8 to 25 horsepower, with a 16 horsepower size being shown as a typical model. This particular engine developed 48 brake horsepower, using a boiler pressure of 135 psi. Aultman-Taylor built their own boilers, and in fact operated a separate division of the company specializing in huge water-tube boilers for power house and processing steam. Advance-Rumely apparently did not acquire this portion of the business in their 1924 acquisition of the farm machinery line.

Advance Rumely

Aultman-Taylor entered the gasoline tractor market in late 1910 or early 1911, shortly after archrival M. Rumely Co. introduced its famous OilPull tractors. The 25-50 model shown here followed the initial 30-60 size in 1915. Later that year a similar but smaller 18-36 model appeared. While Rumely and Hart-Parr were intent on building one or two-cylinder tractors, Aultman-Taylor initialized their models with a four-cylinder gasoline engine, citing its operating smoothness over those with fewer cylinders. Production of the 18-36 and 25-50 models seems to have ended about 1918,

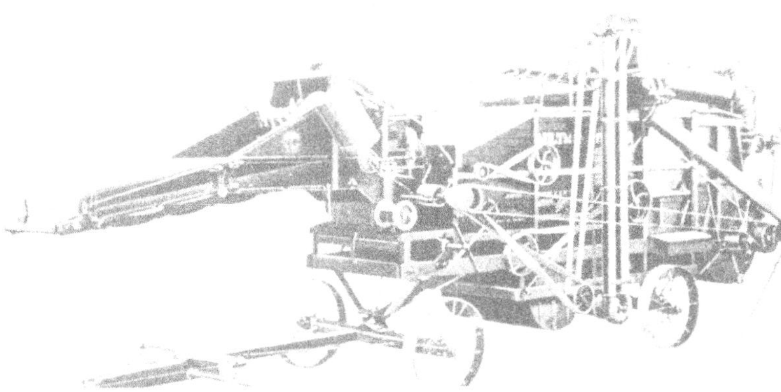

Aultman-Taylor began building threshers immediately after the company's formation in 1867. Henry H. Taylor, a founder of the firm, had earlier been associated with Nichols & Shepard Co., and was successful in obtaining certain manufacturing rights on their "Vibrator Thresher" designs. From these beginnings came the New Century separator offered by Aultman-Taylor in 1900, and built with little change until 1924. The New Century featured what was termed as a "rotary-type shaker" as opposed to the customary straw racks. An open chain raddle carried the grain and straw to the shakers after leaving the cylinder. After purchasing the company, Advance-Rumely continued supplying repair parts for the New Century, a practice which Allis-Chalmers continued until 1935.

The Aultman-Taylor 22-45 tractor was the last of their heavyweight models. It was also their only heavyweight to offer two forward speeds. The general dimensions were the same as for the 18-36, and its design was followed the same lines as all the Aultman-Taylor heavyweights. The 22-45 was tested at Nebraska in 1920 (Test No. 32). A lightweight model, the 15-30, was also tested at Nebraska in 1920 under Test No. 31. Further details of these tests may be found in the book, *Nebraska Tractor Tests Since 1920* (Crestline, 1985). A 1979 Crestline title, *Encyclopedia of American Farm Tractors* gives additional details on the Aultman-Taylor tractor line, while yet another Crestline title, *Encylopedia of American Steam Traction Engines* illustrates a number of surviving steam examples plus gives a historical sketch of the company.

Aultman-Taylor initially offered their 30-60 tractor, continuing its production until bought out by Advance-Rumely in 1924. The 30-60 was a strong competitor for the Rumely OilPull, and in fact, more 30-60 A-T tractors were in use for Iowa road maintenance than any other make. Nebraska Tractor Test No. 30 of 1920 indicated that this tractor was slightly more powerful than the 30-60 OilPull and was also slightly more efficient in the use of fuel. By piecing together numerous oral and written accounts, it appears that after buying out Aultman-Taylor, Rumely built enough additional tractors to use up much of the parts inventory, after which further production was abandoned. Even though the purchase stilled a strong competitor, the high hill of historical perspective shows the purchase to be an exercise in futility at its worst, and of skimpy financial reward at best.

Production of the OilPull began in 1910. During the previous two years there is every likelihood that a number of prototype designs were built and tested. Another early photograph, although unidentified, appears to have been another experimental design. At this point the main frame assumed the same general design as the production model, and the huge flywheel governor of the original design has been modified. Curiously, this model appears to use the exhaust-induced cooling design, but with the stack attached to the back of the radiator itself. Apparently this style pulled the cool air from an open bottom in the radiator sections. Note that this particular tractor carries the exhaust pipe beneath the engine and directly into the bottom of the exhaust stack. Note also that the rear wheel design of this tractor is virtually identical to "Kerosene Annie" of 1910.

Yet another progression in the OilPul design was a prototype which although still crude, shows considerable improvement in design over the early tractors. The radiator has now been refined to nearly its final state, and the tractor now includes side fenders and a useable operators platform. Given the designs of 1908-1910 generally, this prototype might have been built by almost any company, but once again, the rear wheel design is identical to that used in the "Kerosene Annie" version yet to come. Quite possibly the same chassis might have been used for several different prototypes, saving the need for a great deal of new work on each experimental model. Only conjecture will do — no information accompanies any of the photographs, and time has effectively clouded the background of the OilPull experiments, save for occasional newspaper and magazine articles.

With the appearance of "Kerosene Annie," the Rumely OilPull was close at hand. "Kerosene Annie" was a nickname given, probably because of the tractor's ability to successfully use kerosene fuel. This success can be attributed largely to John A. Secor. The carburetor used in the final OilPull design carried the "Secor-Higgins" name, again giving tribute to Secor, and Wm. H. Higgins the Rumely factory superintendent. A close comparison of this tractor with the preceding versions shows that design changes were less dramatic than at first. Noticeable though, is the modified exhaust stack, somewhat higher, and somewhat smaller in area than before. Added has been a rather unrefined canopy consisting of a pipe framework over which is attached a canvas cover. Neither the Rumely nor the OilPull slogans are to be found on this tractor, giving this writer the opinion that "Kerosene Annie" as shown here was not yet a production model. This opinion is reinforced by the simple fact that the Rumely people were not at all shy about using an ample supply of decals to mark their products.

Advance Rumely

Rumely employees went to work in the brand-new OilPull factory on February 21, 1910. The new tractor was known as the Type "B" and was rated at 25 drawbar and 45 belt horsepower. The "B" used two cylinders of $9\frac{1}{2}$ x 12 inch bore and stroke with an engine speed of 375 rpm. Although the "B" was of the same general design as the "E", 30-60 tractor, it used somewhat smaller rear wheels having a diameter of 64 inches with a 26-inch face. Production of the "B" continued into 1912. Rumely introduced the OilPull with a low-tension ignition system using a mechanically actuated igniter in each cylinder. Rumely summed up the advantages of this system by noting that "It gives better spark and the wiring is not as likely to go wrong."

During 1911 Rumely introduced their 15-30 "F" tractor. The engine was simply as single-cylinder version of the 30-60 "E" model. This particular 15-30 is shown at work in the Kankakee Marsh district of northwestern Indiana. As illustrated in this photograph, the soil surface was covered with a dense growth of weeds and vines. The soil itself was a combination of sod, muck, and peat. Gaining the necessary flotation to work effectively required the 22-inch extension rims shown on this tractor. Adding this large amount of wheel surface dropped the soil pressure from an initial figure of 27 pounds per square inch to 14 psi. Other photos of this pictorial series illustrate a five-bottom plow behind the 15-30. Due to the conditions a three-man crew is on hand. An engineer was required of course, and two more men were needed to operate the plows. Chances are that frequent stops were required to clean the plow of entangled vines and brush — a job that must have been very unpleasant.

The Rumely 30-60 Type "E" OilPull tractor entered production in 1910, remaining in production into 1923. None were built during 1914. The 30-60 was built on the same basic chassis as the 25-45 "B" tractor, but used somewhat larger 30 x 80 inch rear wheels. Both models used a main frame of 12-inch steel I-beams weighing $31\frac{1}{2}$ pounds per foot. A huge crankshaft $4\frac{7}{16}$ inches in diameter carried power from the two cylinders of 10-inch bore and 12-inch stroke. The rear axle was $5\frac{7}{16}$ inches in diameter, and the engine flywheel weighed 1,300 pounds. The 30-60 remained popular for perhaps a longer period than any other heavyweight tractor of the time. Even though some sixty years have passed since the 30-60 was produced, it remains as one of the most sought-after tractor models by vintage tractor collectors. The 30-60 was tested at Nebraska in 1920 under Test No. 8. Further details concerning this test and many other OilPull tractors may be found in Crestline's book, ***The Nebraska Tractor Tests Since 1920.***

Newspaper headlines of October 14, 1911 stated, "Traction Engine Plowing Breaks all World's Records." Under the auspices of Purdue University and the Indiana Agricultural College, three 30-60 OilPull tractors were coupled to a fifty-bottom plow built by Oliver Chilled Plow Works. Accounts of this trial show that the three tractors together consumed about 22 gallons of fuel per hour for a fuel cost (based on 4 cents per gallon) of $6\frac{1}{2}$ cents per acre. The three huge plows turned over nearly sixty feet of soil at each pass, or 7 acres for each mile of travel. During the tests, an acre of ground was plowed each $4\frac{1}{4}$ minutes. Working 24 hours per day, 6 days per week, this huge outfit was capable of plowing 2,000 acres. This demonstration was intended to show what power farming could achieve compared to man-and-beast methods. A record-breaking demonstration such as this one caught the attention of almost everyone interested in power farming and probably was very beneficial in terms of additional sales to both the Oliver Plow people as well as to the Rumely organization.

Advance-Rumely

A rear view of the 15-30 with special extension rims is shown in this photograph. Although breaking of marsh lands was difficult at its best, these lands proved to be very productive with proper management. A photographic view such as this often provokes a nostalgic twinge to return to the "good old days," but in fact, operating this or any other heavyweight tractor was hard work. Beyond this, even the perfection of the OilPull design left much to be desired in tractor engineering, so every stage of the day's work was associated with a liberal output of muscle power. Features of the 15-30 "F" included low-tension ignition and a mechanical igniter. Its single-cylinder engine operated at 375 rpm. Unlike the single forward speed of the "B" and "E" tractors, the "F" offered two forward speeds of 2 and 2¾ mph. The "F" weighed 14,500 pounds, compared to 23,800 pounds for the "B" and 25,800 pounds for the "E" tractor.

Rumely's 1912 purchase of Northwest Thresher Co. of Stillwater, Minn., brought with it the Universal 20-40 tractor. A. O. Espe had designed this tractor, and several companies sold it, including the Minneapolis Threshing Machine Co. Its two-cylinder opposed design used a 7½ x 8 inch bore and stroke with a rated speed of 500 rpm. Soon after purchasing Northwest Thresher Co., Rumely christened this model as the GasPull, simultaneously rating it downward to a 15-30 model. The GasPull carried a 50-plus gallon fuel tank at the front of the tractor. Unlike the OilPull, the GasPull was water-cooled, using a thermo-syphon system with a cooling fan. Standard equipment included the canopy and lug equipment. Extension rims were also available, and at extra cost the GasPull could be equipped with rice field equipment consisting of special extension rims and mud scrapers.

Production of the 15-30 "F" ran from 1911 to 1916, with none being built in 1917. For 1918 only, the same basic tractor was offered as the 18-35 "F" OilPull. The slightly higher rating more accurately reflected the tractor's actual output, and to Rumely's credit, the OilPull was consistently underrated by the company. This conservatism was actually a sales builder, since users were always delighted to find their tractor would do substantially more work than had been advertised. Although the 18-35 was virtually identical to the 15-30, this model could be optionally equipped with jump spark ignition in lieu of the standard make-and-break igniter system. The Rumely file photo shows a Model F operating in zero weather, using low-grade fuel. The usual practice in all types of weather was to first start on gasoline, and after the engine had warmed up slightly the switch was made to kerosene fuel. Rumely's induced draft cooling system made temperature regulation virtually automatic. The entire system was so designed and so proportioned that at idle or light loads very little draft was created, thus delivering less air movement over the radiator sections. Under heavier loads, considerably increased air movement gave the required cooling. The use of an oil coolant gave the somewhat higher jacket temperatures which were a part of the OilPull success story. The 18-35 listed at $3,000, a substantial sum for a farmer in the mid-teens.

The 35-70 Advance tractor shown here bears an uncanny similarity to the Twin City design built by Minneapolis Steel & Machinery Co. Research into the Rumely operations shows that Jack Junkin of Advance Thresher Co. designed this tractor, and research into Minneapolis Steel & Machinery Co. indicates that Junkin also designed the Twin City. Junkin came into the Rumely organization when they bought Advance Thresher in 1911. In fact, it was Junkin who finally polished off the OilPull design. In Dr. Rumely's own words, "He [Junkin] experted the OilPull and helped finish off the design." Junkin's big Advance 35-70 tractor shown here was apparently a venture that never left the ground, since the Rumely acquisition of Advance Thresher precluded any need for a design that would compete directly with the OilPull.

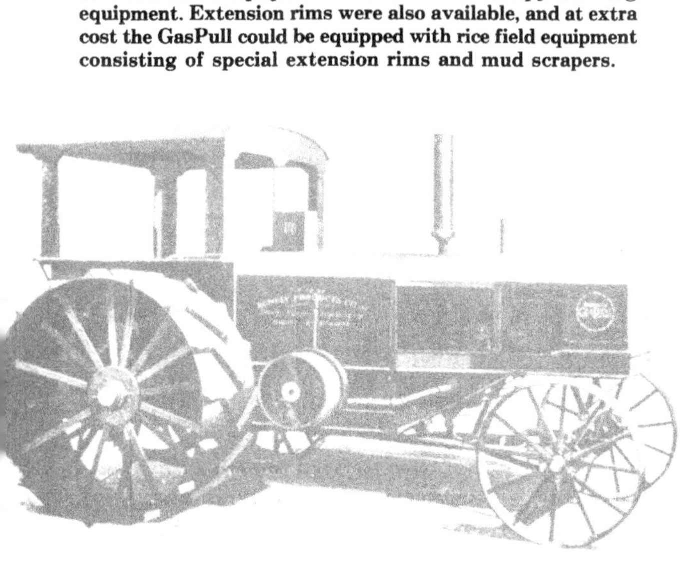

Rumely's test shop used large generators to build a test load. The generated power was then used in the factory. An early photo of the test shed shows half a dozen or more OilPulls in the testing process. Although specific details are unknown, the test probably lasted a day or more, with each tractor-operating a light load, intermittent loads, and of course under full load. The testing process not only insured the customer against defects in castings or workmanship, it also limbered up the engine so that it was ready for work as soon as it arrived at its destination. Ironically, the test process using electric generators to build the load is still practiced by tractor testing laboratories, and is also used by several current farm tractor builders. The technology has improved and the methods have changed, but the goal is consistent with those set out some 70 years ago.

In 1912, Rumely acquired this orchard tractor design from Joshua Hendy Iron Works, of Sunnyvale, Cal. Christened by Rumely as the ToeHold tractor, it had a rating of 14 drawbar and 28 belt horsepower. The ToeHold carried a two-cycle gasoline engine design totally out of character with the heavyweight, low-grade-fuel philosophy of the OilPull. Lighting was available as an option, either of the carbide or electric variety. The unusual design was intended primarily for orchard work, with special guards provided to protect both the trees and the operator from injury. The almost total lack of advertising material on the ToeHold design tends to indicate that this venture was of short duration, and marks Rumely's only attempt to enter the rather specialized orchard tractor market. Curiously, Rumely also announced plans about this same time to offer marine engines. This venture too, never was put into effect.

Initially announced in late 1915, the Advance-Rumely 8-16 tractor did not enter production until the following year. Following generally the new lightweight designs that had entered the market, the 8-16 used a single drive wheel, plus an outside idler wheel, and a rear-mounted support and steering wheel. Shown here with two mounted plows, the design required only to pull two pins and less than two minutes work for removal. Rumely advertising noted that this tractor was equipped with a 4-cylinder engine of their own design and manufacture. Rated at 850 rpm, it carried a 4 x 5½ inch bore and stroke. A single speed was offered, in forward or reverse, of 2.1 mph. Weighing 5,600 pounds, the 8-16 was offered during 1916 and 1917.

LaPorte area newspapers of September, 1921, note that Advance-Rumely had displayed their new line-drive tractor at the LaPorte County Fair. All motions, forward, backward, left, and right were accomplished by a pair of leather lines held by the operator — in other words, a mechanical team of horses. Several companies tried this approach, probably hoping that diehard devotees of animal power would see the line-drive scheme as the best of both worlds. Even the more successful line-drive tractors, such as the LaCrosse, were of limited production, while the Advance-Rumely line-drive tractor seems to have never gone beyond the production of a few prototypes. Although undocumented at this point, there are strong indications that only two of these units were built, with both still existing. At the present time, the line-drive model shown here is owned by a New York collector, while the other is in Michigan.

Rumely's 12-24 tractor was of the same general design as the smaller 8-16 version. Its 4-cylinder engine carried a 4½ x 6¾ inch bore and stroke with a rated speed of 750 rpm. Production apparently lasted only into 1917. Quite possibly the design had its impetus from the Wolf tractor offered about the same time by Leo Rumely Tractor Co. of LaPorte. Leo Rumely was a brother of Dr. Edward Rumely and John Wolf was an uncle of the doctor. Rumely claimed this to be an ideal one-man plowing outfit, but the short-term production of these two models indicates that it, like others of the same general design were not the answer. For 1916, this 12-24 model listed at $975.

Advance Rumely

Advance-Rumely introduced the "K", 12-20 OilPull in 1918, with production of this model continuing into 1924. The 12-20 followed the usual OilPull 2-cylinder design and employed a 6 x 8 inch bore and stroke. Rated speed was 560 rpm. Weighing 6,682 pounds, the 12-20 featured two forward speeds of 2.1 and 2.9 mph. The Secor-Higgins carburetor remained as a standard feature, and a Bosch high tension magneto was factory equipment. The OilPull engine design used ample quantities of water with the fuel mixture to retard preignition. Nebraska Test No. 10 of 1920 shows the 12-20 using 1.855 gallons of kerosene fuel per hour on the rated brake horsepower test. With an output of 20.06 brake horsepower, the 12-20 also consumed 4.81 gallons of water in the fuel mixture under this load.

By the time Advance-Rumely announced the 20-40 "G" tractor in 1918, the OilPull name had gained the reputation of first-class quality and reliability among its peers. The 20-40, following the same general design as its counterparts, provided an intermediate size large enough for most threshing jobs but without the added size and cost of the big 30-60 model. Weighing nearly 6½ tons, the 20-40 was powered by a 2-cylinder engine rated at 450 rpm and using an 8 x 10 inch bore and stroke. Fuel consumption totaled 4¼ gallons per hour under the rated load of 40 belt horsepower. The original OilPull styling carried with it the dependability that farmers required, but by the early 1920s, farmers were looking for lightweight designs that could be used all around the farm. Development of new tractor design brought demand for the heavyweight OilPull design to a halt, so in 1924 Advance-Rumely ceased production of the entire "heavyweight" series. This entire series was available with various options that included various extension rims and an extensive selection of drive lugs to suit specific applications. Also available was a special high altitude cylinder head for use at altitudes of 4,000 feet or more.

The 14-28 OilPull as a production model emerged in 1917 with a list price of $2,400. Weighing 9,600 pounds, it offered two forward speeds of 2.1 and 3 mph. The 14-28 featured a 2-cylinder engine having a 7 x 8½ inch bore and stroke with a rated speed of 530 rpm. A standard feature was the low tension ignition system utilizing a magneto with mechanically operated igniters. The 14-28 could be retrofitted in the field with high tension ignition. Detailed production records indicate that the 14-28 was built into 1919, with production then being superseded by the 16-30 OilPull. The 16-30 was virtually identical except that it was furnished from the factory with high-tension ignition.

The OilPull 16-30 was a very popular model. Its size was ideal for the average farmer — the 16-30 could easily handle a 3-bottom plow or a 28-inch thresher. The 16-30 used the same engine and chassis as the 14-28 noted above. Test No. 9 at the Nebraska Tractor Test Laboratory indicated a rated output of 30.50 belt horsepower. This test likewise indicated a kerosene consumption of 3.070 gallons per hours, together with 3.28 gallons of water in the fuel mixture. The OilPull design was predicated on the higher cylinder temperatures provided by the use of oil coolant, together with a compression ratio increased somewhat over standard design for engines using gasoline fuel. Another important OilPull feature was the Secor-Higgins carburetor which provided excellent regulation under all loads, and did so without the use of springs, floats, and other complicated parts. Carburetor design of the 1920s included a variety of methods to automatically regulate the air-fuel ratio under different loads, with one carburetor design using a series of variously weighted steel balls over small draft-tubes to achieve the goal. By comparison, the Secor-Higgins was the most simple carburetor of the period. Production of the 16-30 "H" ran from 1919 to 1924.

Apparently in an effort to streamline the OilPull design, Advance-Rumely built this 14-28 prototype about 1915 or 1916. A note on the back of the original photograph simply states, "First 14-28 tractor." From all appearances, this experimental model was virtually identical to the production model of the 14-28 except for the addition of the extra-ugly sheet metal work. The radiator continues the same induced draft design so typical of the OilPull, except that the outlet has taken on a full-length pattern. Chances are that the engine exhaust nozzles were distributed under the full length of the outlet stack. The original photograph was located in the Allis-Chalmers Archives, but subsequent searches have failed to locate the original image of a very unusual OilPull design.

Advance Rumely

The OilPull design is best explained with an illustration. An oil collant was used; this provided the higher jacket temperature required by the OilPull design. A circulating pump moved oil from the bottom of the radiator through the cylinder jacket, returning the heated oil to the radiator. Meanwhile, the engine exhaust was directed into the radiator and expelled through a series of nozzles within the exhaust stack. This induced a draft through the bottom and sides of the radiator, pulling cool air over the multiple radiator sections. An expansion tank in the oil system maintained the proper collant level and eliminated coolant loss. Through the specific design of the OilPull system, less cooling was required at light loads, and of course the exhaust being light, less cooling air was pulled over the radiator sections. Under heavy load, maximum cooling was required, and with the heavy load came a heavy exhaust which automatically created a maximum air movement.

Rumely's move to a lightweight design came in 1924. The smallest of this series, the 15-25 "L" tractor carried a list price of $1,350. Weighing only 3 tons, the 15-25 featured totaly enclosed gearing with the entire transmission running on ball bearings. The Secor-Higgins carburetor, although slightly modified remained a centerpiece of the design, as did the unique OilPull cooling system. The new series of Lightweight tractors introduced in 1924 also featured a differential lock — this feature had been patented by Dr. Rumely under No. 1.055,085 of 1913. Despite the importance of the differential lock as a feature of tractor design, this OilPull series was one of the few to utilize it until at least four decades later. The 15-25 featured a 2-cylinder design using a 5⅜ x 7 inch bore and stroke with a governed speed of 755 rpm.

Production of the 25-45 "R" tractor was apparently delayed until stocks of the older 20-40 "G" were sold off. Only one of the "R" tractors was built in 1924, and by the time production ended in 1927, only 761 had been built. Although slightly larger than the 20-35, the 25-45 was substantially more expensive — it carried a 1926 list of $3,200. Like other tractors of this series, three forward speeds were offered and an American Bosch magneto was standard equipment. Rated at 540 rpm, the 25-45 used a 7$^{13}/_{16}$ x 9½ inch bore and stroke. Advance-Rumely continued to stress their design, calling it the Triple Heat Control. In reality the Lightweights used a condensed version of the original design. Some years later it would be revealed that the new lightweight design occasioned many more field difficulties than had been experienced by the original slow-speed heavy-duty design.

The entire series of lightweight OilPulls introduced in 1924 were virtually identical in appearance except for their physical size. All carried the same basic features, and all utilized a lightweight pressed steel frame in lieu of the heavy structural steel members of earlier days. Priced at $2,000, the 20-35 "M" tractor was advertised to handle four 14-inch plows under average conditions, or to operate a 28 x 44 inch thresher. The heavy spring-cushion drawbar was adjustable from side to side, the Alemite pressure oiling replaced the oil fashioned grease cups. Automobile-type steering was featured, with the front wheels running on Timken tapered roller bearings. Rated at 635 rpm, the 20-35 carried a 6$^{13}/_{16}$ x 8¼ inch bore and stroke. Production of this model, like that of the 15-25 "L" ended in 1927.

Advance-Rumely

Advance-Rumely capped their new lightweight series of 1924 with the big 30-60 "S" OilPull. It embodied virtually all the same features as its smaller brothers, but did so on a larger scale. Rated at 470 rpm, it used a 9 x 11 inch bore and stroke. Like the other Lightweights, the 30-60 offered removable cylinder sleeves as a new feature. An extensive variety of lug equipment was available to suit local requirements. Likewise, various widths of extension rims could be secured for added flotation. As with the earlier OilPull models, a 1924 catalog announced that these tractors "are painted a dark green color and varnished. Fenders and spokes of the wheels are striped in red. The top of the exhaust stack is painted a brilliant black." Without a doubt the green color varied from batch to batch, so the exact shade varied considerably from year to year. For 1926 this 30-60 model listed at $4,200. Production ended in 1927.

The Advance-Rumely 25-40 "X" OilPull was simply a conversion of the earlier 20-35 "M" tractor. In fact, detailed production records indicate that a substantial number of the 25-40 tractors were a conversion from 20-35 tractors already built and in warehouse storage. The 25-40 used the same bore and stroke dimensions as the 20-35 but carried a rated speed of 700-725 rpm. This tractor and the smaller 20-30 "W" could be optionally equipped with a power-take-off drive. A belt-drive was used to a cross-shaft mounted beneath the engine. It terminated in a right-angle gear box which carried the final PTO drive at 540 rpm to the rear of the tractor. The fuel tank situated behind the radiator was partitioned so as to offer a capacity of 26 gallons of kerosene fuel and 23 gallons of water for use in the fuel mixture. A rear view of the 25-40 illustrates the roomy platform and rather uncomfortable seating arrangement. The tractor shown here carries the earlier air cleaner system found on the 20-30. Production of the 25-40 ran from 1928 to 1930.

With the announcement of the Lightweight tractors in 1924, Advance-Rumely also made frequent mention of their patented front axle shifting device. A simple rack-and-pinion affair was situated between the front axle and its pivot. For belt work it was simply a matter of offsetting the front axle to move the front wheel out of the path of the drive belt. Since belt-driven farm machinery is now history, this feature seems to hold small merit. However, in a day when virtually all machines were belt-powered this was an important development indeed. The tractor shown here is a 20-30 "W" OilPull. It was nearly identical to the earlier 15-25 "L" except the engine was speeded up another 100 revolutions to 850 rpm. Another obvious design change is the relocation of the coolant expansion tank to the top of the radiator. Although standard equipment for the 20-30 included 46-inch rear wheels, these could be substituted for 44-inch wheels where reduced height was a requirement.

Both the 20-35 "M" and the 25-40 "X" tractors could be furnished as a special OilPull winch tractor. The two tractors were virtually identical, with the difference being primarily an increased engine speed for the "X". The meager advertising material on this unit suggests that its primary application was for oil field work, although it may have found use in many other applications. A 1926 price list shows this unit with a tag of $3,200, when attached at the factory. Other data suggests that the winch attachment could also be installed in the field. During the mid-1920s, Advance-Rumely was experiencing serious financial difficulty, and perhaps the winch tractor was an attempt to find a new application and additional sales for OilPull power. Nothing has so far been located to indicate the total production of this unit, leading ultimately to the conclusion that very few were built.

By raising the crankshaft speed to a top of 635 rpm, the OilPull people made the conversion of this 30-50 "Y" from the earlier 25-45 "R" tractor. As previously noted, another change, this one cosmetic, was the relocation of the coolant tank from behind the radiator to ahead of it. The fuel pumps were built into the carburetor body — their liquid seal eliminated the need for any packing or packing glands. In Nebraska Test No. 145 the 30-50 "Y" demonstrated an output of 63.32 brake horsepower and delivered 10.79 horsepower hours per gallon of kerosene fuel. Total consumption was 4.667 gallons of fuel per hour, along with 3.29 gallons of water added to the fuel mixture. Weighing 13,025 pounds, the 30-50 made a maximum pull of 6,695 pounds, or 51% of its own weight. Only 245 "Y" tractors were built, and these only in 1929. Of these, 125 were converted from earlier "R" tractors still unsold. When Allis-Chalmers bought Advance-Rumely in 1931, five of the 30-50 "Y" tractors were still in inventory, and were eventually sold at a reduced price of $1,350 per copy.

The 40-60 "Z" OilPull saw production only during 1929. Only 215 units were built, and of these, 62 were converted from 30-60 "S" tractors in stock, but were later reconverted back to the original 30-60 specifications. Thus, only 153 units were actually marketed as the 40-60 "Z" tractor. The 40-60 maintained the same engine speed as was used on the earlier 30-60 "S" tractor, and other modifications were slight. A great majority of the Lightweight series introduced in 1924 was equipped with a solid disc flywheel. In itself this represented good design, but the solid surface acted as a sounding board, magnifying normal operating sounds immensely. The final Lightweight series of 1928 and 1929 reverted to a spoked flywheel, although it was of somewhat different design than had been used in the original heavyweight models. Likewise, a great many of the early Lightweights were retrofitted in the field with a spoke flywheel to eliminate the noisy effect of the original equipment.

Like several other tractor builders, Advance-Rumely at least did some experimenting with a crawler tractor design. Obviously using either a 20-30 or a 25-40 chassis, this crawler at least made it from the drawing board to the prototype stage. A close examination of the photograph indicates that modifications to the tractor chassis itself were minimal indeed, with the crawler bull pinions replacing the rear wheels, and terminating through heavy roller chains to the final drives. Possibly the crawler unit was built by another company — no information accompanied the original photograph, nor is any reference to this unit recorded in the production lists. All that seems certain is that this experimental model was built in the 1928-30 period, and was then apparently abandoned. Since Allis-Chalmers had already purchased Monarch Tractors in 1928, it seems certain that they displayed little or no interest in the design.

Advance-Rumely acquired the Toro Motor Cultivator line about 1927. The following year came announcement of the Rumely DoAll tractor. Shown here in its cultivating mode, the DoAll was another of the convertible tractors which attempted to invade the 1920s tractor market. In addition to the cultivator equipment shown here, the DoAll could be adapted likewise to corn planters, grain binders, mowers, and many other farm implements. By so doing, one man could operate both the tractor and the attached implement. The drive wheels were reversible, giving an optional change in tread width when required, although the tractor could be furnished with tread widths of 60, 66, or 75 inches. The cultivator model could be optionally furnished with fenders over the drive wheels.

Advance-Rumely Thresher Co.

Originally priced at $1,160, the remaining DoAll stocks were sold out to dealers at the bargain price of $543. The DoAll in its tractor version as shown here, could also be dressed up with fenders. Other options included a belt pulley attachment as well as a power-take-off drive attachment. Seven different rear wheel lug configurations were available either as original factory equipment or for field replacement. For special applications a special 66-inch front axle was available, as was a furrow guide to use when plowing. The location of the PTO shaft would be a nightmare to today's safety engineers. The output was located just ahead and to the left of the operator; a perfect invitation to tragedy!

Designed especially for lister units, this special DoAll bears a strong resemblance to a conventional row-crop tractor. Curiously however, Rumely made no particular claim in this direction, and this at a time when the Farmall of International Harvester Co. was consuming the tractor market. The DoAll used a Waukesha X Special engine, which although built by Waukesha, was built to Rumely's specifications. Rated at a top speed of 1,400 rpm, it carried a 3½ x 4½ inch bore and stroke. Two forward speeds were provided of 2⅝ and 3¾ mph. Full pressure engine lubrication was featured, as was a dual centrifugal and filter-type air cleaner. Total weight of the standard DoAll tractor unit was 3,000 pounds. Production of the DoAll effectively ended in 1931, although Allis-Chalmers may have assembled additional units from existing parts stocks.

Production of the Rumely 6A tractor began in 1930, ending in 1931. Sales were apparently very poor, since in late 1931 when Allis-Chalmers assumed control, well over 700 of the Rumely 6A tractors were still on hand in the LaPorte factory. Allis-Chalmers continued to offer the 6A until remaining stocks were depleted, thus this model is listed in various tractor directories as late as 1934. The Rumely 6A was equipped with a Waukesha 6-cylinder engine built to Advance-Rumely specifications. Rated at 1365 rpm, it used a 4¼ x 4¾ inch bore and stroke. Its features included six forward speeds compared to the usual three offered by the competition. Weighing 6,370 pounds, the 6A pulled a maximum load of 4,273 pounds under slippage of 9.96%. Thus the 6A managed to pull 67% of its own weight. For further data in this regard see Test No. 185 in *Nebraska Tractor Tests Since 1920.*

The Rumely OilTurn motors were available either in single or in double cylinder design, identical to the engines available in OilPull tractors. This particular style uses the Type "F" single cylinder, 30 horsepower, oil-cooled engine, and in this case it is mounted on I-beam sills. The OilTurn motors, either single or double cylinder models, could also be furnished as portable engines for ease of moving from one job to another. Instead of the single flywheel as used on the tractors, this OilTurn motor was equipped with double flywheels, each weighing 1,000 pounds. Rumely offered the OilTurn completely equipped and ready for service. Standard equipment included a friction clutch pulley, fuel and water tanks, gasoline tank, magneto, and all necessary piping. Various other OilPull engines were sold for stationary service, using currently manufactured sizes from the OilPull tractor line.

Rumely was prepared to furnish the OilPull engine as an independent stationary power unit. Shown here is a big 60 horsepower engine, identical to that used in the 30-60 "E" OilPull tractor. Because of the OilPull design, supplying a power unit required virtually no modifications to the engine itself. This particular unit is apparently belted to an irrigation pump. Lacking however is the familiar radiator system — possibly this engine was cooled with a portion of the pump water. A huge fuel tank is barely visible in the lower right corner of this photograph. The two straight pipes from the 10-inch cylinders must have created some delightful stack music!

Rumely-Olds Type "B" engines were offered in one, two, and four cylinder sizes, all having a 7⅞ x 10⅛ inch bore and stroke. These three sizes had respective ratings of 18, 35, and 65 horsepower. Type B engines were available only as stationary models. Jump-spark ignition was standard equipment, with a magneto included on the 35 and 65 horsepower models, but an extra on the 18 horsepower size. Although a circulating pump came as standard equipment, the cooling tank was an extra which added another $151 to the base price of $2,575 for the 65 horsepower engine. This size weighed in at six tons. The 18 horsepower model weighed 4,120 pounds and listed at $740, while the three ton, 35 horsepower model sold for $1,450.

Type "G" Rumely-Olds engines were only built in the single-cylinder style. The 18 horsepower size carried an 8½ x 14 inch bore and stroke with a list price of $740. The 25 horsepower size used a 9½ inch bore and the 35 horsepower model featured an 11½ inch bore, with both engines using an 18 inch stroke. These two sizes carried respective prices of $1,090 and $1,290. Topping this series was the big 50 horsepower single cylinder engine. It carried a huge 14 x 20 inch bore and stroke, operating at 200 rpm. Weighing 14,000 pounds, it listed at $1,700. Additionally, the Type G engines could be furnished as portables in the 18, 25, and 35 horsepower sizes. Thus equipped they carried respective prices of $915, $1,295, and $1,495.

During 1912, Rumely Products Co. purchased a block of stock in Seager Engine Works, builders of the Olds gasoline engines. Full details of the Rumely-Seager deal are unknown, but from this arrangement came the Rumely-Olds engine line of 1912. The illustrations and descriptions of the engines remained exactly the same, whether from the Seager Engine Works catalog, or from the Rumely-Olds catalog. A 1913 price list from Rumely Products Co. shows this 1½ h.p. Type A engine with a list price of $67. Six other sizes ranging from 3 to 15 h.p. were available in this series — the big 15 horsepower model listed at $618. It carried a 7⅝ x 9 inch bore and stroke. The 1½ h.p. model shown here was available only in a hopper cooled design, while the remaining sizes could be furnished as either open or closed jacket engines.

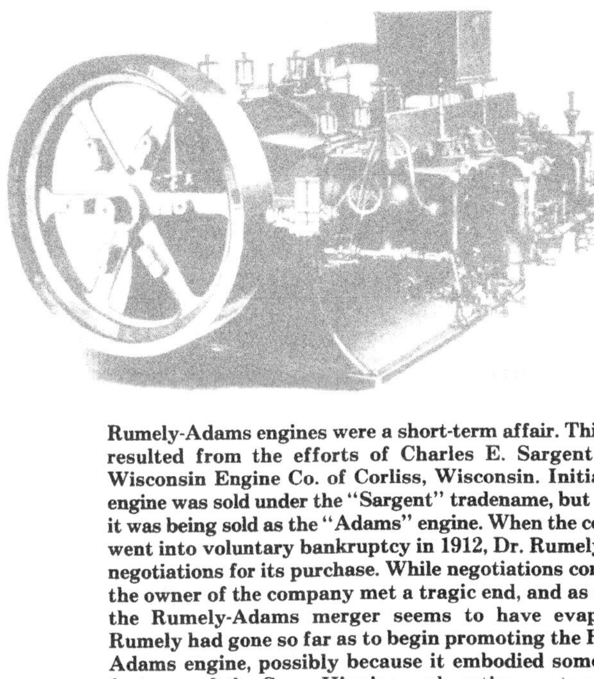

Rumely-Adams engines were a short-term affair. This series resulted from the efforts of Charles E. Sargent at the Wisconsin Engine Co. of Corliss, Wisconsin. Initially the engine was sold under the "Sargent" tradename, but by 1912 it was being sold as the "Adams" engine. When the company went into voluntary bankruptcy in 1912, Dr. Rumely began negotiations for its purchase. While negotiations continued, the owner of the company met a tragic end, and as a result the Rumely-Adams merger seems to have evaporated. Rumely had gone so far as to begin promoting the Rumely-Adams engine, possibly because it embodied some of the features of the Secor-Higgins carburetion system. These engines were built only in larger sizes, with this being the smallest at 50 horsepower. It used a 14 x 18 inch bore and stroke with an operating speed of 270 rpm.

Advance-Rumely

During 1912 Rumely Products Co. gained control of production on the Falk engine built at Milwaukee. It now appears unlikely that the Rumely acquisition included other portions of the Falk operation. Coincidentally, this was the same family from which Otto H. Falk arose — he assumed the presidency of Allis-Chalmers in 1913. Falk engines were of superior design, built to exacting standards, rather than to a price. Since Falk engines were of the volume governing design and built to handle kerosene fuel, they were easily adaptable to the Secor-Higgins carburetor as used on the OilPull tractors. The subsequent demise of M. Rumely Company, together with extreme competition in the engine business brought an end to the Rumely-Falk line by 1915.

The Advance-Rumely No. 1 combine was announced in 1925. Only one was built that year as a prairie combine, but 15 No. 1 Hillside combines were produced. These apparently were shipped to Oklahoma. The No. 1 Hillside was capable of working successfully on a 60% grade, and a 16-foot header was standard equipment. Rumely equipped the No. 1 with a 26-inch cylinder and a 38-inch cleaning shoe. A 35 horsepower Hercules engine operated the machine — it used a 4 x 5 inch bore and stroke. This unit shows an attached 40-bushel grain bin, but this was optional equipment, with a sacking elevator being the standard setup. Production of the No. 1 Prairie Combine continued only into 1927, but the No. 1 Hillside Combine remained in the line through 1929.

Rumely's No. 2 Prairie combine apparently replaced the No. 1 Prairie model. Introduced in 1926, it remained in production through 1930. Weighing 10,500 pounds, the No. 2 used a spike cylinder identical to that in the threshing machine. In this case a cylinder 26 inches long was featured, together with a 38-inch cleaning shoe. Although a 16-foot cutter bar was standard equipment, this machine could also be furnished with a 12-foot header. This machine was sold with neither a bagging elevator nor with a grain bin — these were selected by the customer at the time of the purchase as extra cost options. The No. 2 was not furnished in a hillside version.

Hillside or Prairie versions were available in the No. 3 Advance-Rumely combines. The Prairie version was first announced in 1927, with production continuing to 1930. No. 3 Hillside combines were built between 1929 and 1931. Advance-Rumely equipped the No. 3 models with a 10-foot header. A 20 horsepower Waukesha engine was featured — rated at 1,350 rpm, it used a $3\frac{1}{2}$ x $4\frac{1}{2}$ inch bore and stroke. Due to widely varying conditions, these combines allowed adjustment of cutting height anywhere from 4 to 30 inches. The Advance-Rumely, like other combines of the period carried the cut grain from the side of the machine over traveling canvas aprons into the cylinder. Within a short time the Allis-Chalmers All-Crop Harvester would emerge from LaPorte Works — this revolutionary little machine carried the cut grain directly to the cylinder in a straight line.

Rumely's last combine model was the No. 4 Prairie machine. Using only a seven-foot cut, it was the only Rumely model built to the needs of the small farming operation. Additionally, it was the only Rumely combine to be factory equipped with a tractor power-take-off drive. Other unique features included an auger conveyor to move the grain from the cutting platform into the cylinder. The No. 4 was also the only Rumely combine to use corrugated cylinder bars instead of the customary spiked cylinder. Despite these innovative features, Advance-Rumely was a serious financial difficulty, and the No. 4 was incapable of returning the company to a secure position. Only 100 of these units were built with a production period starting in 1928 and ending in 1929.

Advance-Rumely Thresher Co.

The Rumely-Watts corn sheller is yet another 1913 machine offered by Rumely Products Co. The Watts sheller seems to have been in some way connected with Seager Engine Works, builders of the Rumely-Olds engines, especially since virtually all advertising literature of the period illustrates the machine in conjunction with an Olds engine. The Rumely-Watts was a cylinder sheller; it represented a dramatic improvement over the earlier picker-wheel design used by most manufacturers. The largest size, the No. 6 shown in this photograph, had a capacity up to 500 bushels per hour. Standard equipment included a shelled corn conveyor, cob stacker, and ear corn feeder. In addition, an extension feeder and intermediate sections could be purchased as required. At the small end of the series were the No. 1 and No. 2 shellers having a capacity up to 100 bushels of corn per hour. Either of these shellers could be optionally equipped with a feed grinding attachment that permitted using the machine as a sheller alone, a grinder alone, or as a combination machine that first shelled the corn, then fed it into the grinder in a single operation. The No. 2, complete and ready for operation listed at $50, with the grinding attachment adding another $15 to the cost. Advance-Rumely serial number listings indicate that a corn sheller was again offered during 1925 and 1926. Production was limited to about 300 units, and no historical data concerning these machines has been located.

The Advance-Rumely silo filler was introduced in 1922 with production continuing until 1929. Two sizes were offered — the Type "B", 12-inch model, and the Type "A", 16-inch machine. Both were ball bearing equipped for easy running, and both used cutter knives mounted on a large flywheel that also served as the discharge blower. A variable-speed feed machanism permitted on-the-go changes in length of the cut feed. Rumely insured the strength of the flywheel by building it of 3/4 inch solid steel plate. The fan blades were also of steel, and were mounted to malleable iron brackets which in turn were hot rivited to the flywheel plate. The Type A machine was priced at $400, while the slightly smaller Type B silo filler had a price tag of $350. Extra cutting knives were $5 each, and a special knife grinding attachment added another $15 to the cost.

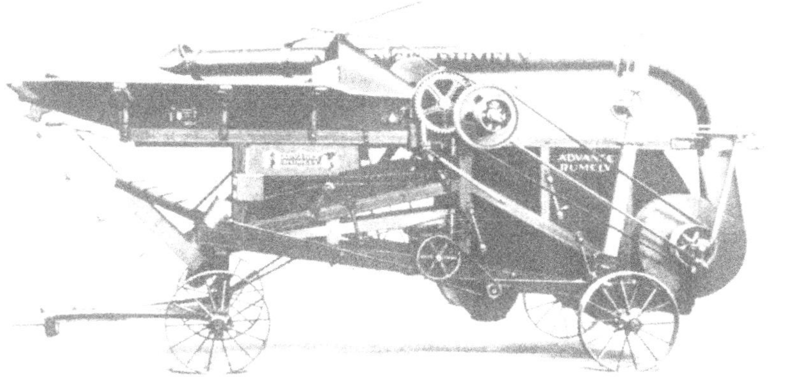

Advance-Rumely serial number listings show that husker-shredders were introduced already in 1901, with production continuing through 1928. Although the earliest offering was strictly a Rumely product, the company seems to have adopted the Advance husker-shredder after acquiring that company in 1911. Mature corn was cut in the field and placed into shocks for curing. During the fall and winter months it was brought to the husker-shredder which snapped the ears and conveyed them to a wagon. Meanwhile the stalks passed through the shredding rolls, finally being delivered to a stack. The shredded fodder could then by utilized for feeding livestock, while the corn itself could either be fed or sold at market. Since these machines were fed by hand, husker-shredders were quite dangerous, and indeed, many serious accidents resulted from their use. Rumely claimed theirs to be a safe feeder, designed to prevent hands from being caught in the machine. For 1926 the eight-roll husker-shredder listed at $1,125.

Rumely's Automatic Baler of 1913 appears to have been built by Robinson & Co. of Richmond, Indiana. Two sizes were available; the Rumely Junior with a 14 x 18 inch chamber and the Rumely Standard which was a 17 x 22 inch press. The Standard could be equipped with the self-feeding device shown here, in which case it was designated as the Rumely Automatic Baler. A Rumely baler catalog of 1913 is from Rumely Products Co. This firm, although a part of the Rumely organization, was primarily a sales arm, doing no manufacturing whatever. This portion of the operation was continued by M. Rumely Co. Rumely Products Co. however, pursued a most vigorous approach to gaining a larger portion of the total farm equipment market. Had not crop failures and uncooperative bankers intervened, Rumely Products Co. might well have been a success story during the 1920s and beyond. Fate however reared its head so that the house of Rumely crumbled at its very foundations by 1914.

Advance-Rumely

During the 1920s, Advance-Rumely introduced a swather or windrower. Probably intended for wheat-growing regions, this unit was also exported to some extent. Little information can be found concerning this machine — leading to speculation that production was of rather short duration. The reel, sickle, and conveyer apron were all ground-driven. Quite possibly this unit was also available with an extension conveyor which permitted the cut grain to be elevated directly into a wagon. Presumably the Advance-Rumely swather was available in several widths, but current information leaves even this basic question unanswered.

Sometime during the 1920s Advance-Rumely went so far as a prototype corn picker. Since nothing further than this photograph has been located, the natural assumption is that after some experiments the idea was abandoned. By 1920, corn picking remained as one of the few farm operations to be completely unmechanized. While some farmers enjoyed picking corn by hand, the job was drudgery for the majority. Although this particular unit was pulled by a tractor, the mechanism was operated by the small LeRoi engine shown here. The experimental machine was of one-row design and used a side elevator not shown in this photograph to convey corn into a waiting wagon.

In March, 1920, Advance-Rumely announced that their new motor truck would be built in a Chicago factory. The first shipment of these arrived in LaPorte during July of that year. Rumely continued production of the truck until about 1927. Features included a Buda CTU engine. Of 4-cylinder design, it used a 3¾ x 5¼ inch bore and stroke. Also featured was a Sheldon worm-drive rear axle with internal expanding brakes. For 1926, the sum of $2,201 bought the chassis with a box seat and cushion. An open cab added $85, or a totally enclosed cab could be purchased for another $140. A platform body with a grain endgate added still another $140. Solid tires came as standard equipment, but a set of 36 x 6 front and 38 x 7 rear tires could be secured as a $275 option. To bring the Rumely truck up to its optimum, the optional electric lighting and starting system was available for $75 when installed at the factory. Rated as a 1½ ton truck, the Rumely is illustrated in the usual Brewster green color already characterized by the OilPull line.

Although Rumely seems to have had a working relationship with Oliver Chilled Plow Company, by 1912 the company seems to have moved into plow manufacturing on its own. For the 1912 season Rumely offered engine gang plows in 5, 6, 8, and 10-bottom sizes. These could be furnished with a variety of moldboards for adaptability to local conditions. Each plow bottom was protected by a wood break pin which permitted the moldboard to turn back away from hidden stones or other unmoveable objects. Each individual plow had its own gauge wheel — these were cast solid to withstand virtually anything coming their way. Individual levers controlled the raising and lowering of each of the plows, with a large platform being provided for the plowman in performance of his duties. Rumely also offered their plows with a power-lift arrangement which eliminated the need for a plowman.

The Rumely Hercules brush breaker plow was designed to tear up and root out heavy roots and brush. Because of the severe conditions under which the brush breaker was used, it was designed with a beam consisting of two 1½ x 6-inch steel plates bolted together to form the beam. A one-piece steel casting formed the frog and the plow standard. The standing cutter was placed between the beam to hold it firmly in position. This plow was built in a single 24-inch size, and could plow any depth up to the beam itself. The two small control handles at the back of the plow were for the convenience of the plowman, although it seems unlikely that he could wield much control of this implement.

Advance-Rumely Thresher Co.

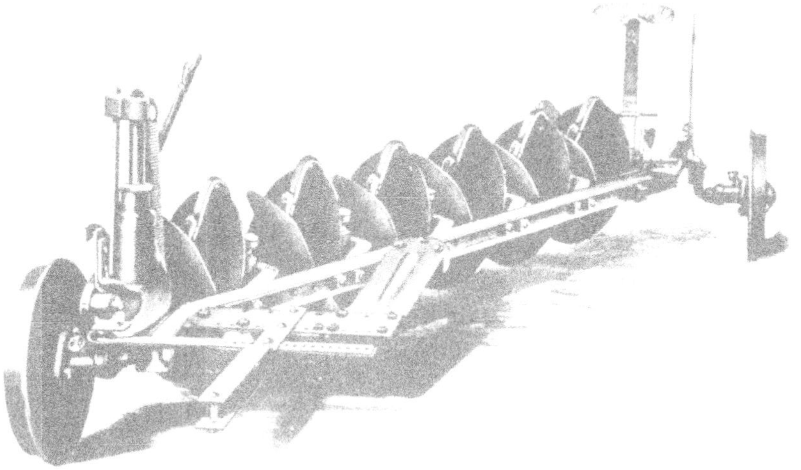

The Rumely Engine Disc Plows were actually built by Newell Sanders Plow of Chattanooga, Tenn. These units were available in 4, 5, 6, and 8-disc frames, based on a 10-inch furrow width. Dropping back to a 6-inch furrow permitted 6, 8, 10, or 13 discs to be used in these respective frames. Rumely recommended that a 20 horsepower steam engine be equipped with two 5-disc gangs, and that a 40 horsepower steamer pull two 8-disc gangs. By comparison, a 20 horsepower gas tractor was recommended to pull a single 5-disc gang. Standard equipment included 24-inch plows for furrows from four to ten inches deep. On special order, 28-inch plows could be furnished where deep plowing was required.

As part of their total offering, Advance-Rumely marketed a line of lubricants under their own trademark. Four viscosities were available to suit various requirements, with OilPull oil being recommended "for tractors, trucks, & automobiles." Undoubtedly the company went to considerable effort in securing the highest quality lubricants, and probably made this move to counter the effects of some rather poor quality oils than on the market. For the benefit of OilPull enthusiasts, this illustration shows one of the OilPull Oil fencepost signs similar to those which formerly adorned many highways.

About 1912 the Rumely Rainbow Grain Grader appeared, at least for a short time. It was designed to separate seed from dirt, weed seed, and foreign matter. After doing so, the clean seed was divided by gravity into four different grades. Two sizes were offered, a small hand-operated version, and a larger, power-driven model. In a drive to broaden their farm machinery trade, Rumely acquired or sold a wide variety of implements and machines during the 1912-1914 period. Also included was the Rumely Cream Separator, which although it carried the Rumely name, was apparently the same unit as supplied by the Seager Engine Works, builders of the Olds engines, and co-partners in the Rumely-Olds engine line.

Steam traction engines required huge amounts of water each day, so the water wagon occupied an important place in the operation. Rumely offered this 12 barrel (378 gallon) tank for mounting on the customer's own trucks, or as a complete unit. The tank alone sold for $120, but the trucks added yet another $100 to the price. This unit is shown with a Meyers pump and suction hose, but again, this was an extra cost option. Steam engineers were generally quite particular regarding the water source, preferring creek or pond water over that from a well. Usually, well water carried with it a great deal of lime, iron, and other impurities that ended up in the boiler as scale which diminished the boiler's capacity and caused endless problems to remove.

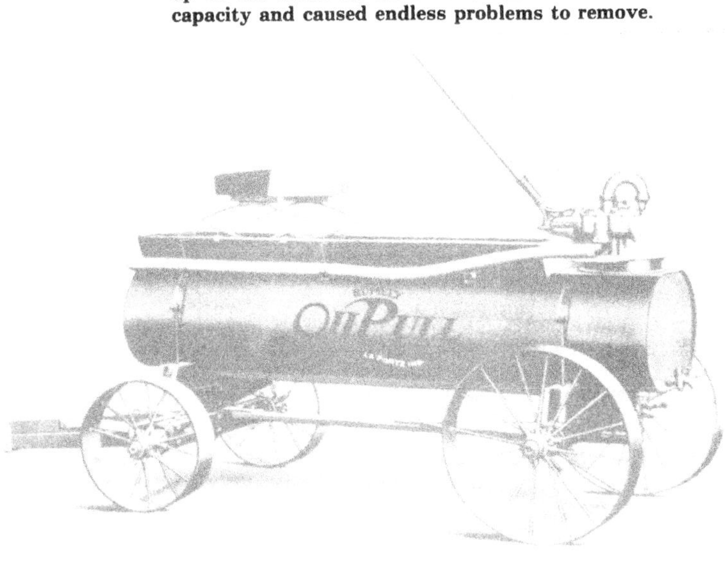

Air Compressors

By purchasing the Christensen air brake patents in 1906, Allis-Chalmers entered the air brake and air compressor business. Christensen had perfected the air brake for electric railway systems, so Allis-Chalmers found a ready market for these units. The compressor itself was operated from the same overhead electrical supply that operated the drive motors. Although a major portion of this enterprise was lost to competitors by 1910, the company continued to offer these enclosed, direct-drive compressors into the 1920s. The style shown here was available in three sizes with capacities ranging from 11 to 50 cfm. Standard equipment included a 550 volt direct-current motor. For stationary service the same compressor units were supplied with the choice of 2-phase or 3-phase motors wound for either 50-cycle or 60-cycle service.

Allis-Chalmers was one of the few companies to offer a combination compressor and pump. Intended primarily as a sprinkler filler pump, it combined both needs in a single unit. The compressor side was used to pressurize the sprinkler mains and maintain pressure on the water storage tanks. The pump was needed to fill the tanks and maintain the water level. Four sizes were available, with the largest machine having an air capacity of 20 cfm, and a water pumping capacity of 50 gallons per minute. All compressors of this period were operated through extra heavy duty herringbone reduction gears, a feature which added substantially to the production cost. Whether any of these innovative and unique pump-compressor units are still in operation is unknown.

During the 1930s Allis-Chalmers began moving away from reciprocating compressors in favor of turbo blowers and axial blowers. Rotary compressors were also commanding some attention. Axial blowers had been developed by Brown, Boveri & Co. of Switzerland in the 1920s, and in 1927 they introduced the centrifugal refrigeration compressor in Europe. Allis-Chalmers incorporated much of this technology into their own designs. The large reciprocating unit shown here was installed in the Allis-Chalmers foundry during 1937. Powered by a 500 hp synchronous motor, it had a piston displacement of 3,078 cfm at a rated speed of 180 rpm. Using a 21-inch stroke, it featured a first stage cylinder 30 inches in diameter and a second stage cylinder diameter of 18 inches. Allis-Chalmers established a Blower and Air Compressor Department in 1935 and built their first turbo blower in 1937. From this point on it appears that reciprocating compressors began the phaseout process in favor of technologically advanced rotary designs.

A 1922 Allis-Chalmers catalog continues to list portable air compressors designed for shops and other applications. These units could be supplied with either alternating current or direct current motors, and could even be equipped with a countershaft and pulley for operation from a lineshaft or a customer-supplied engine. Some of the many features included a specially designed safety valve, automatic unloading valve, and an electrically driven compressor governor. Any of these units could be adapted to use as a vacuum pump by adding suitable piping, stop cocks, and an oil separator. The big D-4 model shown here could deliver 50 cfm with a rated compressor speed of 182 rpm. It used two cylinders, each having an 8½ inch bore and a 4 inch stroke. Weight of the unit shown here was 2,750 pounds.

Air Gauges

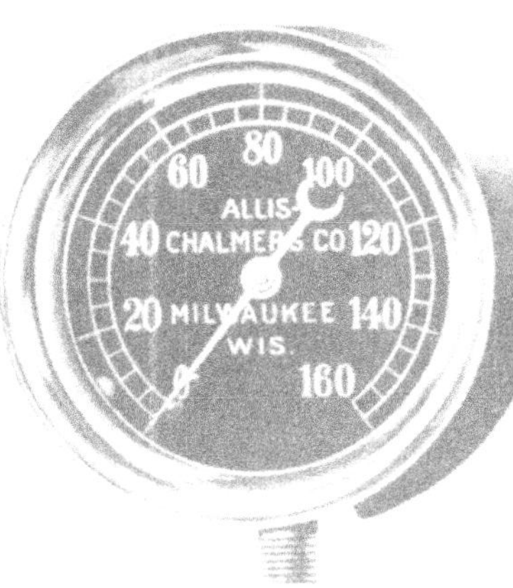

Like most other companies of the period, Allis-Chalmers proudly displayed its name. Even the pressure gauges used on the compressor units carried a gauge with the Allis-Chalmers logo. Although this was a common practice, the vast majority of gauges were built to order by companies such as Crosby or Ashcroft, to name a couple. Exceptional workmanship was displayed in early gauges, from an elaborate and precision-built internal mechanism to the silver plate dial with etched figures. The case was either of brass or cast iron, but the great majority of early gauges used a highly polished brass bezel. Today, these masterpieces of precision are highly sought after collector's items.

Backhoes

During 1950 A-C began selling a modified WD tractor equipped with a C-10 Henry backhoe. It was built by Henry Corp. of Topeka, Kansas. When the WD was replaced in 1953 with the WD-45 tractor, it too was suitably modified to accept the C-10 backhoe. This unit had a maximum digging depth of 8 feet with a standard boom, or 10½ feet with a telescopic boom. Maximum reach with a standard boom was nearly 12½ feet. Standard backhoe buckets were available in sizes ranging from 12 to 24 inches wide, but special grave buckets were available in widths up to 38 inches. The C-10 included a front end loader with a 9 cubic foot capacity and a width of 40 inches. Allis-Chalmers purchased Henry Corp. in 1968.

In 1964 Allis-Chalmers introduced the I-40 and I-60 industrial tractors. The I-40 and I-60 backhoe units were built specifically for these models. For the 1965-69 period A-C built an improved version called the I-600, and from 1966-69 the I-400 was produced. Although the later backhoe series was somewhat improved over early production, the general design was basically unchanged. The I-600 unit shown here was available with buckets ranging from 18 to 30 inches in width. A maximum digging depth of 14¼ feet was possible, as was a reach of 17 feet. The slightly smaller I-400 backhoe had a maximum digging depth of 12½ feet and a maximum reach of 16 feet. It likewise used smaller buckets, with several sizes being available in a width range of 12 to 36 inches. Between 1965 and 1969 Allis-Chalmers also produced the Industrial 64 backhoe designed specifically for the H4 and HD-4 crawler tractors.

An improved backhoe series appeared in 1957. The Mark 2 appeared from 1957 to 1959, and was followed by the Mark 1, 3, and 4 models built in the 1959-1962 period. A somewhat improved Mark 10 was produced between 1963 and 1968. Shown here is a Mark 4 backhoe, and L-100 loader, and a DH-8 hydraulic hammer mounted on a 1961 A-CD-15 tractor. The Mark 4 could also be attached to the D-14 and D-17 models. A Mark 5 unit was built specifically for the H-3 and HD-3 crawler tractors, differing primarily in the mounting methods. These units had a maximum digging depth of 13½ feet and a maximum reach of 15¾ feet. Allis-Chalmers first marketed their hydraulic hammer in 1962, continuing with several different styles through 1971. Barely visible in this photograph is a 66-B, hand-operated hydraulic hammer. Also available were rear-mounted drop hammers which operated with hydraulic power.

Between 1969 and 1975 Allis-Chalmers produced a final series of backhoe units. These included the 650, 710, 714, 715, and 816 models. The 715B shown here typifies the design. Features included an integral frame for both the backhoe and the loader. Industrial front axles were used — they included yoke-mounted spindles for better strain distribution. The arched boom design permitted greater digging leverage without sacrificing reach or digging depth. A three-way taper in the bucket design permitted easier filling and faster dumping. By 1975 however, Allis-Chalmers was moving away from industrial equipment, with production ending entirely during 1976. Additional illustrations of backhoe/loader units are located in the *Tractors* section of this book.

Bale Conveyors

Between 1963 and 1967 Allis-Chalmers offered this bale conveyor system as part of a complete forage harvesting machinery line. This particular one is being powered by an Allis-Chalmers B-10 lawn and garden tractor, with a load of bales being carried on an Allis-Chalmers Power Feed Box. The shorter bale length notes is due to the use of an automatic bale thrower in conjunction with an Allis-Chalmers square baler. The A-C bale conveyor system consisted of a complete package including horizontal conveyors for use within a barn. For close quarters the first conveyor could be built for vertical attachment to the outside of the barn — this particular unit included the necessary attaching parts for a smooth flow of bales from wagon to hay loft.

The 918 Backhoe-Loader was built especially for contractors and pipeliners needing a rugged, versatile unit for digging and loading. An 18-foot digging capacity made the 918 one of the industry's largest wheel-mounted backhoes, yet this machine was versatile for its size. Weighing 18,500 pounds, the 918 had a loader capacity of 1½ cubic yards, while the backhoe bucket was capable of ½ yard per fill. Other features included an 84 horsepower, Model 2800 A-C diesel engine; a single-stage torque converter, and hydrostatic steering. Allis-Chalmers built this unit in the 1971-1975 period, but production was rather limited, totaling less than 250 units. Company production figures are slightly misleading, since the very first 918 unit was built already in 1967, with only 26 machines being built by 1969. The remainder were built in the 1970-75 period. The Author owns a 918 unit virtually identical to the one shown here — it was the second one built.

Bale Loaders

Between 1948 and 1953 Allis-Chalmers built 4,500 bale loaders at LaPorte Works. An ideal companion for the A-C Roto-Baler, it was designed and built specifically for round bales. Roto-Baler design did not permit the attachment of a wagon hitch for direct loading, so the PTO operated bale loader was a viable alternative. As shipped, this unit was designed to mount on the Allis-Chalmers C, WC, or WD tractors. Spring loaded prongs at the base of the conveyor straightened each bale for proper delivery up the conveyor. An adjustable delivery platform at the rear of the conveyor could be raised or lowered to the convenience "of the man on the wagon" to quote an A-C advertisement. Surprisingly, production of the bale loader was less than 6% of total Roto-Baler sales.

Balers

Although Allis-Chalmers began production of the famous Roto-Baler in 1947, the idea went back to the early 1900s and Ummo Franklin Luebben of Lincoln, Nebraska. During 1909 Luebben applied for a patent on a "Baling Press" and was given patent protection on October 18, 1910 under No. 972,884. Luebben's early design embodied many of the principles which A-C adapted to the Roto-Baler nearly 40 years later. Along the way though, Luebben continued his efforts, gradually improving and refining his machine. Not however, until Allis-Chalmers put the Roto-Baler into production did the Luebben design make noticeable progress.

The new Roto-Baler was simple and economical. At thousands of country fairs, Allis-Chalmers set up their familiar display. Within a steel framework were mounted six round bales which rotated slowly. Water dripping on the tightly wound bales nearly all ran off, much like a duck's back sheds water. Farmers were quick to see the advantages here — now they didn't have to worry about losing a hay crop if the bales were caught by a sudden thunderstorm. By comparison, rain on a field of square bales was disastrous!

Sales swelled to a mighty crescendo! During the 1947-1974 period, LaPorte Works built 75,301 Roto-Balers. Granted, the last years of production saw some improvements in design, but the basic operation was nearly identical to that envisioned many years before by Ummo F. Luebben. With the success though came problems. The difficulties arose not so much from defects with the machine but from the carelessness and risk taking of farmers in too big a hurry. As with any machine, the Roto-Baler plugged up occasionally, not necessarily because of the design, but usually because the windows were not properly raked. Rather than follow the large, printed warning signs to keep away from moving parts, a few were tempted to kick loose hay into the conveyor, or attempt to kick it on its way to the bale chamber. Getting caught up in the mechanism by this means often meant serious injury. Therein came the problems for Allis-Chalmers — operator negligence was being overlooked while the company was being castigated for a poor or unsafe design.

While interviewing a company official several years ago, this writer asked why Allis-Chalmers failed to get into the "big bale" market, referring to the large round balers now in production. After a measured pause and a carefully worded reply the message came through that the difficulties experienced with the Roto-Baler were sufficient to dismiss any thought of entering yet another situation of a similar nature. This writer sees the situation as a hopeless one under the current trends. To sum up, it seems incredulous that a major corporation, having the experience, the background, the patents, and the technology to enter a new endeavor would opt to remain out of a huge potential market simply because of a virtual certainty of adverse litigation. Equally incredulous is the notion that inanimate, unknowing, and unsympathetic machines, however well designed are usually to blame, when in fact, careless and negligent operation of a machine is the probable cause of accidents in the vast majority of cases.

Recognizing that some farmers wanted a conventional square baler, Allis-Chalmers began production of these machines in 1961, continuing with several different models until 1974.

After acquiring the Luebben patents, Allis-Chalmers modified and re-modified the design until in 1947 the A-C Roto-Baler appeared. While this machine scarcely resembled the original efforts of Luebben, its basic principles were intact — rather than press the hay into a bale, the hay was rolled into a neat, compact, and virtually waterproof bale. The Roto-Baler offered low first cost, one-man operation, independence from custom operators, and from the Allis-Chalmers point of view, better bales. The bales could be unrolled into windrows for feeding or bedding. A built-in regulator controlled finished bale size anywhere from 14 to 22 inches in diameter. A density regulator permitted a finished weight anywhere from 40 to 100 pounds. Any two-plow tractor provided the necessary power through a PTO drive. Baling capacity ranged from 4 to 8 tons per hour. Tandem wheel equipment was available as an extra-cost option for contoured or terraced fields.

Shown here is a Roto-Baler with a WD-45 tractor. Obtaining well-formed bales was predicated on using a double windrow 3½ to 4½ feet in width. A single windrow entering the center of the machine produced bales that looked like an oversize cigar with a large center and tapered ends. When a bale was formed in the chamber, a trip mechanism operated the twine wrapping mechanism. This required the operator to stop until the bale was ejected, after which forward motion was again resumed. Standard binder twine was specified for wrapping the bales. For 1956, the Roto-Baler listed at $1,320, plus another $85 for the PTO attachment. In lieu of the PTO drive, the Roto-Baler could be equipped at extra cost with an A-C engine.

Balers

By 1954 Allis-Chalmers had developed a continuous feed system that eliminated the need to stop for each bale to be wrapped and ejected from the Roto-Baler. The continuous feed mechanism simply shunted the loose hay over the upper conveyor until the chamber was ready to accept it. This operation was entirely automatic. Although it would appear that this new improvement would have been very popular, the No. 10 continuous baler saw sales of only 1,500 copies, after which production was discontinued. Although current research fails to disclose any pricing information on this specific unit, the standard Roto-Baler could be equipped with options that included a bale counter, tongue support jack, and a soft-center bale attachment. Also available was a fitted canvas cover for $36.

A September, 1954 file photograph illustrates the Roto-Baler as equipped with the continuous conveyor and the rear twine mechanism. The Roto-Baler used a total of twelve forming bands or belts, six above and six below. The bale was compressed under pressure by the forming bands, and after attaining a preset size an automatic trip mechanism actuated the twine tying mechanism. The twine secured each bale, keeping it under tension, and minimizing weather damage. The heavy springs shown on the machine were part of the bale tension mechanism. As previously noted, the Roto-Baler embodied many basic principles set out many years before in the Luebben patents. It remained however for Allis-Chalmers to modify the design so that the "round baler" became a truly practical machine. Curiously, Allis-Chalmers neglected to apply for trademark protection of the "Roto-Baler" title until March of 1948, even though they claimed first use of the term on February 3, 1947. Trademark No. 552,559 was issued to the company on August 22, 1950.

The Roto-Baler remained in production from 1947 to 1960, although it appears that production had diminished to a very low level before finally being suspended entirely. The late-model machine shown here offered extensive shielding from the operating mechanism, but mechanical improvements were such that the basic design was virtually unchanged from the earliest models. Allis-Chalmers noted that the Roto-Baler design provided better quality hay, since their method of compression handled hay more gently for less leaf shatter. Likewise noted was the weather-resistant outer surface, making the hay safer from weather damage than square bales. The twine wrapping mechanism was very simple compared to complicated knotters or needles to keep in adjustment. By 1974 however, farmers were looking to round bales, but of much larger dimensions — instead of a 75 pound bale from a Roto-Baler, these new bales often weighed as much as a ton apiece! During a 13 year production run, Allis-Chalmers built over 74,500 Roto-Balers at LaPorte Works.

Beginning in 1961 Allis-Chalmers offered a conventional square baler, with the initial model designated as the No. 200. Production records indicate this model being built only during 1961 and 1962, but a 1967 price book continues to list the machine at $1,800. The 300 series, built from 1963 to 1968 offered the 302 and 303 models. Both were of the same general design as illustrated here, but the 303 was perhaps better equipped, offering a few features not found on the 302. A-C offered the 300 series with the choice of a twine model shown in this illustration, or as a wire-tie machine. The 303 Bale-Chief used a standard 14 x 18 inch chamber, a stroke length of 28 inches, and an operating speed of 70 strokes per minute. This twine model carried a 1967 list price of $2,000.

Balers

Designed for areas where hay was grown commercially and bales were to be handled frequently, the 302 and 303 wire-tie models were also produced during the 1963-1968 period. Wire-tie balers were nearly identical in price to the twine-tie models. All sizes could be equipped with an auxiliary engine instead of the usual PTO drive. Other optional attachments included the Model 20 bale thrower, complete with its own 5¾ h.p. engine to completely mechanize the baling operation. A speed-lube system for the knotter system provided one-shot greasing of twelve locations. Another innovative feature was the hydraulic bale tensioning system which automatically maintained bale tension at a predetermined level, and which could be altered without leaving the tractor seat.

A 1967 photograph illustrates a Model 303 baler complete with a PTO-driven bale thrower. Also illustrated is an Allis-Chalmers Power Feed Box to receive the bales, and leading the parade is an A-C D-17 Series IV tractor. During the 1960's Allis-Chalmers made a concerted effort to provide a complete series of hay and forage equipment, doing so in admirable fashion. Changing times and changing demands, coupled with corporate management decisions saw a gradual withdrawal from the hay and forage lines in the early 1970s. The 303 baler shown here, complete with bale thrower and optional wagon hitch listed at $2,605. Adding the hydraulic bale tensioning system added yet another $151.

For the 1969 season, Allis-Chalmers offered the Model 442 baler as a replacement for the Model 300 Bale-Chief series. This model was apparently an interim machine which bridged the gap between the 300 Series and the still-to-come 443 and 444 balers. The unit shown here was equipped with a bale thrower — the PTO drive for this unit is distinctly visible in the photograph. Other features included a safety guard over the pick-up area to prevent operator injury from reaching into or being pulled into the mechanism. Allis-Chalmers appears to have been very safety conscious. In addition to the usual warning decals and protective guards, instruction manuals which accompanied each machine carried what might be termed a short course in operating safety.

Between 1972 and 1974 Allis-Chalmers offered the Model 443 baler. It was virtually identical to the Model 444 baler except that the 443 used a two-joint power line, compared to a three-joint power line on the 444. This particular unit is shown with the conventional bale chute which delivered finished bales to a towed wagon. At an operating speed of 540 PTO rpm, the 443 gave 80 bale strokes per minute. A standard 14 x 18 bale chamber was used, together with a 28-inch bale stroke. Bale length was adjustable, up to 42 inches. Safety features abounded in this machine. Included was a flywheel shear bolt for ultimate protection of both the baler and the tractor PTO drive. Also featured was an overrunning clutch, spring loaded slip clutches, and a ram stop which prevented the ram from hitting the needles in case of trouble. Instead the ram hit the stop, thus shearing the flywheel pin.

Closing out the line was the Model 444 baler built during 1970 and 1971. The 444 used a three-joint drive line, but was otherwise the same as the 443 baler. Even these late models were available with wire-tie mechanisms in lieu of the standard twine-tie system. Allis-Chalmers offered a unique steering system for the bale thrower. The conventional wagon tongue control forced the thrower to follow the wagon — an important feature on contoured or hilly ground. Optionally available was a hydraulic steering mechanism controlled from the tractor seat. Although certain A-C production records indicate that Roto-Baler production had ended in 1971, advertising materials suggest that this machine was still available in 1974. Possibly the Roto-Baler resumed production for a time, or perhaps a number were built from existing parts stocks. Regardless, the total production by 1970 appears to have been very limited.

Beach Sanitizer

Designed specifically for the park and recreation industry, the Allis-Chalmers Beach Sanitizer presented an efficient means of reducing beach upkeep while satisfying the public demand for clean beaches. The Beach Sanitizer, operating at a depth of six inches was capable of covering from 1.2 to 3.75 acres per hour. A reel mechanism at the head of the machine moved sand onto the conveyor apron. Long-wearing brushes were used to move heavy, wet sand through the apron, but stones and trash were carried to a trash receptacle at the back of the machine. When full, the receptacle could be emptied hydraulically into a waiting truck. The Beach Sanitizer was intended specifically for use with the Beachmaster tractor. This was an industrialized version of the current 190 tractor, and was available in either a gasoline or diesel-powered version. The Beachmaster tractor used a hydraulically actuated clutch and offered eight forward speeds ranging from just under 1 mph to nearly 14 mph. This system provided on-the-go shifting from low to high range with no clutch wear. High-flotation tires were standard equipment. The Beach Sanitizer has a listed production period of 1964 through 1970.

Bean Cutters

Production of the bean cutter and windrower attachment approximately paralleled the Model B tractor. Built specifically for use with this tractor, the bean cutter could be purchased with the necessary frame and brackets as a separate item, or could be attached to the frame of the No. 65 and No. 66 cultivators. This unit was designed to cut and lift two rows of beans at a time, preparing them for final harvest. The tapered cutting knives were adjustable for virtually any type of bean crop and to almost every crop condition. Despite the value of this unit for specific crops, especially edible beans, production was rather limited, and appears to have ended entirely by 1967.

Betatron

Already in 1944, Allis-Chalmers had perfected the Betatron, undoubtedly because of its involvement in the Manhattan Project which produced the first nuclear bomb. During 1952, a 24 million volt Betatron was installed at New York Memorial Center to be used in cancer research. It was the first such machine to be designed specifically for the treatment of cancer. The Betatron is an instrument which produces high energy electrons. The flux of a magnetic field accelerates the electrons to nearly the speed of light — 186,000 miles per second. In operation the Betatron functions similar to an alternating-current transformer, but a large doughnut-shaped vacuum tube replaces the transformer secondary, with the electrons being held in place by a magnetic field instead of a wire coil. When X-rays are produced, the beam of free electrons is directed into a tiny platinum target, and from this target, a sharply defined X-ray beam is produced. The unit shown here was built in April, 1946.

Shown in this photograph is the first completely portable Betatron being loaded on a waiting truck. This unit was designed to test castings for internal flaws, and was capable of penetrating castings 22 inches in thickness! The trailer unit housed the control room, and a similar trailer carried the diesel-electric unit to provide operating power to the 25 million volt Betatron. This unit was shipped to Belgium to serve three private companies for radiographic inspections. A news release of November 27, 1973 announced the sale of the Betatron Division to ATC Betatron Corp. headquartered in West Allis. This news release also indicates that up to the date of the sale, Allis-Chalmers had built 80 units, or nearly one-third of those in use around the world. Reasons for the sale, or the ATC purchase price were not disclosed.

Blades

In 1953 Allis-Chalmers announced the "Snap-Coupler" hitch for the WD and CA tractors. Following this announcement came the gradual introduction of implements centered about this draft arrangment. Included early in the process came an "earth and snow blade" as shown here on a WD-45 diesel tractor. Light in weight and low in cost, tractor implements such as the earth and snow blade were great labor savers and found extensive use. In the following years, Allis-Chalmers offered an extensive variety of blades to fit the Snap-Coupler hitch, eventually building others to fit the standard 3-point hitch designs. While in itself of slight historical significance, implements such as this were indeed a part of the transition from manual to mechanized farming.

Brenneis Mfg. Co.

During 1898, Louis Brenneis established a blacksmith shop at Oxnard, Cal. Initially it was a general repair shop, but within a couple of years Brenneis was spending much of his time modifying farm implements to suit local needs, especially those implements associated with the beet industry. Brenneis concluded that there was a definite need for beet planting and cultivating machinery built to suit the job, rather than modified from other equipment, so by 1900 the first Brenneis beet cultivator appeared. During 1910 the first Brennis beet planter appeared. Although these implements were quite successful, their application was limited primarily to the beet industry.

By 1915 the Brenneis line included an extensive offering of orchard tools, including chisel cultivators, tandem discs, and subsoilers. A 1914 price list of Brenneis orchard, field and road tools lists beet cultivators, bean cultivators, orchard cultivators, beet drills, bean planters, fertilizer drills, and bean cutters.

By the 1930s Brenneis had heavy competition with Killefer Mfg. Co. of Los Angeles. This firm held numerous patents on offset disc harrows, as did Towner Mfg. Co. of Santa Ana, Cal. The latter firm, while listed as a farm equipment builder, appears to have had the primary role of a patent holding company with manufacturing being in a distant second place. Because of numerous patent interferences, the three companies agreed to mutual use of each other's offset disc harrow patents in early 1938. Killefer and Towner had made a licensing agreement with Deere & Co. the previous year, and this complicated matters even further.

Allis-Chalmers was by this time looking for a line of deep tillage implements to bolster sales of their crawler tractor line. Even though Brenneis employed but 23 men, the quality of Brenneis products was such that Allis-Chalmers was soon able to expand their new "California Works" considerably. This title was soon changed to Oxnard Works, remaining so during its entire life with Allis-Chalmers. Purchase price of Oxnard Works from Brenneis was $103,000.

Allis-Chalmers announced a major expansion of Oxnard Works in 1949. This $250,000 project followed a $100,000 expansion of the previous year. Offset disc harrows and much of the deep tillage implement line was built at Oxnard Works.

Although little pictorial data is currently available regarding the early Brenneis products, a catalog from the 1920s illustrates a small offset disc harrow designed for orchard cultivation. This particular one is being pulled by a late style Fordson tractor. Orchard cultivation called for low-profile machines as a major requirement. Keeping the implements small in size gave better maneuverability — another requirement for orchard and vineyard work. Although Brenneis started into the implement business with little competition, Killefer and Towner gave it plenty during the 1920s and 1930s. Shortly before selling out to Allis-Chalmers in 1938, Brenneis entered into an agreement with Killefer and Towner that permitted all three companies to jointly use their collective offset disc harrow patents.

Brenneis Mfg. Co.

The Brenneis subsoiler was an important addition to the line. In many areas a hardpan existed beneath the soil surface. This became very impervious to water, with the subsoiler being the most practical way of breaking up the hardpan to promote soil drainage. Considerable power was required, especially when the hardpan was quite deep, so this machine was built extra heavy to withstand the abuse to which it almost surely would be subjected. Two heavy steel plates form the beam and extend to the tractor hitch. Because of the power requirements for this machine, crawler tractors were often used. Possibly this was one of the machines that attracted Allis-Chalmers to Brenneis initially. In 1938 Allis-Chalmers acquired Brenneis, and from this purchase came the Oxnard Works.

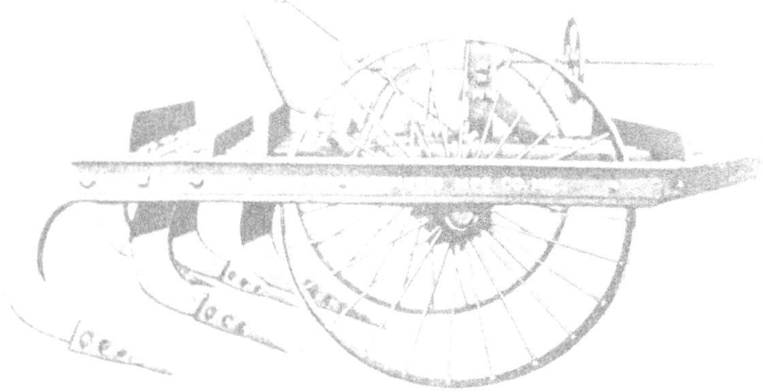

The Brenneis Giant Road Ripper must have indeed been an impressive sight when in operation, but probably more impressive was the performance of the tractor to which this outfit was connected! The tooth design was such that once penetration was achieved, a pronounced lifting effect and the subsequent rolling action of the material should have been very effective in reducing a hard-packed road bed into loose material. As road men know, macadam roads, once filled with pot holes, are virtually impossible to repair from the top side. The only effective way to remove them is to tear up the roadbed deeply enough to get a fresh start. Certainly the Brenneis Giant Road Ripper must have been equal to the challenge.

Buda Mfg. Co.

In November, 1953 Allis-Chalmers purchased Buda Co. of Harvey, Ill. Prior to this time, Allis-Chalmers had been using General Motors diesel engines in their crawler tractors, while using their own gasoline engines in the farm tractor line. General Motors purchased Euclid Road Machinery Co. in August, 1953 and Allis-Chalmers apparently saw this move as a direct competition with their own road machinery business. Buda had a long and proven track record in diesel engines, and indeed, Allis-Chalmers had made limited use of Buda engines in previous years. Thus, the purchase of Buda dovetailed into the Allis-Chalmers plan for extensive involvement in the engine business, not only for their own manufacturing needs, but also for the OEM and military markets.

George Chalendar organized Buda Foundry & Mfg. Co. at Buda, Ill., during 1881. The original plant included a machine shop, foundry, and office within a single building. During 1891 the Harvey Land Assn. was instrumental in getting the company to move into new facilities in Harvey, Ill. Up to 1910, production included an extensive offering of railway equipment and supplies — everything from handcars and switch stands to cast iron heating stoves. An 1889 Buda catalog lists the "Giant Heating Stove," of the Round House Pattern. Weighing 1,100 pounds, it had a gigantic 26-inch fire pot.

During 1907 Buda expanded its line to include a series of motors and transmissions, and by 1910 Buda gasoline engines accounted for a major portion of annual sales. Although Buda soon gained a reputation for truck and tractor engines, a few Buda engines made their way into automobiles. One of the earliest examples is the McKay, built at Kentville, Nova Scotia by McKay Bros. Originally known as Nova Scotia Carriage & Motor Car Co., this firm produced 200 cars before going bankrupt. Of special interest to this historical sketch is the fact that a Buda engine was used.

Buda continued in the railway supply business, offering a series of motorized inspection and hand cars during the 1911-1925 period. This time frame also saw the introduction of an extensive engine line, with Buda taking the innovative step of offering "re-power" units for many truck models including the Acason, Clydesdale, Diamond T, F.W.D., G.M.C., Indiana, International, Mack, Packard, Pierce-Arrow, Signal, Velie, and White, among others.

The 1926-40 period saw further expansion of the engine line, and engine-generator sets became a major part of the business. Railroad equipment was still manufactured, but this was no longer the primary emphasis within the company. During this period Buda developed and marketed a new series of lifting jacks in over 100 models. These were widely used by railway, construction, and industrial businesses. Mechanized material handling got a big boost with the 1938 introduction of the Chore Boy industrial truck. Powered by a small air-cooled engine, this little outfit permitted one man to do the work that formerly required three or more people.

During 1926 Buda took another innovative step in engine design by introducing their first diesel engine. This unit was built under design from M.A.N. *(Maschinenfabrik-Augsburg-Nuremburg)* of Germany. The M.A.N. design as originally used by Buda featured two injection nozzles placed horizontally in the side of the cylinder head, and diametrically opposed to each other. The head itself was recessed to form the combustion chamber. Fuel for each cylinder was supplied by one plunger of the M.A.N. pump — a single injection line was split with fuel going to each injector. The injector units were not spring loaded, and in fact, were virtually an open line terminating in the cylinder head. Each line had a "backfire valve" to prevent the pressure of compression and ignition making its way into the pump unit. Fuel injection timing remained constant, and the end of injection cycle varied with the load. Buda modified the engine in line with M.A.N. developments, occasionally adding refinements of their own. By 1933 the Buda diesel of M.A.N. design was available in a high-speed automotive design, compared to the original models which were suitable only for marine propulsion, stationary power units, and engine-generators.

In 1938 Buda introduced an entirely new diesel engine series which embodied the Lanova design pioneered by Franz Lang of Germany. Lang had earlier introduced the Acro system which employed a specially designed piston head, hollowed out so as to form a combustion chamber. The drawbacks of the Acro system included a distinct diesel knock inherent in the design. Likewise, the Acro was unsuited to very large cylinder bores due to excessive heating of the piston head.

Lanova engines used a new approach of directing the injected fuel toward an air cell or "energy cell." For cold weather starting the cell could be closed by means of a valve, thus raising the compression ratio temporarily. In ordinary operation, the combination of the air cell and the unique cylinder head design yielded an engine of exceptional economy and dependability. By taking on the Lanova system, Buda had the backing of Lanova Corp. and a myriad of patents covering the design. Except for occasional refinements, the Buda-Lanova diesels remained almost unchanged until Allis-Chalmers took over in 1953. Even after the A-C purchase, Lanova designs were still prevalent at Harvey Works.

Purchase of the Buda plant seems to have been an important milestone in Allis-Chalmers progress. Though the company had built heavy stationary-type diesel engines already in the early 1920s, their efforts to build the high-speed diesels seem to have made poor headway in the ensuing years. Allis-Chalmers had been building its own gasoline engines for the tractor line at West Allis for many years, but the Buda purchase now made it possible to obtain the research and development of an engine manufacturing specialist. Yet another key factor was a well-developed line of engine-generator sets. Here again, Allis-Chalmers was in a position of simply taking up the market already established by Buda.

From the Chore Boy industrial platform truck of 1938 eventually came an entire series of Buda lift trucks, and the Shop Mule tractors started an extensive series of machines designed to handle heavy loads, including aircraft.

With the formation of Deutz-Allis in 1985 came the phaseout of numerous plant operations. Harvey Works, formerly the Buda Co., closed its doors in mid-1986.

Although Buda Foundry started out in a small building that served as foundry, factory, and office in 1881 it seems obvious that the company grew by leaps and bounds during the next few years. In fact, an early catalog shows a rather extensive group of buildings at the Buda, Ill. location. The reasons for moving from Buda to Harvey, Ill. in 1891 are somewhat unclear — perhaps a strong inducement from the Harvey Land Assn. was involved, and most likely there were other factors of importance. Regardless of the reasons, Buda Foundry & Mfg. Co. moved out of Buda in 1891, and immediately occupied new facilities at Harvey, Ill. From this humble beginning came the huge Harvey Works of Allis-Chalmers.

An early Buda catalog devotes the majority of its space to the company's main endeavor — railway supplies. This included an extensive line of hand cars, including the No. 4 unit shown here. Buda gave considerable space to advertising its wood-wheel construction as being the "lightest, the strongest, the most durable, and the safest" on the market. The steel tire was fitted with a wood center — it was pressed into place. The frame of Buda handcars was made of ash, locked and bolted at each joint, and furnished with iron truss rods for reinforcement.

Buda Mfg. Co.

For those desiring an iron spoke wheel construction, Buda offered this hand car for 1882. Although the hand car was a great improvment over walking to a repair or construction site, operating one required a considerable effort. The occasional escape scenes of Western movies in which a pair of desperadoes gamely pump their way to freedom via a hand car really do not do justice to the more compelling fact that virtually everything connected with railway construction and repair was hard work. In addition to the two hand car styles shown here, Buda offered an extensive variety for specific requirements, including special cars for use in mines. Additionally, Buda was equipped to supply hand cars built to the buyer's specifications.

From its earliest beginnings, Buda offered not only an extensive line of hand cars, but also provided railroads with an excellent choice of switch stands and other railway necessities. The "Empire" switch stand shown here was one such variety offered during the 1880s, with half a dozen other varieties available to suit the requirements of individual railway companies or specific situations. In addition, Buda offered switch rods, rail frogs, "T" crossings, and semaphores.

Railway construction and maintenance required a great amount of heavy equipment, and this included a pressing need for compact, lightweight, and powerful jacks. Shown here is a typical example of a Buda model. The company offered dozens of different styles and sizes, each built to exacting specifications, and tailored to specific needs. While Buda jacks were originally intended for railway maintenance, they soon found their way into virtually every phase of construction and industrial work all over the world. With constantly changing technology, Buda began offering an extensive line of hydraulic jacks during the 1930s.

The majority of foundries in the 1880s offered heating stoves for sale. Buda was no exception, with their early catalogs listing a variety of "volcano" stoves, way car stoves, station stoves, box heaters, and this "Giant" heating stove. This particular stove was termed by Buda as their "Roundhouse Pattern" and had a huge 26-inch firepot. Weighing 1,100 pounds, it most likely was able to break the winter chill in a large building such as a roundhouse. Considering that stoves of this type were to be found virtually everywhere in cold climates, the market was endless. As Buda moved into other manufacturing areas, stoves disappeared from their offerings.

The 1940s saw still further excursions into the railway business, with Buda introducing a compact car that could be furnished either as a passenger unit, or as an inspection car. The passenger style could carry up to 12 people including the operator, but the inspection style was limited to 8 passengers. This attractive unit was of course equipped with a Buda engine. Standard equipment included 20-inch chilled cast iron wheels, 3-inch axles, and Timken roller axle bearings. The car was carried on semi-elliptical leaf springs and radius rods with Buda's patented thrust plates. For exceptionally smooth riding, rubber cushioned wheels were available.

Buda Mfg. Co.

By 1911 Buda was offering an extensive series of motorized handcars. Their Model M railroad inspection vehicle included a landau top and upholstered seats, but the chassis was essentially the same as the section car shown here. Coincidentally, the Model M used the same engine as the McKay automobile built in Nova Scotia, for which Buda supplied the engines. Never content to remain static, Buda was a forerunner in developing motorized railway baggage wagons, and these in turn led to the development of the lift truck. Buda's entry into engine manufacturing about 1910 was a fortunate move — within a few years this division accounted for a major portion of company operations.

Buda pioneered the idea of motorized baggage carriers and eventually from this came the forklift truck. During the 1941-1955 period, Buda introduced an extensive series of fork lift trucks, with industry seizing them by the thousands. The fork lift made dangerous hand-lifting obsolete. It was able to do the same lifting jobs much faster and more efficiently than before. With the forklift came development of palletized systems of storage and utilization, saving industry even more money in terms of warehouse space. Shown here are some typical Buda forklifts of the 1950s.

During World War II, large planes required heavy-duty moving equipment, and from this requirement came the Buda Shop Mule. Although industrial plants used the Shop Mule extensively, it and its descendants were and still are commonly seen in airports. This Shop Mule is linked to a plane in front of a United Airlines terminal, waiting for passengers to board. After boarding is completed the Shop Mule pushes the plane backward so that it can then taxi toward the runway. The Shop Mule Division was sold prior to the purchase of Buda by Allis-Chalmers in 1953.

The year 1938 marked a new milestone for Buda with the introduction of the Chore Boy industrial platform trucks. Capable of carrying a half-ton load, the Chore Boy was a welcome addition to industrial plants which heretofore had used men with balance trucks and push carts to move parts and equipment from one place to another. The Chore Boy appeared just prior to World War II, and when women replaced men in the factory while the men were off to war, equipment like the Chore Boy was essential to maintaining peak production. Thus in its own way, the Chore Boy contributed greatly to the war effort.

Buda began its meteoric rise as a renowned engine builder in 1910. Their early engines used an integral block and cylinder head. This eliminated the need for head gaskets, often a troublesome flaw in early engine designs. From Buda's earliest beginnings came styles like the well-known "HU" style shown here. A number of early motor trucks used the Buda engine, as did several farm tractors. A 1918 issue of *Cooperative Tractor Catalog* indicates that Buda tractor engines were available in six different sizes ranging from 22.3 to 32.4 S.A.E. horsepower. In addition to auto, truck, and tractor applications, Buda engines were extensively used in marine duty and for numerous industrial applications.

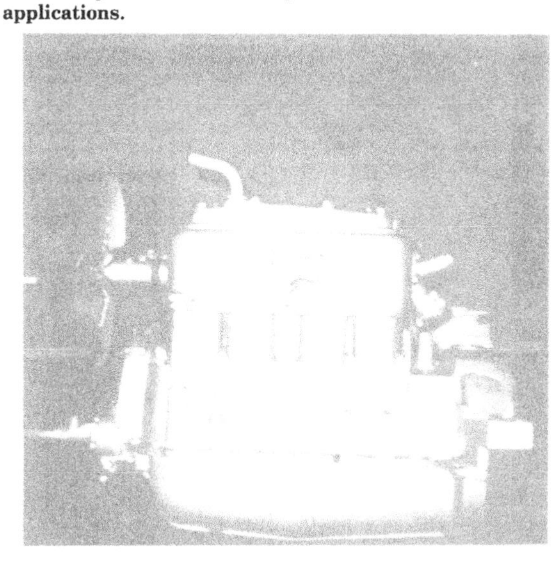

Buda Mfg. Co.

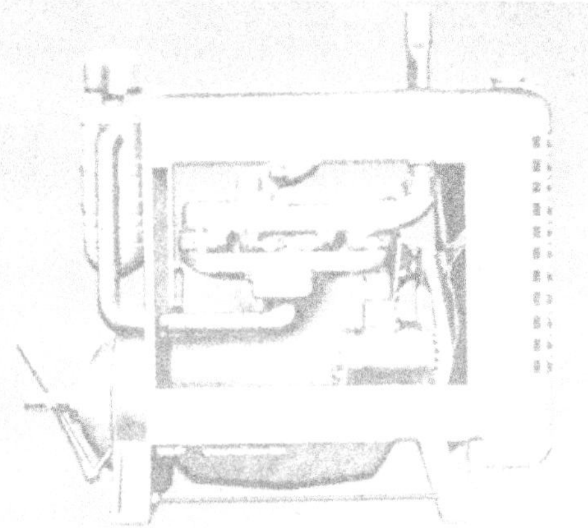

Diesel engines were introduced by Buda in 1926. Initially they were for stationary duty and used the M.A.N. design. More than likely, sales were not exceptionally high for these early models, yet the end-of-year totals were obviously good enough for continued research and refinement. By 1933 Buda had announced a high-speed automotive design using the M.A.N. fuel system, but this was replaced in 1938 by a completely new design featuring the Lanova system pioneered by Franz Lang of Germany. From this time onward, Buda diesel engines used the Lanova system. Shown here is a typical design of the 1940s. The little 1-BD-38 industrial diesel weighed 670 pounds. Its 38 CID resulted from a $3\frac{7}{16}$ x $4\frac{1}{8}$ inch bore and stroke. At its rated speed of 1,800 rpm, $6\frac{1}{2}$ horsepower was available. Conversations with former Buda employees indicate that the majority of these engines were sold, not for specific power uses, but to major oil companies for purposes of testing caloric values of fuel, along with other fuel testing procedures.

Following the single cylinder engine of the BD series was a 2-cylinder model using the same bore and stroke, but with an output of 16 horsepower. Likewise, this series included the 4-cylinder, 4-BD-153 engine. The model designation was simple indeed — the first digit indicates the number of cylinders, the following letter code denotes the engine style, and the latter number gives the cylinder displacement in cubic inches. Rated at about 30 horsepower, this engine was accompanied in the BD series by another 4-cylinder style, the 4-BD-182 with a $3\frac{3}{4}$ x $4\frac{1}{8}$ inch bore and stroke. The BD diesel models had a design which closely matched that of the Model B gasoline engines built concurrently. Topping out the BD series were two 6-cylinder styles, the 6-BD-230 and the 6-BD-273. The latter size topped out at 75 maximum horsepower.

Buda's engine line for 1953 included the DA and DAS series. The latter designation simply indicated that a supercharger was used. Shown here is a 6-DAS-844 Industrial Diesel of supercharged design. The blower is obvious — it was a positive displacement type with helical form rotors "to obtain quietness." This complete package unit weighed 5,400 pounds. Like the naturally aspirated model, the 844 was a 6-cylinder model using a $5\frac{1}{4}$ x $6\frac{1}{2}$ inch bore and stroke. At a rated 2,000 rpm it delivered 280 horsepower, compared to 215 horsepower on the naturally aspirated version. The DA series also included an 8-cylinder model — in the supercharged design it cranked out a maximum of 350 horsepower.

Buda's 1953 engine line climaxed with the huge 8-DCS-2505 series, available in a 388 horsepower naturally aspirated design or equipped with a supercharger for a maximum output of 516 horsepower. This huge in-line engine carried a $6\frac{3}{4}$ x $8\frac{3}{4}$ inch bore and stroke and was rated at a maximum speed of 1,300 rpm. Five v-belts were required for the huge positive displacement supercharger. Buda's advertising noted that the 2505 was large enough to operate extra large equipment including $3\frac{1}{2}$ yard shovels and similar machines. While the Buda gasoline and diesel engine series was far more extensive than shown in this volume, it at once becomes obvious that the company was serious about the engine business by offering such a wide range of sizes. When Allis-Chalmers took over in 1953 the Lanova design continued in production for several more years, eventually introducing new designs. The D-262 as built by Allis-Chalmers was however, basically of the Lanova design.

By the mid-1920s Buda included marine engines as part of the full line. Already with the introduction of Buda's first diesel engine design using the M.A.N. patents, Buda offered these engines equipped for marine duty. After changing over to the Lanova design, Buda's marine engine series was extensive indeed, running the full gamut of sizes up to 500-plus horsepower. This photo from the old files of Buda Company shows a fishing boat called "Leif the Lucky". Although information with this photo indicates that a Buda Diesel engine is on-board, specific data of the size or style is unavailable.

Buda

Early to recognize a market for engine-generators, Buda began offering these units in the 1930s and perhaps earlier. By the 1950s Buda had developed an extensive series of engine-generator sets, ranging all the way from 2½ to 250 kilowatts. Both gasoline and diesel engines were available, with this particular unit also carrying an auxiliary drainage pump on one side and an air compressor on the other. This particular unit was designed specifically for application on board ship and on shore. The engine is a 2-BDG-77 gasoline engine of 2-cylinder design. Its 77 CID is sufficient to operate a 3kw, 32-volt DC generator. The bilge pump has a capacity of 90 gpm, and the air compressor delivers 9.25 cfm at 250 psi. During World War II a great many of these units were pressed into service all over the world. The postwar period saw new markets as power needs increased in areas that had limited power sources, or in some cases, no "high-line" power at all.

Buda engines also saw extensive use in pleasure boats, including the Tradewinds Kingfisher shown here. Once again, this photo from the Buda files indicates that a Buda engine was installed, but fails to state the specific size or style in use. Buda marine engines were equipped with a heat exchanger system making a radiator unnecessary. Larger sizes were equipped with a shell-and-tube oil cooler. Electric starting equipment was standard, but an air-operated starting motor could be furnished on special order. Also of interest, Buda offered a series of engines specially tailored to the needs of the oil fields, and went so far as to offer specific catalog data on their oil field engines.

A 1940 catalog illustrates this Ford truck equipped with a Buda 6-DT-317 diesel engine. Maximum torque was achieved at 1,800 rpm, and at this speed the 317 gave an output of approximately 75 horsepower. The truck shown here pulled a 20-ton payload with a gross weight of 60,000 pounds. On a 165 mile trip, empty one way, and loaded the other, it had a total fuel consumption of 25 gallons over the 330 mile round trip, for an average of 12 miles per gallon, compared to about 5 miles per gallon with the original equipment. Buda offered engine styles to fit conventional or cab-over bodies, and went even further by making installation so simple that "it can be handled by almost any mechanic." The company's innovative approach probably led to a great many sales of their diesel engines at a time when diesel engines were but a tiny minority in over-the-road vehicles.

Although Buda engines were used in virtually every application from autos to tractors to shipping vessels, the company seems to have devoted an extraordinary amount of effort in promoting their diesel designs for use in trucks. Already in the late 1930s Buda was offering retrofit engines for many current truck models. One such engine was the 8-DA-1125 automotive diesel shown here. Its 1125 CID could deliver a maximum of 300 horsepower. Although the early styles were much smaller, this one was well suited to hauling heavy loads over mountain roads as one example. The flywheel housing was so constructed that the electric starting motor could be mounted on either side of the engine.

A 1950s photo illustrates a Diamond T Model 950 logging truck making a haul from the landing to a re-load station. The 950 is equipped with a Buda DAT-779 turbocharged engine. Working in an Oregon-based logging operation, this truck ran 11 hours per day, 5 days a week, hauling up to 75,000 board feet per day in five trips. The DAT-779 seems to have been used extensively in log hauling operations, as was its supercharged version, the DAS-779 model. By the early 1950s several truck builders were regularly using Buda engines — included were Kenworth, Hendrickson, Peterbilt, and F.W.D. After Allis-Chalmers took over in 1953 the all-out effort to become a major force in the truck engine market seems to have faltered, even though Allis-Chalmers apparently equipped a portion of its own truck fleet with Buda diesels.

Bullock Electric & Mfg. Co.

In his book *An Industrial Heritage*, Walter F. Peterson comments that "Allis-Chalmers was forced into the electrical field to maintain its existence." At first glance, this comment seems to overstate the case, but closer investigation reveals a great deal of truth in this statement. Edward P. Allis Company had by 1900 built an excellent reputation for its Reynolds-Corliss steam engines, and after the formation of Allis-Chalmers Company, was eager to move into the hydraulic and steam turbine fields. Gaining the engineering know-how was achieved with little difficulty, and A-C already had most of the facilities required for building prime movers. The problem was in matching their machinery to electrical generators.

A-C appears to have made overtures to General Electric Co. for an arrangement whereby Allis-Chalmers would supply the prime mover with General Electric building the electrical equipment. Since General Electric was already in the turbine business, they indicated no interest whatever. Likewise, Westinghouse Electric was quite content to operate on a completely independent basis. Therefore, if Allis-Chalmers was to have any role whatever in the electrical power industry, it would have to either build its own factory or find one that could be leased or purchased. Thus in 1904, Allis-Chalmers entered into a complicated lease arrangement with Bullock Electric & Mfg. Co. of Cincinnati.

Bullock had a history going back to 1887 and the Geo F. Card Mfg. Co. Although Westinghouse Co. had installed the first large hydroelectric alternators at Niagara Falls in 1895, Bullock followed close behind by building their first alternators in 1898. For the St. Louis World's Fair of 1904, Allis-Chalmers supplied a 5,000 horsepower Reynolds-Corliss engine of horizontal-vertical cylinder design. It was coupled to a 3,500 kva Bullock alternator. Operating at 75 rpm, it delivered 2,200 volts and 25 cycles. This huge machine supplied all the electricity required for decorative lighting at the St. Louis Fair. The complete display received three Grand Prize awards, one for the 5,000 horsepower engine, one for the Bullock generator, and still another for the exhibit that included circuit breaker equipment.

To better understand the company's intense desire to get into the electrical power industry requires an introspective look at the early 1900s. America was emerging as an industrialized nation. Until this time, steam engines or gas engines were the primary source of reliable power, and these prime movers had been developed to reasonable success only during the previous quarter century. Prior to that, crude and inefficient waterwheels harnessed but a tiny fraction of the natural energy available from many of America's rivers and waterfalls. The potential was enormous. Allis-Chalmers and its predecessor companies, especially Edward P. Allis Co., had built much of their reputation on being able to supply a complete, ready-to-go operation, whether it was a flour mill or a sawmill. Being able to build a hydraulic turbine or a steam turbine meant nothing if there was no generator to go with it. True enough, there is no doubt that A-C had the resources to build an electrical factory from the ground up if necessary, but that would delay the entire operation for two or three years. Thus, the A-C operating arrangement with Bullock would serve to launch the company into one of its greatest manufacturing divisions.

Bullock had its birth back in 1887 as the George F. Card Mfg. Co. In 1890 the company name was changed to Card Electric Motor & Dynamo Co., a title that more accurately reflected its primary function. The struggling company probably found a very limited market for small motors and dynamos, and as a result it was reorganized in 1897 as the Bullock Electric Mfg. Co. Until 1899, Bullock operated from a plant in Cincinnati, but that year a new plant was occupied in the Cincinnati suburb of Norwood.

Bullock built its first alternating current generators in 1898, and in 1900 Bullock secured manufacturing rights for a-c electrical equipment from Oerlikon Co. of Switzerland. While still functioning as an independent company, Bullock built the huge generator that was coupled to an Allis-Chalmers 5000 horsepower steam engine at the St. Louis World's Fair of 1904. The company secured the Grand Prize for this generator, and in addition received a Gold Medal for their multiple voltage balancing sets and a Silver Medal for their street railway apparatus. If nothing else, Bullock gained excellent name recognition from these achievements — the company was no doubt very proud of their position in the industry at this point. After all, the boy had successfully defeated the electrical giants of General Electric and Westinghouse. For Bullock and Allis-Chalmers to join forces in September, 1905 should therefore have surprised no one. Even though Allis-Chalmers gained complete control of Bullock in 1905 under a leasing-buy arrangement, the takeover was not completed until 1914. Shortly after acquiring Bullock, A-C renamed it as the Norwood Works, and from this factory emerged countless thousands of motors, generators, transformers, and other electrical equipment.

The Bullock operation was a complete one, all the way from the foundry to the shipping room doors. An illustration of the erecting floor shows all sizes and styles of generating equipment in process, and even includes a large flywheel in the foreground. Due to the uneven action of slow-speed reciprocating engines, the stored energy of a large flywheel was imperative to maintain an absolutely constant speed. This alone was a major problem with alternating current generators, since the frequency is speed-dependent. Bullock, like other manufacturers of the day, tended to overdesign and overbuild their machines, all in an attempt to gain and maintain the ultimate in reliability. By placing today's technology against that of eighty years ago, they did their job remarkably well, considering that the available materials were far less developed than at the present time.

Each Bullock motor and generator was thoroughly tested before leaving the factory. This early photograph illustrates several different generators being tested at once. In the foreground a rheostat and several grid banks are observed. These cast iron plates were probably used to dissipate the generated power, or in other words, served as a dummy load. Obviously, motor testing required only to load a generator up to the motor's full load capacity. An operator stands at a large switchboard in the background, monitoring the readings, and to the right, two men are observing meter readings as the testing continued. Thorough testing provided the customer with greater reliability, and this point alone was an important sales pitch.

As alternating current grew in importance, power lines were soon to be found almost everywhere. The concept of electric railways, so popular in large cities, now made its way to smaller cities and towns in the form of street car lines. A further refinement was the construction of numerous electrically operated interurban lines which ran between cities and towns. These required converter sets in which the a-c power was converted to direct current for the traction motors. Many electrical applications required direct current power, but since it cannot be economically transmitted over long distances, alternating current is used, and in the early 1900s, motor-generator sets were the only feasible means of making the conversion from one form to another. This scene from the Bullock shops illustrates several large motor-generator units which are either completed or in various stages of construction. Note the extensive test equipment in the lower right-hand corner of this photograph.

In addition to motors and generators, Bullock also introduced an extensive line of transformers. A view of the erecting floor shows a number of very large units under construction, as well as several smaller transformers. At this point in time the great majority of this work was done by hand, and for the larger units, handling the heavy coils was no easy task. Beyond this, Bullock was very energetic in its promotion of electric motor drives for all sorts of industrial machinery. Company advertising illustrates many different machines from newspaper presses to turret lathes converted to electric motor drive. The conversion to electric motor drives from the leather belts and lineshafts of the past opened an entirely new and potentially profitable method of selling electrical equipment, and in this respect, Allis-Chalmers closely followed the trend established by Bullock. Further information on this equipment is found in the *Electrical Equipment* section of this volume.

A heavy lift crane at the Port of Milwaukee's open dock raises one of the twenty-plus cement kiln sections bound for Loma Negra in Argentina. In 1960, this was the largest such kiln in the western hemisphere, with each of the giant sections weighing 19 tons. Two separate voyages were required for delivery of this unit, including the auxiliary machinery such as crushers, grinding mills, and motors. The assembled kiln is 610 feet long. Portland cement dates back to an English patent of 1824. The name comes not from its proximity to a town or county by the name of Portland, but instead was derived because of the finished material's similarity to a unique sort of oolitic limestone found on the Isle of Portland. Although the process itself has been improved and the manufacturing methods are less costly, Portland cement is still made under basically the same processes invented decades ago.

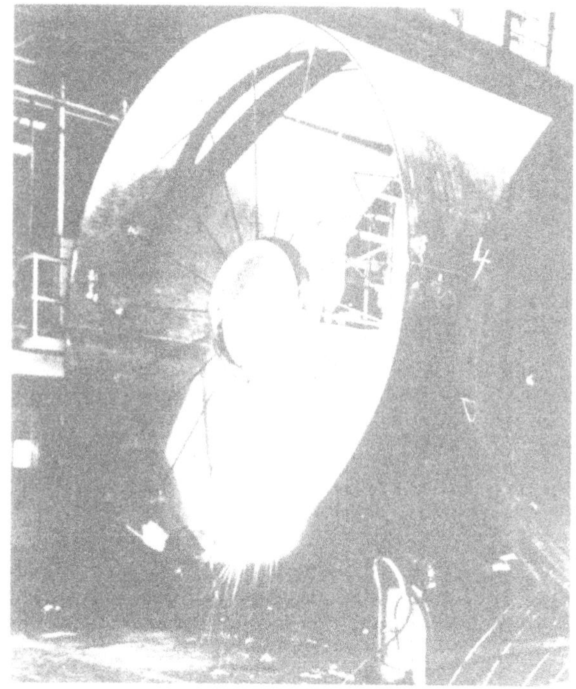

A 1960 photograph shows a dazzling stream of sparks cascading out of this giant kiln section. Taken at the Allis-Chalmers tank and plate shop in West Allis, a welder is tacking stay rods in place prior to shipment. The extra bracing is necessary for the 6,000 mile voyage to Loma Negra, Argentina's largest cement producer. This unit, 610 feet long, was in 1960, the largest in the western hemisphere. It was shipped via the St. Lawrence Seaway and the Atlantic Ocean. This particular kiln, 19 feet in diameter, commands special attention because of its size, and because of its faraway destination. Allis-Chalmers has built an extensive line of cement machinery over the years, and further examples are found in the section entitled *Ore and Mineral Processing*.

Candy Puller

Except for a 1926 advertisement in the *Candy Factory Jobber*, virtually nothing is known about the American Candy Puller offered by Allis-Chalmers. Presumably A-C bought out this company, since it would be out of character for them to sell the products built by others, at least in the 1920s time frame. This particular advertisement points to the simplicity of the American Candy Puller, noting that "the big figure '8' traced by three floating pins, tells the story of the principle of the American Candy Puller. It has been known for years that the figure '8' motion is the true principle upon which efficient candy pulling is based, and now this true principle is available." A-C went on to note that "to operate, simply hang a batch on one of the pins, turn on the motor and the candy will wind itself onto the three floating pins — a moment's job." Despite the simplicity, and despite what must have been at the very least a limited market, current research has unearthed no further information on a most interesting and unique machine, not only in terms of its application and use, but also in terms of the company's departure into a field totally unrelated to sawmills, flour milling, rock crushers, or hydroelectric machinery.

Carriers

Allis-Chalmers built an extensive line of attachments for their tractors, especially during the 1950s and 1960s. All were aimed at operator comfort, convenience, and efficiency. This photograph of about 1959 illustrates a convenient carrier attachment which was readily adaptable to the hydraulic lift arms. In this instance the farmer is about to pick up several milk cans for transportation to a waiting truck. This simple, sturdy unit was not built for industrial duty, but was designed to ease farm tasks. In itself, this utility carrier did little to change the direction of mechanized agriculture, but taken with countless other mechanical conveniences it certainly eased many jobs that had been done entirely with manual labor. The tractor is a D-14 of about 1958 vintage.

Chain Saws

A-C production records indicate that chain saws were offered in the 1971-1979 period. Information regarding this series is rather sketchy in the A-C Archives, but an instruction manual indicates that certain special tools could be obtained from Beaird-Poulan Inc. at Shreveport, LA. Our research has so far failed to reveal whether Poulan built this series, although this appears to be likely, at least for a portion of the line. A late 1970s advertisement shows five models ranging from the small Model 75 up to the big Model 295. Sometime earlier a Model 65 was offered. Its little 1.86 CID engine operated at a maximum of 8,500 rpm, giving a chain speed of 3,190 feet per minute. Needle bearings were used 100% in the Model 65.

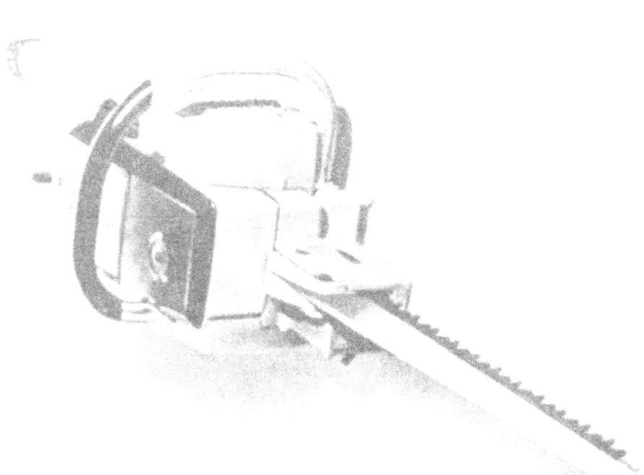

Allis-Chalmers offered the distinctive model 85 blade saw as a part of its chain saw line. This versatile woodcutter used the same natural cutting action as a hand saw, making it ideal for pruning limbs, cutting small trees, brush, and firewood. A special blade design prevented clogging of the teeth. A two-cycle engine was used, and the total weight of this unit was only 15½ pounds. Except for some occasional magazine advertisements, the current research has yielded virtually no detailed data on the Allis-Chalmers chain saw line. Other distinctive and unique A-C products can be found in the *Outdoor Equipment* section of this volume.

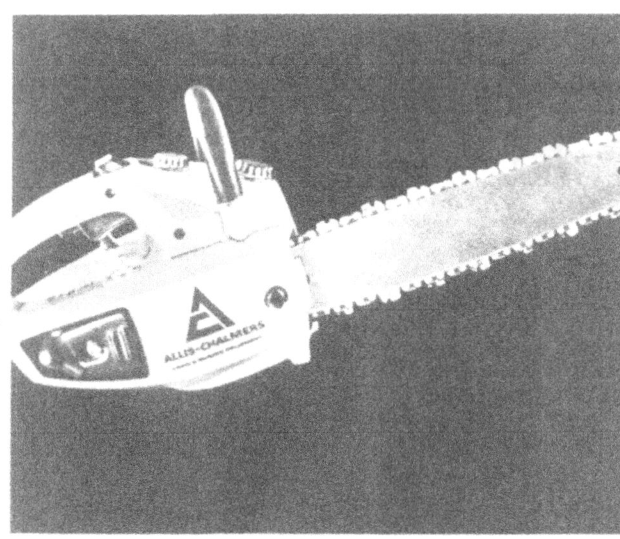

A closeup view of the Allis-Chalmers chain saw reveals the forward mounted fuel and oil tanks. This location provided a slightly better factor of safety for the operator, and the large size of the tanks permitted longer operating cycles between refueling. The choke button could be operated by either hand for starting, and the tuned, baffle plated muffler was designed to increase operator comfort. During the 1970s Allis-Chalmers offered an extensive variety of "Outdoor Equipment," with a number of these items actually being built by other companies, but to Allis-Chalmers specifications, and were of course equipped with the Allis-Chalmers trademark.

Chisel Plows

When Allis-Chalmers bought out Brenneis Mfg. Co. in 1938, a series of chisel plows came along as part of the package. Until the late 1940s however, these were used primarily in conjunction with large crawler tractors, and were apparently not widely adapted to the average midwestern farm. In the early 1950s Allis-Chalmers built a mounted spring-time chisel plow for direct attachment to the then-current WD tractor. At this point in time it appears that implements of this category were built at the Oxnard, Cal., Works formerly occupied by Brenneis. A-C production records are hazy on this particular unit, leaving some doubt as to whether this particular machine was produced in any substantial numbers. Even if it properly falls in the experimental category, this unit was a very early demonstration of future developments in direct-mounted farm equipment. The number of teeth and configuration would, in terms of today's terminology, give this particular unit the designation of a field cultivator rather than a chisel plow.

During the 1960s Allis-Chalmers developed a number of field cultivators — these were built in various widths to match various horsepower classes, and were also designed in direct-mounted and towable versions. Shown here is a D-15 tractor with a tool bar-type machine. Although the original photograph designates this unit as being a field cultivator, its eight spring shanks seem to indicate that the chisel plow designation might be more appropriate. In fact, the tool bar design was such that various equipment could be used with the same tool bar, much as different shovels could be used on the spring-coil shanks with which this unit is equipped.

During the 1970s and early 1980s Allis-Chalmers offered an extensive variety of chisel plows. Toward the end of A-C implement production came the 1500 Min-Til conservation tool which combined a gang of 20-inch coulter blades ahead of heavy-duty chisel shanks. Conventional chisel plows were offered in mounted, rigid pull-type, short-wing pull-type, and wide-wing pull-type units. Shown here is a rigid pull-type chisel plow behind an A-C D-21 tractor. Of the A-C 600 Series, this unit was available in widths from 7 to 16 feet, while the mounted type could be secured in seven sizes ranging from 8 to 20 feet in width. Options included tool bar extension kits and an extensive variety of shanks and shovels to suit specific areas and conditions.

By the early 1970s Allis-Chalmers was offering the 600, 700, and 800 series of chisel plows ranging all the way up to an operating width of 34 feet. The largest machines were coupled to crawler tractors and used in large operations, while the smaller sizes found their application on small to medium-size farms. The A-C 1600 Series of chisel plows was offered in sizes up to 30 feet, using up to 30 shanks. Six different shank styles were available, with an extensive number of shovel configurations also being offered. Shown here is a D-21 Allis-Chalmers tractor pulling a 28-foot chisel plow. Additional data on A-C tillage equipment may be found under specific headings.

Clover Hullers

Allis-Chalmers bought out Advance-Rumely Thresher Co. in 1931. An obvious advantage to this purchase was that A-C now had a ready-made dealer organization for its tractors. This logic does not explain however, why Allis-Chalmers bought out the Birdsell Mfg. Co. that same year. Perhaps the purchase was intended to further enhance the farm equipment line, but existing sales literature of the period seems to give little attention to clover hullers. Perhaps the Birdsell factory at South Bend, Ind., was considered to be the site of other A-C operations. Whatever the reasons, all seem to be now obscured.

It appears entirely likely that the Birdsell family wished to dispose of the business. By the late 1920s, only two of John Comly Birdsell's sons remained, and they were in advancing years. Their probable desire to sell does not seem to match with Allis-Chalmers' usual procedure of buying carefully, cautiously, and as cheaply as possible while under the presidential leadership of General Falk.

John Comly Birdsell (1815-1894) spent most of his adult life perfecting the clover huller. Although clover was a most desirable crop, early methods of saving the seed were so laborious that the effort was scarcely worth the time. By bringing the clover huller to a high state of perfection, Birdsell contributed mightily to the production of clover as an important hay crop.

Patent No. 5,599 of 1848 appears to be the earliest of the Birdsell patents, with many more following in later years. Between 1855 and 1863 he built ten or twelve hullers per year at his farm near West Henrietta, N.Y. After this time he moved to South Bend, Ind., and set up an entirely new factory — in only a few years the Birdsell name was synonomous with clover hullers all over the United States, and in many parts of the world.

Mr. Birdsell seems to have gone through life plagued with adversity. Securing patents was one thing, but maintaining rights to them was quite another. Birdsell spent much of his time and money defending his many designs. After certain of the patents expired, competition sprang from everywhere to gain a share of what was a rather limited market. Despite these many problems, John Birdsell prevailed, and is now generally recognized as having invented and perfected the clover huller as a single machine capable of threshing, hulling, cleaning, and re-cleaning the clover seed in a single operation.

Combines

Recapping the history of Allis-Chalmers combine development consists of two separate and distinct endeavors which will be treated as such. The work of A-C engineers and the subsequent All-Crop Harvester represents one area, with the other being the separate and competing efforts of Gleaner Harvester Corp. A-C acquired Gleaner in 1955, and from that point onward the Gleaner machines are covered in this section. Early Gleaner developments are of such significance however, that a separate section entitled, *Gleaner Harvester Co.* may be found on Pages 157 to 160.

Numerous indications exist showing that Allis-Chalmers had decided to pursue the development of a new lightweight combine in the late 1920s. From a study of the Allis-Chalmers sales bulletins and other available documents it would appear that these developments had not even reached the embryo stage by 1930. At this point however comes the Fleming design under the auspices of its promoter, Guy H. Hall.

Robert Graham Fleming of Chico, Cal., developed a revolutionary cylinder design using flexible bristles in the late 1920s. The Fleming machine also featured a straight-line, straight-through design that simplified the process of carrying cut grain to the threshing cylinder. Rather, the Fleming machine used a cylinder the same width as the cutter bar, and in this feature Fleming forever changed the design of the combine-harvester. Fleming filed for a patent on this machine in January, 1930, and was granted Patent No. 1,814,305 on July 14, 1931. When this application was filed, Guy H. Hall of Evanston, Ill., was named as the assignee. Hall demonstrated the first machine at various midwestern locations during the 1930 harvest. As a result, Harry Merritt of A-C paid $25,000 for a shop license to build the machine. A-C announced through its *Sales Bulletin* of Nov.-Dec., 1930, that "a new combine" had been added, noting that a limited number of these machines would be produced in 1931.

This series of events seems likely to have had a positive effect on the continuing negotiations with Advance-Rumely Thresher Co. of LaPorte, Ind. Talks between A-C and Advance-Rumely had gone on for an extended period, but no agreement had been reached. Therefore it seems entirely plausible that the purchase of a shop license for the Fleming machine might have been the catalyst which moved the two companies closer to agreement. Purchasing Advance-Rumely would give A-C an intact dealer and branch house organization, almost overnight. Furthermore, purchase of Advance-Rumely would give A-C a manufacturing site for their proposed new combine. While all of this was taking place, A-C could continue to sell the already established Advance-Rumely combines of conventional, heavy-weight design. For several years following the 1931 purchase of Advance-Rumely, A-C continued marketing the No. 2 Prairie, the No. 3 Prairie, and the No. 3 Hillside combines originally designed by Advance-Rumely. A rather rare piece of A-C advertising literature of 1933 illustrates A-C "35" tractor and the No. 3 combines. Thus, the A-C purchase of Advance-Rumely probably

Combines

had far more long-term implications than the acquisition of a ready-made dealer organization. For the residents of LaPorte, the impact of the A-C takeover was soon evident — little orange combines were shipped out by the thousands, and all of this manufacturing activity meant lots of jobs for many people — getting and keeping a job during the Great Depression of the 1930s was indeed an achievement.

The Fleming design used a 5-foot long cylindrical brush — it was the same width as the cutter bar. Other features of this machine included a double-lip screen which permitted grain to fall through, but which retained heavy chaff and short straws. Narrow channels on the grain pan prevented grain from piling up against the side of the machine when operating on a slope. An automatic fan governor maintained the air blast through a moveable shutter system. Thus, when the tractor speed dropped, the governor opened the fan shutter proportionately, maintaining the air blast at a reasonably constant level.

The new machine weighed only 3,000 pounds. Despite this and other advantages, the cylindrical brush presented continuing problems. For one thing, the bristles would break off, ending up in the grain, and causing serious health hazards for livestock ingesting the tiny steel bristles. Harry Merritt assigned the task of redesigning the machine to Walter Dray at LaPorte Works. The A-C *Sales Bulletin* for Oct.-Dec., 1934, announced that a new Corn Belt Harvester would be ready for production in 1935. The new machine featured angle iron bars on the threshing cylinder. Attached to the cylinder spiders with a slight spiral, these bars were faced with rubber.

Following Dray's design came further and somewhat dramatic improvements by Charles J. Scranton Jr. Dray had replaced many of the chains on the original machine with Texrope (v-belt) drives. After all, Allis-Chalmers was pioneering the Texrope multiple v-belt drive in another division, and had a number of eminent engineers on this single project. Scranton came to LaPorte Works in 1935 and moved quickly to redesign the housing within which was carried the cylinder, straw racks, and separating mechanism. This design was tailored to use automotive production line and manufacturing methods as much as possible. The design further lightened the machine and extensive use of pressed steel sections joined together by spotwelding gave an exceptionally strong machine that could be produced at far less cost than the conventional angle iron frame with riveted steel panels.

Scranton also began design work for the Model 40 All-Crop Harvester which appeared in 1938. This little machine was assembled entirely with spot welds. Curiously, A-C did not apply for a patent on the Model 40 design until 1940, with Des. Pat. 123,896 being issued that year. An improved Model 60 All-Crop design was later protected by Des. Pat. 143,098 of 1945. In addition to the above design patents, all important aspects of the All-Crop Harvester were under patent protection wherever possible. A lion's share of these were issued to Charles J. Scranton. His important efforts, along with those of Harry C. Merritt brought each of these men the coveted Cyrus Hall McCormick Medal of the American Society of Agricultural Engineers. Mr. Merritt was awarded the Medal in 1941, with Mr. Scranton being similarly honored in 1952.

Production of the Model 60 All-Crop Harvester continued through 1949, with 176,916 of these machines being built over the production run beginning in 1935, and including 10 pre-production machines built in 1934. A somewhat improved Model 60-A machine appeared in 1950. By the time production ended in 1952, Allis-

Allis-Chalmers actively promoted the "A-C Rumely" combine during the 1930s, giving most of their sales push to the wheat growing regions, but likewise paying some heed to the small grain areas as well. This photograph of the mid-1930s shows an Advance-Rumely No. 3 combine behind an Allis-Chalmers WC with rubber tires. The combine has likewise been equipped with pneumatic tires, eliminating much of the draft created by steel wheels. Since the No. 3 combine carried its own Waukesha engine, no PTO power was required of the tractor, leaving all of its horses available to pull the machine over hill and dale. Further data on these machines may be found under the *Advance-Rumely* section of this volume.

In the book, *The Grain Harvesters*, Quick and Buchele note that the Advance-Rumely combine design originated with the Savage Harvester Company machine of the 1920s. Ironically, C. C. Baldwin was competing for this machine, but lost out to Advance-Rumely. Eventually however, Advance-Rumely and Baldwin would both come under the A-C corporate umbrella. After acquiring Advance-Rumely in 1931, Allis-Chalmers continued marketing the Rumely combine until 1938. The Model 2 Prairie sold at $1,400; it had a 16-foot cut. The somewhat smaller model 3 Prairie combine had a 10-foot cut — it sold for $1,200. The Model 3 Hillside combine offered a 12-foot cut — its complicated leveling mechanism raised the list price to $1,950. These machines appear to have followed some of the Baldwin designs, some of which had been incorporated in the Savage harvester which Rumely bought out in the late 1920s.

Combines

Chalmers had built 75,000 of these machines, bringing the grand total for all Model 60 machines to well over 250,000 units. This figure also includes a few Model 60 Hillside combines produced only in 1940.

The Model 66 All-Crop Harvester introduced for the 1953 harvest used a 5½-foot sicklebar, but depended on the divider tips to actually take a 6-foot cut of standing grain. The PTO-drive model had a 1955 list price of $1,195, while the same machine with an auxiliary engine drive carried a price of $1,682. Allis-Chalmers also built a "big bin" series for the Model 66 from 1955 to 1958. This one had a price of $2,314 with an auxiliary engine drive, or $1,675 with a PTO shaft. The Model 66 retained many features pioneered in the Model 60, but added a variety of new features and new attachments to better conform with rapidly changing farming methods and radically different tractor designs and capabilities. Included was a new option of hydraulic control for lifting the grain header — a job that formerly had no option except that of hand operation.

Already in the 1940s Allis-Chalmers had experimented with a cornhead for its pull-type combines. Between 1957 and 1959 A-C built 850 of these units for attachment to the Model 66 All-Crop Harvester. Similar happenings occured with the Model 72 and Model 90 harvesters.

A-C continued building pull-type combines for several years following their acquisition of Gleaner Harvester Corp. These included the All-Crop 72 produced from 1959 to 1962 and the All-Crop 90 built between 1957 and 1960.

By the late 1940s Allis-Chalmers began looking at self-propelled combine designs. This resulted in the All-Crop 100 marketed between 1953 and 1957 — it was followed in 1958 only with the Super 100 All-Crop machine. With 1958 came an end to All-Crop self-propelled harvester production — the model was superseded by a far more extensive line of Gleaner combines built at the Independence Works formerly operated by Gleaner Harvester Corp.

Gleaner had its earliest beginnings back in 1911 with Curtis C. Baldwin and The Baldwin Company at Nickerson, Kan. From that time onward came a series of companies, failures, partnerships, and new companies. Somehow or other the indefatigable and assiduous Baldwin remained afloat despite untold adversity. By acquiring Gleaner Harvester and thus creating the Independence Works, A-C was able to not only maintain their industry position, but also to enhance it considerably because of existing designs and those on the drawing board. To better delineate the Gleaner-Baldwin story, and to avoid redundancy, further information on these designs may be found under the *Gleaner Harvester* heading elsewhere in this volume. Due to a certain amount of overlap with the All-Crop and Gleaner designs in the early stages, the Gleaner combines built by Allis-Chalmers are pictured and captioned within this section.

Acquiring a shop license for the Fleming design was a forerunner to one of the greatest Allis-Chalmers triumphs, the All-Crop Harvester. The Fleming patent covered the wire brush cylinder design primarily, while at least a portion of the straight-line feed design appears to have come from Patent No. 1,765,582 issued to William L. Hanlon of Sterling, Kan., in 1930. Whether A-C bought the patent outright or secured a shop license for it is presently unknown, and is rather unimportant. What is important is that this lightweight machine was about to revolutionize the whole business of the combine-harvester as a method of harvesting grain. The wire-brush cylinder was 5 feet long, the same length as the cutter bar. Operating at 1,000 rpm, it did an efficient job of removing grain from the heads, even on crops previously considered difficult or impossible to thresh. This particular machine is operating in an Illinois soybean field and is being pulled by the new "All-Crop" Allis-Chalmers tractor. The original photograph was taken during the 1930 harvest season.

The A-C *Sales Bulletin* for December, 1934, notes that "the development work that has been going on the past four years on the small five-foot cut combine . . . will be placed on the market in 1935." At this point in time the new machine carried the title of "Corn Belt Combine" with the "All-Crop Harvester" title not appearing until late in 1935. The Trademark Register of the U.S. Patent Office shows that the "All-Crop" designation was not filed until September 17, 1947, and was published for opposition on February 1, 1949 under Ser. No. 534,717. However, in its application Allis-Chalmers claimed first use of this title on November 25, 1935, "for agricultural machines, namely, harvesters." The All-Crop title was however used earlier in connection with the original UC tractor — it carried the All-Crop designation for a short time. Curiously though, the Corn Belt Combine title was followed for a very short time with the "High Speed Combine-Harvester" logo, probably during the latter part of 1935, although our research has failed to *precisely* establish the time period of this designation. Immediately following the "High Speed Combine-Harvester" logo came the "All-Crop Harvester" designation, and this name stuck. Since the company claimed first use in November, 1935, it would appear that the 1936 and subsequent production carried this title.

The Corn Belt Combine was a small lightweight machine that embodied the newest in combine design and attempted to follow the lightweight framework that was beginning to see some success in the automotive field. This machine however was built by then-used methods of bolting and riveting various sections together, with most of the sheet metal being fabricated by hand or crude power equipment. This 1935 photograph illustrates the rather unrefined straw delivery chute in the immediate foreground — by itself, it probably required several manhours to build and attach to the machine. A close look at the photograph reveals an open PTO shaft running over the tongue and back to the machine.

A closeup view of the Corn Belt Combine illustrates the bolts-and-rivets construction alluded to in another caption. Walter Dray at LaPorte Works had brought the machine to this stage from its rather primitive state of the original brush-type cylinder of the 1930s. Now it was up to Charles J. Scranton Jr. to redesign the machine so that much of the fabrication could be done with large presses and assembly could be completed with multiple spot welds. From Scranton's efforts emerged the All-Crop combine as the heir apparent to the Corn Belt Combine. This machine, as others of its type, carried the cut grain into a cylinder as wide as the cutter bar. After threshing, the grain and straw dropped to the straw racks which carried this material out of the machine at a right angle to the cylinder delivery. This system eliminated the troublesome cross conveyors required on conventional designs of the day, including the Advance-Rumely machine illustrated elsewhere in this volume.

A field scene of 1935 illustrates the High-Speed Combine at work. Due to Charles J. Scranton's redesign of the machine, the heavy steel straw hood has been replaced with a canvas chute, lighter in weight and less difficult to build than its predecessor. The new housing style is obviously different than in the Corn Belt Combine — this new design used formed steel sheets to make up the housing, and indeed to carry most of the working elements within. Scranton was not at all hesitant about adapting automotive manufacturing methods to combine manufacturing and did so with amazing skill. The machines built in 1935 sold for $595 each f.o.b. LaPorte, Ind. This machine was so impressive that during the next year sales jumped ten times to a total of 5,500 units.

Another 1935 photo illustrates a High-Speed combine behind a McCormick-Deering 10-20 tractor. Since ASAE pto and drawbar standards were not then in wide use, Allis-Chalmers could furnish the proper equipment for coupling to almost any make of tractor which was equipped with a pto shaft. Suitable drawbar equipment and recommended alterations were likewise available to accommodate varying drawbar height and length to suit the combine. While the High-Speed combine would undergo further cosmetic changes, along with a few internal improvements, the design was virtually fixed and would remain for another 20 years — not a bad achievement in an age when almost every new machine reached obsolescence in a few seasons.

While the original photograph fails to give any details, this machine is obviously parked just outside of the A-C factory, and is likely one of the first machines to bear the High-Speed Combine title. From all appearances this particular designation was short-lived — of far greater importance is the overall design that included a variable speed drive for the cylinder, and v-belt drives wherever possible. Rubber tires made the combine easier to pull, permitted higher transport speeds on the road, and minimized jarring on rough ground. At this point in time the PTO shaft is entirely unshielded and virtually cries out aloud for the operator to beware of its dangers! Even as this title is being written in the late 1980s, equipment operators still fail to heed this warning occasionally, and often with tragic results!

Combines

By 1936 the All-Crop combine was well under way, and as previously noted, sales for that year totaled 5,500 machines. This was but a prelude to the 1937 production which again doubled to well over 10,000, and climbed to an unbelievable 16,500 units for 1938. It is well to bear in mind that these were tough years for farmers — land values had dropped dramatically, and commodity prices were far below production costs. Yet there was hope, and the low-cost, but very efficient All-Crop Harvester attracted many farmers.

A letter written by Harry C. Merritt to General Falk, Chairman of the Board, and dated November 11, 1937 explains this photograph. Mr. Merritt writes in part, "I am attaching a picture showing the attachment that we are experimenting with for combining corn, which will be attached to the ALL-CROP combine. It is now being tested out near Waterloo, Iowa, on Frank Zink's farm. We tried to use the regular header and found that we couldn't get the stalks back into the cylinder; but when we did get them back, we did an excellent job of shelling and saving the grain, and at the same time, saving all of the stover which was blown into a wagon pulled behind the combine. If we can work out a method of picking up the stalks and cutting them, it will vastly increase the market for the ALL-CROP. It shows considerable promise at this time." As it turned out, Merritt's hopes had another 20 years to go, for it was not until 1957 that Allis-Chalmers offered a corn head attachment for the "60" and "66" combines. Weighing 1,800 pounds and priced at $1,510, it was taken out of production the following year.

NEW 1940 MODEL "60"

The Model "60" for 1940 illustrates a constant cosmetic refinement, and included numerous mechanical refinements as well. By this time the pto shaft was completely shielded for operator safety. A hallmark of simplicity, the "60" depended on v-belt drives wherever possible, and completely eliminated troublesome flat belt drives. The cylinder carried eight spiraled steel angle bars with a vulcanized rubber face. Two bar-type block rubber, interchangeable concaves were furnished, as was an adjustable, rubber-faced shelling plate. The cylinder itself was 60 inches long and 15 inches in diameter. Through a variable speed drive, the cylinder speed could be varied anywhere from 450 to 1,600 rpm. In addition, ball or roller bearings were used on every important drive. The grain bin had a holding capacity of 18 bushels — it could discharge its load in 90 seconds.

A-C offered an extensive and rather impressive list of attachments for the All-Crop combine. These included the tandem wheel attachment shown here. Recommended especially for fields with ditches or gullies, this attachment minimized torsional and twisting strains on the machine. The tandem wheel attachment carried a price of $85. Bolt-on dual wheels were available where extra flotation was required as in the case of very sandy or muddy fields. For 1947 the "60" All-Crop carried a list price of $885. Since this machine was capable of successfully harvesting well over 100 different seed crops, a large selection of screens was available at an average cost of $3.40 each. Special finishing sieves, special flax rollers, and a variety of other items were also available at extra cost.

Built only in 1936, 1939, 1940, and 1941, the 60-H Hillside harvester represented the company's first attempt at such a machine. This illustration, taken from the front page of the applicable parts book, fails to show the machine clearly, but a study of the manual indicates that the separating mechanism was suspended pendulum style so that it remained relatively level, regardless of the operating slope. Crawler tractors were best suited to hillside work due to their low center of gravity, and the 60-H shown here was being pulled by an A-C crawler. Whether from a simple lack of demand, overwhelming competition, or an imperative reduction of activities due to World War Two, this machine only saw production of some 360 units during this period of time. Virtually nothing has been thus far located regarding specific details of this unusual, and very innovative machine.

Allis-Chalmers shot literally thousands of photographs illustrating not only the machine itself, but also showing the "60" at work in all kinds of crops and conditions. The extensive variety of special screens included one for alfalfa, another for red and white clovers, yet another for buckwheat, and so forth *ad infinitum*. This particular machine is at work behind an Allis-Chalmers WC tractor which not only pulls the "60" but powers it as well through the PTO shaft. A-C actively promoted this method of machine operation as being more efficient and less costly than using an auxiliary engine drive. In this photograph the unloading auger is tipped into its operating position, although the discharge spout is turned back out of the way. A-C claimed that it was possible to harvest from 14 to 18 acres per day with this machine, although with its five-foot cut, this would indeed be a long day!

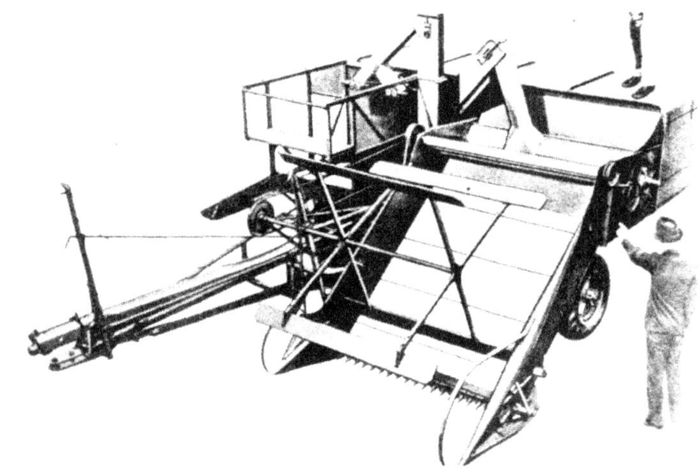

For grass seed, legume seed, and other special crops, the "60" could be furnished with a sacking platform in lieu of the customary 18 bushel grain tank. The machine shown here with the sacking platform and bagging spouts carried the same 1947 list price of $885 as the standard model with a grain bin. Machines already equipped with a grain bin could be equipped with an auxiliary bagging platform for an extra cost of $54.50, but this equipment could only be used on the PTO equipped machines. With this, the bagging attachment could also be purchased separately at a cost of $49.50. Although this machine is shown with the standard cutting bar, several pick-up attachments were available to accommodate windrowed crops.

A nighttime scene shows the Model 60 All-Crop at work in maize or a similar row crop, since the original photograph fails to name the crop and an inspection of the photo gives no conclusive evidence. However, the unique nature of this illustration lies not only in the type of crop being harvested, but also gives evidence of a new phenomenon for the 1940s farmer — that of being able to work past dark by the use of electric lights on the tractor. Although the majority of farmers attempt to complete their day's work in daylight, the threat of adverse weather and other considerations often makes it imperative to continue past sunset in order to save valuable crops from damage by the elements.

This 1940s scene typifies the Model 60 All-Crop at work in a midwestern setting. Although this machine was typically equipped with rubber tires, wartime shortages during World War Two necessitated the use of steel wheels for several years. An Allis-Chalmers catalog of 1943 illustrates the Model 60 so equipped. Of all attachments for the "60" however, one of the most unique was the bundle topping attachment. The regular sickle bar was hinged into a vertical position with the bundles being fed over the cutter bar by hand. Designed specifically for bundled kaffir or maize, it was also used to a minor extent for harvesting shocked oats and other small grain. This attachment was available at least into the late 1940s for a cost of $18.

Combines

A portion of the Model 60 assembly line at LaPorte Works illustrates the main housing assembly on the line. At this point only a few components have been installed, including the cleaning fan housing. The traveling assembly line moved from one end of the plant to the other making a huge U-turn in the process. Charles J. Scranton of LaPorte Works designed much of this system, and to Scranton must go the credit for introducing automotive assembly techniques to the farm equipment business, at least so far as Allis-Chalmers was concerned. Because of Scranton's ingenuity, production costs were so much reduced that initially the Model 60 carried a list price of $595, well within the reach of most farmers.

An overhead view of the Model 60 assembly line shows the moveable support frames in the right foreground. Following the righthand line, the bottom steel is soon in place, followed by the top housing. Small components are installed at each station as the unit makes a U-turn and returns. Soon the grain elevator and return elevator appear, and by the time the machine reaches the station in the lefthand foreground the axle is in place. At this point little is left except to mount the cutting and heading attachment and install the tongue. This line had a capacity of 150 Model 60 combines per day. Production reached a peak of 20,825 units in 1947, or about 140 working days at full capacity.

Another view of the Model 60 assembly line illustrates stacks of component parts awaiting their turn on the assembly line. Although Henry Ford had been using this method in general terms for many years, most farm equipment of the 1930s was still being built on a one-at-a-time basis. This added immensely to the final cost, but also required precise fitting of all parts to be successful. Machine-made stampings were used extensively in the Model 60, and once these were correct, each and every part would fit together with ease. A close look at this photograph shows workmen scattered along the entire assembly line, each doing his own task as another unit moves by.

Well on their way to completion, these Model 60 Harvesters are now equipped with the tongue, axle, and PTO drive. The elevators are knocked down to prevent shipping damage and will be assembled by the dealer on arrival. The header is not yet installed, and likely will not be attached to the machine at this time. Like the elevators, it will be completed in the dealer's shop. With a capacity of 150 units per day, a number of men were required to make the necessary tests on each machine prior to loading it out. This scene within LaPorte Works illustrates some of the factory personnel making the final adjustments and performing the final testing prior to acceptance.

A photo of the load-out area shows dozens of Model 60 combines ready for shipment, and in fact the big overhead crane is in the process of loading yet another one onto a waiting flat car. Shipping farm equipment by truck is a relatively recent development — when this photo of the 1930s was taken, virtually all farm equipment was shipped by rail. Dealers would attempt to set up orders with their territory man in such a way that they could fill a complete rail car to obtain the most favorable rates. Oftentimes this would mean a mixed load of tractors, combines, and other machinery, but the great popularity of the Model 60 All-Crop led to frequent shipments of several carloads to a single dealer.

Unfortunately this photograph is undated, but was probably taken sometime during 1950. Allis-Chalmers was obviously proud of building 200,000 Model 60 All-Crop Harvesters, and took time off for a photo session of the plant personnel, together with this machine. More unfortunate yet than not knowing the date is not having a list of those individuals pictured — it would seem likely that many of those who were so instrumental in launching the Model 60 as a viable means of harvesting were included. Despite this milestone however, A-C continued building the slightly modified Model 60-A with its production totaling 75,000 units before production ended in 1952.

When Allis-Chalmers purchased the Advance-Rumely factory in 1931, they bought not only the stock-in-trade, but also secured for themselves an immense factory. An aerial scene of the Allis-Chalmers LaPorte Works illustrates factory buildings spread over several acres, and bounded on the right by a major highway, with a large lake on the left. From this facility Advance-Rumely had produced the huge OilPull tractors, and from this same plant A-C built All-Crop Harvester's numbering into the hundreds of thousands. As various machines built at LaPorte Works were phased out, less and less significance was attached to this plant, and it was finally closed in 1985.

Prior to the days of mass media as we know it today, the county fair was a favorite spot to display new farm equipment. Allis-Chalmers generally played a prominent role in these displays, with this particular one taking place at the LaPorte County, (Indiana) Fair under the sponsorship of the Rumely Implement Center Inc. Several Allis-Chalmers tractors are in evidence, as is a forage harvester in the distance. Barely distinguishable to the backside of the tent is of course, the venerable All-Crop Harvester. These machines were so very popular that All-Crop virtually became a household word among farmers.

The Model 40 All-Crop Harvester appeared in 1938, with production ending in 1940. This little machine had a 40-inch sickle, with 45 inches between the divider boards. Weighing only 1,730 pounds, it offered a straight-through design ominously comparable to competing designs of future years. Like the Model 60, it used sawtooth wind valves behind the underblast cleaning fan to give better air control and distribution. The grain tank had a capacity of 11 bushels. Like the larger Model 60, this machine was capable of harvesting well over 100 different seed crops, although special attachments were required to facilitate the handling of certain crops. The small power requirements of the Model 40 permitted the use of rather small tractors including the Allis-Chalmers Model B shown in this illustration.

Priced at only $375, the Model 40 All-Crop was within the financial reach of almost every small farmer. This machine was especially well suited to farms with small patches of tillable land and to the harvesting of special seed crops including turnip seed, cabbage seed, and others equally unusual. The housing design was that of Charles J. Scranton who was likewise responsible for much of the Model 60 design. For his efforts, Scranton was awarded the Cyrus Hall McCormick Medal by the American Society of Agricultural Engineers in 1951. Built only during 1938, 1939, and 1940 the Model 40 saw total production of 15,100 units during that period. Parts service for this machine was terminated in 1963.

Combines

Assembling the Model 40 All-Crop Harvester first required the necessary stampings to be formed in a huge press. The next step in the assembly was assembling these pieces. For this purpose Allis-Chalmers used a specially designed spot welding machine that welded the entire housing in less than one minute's time, with no rivets being required in the entire machine. This manufacturing method, borrowed from the automotive industry, gave genuine production line techniques to the Model 40 and served to greatly reduce its final cost compared to earlier hand assembly methods. Here a Model 40 housing has just been welded and will now be lifted to the next work station while another set of housing parts is located for yet another machine.

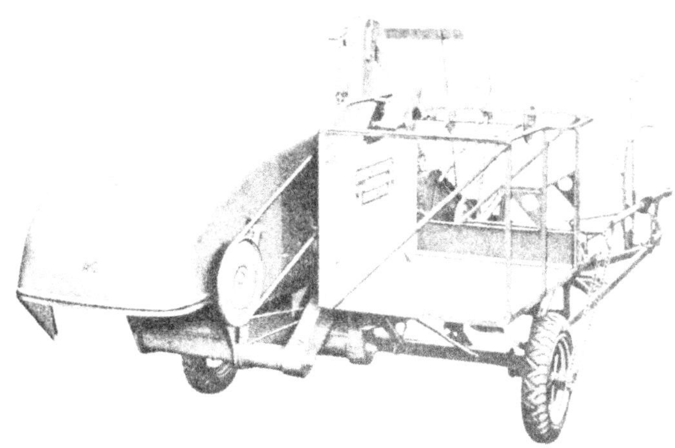

Like the larger Model 60, the little Model 40 could be equipped from the factory with a bagging platform in lieu of the 11-bushel grain tank. This was a particularly valuable asset when harvesting relatively expensive seed crops such as legumes or difficult crops like parsnip or carrot seed. Also like the Model 60, this machine could be equipped with the Hart Scourkleen, a $50 option that went a step further in the cleaning process by removing virtually all unwanted weed seed and other debris. Allis-Chalmers claimed this machine could harvest about one acre per hour, so it must have been disheartening indeed to pull into a 40 acre field knowing that it would take at least four 10-hour days to get the standing grain into the bin!

Just as previously illustrated for the Model 60 All-Crop, the Model 40 was built using the latest in production techniques, including an automatic conveyor line. These machines appear to be moving from right to left during the assembly process, and by the time they return to the opposite end of the plant, virtually all the assembly will have been completed. These extremely rare photographs illustrating the Model 40 during its actual manufacturing processes are seldom found in the archives of any company — many times they are never released to the public, but more often they are disposed of periodically, simply for lack of storage space.

Building the Model 40 and Model 60 All-Crop machines at LaPorte required an immense amount of stock just to carry on daily operations. The LaPorte warehouses literally bulged with materials for these machines. While some items were purchased from outside sources, many of the parts were built within the LaPorte Works and stockpiled in warehouses for future use. While this required more storage area, the storage costs were more than offset by the manufacturing efficiencies realized by building a large number of components at a single setting, rather than building them a few at a time.

Although no definite data has yet been located, a photograph in the Archives of the LaPorte County (Indiana) Historical Society indicates that experimental work on a new series of combines was in progress by 1945. The original photograph is marked "Model 44 & 66 Combine" and is dated February 9, 1945. This particular machine has the tailings conveyor relocated to the rear of the machine, and offers new features such as an enclosed and vented housing over the cleaning fan, and a large inspection or cleanout door is obvious at the head of the straw racks. Presumably this particular machine was not marketed, at least to any extent. Nowhere has any notation of a Model 44 or Model 66 combine been located in the production records until a Model 66 All-Crop made its public debut in 1953.

Combines

The "Big Bin" 66 was virtually identical to the earlier Model 66 except that a 25 bushel grain bin replaced the original 18 bushel size. This meant fewer stops for unloading, and a high capacity unloading auger with an extra long reach aided this process. To accommodate the additional weight of grain in the 25 bushel bin, an eight-ply tire was used on the right side to carry the load. Hydraulic or lever-lift header control was available at the buyer's option, as was bin or bagger grain handling equipment. In addition to the standard sickle head, three different pick-up attachments were available for windrowed crops. The standard rubber cylinder bars and concaves could be replaced with a special peg-tooth cylinder for use in heavy vine crops such as navy beans. The Model 66 combines offered a width of 6 feet between the divider points, along with a 66-inch sickle. Weighing about 3,070 pounds, the Big Bin 66 was offered between 1955 and 1958. The standard PTO model carried a list price of $1,675, compared to $2,314 for the engine-powered model.

Allis-Chalmers offered one of the industry's most extensive lines of attachments for their All-Crop Harvesters. A vast majority of these remained available, even for some years after the specific harvester model had gone out of production. As an example, A-C offered this special straw chopper attachment. A 1956 price book indicates that it carried a list price of $311. The same price book shows that the 2-row corn head for the Model 60 and Model 66 harvesters carried a price tag of $1,400, including all necessary attaching parts. Going "the extra mile" on attachment parts certainly must have benefited the company, at the very least, this extraordinary service must have enhanced the company's image from a public relations viewpoint.

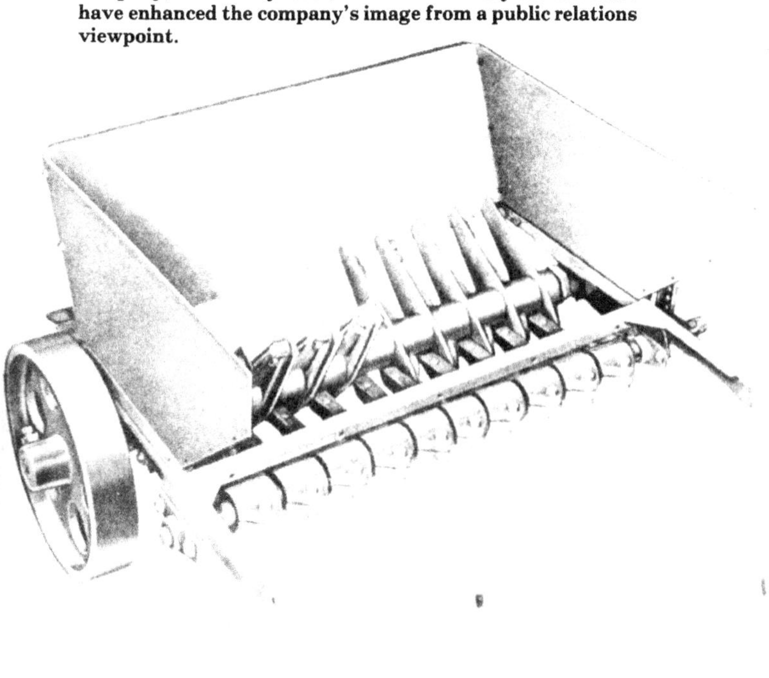

With the close of 1952 came the end of the Model 60 All-Crop Harvester. It had been a reliable workhorse for many years, but higher tractor horsepower and larger farm units demanded a machine that not only had greater capacity, the new machine also had to incorporate improved operating designs and manufacturing methods. The Model 66 introduced for the 1953 crop season offered many of these new features, even through the basic design remained virtually identical to the original All-Crop Harvester. As with its predecessor, the Model 66 was usually equipped with a PTO drive, but an auxiliary engine could be furnished as an extra cost option. Weighing 2,890 pounds, the Model 66 PTO model carried a list price of $1,195, while the same machine with an auxiliary engine drive carried a price tag of $1,682. Production of this machine with an 18 bushel grain tank continued into 1955.

Between 1959 and 1962, Allis-Chalmers offered the Model 72 All-Crop Harvester. Although the Model 66 boasted total sales of 72,839 units, the Model 72 tally sheet only yielded 4,532 machines during its production run. However, with the end of the Model 66 came also the end of the canvas-type conveyor for cut grain. The Model 72 used an auger-type grain conveyor and offered various other design changes. Reflecting a change in the A-C tractor line, the D-14 Auxiliary Engine was now supplied when this type of drive was required, although the PTO drive was by far the most popular style. This particular photograph illustrates a Model 72 all-Crop and an Allis-Chalmers D-17 tractor at work in an Illinois soybean field.

A comparison of specifications for the Model 66 and the Model 72 combines indicates that these machines were virtually identical regarding threshing width and cleaning area. The differences seem to have been in the feeder house design and various cosmetic changes which made the Model 72 more convenient to the operator. This particular machine is equipped with a Hart Model K ScourKleen unit mounted above the grain tank. These units were very effective in removing weed seed from the grain — a feature especially important when harvesting grain which would be again used as seed rather than being fed to livestock. The ScourKleen carried a 1960 list price of $93. An extensive variety of special screens were also available at $11 each.

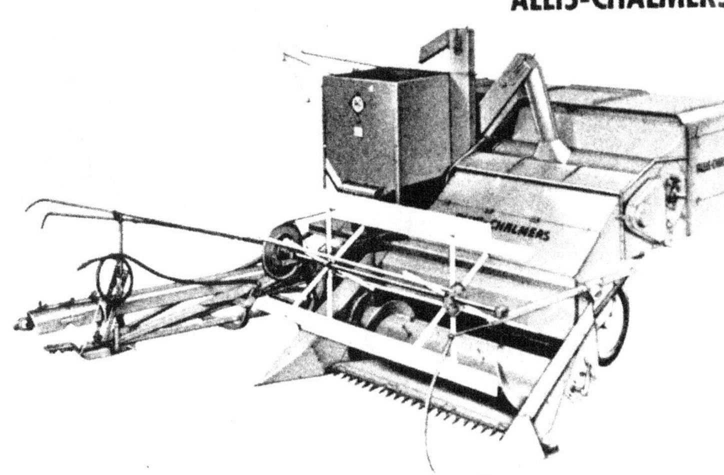

For 1960 the Model 72 All-Crop was priced at $1,898. When the PTO drive was desired, another $96.50 was added to the base price. Auxiliary engine drives required the D-14 Auxiliary motor. Complete with remote control and clutch, it was priced at $685, but required another $95 for the motor mounting and power drive. Electric starting for this engine added yet another $98. Spanish peanut equipment was a $90 option — the same option was used for harvesting navy beans. The two-row cornhead was priced at $1,675. By 1960 the self-propelled combine was becoming increasingly popular, and this fact alone seems to account for diminishing sales of pull-type machines.

The Model 90 All-Crop Harvester, introduced in 1957 remained in production through 1960. Its 7½ foot header made it the largest of the A-C pull-type combines. The 42-inch wide cylinder offered the same type of rubber-on-rubber shelling that had been a cardinal feature of All-Crop design since the 1930's. In addition to hydraulic control of the header unit, the Model 90 also featured hydraulic control of the reel height — a very important feature when working in almost any crop. Over 4,500 square inches of separating area was available, and the saw-tooth wind valve design remained almost identical to that introduced many years before. Other features included a 30 bushel grain bin, several different pickup attachment for windrowed crops, and the Model 90 corn head attachment for field shelling with the combine. Priced at $2,476 with the PTO drive, this machine offered all of the options available for other All-Crop machines, with the Model 90 corn head being priced at $1,653.

Built from 1957 to 1960, the Model 90 All-Crop was also offered as a Model 90 "Peanut Special" for a short time during this period. Of the total number of 540 Model 90 machines built, 51 were set up as shown here for harvesting Spanish peanuts. Despite what appeared to be a promising market, sales of these machines were disappointing, and in fact, it appears that at least some of these 51 units were later reconverted to standard Model 90 machines. The machine modifications appear to have been confined primarily to the feed house and pickup design, with peg-tooth cylinder equipment being substituted for the usual rubber-on-rubber threshing mechanism. This conversion could however, be easily made in the field. Note also that this machine is equipped with the D-14 auxiliary engine. With this machine came an end to the A-C pull-type combines. Allis-Chalmers had been experimenting with their own self-propelled machine for several years, and sometime prior to introduction of the Model 90 they had purchased the big Gleaner factories at Independence, MO.

By 1948, and perhaps earlier, Allis-Chalmers had brought their concept of a self-propelled combine to the point that at least one working copy was built. This photograph of May, 1948, illustrates a left-hand rear view of the machine. Surprisingly few external changes were made between this machine and the Model 100 production machine introduced in 1952. The engine was mounted to the extreme rear of the machine, and in this early example, the rear steering wheels appear to have been borrowed from a tractor, probably an A-C production model. A rather large grain tank is mounted over the engine, and the separator design follows that of the earlier All-Crop machines almost identically except perhaps for physical size.

A front view of the Model 100 machine of 1948 illustrates its great similarity to the original All-Crop design. It appears that A-C engineers and executives alike were intent on retaining this design, even in machines somewhat larger than the original All-Crop Harvester. Not until A-C purchased the Gleaner factory in 1955 was there a change in combines, and this one resulted not from a change in A-C's own design but from the acquisition of an already established design with proven success. Thus, it can be truly said that Allis-Chalmers never changed its own All-Crop Harvester design substantially from that of the 1930s. Despite what appears to have been a substantially complete design, the Model 100 never went into production until 1952.

Between 1952 and 1958, Allis-Chalmers built 5,500 Model 100 machines. Of these, the last 1,000 were the Super 100 design, a slightly enhanced version. This 1952 photograph illustrates the first Model 100 self-propelled machine as a production model. Priced at $4,615, the Model 100 used 10-24 front and 6.00-16 inch rear tires to carry its overall weight of 6,760 pounds. This machine could be equipped with either a 9-foot or a 12-foot header. Power was provided by an Allis-Chalmers W-226 engine with 60 maximum brake horsepower at 1,800 rpm.

A 1953 field scene shows the Model 100 at work in standing grain. Compared to designs by other manufacturers, the Model 100 used a rather peculiar header and feeder house with the latter being situated at one end of the header rather than having the feeder house in the center of the header. This was of course necessary due to the A-C crossmounted thresher design. Likewise, the crossmounted thresher offered its own set of disadvantages, especially in a large machine. When opening fields, either the straw was discharged in the fenceline or was dropped over uncut grain. By comparison, the Gleaner design which A-C acquired in 1955 deposited the straw directly behind the machine.

A rear view of the Model 100 combine plainly shows the elaborate ducting required for the engine cooling system. By extending the radiator air intake above the machine, relatively dust-free air could be obtained, thus minimizing radiator plugging and subsequent engine overheating. The rather long discharge auger had the advantage of permitting on-the-go unloading if desired, although the side discharge of straw and chaff must have made this unpleasant at times. To obviate the disadvantages of straw delivery from the side of the machine, A-C offered a rear-spread straw attachment as an extra-cost option. It included a 3-blade rotary knife which chopped and spread the straw and chaff behind the machine.

For 1958 only, Allis-Chalmers offered the Super 100 self-propelled All-Crop Harvester. If offered several new features over the original Model 100 including an inclined steering wheel, larger tires, and a new retractable finger header auger as standard equipment. Other new features included a new control panel, folding discharge auger, high clearance reel drive, remote control operation of the saw-tooth wind valves, a new radiator, and a higher intake screen. Optional power steering was now available for all Model 100 machines. For 1958 the Super 100 with grain bin and 9-foot header carried a price of $4,795. The offset feeder house design was not however, well suited to corn combining, and no corn head attachment was listed for these machines. This obvious disadvantage at a time when the corn combining was making major inroads into harvesting methods probably spelled the end for the All-Crop design. This harvest scene was photographed near American Falls, Idaho.

Combines

With the 1955 purchase of Gleaner Harvester Corp., Allis-Chalmers acquired a proven combine design, virtually overnight. The Gleaner-Baldwin Model A had been placed into the 1952 harvest and had proven to be a formidable competitor to all comers. Gleaner originally equipped this machine with a Hercules QLXD engine, a 6-cylinder design with about 55 brake horsepower. The engine itself was well proven — it had been introduced in 1946. With the A-C acquisition of Gleaner came the Allis-Chalmers 6B-230 engine of virtually the same displacement as the Hercules. The Model A used a conventional straw walker design with over 13 feet of separation length. The Model R, built simultaneously with the Model A was identical to the Model A in every respect except that it used a raddle type separator instead of straw walkers. This design was tailored to the harvesting of headed wheat and milo maize.

Originally introduced by Gleaner Corp., the Model R was identical in every respect to the Model A Gleaner, except that where the Model A used straw walkers, the Model R used raddles to carry the straw from the machine. The raddle-type separator was specially adapted for maize and kaffir, as well as other grains harvested in a similar manner. In the raddle design, the front raddle carries material to one picker and two rod beaters, allowing the grain to drop through an air blast to the grain pan. The picker and the rod beater thoroughly agitate and shake the straw so that any remaining grain will fall through to the cleaner shoe. Despite the belief that the raddle-type separator is of relatively recent origin, it is in fact, one of the oldest of all separator designs, antedating the straw walker system by many years.

After acquiring the Gleaner line, A-C began extensive testing of new designs tailored to meet rapidly changing farming methods. Shown here is a Model A with a 3-row experimental corn head working in an Illinois field. The 3-row design was intended for use with the relatively recent concept of 6-row planters. In addition to a great many accessory items for specific crops, Allis-Chalmers also adapted the Model A into a hillside version, producing these machines between 1955 and 1962. Thus the Model AH was the first A-C Gleaner to be shipped as a hillside version. These machines could also be equipped with a butane fuel attachment as an extra cost option, and in fact, could be equipped with a diesel engine at additional cost. For 1958, the Model A Gleaner with a 12-foot grain table carried a price of $6,800. A 2-row corn head was also available for the Model A.

The Gleaner Model B Rice Special machine appears to have been the standard Model A, suitably modified for this specialized crop. Obvious changes were the substitution of 18-26 tires for the 13-26 size used on the Model A. Built between 1953 and 1963, the Model B appears to occasionally have the title of "AB" on certain photographs and other company data. Every indication is that a great many parts were interchangeable between the two machines. Standard equipment for the Model B included a 12-foot header, specially equipped for handling the standing rice. This particular scene dates from October, 1962, and was photographed in a field of Nato rice near Jonesboro, Ark.

Another view of the Model B Rice Special combine illustrates the huge flotation tires required for this particular crop. For special conditions such as soft, muddy fields this entire series, Models A, B, and R could be equipped with individual hydraulic steering brakes on the drive wheels to permit shorter turns and closer control of the machine. These were available either for factory or field installation. Another important option was an auxiliary low speed transmission which provided a complete range of speeds for 0.3 mph up to 15 mph. This feature was available for factory or field installation on Models A and R, but curiously was not listed as an available option for the Model B. At this point in time, power steering was optional — within a very few years, demand for power steering was so universal that it became standard equipment. The Model B Rice Special shown here is working in rice near Stuttgart, Ark.

Following on the heels of the Gleaner Model A came the Gleaner A2. Built from 1963 to 1967 it was an improved version of its immediate ancestor and carried much the same design. Designated by Allis-Chalmers as the "Flagship" of their "Great Silver Fleet," the A2 had an increased capacity over the Model A. Rasp-bar cylinders were now standard equipment, although the rubber-faced bar cylinder was optionally available, as was the conventional spike-tooth cylinder. The metal-bat reel was a new feature. Five different headers were available, ranging in size from 10 to 16 feet. Two and 3-row corn heads were available. This 13-foot A2 Gleaner is shown in a 1964 photograph working in soybeans near Greentown, Ind.

To meet constantly changing demands, Allis-Chalmers built various cornhead styles. In this particular instance an A2 Gleaner is equipped with a 4-row narrow cornhead, combining 20-inch rows. This particular machine was owned by Allis-Chalmers, but was operating on a farm near Mowequa, Ill., in October, 1965. The A2 Gleaner carried a list price of $8,513 when equipped with a 10-foot header and a gasoline engine. A diesel engine added $757 to the base price, with a cab being another option at $577. The 3-row cornhead for the "A" series combines was priced at $3,250.

The popular Model C Gleaner first appeared in 1960, remaining in production with little modification through 1963 when it was slightly redesigned and again entered the market as the Model C-2 Gleaner. This photograph taken in Texas illustrates two Model C's combining milo maize. The Model C weighed in at 10,366 pounds when equipped with a 16-foot header. Its 60 bushel grain tank could be equipped with an extension for still greater capacity. Allis-Chalmers equipped this model with its own G-262 engine, although an A-C diesel engine could be supplied at extra cost. For 1963, the machine with a 16-foot header was priced at $8,995.

The Model C Gleaner was available in a "CH" Hillside version from 1965 through 1968. Priced at $17,575 it was built with a 16-foot header. Weighing 12,920 pounds, the CH Gleaner was equipped with 23.1-26 drive wheels. The "C" Series Gleaner combines offered "fingertip" hydraulic controls as standard equipment, using one lever to control the ground speed anywhere from 1 to 12 mph. The reel and header were also controlled hydraulically and with one hand. Power steering and individual hydraulic turning brakes were supplied as standard equipment. This Model CH Hillside Gleaner is combining wheat near Starbuck, Wash., in August, 1964.

Combines

A 1961 photograph illustrates an early Model C Rice Gleaner at work near Yuba City, Cal. Due to the swampy condition of the field, this particular machine appears to be equipped with a tracklayer device rather than the customary flotation tires. Despite the early date of this photograph, various farm implement guides indicate that the Model C Rice machine was built only during 1967 and 1968, and indeed, there are indications that this 1961 machine might have been an experimental model. Due to the limited demand for this specialized equipment, implement manufacturers were reluctant to equip each model of their full line with the necessary changes required for the rice harvest. In the case of the Gleaner line, the Model B Rice Special might have handled this requirement adequately for several years.

Beginning in late 1963, Allis-Chalmers slightly revamped the Model C Gleaner, sending it out as the Model C2. A-C dubbed this one "Giant of the Harvest", noting that custom operators and big farmers had attached this name to the machine. Seven different headers were available, ranging in size from 12 to 24 feet. In addition, a four-row corn head was available, and the "quick detachable" header design permitted the changeover from grain header to corn head in a matter of minutes. This system made use of hydraulic power to bring the hook-type latches into proper alignment with their mating slots. This feature alone was a valuable one for the farmer since it eliminated the difficult and sometimes hazardous job of changing the heads by hand methods. Built from 1963 through 1968, the C2 was priced at $17,525.

Allis-Chalmers made frequent mention of their "true center-line design" on the Gleaner combines. All major units, including the header, grain bin, and engine were centered along a line from front to rear of the machine. This design helped promote even feeding to the cylinder and gave equal weight distribution on each side of the center line. Likewise, the fore and aft weight was proportioned so that the front drive wheels gained proper traction while the rear steering wheels gave easy handling. This C2 Gleaner is unloading on-the-go. The photograph, taken in July, 1964 near Holyoke, Col., shows the C2 combining rye.

Current research has been inconclusive regarding the exact production dates for the Model CR Gleaner combine, a machine built specifically for the rice harvest. This 1964 photograph taken near Ethel, Ark., shows a Model CR equipped with a 13-foot header. In addition, 14-foot and 16-foot header sizes were available, but the standard machine equipped with the 13-foot model carried a price of $13,425. Important features of the "C" series Gleaner included a center-mounted platform, giving the operator maximum visibility at all times. At this point in time an operator cab came only as optional equipment.

Although the four-row corn head was a standard item, the "820" cornhead fitted for eight 20-inch rows made its appearance in 1966. This photograph, taken in January of that year shows a C2 thus equipped working in a corn field near Mowequa, Ill. The corn head alone carried a list price of $7,550. By the mid-1960's Allis-Chalmers offered a variety of corn heads to fit their Gleaner combines. These included row widths of 20, 30, 35, 38, and 40 inches, along with various numbers of gathering units to match the capacity of the machine on which they were to be mounted. The C2 series was also built as a "Corn-Bean Special" equipped with particular emphasis on harvesting these two crops.

Combines

Allis-Chalmers built the Model E Gleaner combine from 1962 to 1967. Standard equipment included a 10-foot header. Weighing 6,905 pounds, the Model E carried a list price of $6,863. Designed not for the large operator but for the smaller farmer, the Model E could also be equipped with a 2-row corn head, thus permitting maximum utilization of the machine. Features were much the same as on larger Gleaners, including a single lever control which permitted control of both the header height and the ground speed. The visual tailings return was another feature that permitted a quick check of what was happening at any time. This Model E is working in windrowed oats near Merrill, Iowa.

This Model E Gleaner is equipped with a 2-row corn head and is operating near LaPorte, Ind. The Gleaner corn head featured adjustable stripper plates to reduce shelling losses, and included grass shields under the rolls to prevent wrapping — this meant less chance for plugging. A-C advertising noted that the Model E Gleaner had the largest separating capacity (4580 sq. in.) of any combine in its class. Power for this machine came from a G-226 Power-Crater engine, the same engine as was used in the D-17 A-C tractors.

A head-on view illustrates the Model E Gleaner equipped with a Hume reel (an extra-cost option) working in soybeans. The Hume reel and other pick-up reels were designed to pull the standing grain onto the feed table, and through a unique but adjustable mechanism, the teeth would retract as they accomplished the job. This view also illustrates the obvious advantages of the centralized operating platform, making it much easier for the operator to observe the entire width of the grain table. Certain crops such as soybeans, when planted in rocky ground, require extreme vigilance on the part of the operator. Carrying even a small rock into the threshing cylinder can spell disaster, and at the very least can take several hours of dismantling to release a stone which has wedged itself in the cylinder.

During 1968 and a portion of 1969 Allis-Chalmers offered the Gleaner E3 as a replacement for the Model E Gleaner. This machine saw considerable change over the earlier model, including an improved platform and operator control design. Hydrostatic steering was a $180 option, and a pressurized cab added another $625 to the price. As previously indicated, this cab was pressurized to minimize the entry of dust, but was not equipped with either a heater or an air conditioner — these were also extra cost options. A wide selection of corn heads was available to match virtually any planting width, and a reversible straw chopper was also available. For hauling or storing, the cab could be tipped forward to lower the overall height. The E3 carried a list price, including a 10-foot header of $8,232, while the E3 Navy Bean Special combine with a spike tooth cylinder came in with a price of $8,707.

Combines

For 1968, Allis-Chalmers introduced the Model F Gleaner combine. When equipped with a 13-foot grain header the Model F was priced at $21,080 and this did not include a cab at $1,190 or air conditioning for the cab at another $950. Six different headers were available, including the 13-foot size, and ranging up to a maximum of 20 feet. In addition, six different corn heads were available in either 3-row or 4-row configurations and various row widths. In addition to this model, A-C offered the Model F as a Navy Bean Special, with yet another style being the Model FKS. The latter model was Gleaner's Corn-Soybean Special. This machine was equipped with a hydraulic reel lift and variable speed reel drive, along with other slight modifications. Production of the Model F series continued into 1976. The Model FR Rice Combine was built during a similar time frame, as was the FR Rice Special. The latter model was equipped with a hydro-traction hydrostatic transmission as standard equipment. All Model F Gleaners could be furnished with the choice of an A-C Model 262 gasoline engine or an A-C Model 2800 diesel engine.

A 1968 photograph illustrates the Model G Gleaner. Built through 1972, this popular model was built in a standard configuration, but could also be purchased as a Navy Bean Special, or as a Corn/Soybean Special. When equipped with a 13-foot grain head, this model carried a list price of about $16,000. This price was however, subject to various company and dealer discounts. List prices, where given, should be referred to as a relative value in comparison to other machines. Initially, the Model G was offered with the choice of an A-C Model 2800 engine, set up for either gasoline or LP-gas. Optionally available was the Model 2900 A-C diesel engine. Later production however, used a Chevrolet 350 V-8 engine in the gasoline version. This machine in all its various versions was equipped with a 100 bushel grain bin and hydrostatic power steering.

Making its debut in 1977, the F2 Gleaner remained in production through 1982. The F2 offered many new features over the Model F Gleaner, and many items offered previously as options now became standard equipment. This included the operator cab, analog tachometer, and a revised separator drive. When the gasoline engine was specified, a Chevrolet 292 engine was used, but if the diesel engine was installed, it was an A-C 433-I model. For narrow-row corn, header spacings for 3, 4, and 6 rows were available, and for wide rows, the headers could be secured in 3, 4, or 5-row spacings, and of course, in varying row widths. Weighing 11,606 pounds with a 13-foot header, the F2 carried a list price of $49,682 — the hydrostatic transmission if desired, added another $3,669 to the cost. During 1982 the F2, M2, L2, N5, N6, and N7 Gleaners were all available in a "Golden Harvest Edition." This designation carried a special exterior cab paint scheme and a "Golden Harvest" decal. Also featured was a special steering wheel cap and sport steering wheel, and a solid brass identification plaque engraved with the buyer's name and a special serial number. The "Golden Harvest" models were supplied with an extended warranty over and above the standard two-year engine warranty.

The F3 Gleaner, was introduced by Allis-Chalmers in 1983. Available in either a standard or a Corn-Soybean Special configuration, the F3 is equipped with an A-C 433-I diesel engine. This 4-cylinder design has a 200 CID and an output of 95 horsepower. A 5-speed torque sensing variable speed drive is standard equipment, with the hydro-traction hydrostatic drive being a $4,121 option. Factory list on this machine was $53,412 for 1985. Electro-hydraulic controls are standard equipment, with other standard features including a monitoring system and the industry's widest choice of headers and corn heads. Numerous cab options are available, with other options including automatic 3-position header height control and automatic reel/corn head speed control. Further data on the "F" series Gleaner is listed in the *Deutz-Allis* section in the Epilogue to this volume.

The Model GH Hillside Gleaner was produced during the same period as the Model G previously noted. This machine was virtually identical in design and operation, save for the necessary alterations to convert it into a hillside machine. Automatic, multi-rate leveling was featured to get the fast, positive leveling required. A load leveler in the grain bin provided automatic load leveling. Another important feature was the uphill weight shift mechanism which actually shifted body weight uphill for extra stability and safety. This design actually placed more weight on the uphill wheels, with less on the downhill wheels, thus assuring that the uphill wheels would maintain proper traction characteristics. Model GH Gleaners used a swivel-mounted, quick-detachable header. The GH Hillside Special carried a list price of $24,544. Also available was a Gleaner GRT-HY Rice Special, again using the same basic design, but suitably modified for the special conditions encountered with this crop.

Replacing the earlier Model K Gleaner was the Model K2 machine. Introduced in 1976, it remained in production until mid-1980. It offered a specially designed concave door which opened automatically if rocks or other potentially damaging objects entered the cylinder and concave area. A monitor light in the cab instantly warned the operator if the concave door was open, and it could be reset to normal position very easily. This feature was exclusive to the Gleaner combines. Standard engine equipment was Chevrolet's 250 CID 6-cylinder engine, although an A-C 433-I diesel was optionally available. The K2 was offered with a diesel engine for a factory list price of $33,800. Also available was a K2 Corn/Soybean Special for $34,335; the KK2 Corn Plus at $32,913, and the K2 Corn Plus for $40,617. These various styles differed from one another primarily in the type and amount of accessory items designed either for operator comfort or for special crop requirements.

Offered from 1968 through 1976, the Model K Gleaner was quite similar in design to the somewhat larger Model F built concurrently. Six different grain headers were available, ranging in size from ten to 15-feet. Also available were two and three-row corn heads. The grain bin capacity was 66 bushels, compared to 77 bushels for the Model F. Standard equipment included a Chevrolet 250 cubic inch, 6-cylinder in-line engine. Allis-Chalmers did not offer a diesel engine option for this machine. Equipped with a 10-foot grain table, this machine carried a factory list price of $14,980. The Model K was also available as a Navy Bean Special with a factory list price of $15,799.

Seven different grain headers ranging in width from 13 to 24 feet were available for the Model L Gleaner. Continuing their practice of offering the widest selection of corn heads in the industry, A-C offered corn heads for the Model L in 4, 5, 6, and 8-row configurations. Several tire choices were available, and a choice could be made between a Chevrolet 350 V-8 engine or an A-C 2900-II turbocharged diesel. Those machines equipped with a hydrostatic transmission could be furnished with a big 426 CID turbocharged diesel. In addition to a host of standard features, the Model L could also be supplied with many different options — these included a hydraulic accumulator to absorb the shock of the header load on the hydraulic system when operating over rough or uneven ground. A grain bin extension was optionally available which permitted a maximum load of 185 bushels. In addition to the standard Model L Gleaner, A-C also built the Model LKS machine, a corn/soybean model with a factory list price of $35,632.

Beginning in 1976 the Gleaner L2 machine replaced the earlier Model L. At this point the gasoline engine was dropped as either standard or optional equipment, with the A-C 2900 Mark II turbocharged diesel being standard equipment on the L2 machine. Cab, lights, hydrostatic steering, and a 200-bushel grain bin were all standard equipment on the L2 — it carried a factory list price of $45,335. Individual hydraulic caliper disk brakes were featured, as was an elaborate monitor system which warned of various harvest problems or component failures. Electro-hydraulic controls were a standard feature, but the full-floating cutter bar was an extra-cost option. Production of the L2 ended in 1982. This style was also available as the L2-HY Rice/Soybean Special shipped from the factory with the necessary modifications for harvesting these two crops. The L2 could be equipped with specially designed tracks replacing the rubber tires for harvesting rice or working in muddy fields.

Production of the Model M Gleaner began in 1973, a year later than the somewhat similar but larger Model L. In fact, except for size and capacity, the Model L and the Model M machines were virtually identical in design and operating characteristics. The same standard and optional features were almost identical, except of course that the smaller Model M could be equipped with a Chevrolet 350 V-8 gasoline engine, or optionally with the A-C 2900-I diesel. The Model M of course offered standard Gleaner features like open bottom straw walkers, transverse flow, dual outlet cleaning fan, and reversible rasp-bar cylinders. Also available was a Model MKS Corn/Soybean Special.

A 2-speed electro-hydraulic system with a parallel linkage beam was the heart of the MH Gleaner Hillside model. Built concurrently with the Model M Gleaner, this design was intended as the designation implies, for operation on sloping ground which would either be too dangerous for safe operation with a conventional machine or would at best be very inefficient in its work. Like the standard Model M, the MH featured a big 12-inch unloading tube for super-fast unloading of the 105 bushel bin. A pushbutton control in the operator's compartment swiveled the unloading auger into position, and after emptying the bin, pushing the button again retracted the auger to a closed position alongside the machine.

Introduced in 1983, the L3 Gleaner embodied most of the features found in earlier "L" models, plus some additional improvements which resulted from improved technology and field experience. Powered with an A-C 649-I turbocharged diesel engine, the L-3 boasted 145 horsepower. A 5-speed transmission was used. Other standard features included Gleaner's 3-stage separation system and 2-stage cleaning of the grain. The Tattletale II monitor system alerted the operator of virtually every major machine or engine difficulty that might arise. The L3 corn/soybean Gleaner had these two crops as the major use of the machine — one change included intercooling of the 649-I engine. Beyond this, A-C offered the L3 "Windrow Plus" model, again with the 649-I engine. Two versions were available with a hydrostatic transmission as standard equipment; the L3-HY Rice Special and the L3-HY Corn PAC. Both were virtually identical to the standard L3 models except that each of these units was equipped with an A-C 670-T turbocharged engine of 158 horsepower.

M2 Gleaner combines, built from October, 1976 through June, 1982 were part of a series that also included the F2, K2, and L2 models. New features of this series included a new cab interior as standard equipment, and an optionally available Power-Fold ladder. Both the L2 and M2 machines offered an exclusive bin loading system which featured a floating center-loading bin auger and automatic folding bin extension flaps. As the bin fills, the auger floats up on top of the grain simultaneously raising the folding bin flaps for extra bin capacity. Several variations were available in the M2 series. These included the M2 Grain model with headers available in seven sizes ranging from 13 to 24 feet in width. The MKS2 corn/soybean model was tailored to these two specific crops, although it was capable of handling various others as well. Also available was a special M2 Gleaner for rice. The MH2 was a hillside version.

Gleaner's MH2 Hillside combine was built concurrently with the standard M2 model and carried all the Gleaner combine traditions of the past — dependability, accessibility, easy adjustment, and ease of operations. The Gleaner MH2 had the power and capacity to master steep slopes. The multi-rated leveling system automatically matched rate of leveling to the degree of slope. Sudden changes in slope gave sudden changes in the leveling system, but slight changes in slope gave a gradual change in leveling. With this, the 140-inch wide-track design was built so that weight was automatically shifted uphill for extra stability, safety, and traction. The swivel-mounted header conformed to the slope, and Gleaner's exclusive "Esca-Ladder" pivoted with the drive axle, staying one step above ground level whether on the uphill side or the downhill side of the combine. A great number of options were also available for MH2 machines, including a straw chopper, rock guards, and a floating cutter bar.

Introduced in 1979, the N5 Gleaner was part of an entirely new combine line using a rotary threshing system instead of the traditional rasp-bar cylinder design. Although other builders mounted the cylinder parallel with the machine, the Gleaner rotary design situates the cylinder transversely. This was done for several reasons — it requires less space, provides optimum weight distribution, and allows for straight-in feeding of crops. The Gleaner rotary design threshes by a combination rubbing action and by centrifugal force. As the grain moves in a spiral path around the cylinder, threshing is completed, with the grain being removed by centrifugal force. The N5 is equipped with an A-C 670-I turbocharged and intercooled diesel engine rated at 190 horsepower.

For Allis-Chalmers, the M3 was the final machine of the series that they would introduce under their own name. During early 1985 a merger with Deutz of Germany gave a new title, that of Deutz-Allis. Introduced in 1983, the M3 offered additional features over its predecessors, with many of these being aimed at improved operator convenience, and greatly enhanced monitoring systems using solid state electronic modules. M3 Gleaner machines continued to use the A-C 649T turbocharged diesel engine in the standard models, but used a slightly larger 670-T engine in the MH3 Hillside version. With the M3, A-C promoted their 2-Season warranty with a program called TQA — Total Quality Assurance. This program was unprecedented in the industry, but certainly suggested that reliability was a hallmark of Gleaner design.

The N6 Gleaner shown here is equipped with a huge 12-row corn head, giving it tremendous harvesting capacity. This corn head alone carries an operating weight of 7,690 pounds, over and above the base machine weight of 19,597 pounds. Grain platforms are built in sizes up to 30 feet, and several modified versions of the N6 are built to accommodate special crops or conditions. Despite its size however, the N-Series Gleaner's have a short turning radius of only 19.9 feet, making these machines very adaptable to small fields as well as large ones. An A-C 670-HI engine is used in the N6 Gleaner. Rated at 2,200 rpm, it uses a 4.25 x 5.0 inch bore and stroke, yielding 220 horsepower. After the Deutz-Allis merger of 1985, production of the "N" series combines has continued.

Allis-Chalmers announced the N-Series Rotary combine in a news release of February 1, 1979. Initially offered were the N5 and N6 models, with the N7 being announced in a news release of October 17, 1979. The above three rotaries, plus five conventional rasp-bar machines gave A-C a total of eight different combines available — one of the largest selections in the industry. The N7 is virtually identical in size and design to the N6 Gleaner, except that this particular model is equipped with a somewhat larger A-C 685-I diesel engine of turbocharged and intercooled design. Information on Gleaner combines built after 1985 may be found in the *Deutz-Allis* section given as an epilogue to this volume.

Comfort Chair

During the late 1970s, the Luxury Allis-Chalmers Comfort chair was offered to A-C dealers. The seat itself was virtually identical to that used in Gleaner combines such as the M2 model. Covered with seamless vinyl and cloth fabric, it was designed to be the ultimate in office comfort. Features included tilt-up armrests, swivel-and-tilt base, chrome-plated legs, oversize casters, and lumbar adjustment. Apparently the Luxury Comfort Chair was a promotional item intended primarily for A-C dealers rather than for the general office equipment market. However, even the promotional concept appears to have been a short-lived venture.

Condensers

Already in 1877, Edward P. Allis Co. was building steam condensers. Edwin Reynolds, like many other engineers of the time, recognized the greatly increased operating efficiency brought about by the use of the condenser — the improved efficiency meant lower costs for fuel. While the early condensers were rather small, ever increasing sizes of steam engines, and later steam turbines, demanded huge condensers to handle the load. The condenser took the exhaust steam and cooled it back to a liquid state, ready for another pass through the boiler. This particular unit of 1937 was of the welded shell type, offering divided water boxes so that half of the unit could be shut down for cleaning while the other half remained in operation.

A 1937 photograph from the Allis-Chalmers *Annual Engineering Review* illustrates a newly installed condenser with its auxiliary equipment at the Granite City Steel Co. plant. Operation and control of these units required circulating pumps, motors, valves, and other accessories which Allis-Chalmers also furnished as required. A 3,000 kilowatt steam turbine required a 9,000 square foot condenser with dimensions approaching 96 inches in diameter and a length of 18 feet. Auxiliary equipment might include a circulating pump, two condensate pumps, two steam jet air pumps, along with electric motors, valves, and necessary controls.

This 1930s photograph shows a huge 31,000 square foot, two-pass condenser. A welded steel shell is featured in this unit. Allis-Chalmers was able to completely manufacture this unit at West Allis, including everything from rolling the huge steel plates to machining the matching surfaces of the shell and the end plates. This particular unit was designed for condensation of exhaust steam, with the inlet visible to the rear of the unit. The ladder standing next to this condenser appears to be about 16 feet high. A workman, barely visible in the lower right hand corner of this photograph is completely dwarfed by this huge unit. A great many Allis-Chalmers condensers were sold for use with steam turbines built by competing turbine manufacturers.

Condensers

The Allis-Chalmers shops boasted a huge 40-foot vertical boring mill, the largest in the country. This 1940 photograph shows a condenser shell mounted on the big 40-foot mill to machine the upper bolt pads and the center manhole flange. Since Allis-Chalmers owned very large shop equipment, they were able to handle large projects out of the reach of most manufacturers. This particular condenser is in its preliminary stages — the tube sheets are yet to be installed, and of course, the next step will be the installation of hundreds, perhaps even thousands, of tubes. In the early 1950s Allis-Chalmers built a unique experimental condenser at West Allis Works to implement long-term studies in heat transfer within condensers. The model was designed to permit use of almost any tube bundle and baffle arrangement, as well as a variety of tube sizes and materials.

During World War II, Allis-Chalmers contributed heavily to the war effort, including the production of condensers for use in ships. The size and productivity of the West Allis Works is illustrated by the fact that from these doors came enough condensers to equip a single ship each working day. Achieving this goal required exceptional coordination of the manufacturing and assembly operations, since so many separate components were required for a complete unit. This 1943 photograph shows workmen rolling condenser tubes into the tube sheet on the first unit. A nearly completed condenser follows, with a third condenser in the background being ready to receive the tube sheets. Allis-Chalmers was heavily involved in manufacturing various items for ship propulsion during the Second World War.

This closeup view shows a workman installing condenser tubes in a two-pass unit. Water circulated through the lower half of the unit, through the condenser head, and returned via the top set of tubes. Hundreds upon hundreds of tubes were required, with each one being carefully rolled into the tube sheet and made pressure-tight. While this task was done by hand in early days, the introduction of pneumatic tools simplified the task. By 1953, Allis-Chalmers condensers averaged 60,000 square feet of cooling surface, with two 115,000 square foot units scheduled for completion during that year. For 1953, the largest completely tubed condenser had a cooling capacity of 27,500 square feet. Weighing 74 tons, it was designed, manufactured, and shipped within six months time.

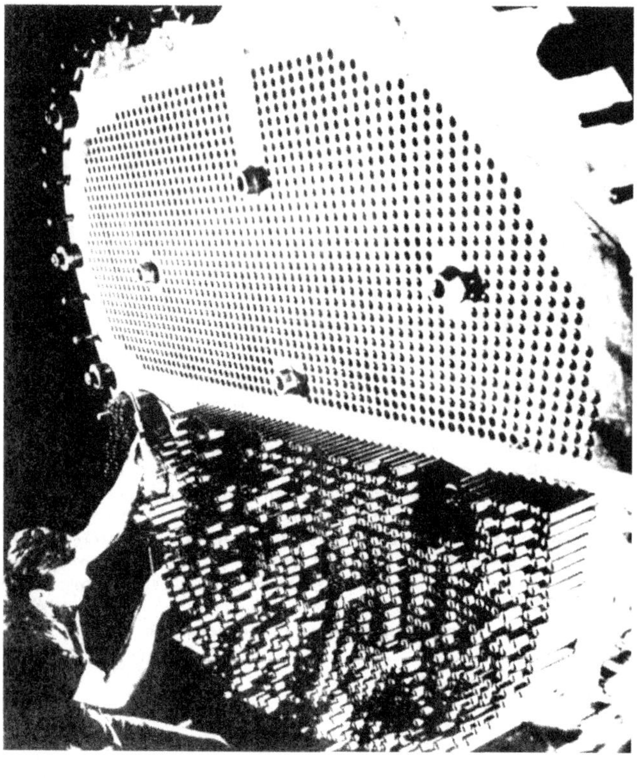

By the 1950s Allis-Chalmers was offering various condenser designs, including this sectional unit with 75,000 square feet of surface. Shown here is the center section in the foreground, with one head section loaded on another rail car. Allis-Chalmers was also active in the design and manufacture of deaerating units, recognizing the need to control corrosion in a steam-water cycle. Allis-Chalmers engineering in the relatively new field of nuclear power further goaded this effort, and during 1953 the company was able to build hot zeolite units capable of treating over 120,000 gallons of water per hour. The company was also active in the engineering and design of ion exchangers for demineralizing and removing silica from the feedwater for high pressure boilers.

Corn Pickers

Allis-Chalmers began experimental work on a mounted corn picker in the late 1930s. This resulted in their first production machine during 1941. Due to the onset of World War II, production was almost immediately suspended, resuming again in 1947.

The A-C mounted picker design was indeed unique. Mounting was simplified by actually driving over the rear mechanism prior to attaching the necessary pins. The two-row design was intended for mounting on the WC tractor, and the single-row version was designed for the Model B tractor. Since the Model B was equipped with a wide-tread front end, A-C engineers mounted the gathering unit on the centerline of the tractor, just ahead of the radiator. Before long though, this design was scrapped in favor of a sidemounted gathering unit.

Early A-C corn picker designs featured adjustable stripper plates and so-called "rubber-on-rubber" husking rolls. The replaceable rubber buttons were designed to accomplish husk and clean the corn in a single operation, doing so with great efficiency when the leaves were damp, but doing so with great inefficiency under dry conditions.

Production data on the early A-C corn pickers is rather sketchy, although it appears the mounted style of 1941 was built with minor changes until the early 1950s. During 1949 however, a special sweet corn harvester was produced to meet the specific needs of the canning industry. A set of shielded rotating knives actually clipped the stalks above the ear, casting them to one side, and thus preventing excessive trash from passing through the machine. This and other modifications were made to the standard corn picker, and it appears that most of these modifications could be made in the field, although at some considerable amount of effort.

The Allis-Chalmers Model 35, one-row, pull-type picker was built between 1950 and 1960, with 300 of these being sold as snappers, having no husking bed whatever. Concurrently, the Model 33 mounted picker was marketed, it being designed specifically for the WC, WD, and WD-45 tractors.

From 1958 to 1961 the Model 14 two-row mounted picker was offered — it was designed for use with the D-14 or D-15 A-C tractors. Also available during this period was the Model 17, built for use with the D-17 tractor.

Finishing out the Allis-Chalmers corn picker line were the 150 and 170 models, each being introduced in 1961, with both of them ending production the same year. The 190 picker could be mounted on the WD-45, D-17, and D-19 tractors.

A head-on view shows the Allis-Chalmers corn pickers of 1940. A note on the back of the original photograph reads: "Allis-Chalmers direct-attached corn pickers on the "WC" and "B" tractors. These are still experimental models. Regular production will probably start next year. In these machines Allis-Chalmers has strived to produce a unit that will sell for about half the price of those now on the market, yet they will work better." Unfortunately for Allis-Chalmers, World War II began just when these machines first entered the market. Although factory production records indicate that a few were built during the war, production remained rather limited until 1947. By that time, many competing companies, fortunate enough to get their designs on the market prior to the war, had gained a distinct sales advantage over the A-C machine.

The Allis-Chalmers 1940 experimental corn pickers are shown here in a sideview. The single-row version is mounted on an Allis-Chalmers Model B tractor. Due to the wide-front design of the tractor, A-C engineers placed the gathering unit directly over the tractor centerline. Even in the days of open pollinated corn, the immense amount of dry material passing in front of the radiator surely must have created immense cooling problems. To protect adjacent corn rows, protective shields were mounted in front of each rear wheel. Note that while the two-row version is equipped wih screen shields for the engine, the single-row model has no protective screening.

A rear view of the early A-C corn picker design illustrates the low-profile of this unit. While sometimes an advantage, this frequently proved to be a disadvantage, since clearance beneath the machine was quite limited. This tended to retard the material passing through the snapping rolls.

From all appearances, the single-row mounted picker for the Model B tractor was put into production as a sidemounted machine instead of the center-line version shown for the 1940 experimental model. This particular unit illustrates a rather elaborate screen to protect the radiator and permit an ample airflow at all times. Electric lights actually permitted harvest to continue past sundown, thus increasing productivity. Compared to hand picking of the past, even a one-row machine represented a substantial time savings, and probably improved the disposition of the entire family by eliminating the need to be in the field "at sun-up" to start the days task of "hitting the bang-boards."

Because of its low-profile design, the Model 33 picker was easy-on, easy-off, requiring no hoists. This was a unique feature at a time when mounting a corn picker to a tractor usually required a couple of men, a few boys, and lots of muscle power. By removing the picker snouts, the entire assembly was low enough so that the tractor could be backed directly over the picker and then mounted to it. A-C used a combination system of stripper plates and rubber-on-rubber husking, citing this arrangement as less damaging than the customary steel-on-steel husking rolls used by other companies. One satisfied customer wrote that he averaged about 1½ acres per hour without and trouble, and in the previous season had picked 200 acres of corn yielding 80 bushels per acre.

Another view of the Model 33 mounted picker illustrates one of the high-capacity cleaning fans. These were driven directly from the tractor belt pulley. This machine was built with 41-inch centers on the gathering unit, making it compatible with row widths of 38 to 44 inches. Great attention was given to shielding, with A-C advertising noting that the Model 33 was the "safest picker built." The two-row corn harvester of 1947 carried a factory list price of $745, with an $8 motor shield being the only extra to buy over and above the list price. For 1957, the Model 33 base machine, a two-row mounted style without the mounting equipment listed at $1,230.

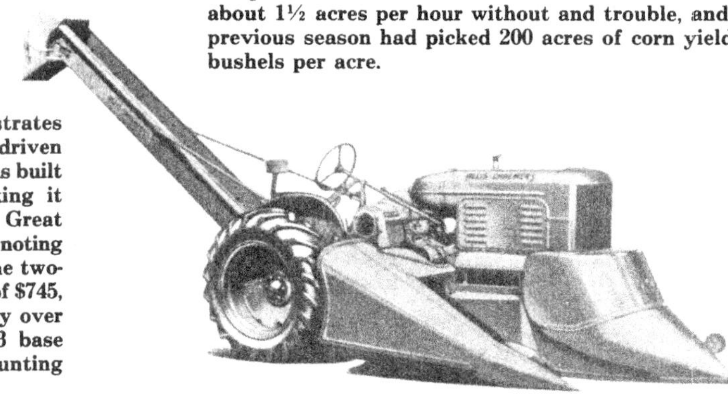

The Allis-Chalmers Sweet Corn Harvester was built especially to the needs of the canning industry. Introduced about 1950, its design included a special corn topping device which out off the stalk just above the ear, throwing it aside and keeping this additional material from entering the machine. Designed for use with the Allis-Chalmers "WD" tractor, it also featured steel tube rollers running with knife rollers to replace the usual snapping rollers. Different stripper plates were used — in this case the stripper plates were mounted over the snapping rolls to prevent injury to the ears by direct contact with the snapping rolls. The sweet corn harvester could then be converted to a standard corn harvester in the fall, with all necessary conversion equipment being included as part of the original purchase price.

Another view of the Allis-Chalmers Sweet Corn Harvester shows a closeup view of the unique stalk-topping mechanism ahead of the machine. A special fine-mesh screen was used over the radiator grille to minimize the plugging effect of pollen and green trash dislodged by the harvesting process. A set of shields was placed over the husking rolls to render them inactive, and special gathering chains were installed to assist in moving large volumes of green materials through the machine. While a unique and very useful machine, it appears that sales of the Sweet Corn Harvester were insufficient to pay for the added engineering and manufacturing costs, so production ended by the mid-1950s.

Corn Pickers

For those farmers wanting a pull-type corn picker, Allis-Chalmers offered the Model 35 design, noting that it was suitable for all tractors with ASAE standard hitch and PTO shaft. Built from 1950 through 1957, the Model 35 saw total production of 6,500 units of which some 300 were built as the Model 36 snapper. It was identical to the Model 35 but lacked a husking bed. For 1957 this unit listed at $905. Design features were virtually identical to those found in the Model 33 mounted corn picker except of course for the design changes necessitated because of the pull-type design. Approximate operating weight of the Model 35 was 1,490 pounds.

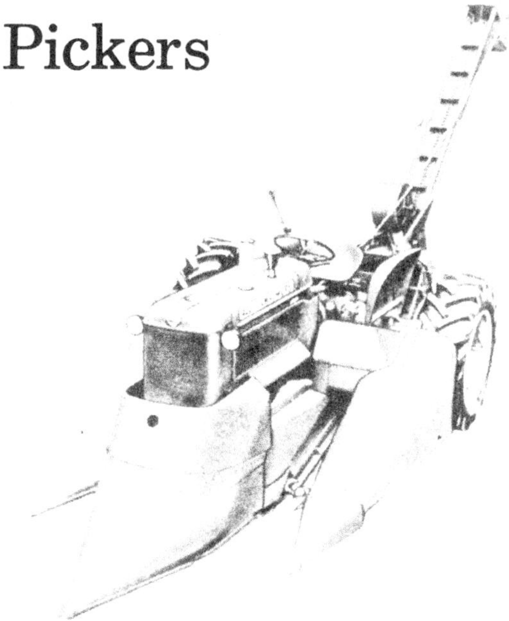

During the 1958-1960 period, Allis-Chalmers built 200 Model 14 Corn Harvesters. Of these, the last 50 were converted from the concurrent Model 17 machine during August, 1960. Matched specifically to the D-14 tractor, the Model 14 picker offered various improvements in design, including a relocation of the cleaning fan to a position over the husking bed. Since production of this machine was very limited, few specific details are now available.

Between 1958 and 1960, Allis-Chalmers built 2,150 copies of the Model 17 2-row corn picker. Although none were built after 1960, this machine appears in the A-C Price Book as late as 1962, indicating that they were still available from existing stocks. Priced at $1,775, the Model 17 was built specifically for the D-17 gasoline or diesel tractors. The rubber-on-rubber husking design was still featured, as were the rotary feeders and other items unique to the A-C design. The question of why sales were disappointingly low for this machine could cover considerable territory — in this time frame, many farmers were beginning the conversion to the corn combine, thus eliminating the need for conventional corn pickers. Of those still using the corn picker, farmer preference was probably a major factor.

An entirely new A-C corn picker design emerged in 1961. The Model 170 picker shown here was built specifically for the concurrent D-17 tractor. During this, its only year of production, 750 units were built. A complete redesign of the ear corn elevators is obvious — this design included a trash ejector mechanism which closely followed some competing designs in its principles of operation. Priced at $2,695, the Model 170 weighed 3,400 pounds. The same picker, when purchased for use with the D-14 or D-15 tractor was known as the Model 150. The primary difference in these machines was in the necessary mounting hardware.

Like the 150 and 170 models, the A-C Model 190 picker offered a 4-roll husking bed featuring rubber-on-rubber husking. Three gathering chains were used on each row. Priced at $2,375 for the base machine, an additional $375 fee was levied for the attaching parts, and these could be furnished for the WD-45, D-17, or D-19 tractors. In fact, the Model 190 appears to be virtually identical to the 150 and 170 models except for this very point — while these models were designed for specific tractor models, the 190 could be equipped to fit a range of three different Allis-Chalmers tractors. Built only in 1961, the Model 190 marked the end of all corn picker manufacturing at the Allis-Chalmers LaPorte Works.

Cotton Pickers

Cotton pickers became part of the A-C agricultural line about 1949. The Two-Row Cotton Harvester shown here was mounted over an Allis-Chalmers WD tractor, with later production using the WD-45 tractor as a base. Row spacings were adjustable from 36 to 42 inches. The basket had a capacity of 188 cubic feet, large enough for about 1,200 pounds of seed cotton. The A-C design used spindles which began spinning as they entered the picking area, with each spindle extending entirely through the cotton plants, wrapping lint from the open bolls. The cotton-laden spindle stopped spinning after leaving the picking area and entering the stripper. Here the cotton was gently removed and elevated to a blower for delivery to the basket.

Basket unloading was a completely power-driven operation which permitted the operator to drive alongside a waiting wagon, empty the basket, and return for another load. The unloading lever hydraulically elevated the basket above the height of the wagon sides. Then a shift lever lowered the floor extension and the entire load was carried by a chain raddle into the waiting wagon. When A-C introduced the WD-45 tractor in 1953, it became the foundation for the Two-Row Cotton Harvester, adding extra horsepower as well as other standard WD-45 features. At this point in time, only the Allis-Chalmers cotton harvesters elevated the cotton to the Flower, a feature that gave cleaner lint with less pin trash, and as a result gave a better grade of cotton. This machine sold for approximately $17,500.

The Model 116 and Model 122 cotton harvesters built between 1962 and 1964. Both were of one-row design, with the 116 having 1,280 recessed barb spindles, and the 122 using 1,760 spindles. Except for this difference, the two machines were virtually identical. The design was such that row spacings of 38, 40, or 42 inches could be accomodated. From all appearances, sales of this one-row machine were disappointing at best. Prior to this style, A-C offered the Model 500 cotton picker only during the 1960 harvest, with production ending after only 225 units.

Introduced in 1961, the Model 616 cotton picker remained in production until 1966. During this period, 244 units were built at LaPorte Works. The 616 was of two-row design, using a 16-spindle slat with 2,560 spindles. The 616 was built for medium height cotton, and the recess-barbed chrome plated spindles gave the necessary aggressiveness, while the chrome plating kept the spindles free of rust, making them easier to strip. Rust-free, oil-impregnated bronze bearings were used on the spindles, requiring no greasing, and preventing oil and grease stains on the cotton. The hydraulically operated picking units were designed for individual raising and lowering to meet existing conditions. Row spacing was adjustable for 38, 40, or 42 inch rows. This 1961 photograph, taken at the annual County 4-H Club Fair in Milwaukee, illustrates some of the award winning dresses modeled by 4-H representatives. The A-C 616 cotton picker provided a fitting background for their "King Cotton" theme.

Cotton Pickers

A 1961 photograph taken near Fort Stockton, Texas, provides the background for this Model 622 cotton picker. Built from 1961 to 1966, it, like the Model 616 used an Allis-Chalmers D-17 tractor as the foundation. While the 616 could be furnished with either the 226 CID "Power-Crater" gasoline engine or the 262 CID diesel engine, the Model 622 offered two additional options — a 262 CID diesel, or a turbocharged version of the 262 engine. Increased use value was provided by converting the tractor to field use after the cotton harvest. A conversion package, available as extra equipment, converted the tractor from under the picker to a conventional D-17 tractor without the complete cost of another tractor. Power Shift rear wheels, while not standard on the cotton picker, could be purchased when making the conversion. Allis-Chalmers recommended using either the 226 cubic inch gasoline engine or the standard 262 cubic inch diesel model for tractor conversion — the turbocharged 262 CID engine was NOT recommended for regular tractor use. The Model 622 carried an approximate list price of $16,700.

From 1966 to 1969, Allis-Chalmers built the Model 801 and Model 802 cotton harvesters. These machines were built as single-purpose units rather than the past practice of mounting the machine over a tractor chassis with the possibility of later converting the tractor to other duties. The 801 was known as a Lo-Unit machine designed for low to medium height cotton, while the 802 was a Hi-Unit design intended for medium to tall cotton. In addition, the 802 carried additional spindles. Also featured was a standard gasoline engine, suitably converted to use propane fuel. Weighing 1,2195 pounds, the Model 801 carried a 1969 list price of $15,795, while the Model 802 was tagged at $17,400. The Model 860 Cotton Harvester appeared in 1973, remaining in production until 1978. The last model of the line was the Model 880 Cotton Harvester. Introduced in 1980, it was built as late as 1984. The 880 offered all the latest features, including a 95 horsepower diesel engine. Factory list on this machine was over $66,000.

For those farmers wanting a one-row machine, Allis-Chalmers offered the One-Row Cotton Harvester between 1952 and 1954. It was built specifically for use with the WD, WD-45, or CA Allis-Chalmers tractors. Hydraulic control eased the raising and lowering task, and full shielding protected the plants and green bolls from damage by the machine. This single-row design used 1,024 rotating picking spindles mounted in vertical slats with each slat holding 16 spindles. The A-C design made extensive use of ball and roller bearings, needle-bearing universal joints, and v-belt drives. One apparent advantage of this machine was that it could be mounted or dismounted from the tractor with relative ease, permitting the use of the tractor for other jobs after the cotton harvest. List price was about $12,800.

The cotton stripper provided "once-over" harvesting of the crop. For this purpose, Allis-Chalmers offered a series of cotton strippers, beginning with the 2-S, built from 1960 to 1965. Intended for direct-tractor mounting, the 2-S provided an economical means of getting the bolls from the field, leaving the remainder of the process to later operations. Prior to harvest, a chemical defoliant or a desiccant was often used to removed as much green foliage as possible.

Between 1966 and 1969 A-C offered the 707 and 707 XTB cotton strippers. The machines were identical except that the machine on the right was equipped with an extractor, while the machine on the left was without this device. The extractor functioned to remove cotton burrs, sticks, leaves, trash, and other foreign material from the harvested cotton. For 1967 the Model 707 carried a list price of $9,650, while the 707XTB machine listed at $12,950. The 1968-1972 period saw the production of the improved Model 760XTB self-propelled stripper — it used a 6-cylinder engine rather than the 4-cylinder style found in the Model 707 series. In most other respects it was the same machine as before. The 760XTB carried a somewhat higher list price — approximately $18,000.

Crushers

Although crushers and various types of reduction equipment appear in the "Ore & Mineral Processing" section of this book, crushers as such, represent an important part of industrial development, and for the purposes of this volume, illustrate their important part of industrial history, not only for industry world-wide, but in particular for some of the Allis-Chalmers predecessor companies.

In 1858, Eli Whitney Blake invented a jaw-type crusher, subsequently receiving Patent No. 27,882 for his efforts. This design was adopted by Gates Iron Works of Chicago, founded in 1842 by H. H. Scoville and Philetus W. Gates. Shortly after the company was organized, Gates married one of Mr. Scoville's daughters. Of their children, P. W. Gates II was one of the organizing members of Allis-Chalmers Co. in 1901. David Fraser married another of the Scoville daughters in 1851 — he was a founder of Fraser & Chalmers Co. in 1872. Thomas Chalmers had joined the Gates Iron Works in 1844.

During the entire period from the 1840s until the Allis-Chalmers merger, the Gates, Scoville, Fraser, and Chalmers activities are so intertwined because of family relationships that separating the various business activities is difficult at best. Suffice it to say however, that these various families, individually and collectively controlled the major portion of the crusher business for many years. Gates Iron Works was renamed Eagle Works Mfg. Co. in 1862 with Gates, Fraser, and Chalmers listed as partners. About this time they began selling the Blake crusher. Finally in 1872, David Fraser and Thomas Chalmers organized the firm of Fraser & Chalmers, announcing themselves as "successors to Eagle Works." W. J. Chalmers, son of Thomas, was bookkeeper of the company — he is the person from whom the "Chalmers" portion of the Allis-Chalmers name was derived, with the other name of course being Edward P. Allis.

Returning again to the Blake patent of 1858, it appears that Gates Iron Works began selling these machines in the early 1860s. A study of these machines tends to support the notion that the Blake machine was the first successful jaw-type crusher. Following this came the gyratory crusher in 1878. In June of that year, Charles M. Brown filed for letters patent on his design, receiving Patent No. 201,646 in 1878. The Brown machine used a two-bearing shaft design, with the upper bearing being spider-mounted, while the lower bearing was eccentric-mounted. A separate conical breaking head called the head center was fitted on the main shaft. An eccentric motion to the main shaft was derived from a bevel gear drive to the power source. Raising or lowering the main shaft altered the size of the product. This design remains virtually identical in principle to the modern Superior and Hydro-Cone crushers built by Allis-Chalmers in recent years.

During 1880, Gates Iron Works sold a gyratory crusher to Buffalo Cement Co. of Buffalo, N. Y. This No. 2 Gates crusher marked one of the first recorded transactions between the cement industry and a parent company of Allis-Chalmers. Some years later, the cement industry would be a major sales arena for Allis-

Designed for finer reduction than the Blake crusher, Dodge crushers used a slightly different design which permitted more accurate product sizing. Dodge crushers were often used to further reduce the product coming from the Blake crusher. Fraser & Chalmers sold a great many Dodge crushers, with the 1901 offering including four sizes. The smallest machine used a 4 x 6 inch feed opening and required only 2 to 4 horsepower. The largest machine weighed 13,500 pounds, had an 11 x 15 inch feed opening, required 16 to 22 horsepower, and had a capacity of 5 to 8 tons of rock per hour.

During the late 1850s Eli Whitney Blake developed a jaw-type crusher for which he subsequently received Patent No. 27,882 in 1858. While some design changes were made in the following years, Blake crushers remained reasonably close to the original Blake concept. Gates Iron Works began building the Blake machine during the 1860s, and when Fraser & Chalmers was organized in 1872 they built the Blake crusher in addition to a wide variety of machine tools, mining equipment, and supplies. This example of a Blake crusher was built by Fraser & Chalmers during 1901. Five sizes were offered that year, ranging from a tiny Laboratory Machine up to the Model No. 8 with a 20 x 10 inch feed opening. Weighing 22,000 pounds, it could handle pieces weighing up to five tons, reducing them to product of 1 or 2-inch size.

Crushers

Chalmers crushing and refining equipment.

Holding several gyratory crusher patents, especially the Charles Brown patent, Gates Iron Works was able to capitalize on the demand for several years with little or no competition. However in 1884, Fraser & Chalmers began marketing their own version known as the Comet Crusher. As noted in previous paragraphs, Fraser & Chalmers was related to Gates Iron Works through marriage if through no other means. The fierce patent infringement suit brought by Gates against Fraser & Chalmers in 1890 must surely have made family gatherings a pleasurable affair! The evidence presented in the case seems to have definitely established the Brown patent as the first successful gyratory crusher. However, it was argued that Gates did not actually have patent protection, since the design features enumerated in this and other gyratory crusher patents in his possession had already been covered by earlier ones. Gates lost the case, and Fraser & Chalmers became even stronger competitors to Gates Iron Works.

By 1896 Gates Iron Works had sold over 4,000 gyratory crushers, and by this time the firm had gained international status. Fraser & Chalmers meanwhile seem to have concentrated their sales efforts on the mining industry, while Gates seems to have devoted more effort to the cement and aggregate industries. Pulling these two firms into the Allis-Chalmers merger of 1901 placed the new company into an almost unreachable market position.

The original Gates Model B crushers were built until about 1895, at which time the Model D crushers were introduced. The Model D was quite similar to its immediate ancestor. By 1901 the Model K crusher was on the market — it was built in larger sizes than the previous machines, ranging all the way up to a 21-inch feed opening.

With the development of large power shovels came a parallel demand for larger crushers. Hand feeding was no longer practical, and in response to the demand, Allis-Chalmers offered a 36-inch machine in 1907, raising this size to 42 inches in 1909. During 1917 the feed opening was raised up to a mammoth 54 inches.

During 1924 Allis-Chalmers purchased the ore and mineral processing lines of Power & Mining Machinery Division of Worthington Corp. With this purchase came the McCully gyratory crusher designs, probably the greatest competitor to the Gates crushers. Two years later, Allis-Chalmers designed, built and shipped two Superior McCully crushers with 60-inch feed openings to South America. For some years to come, these were the largest gyratory crushers in the world.

During 1929 Allis-Chalmers introduced their first "Oil-Adjusted" crusher with the main shaft being supported on an oil-operated hydraulic jack. This early design eventually was refined into the patented Allis-Chalmers "Hydroset" control, and was adapted to all sizes of Superior and Hydro-Cone crushers. During the 1930s Allis-Chalmers introduced the Type R crushers utilizing Hydroset control. This feature enabled instant changes in product size while offering safety protection against tramp iron entering the machine by accident.

Improved manufacturing methods led to improved crusher designs. In this area Allis-Chalmers excelled, capping their previous achievements in crusher size by building the first of their huge Superior Gyratory machine in 1952. These were capable of crushing 6,600 tons of rock per hour, making them the largest gyratory crushers in the world.

A final note regarding the Blake jaw-type crusher. Allis-Chalmers continued building the Blake design in large and small sizes until 1924. With their acquisition of Power & Mining Machinery Division of Worthington Corp., it appears that Worthington's "Superior" jaw crushers moved into an important role with A-C, and from the Superior jaw crushers of 1924 eventually came the A-C Fine Reduction Jaw Crushers with welded steel frames and anti-friction bearings. Canadian Allis-Chalmers began building large sizes of the welded steel Blake crushers in 1938.

Additional crushers and other equipment including stamp mills may be found under the "Ore & Mineral Processing" heading elsewhere in this volume.

Gates Iron Works introduced the C. M. Brown gyratory crusher design in 1878, assuming protection under Patent No. 201,646. During 1884 Fraser & Chalmers introduced their "Comet" gyratory crusher. It closely resembled the Brown machine being built by Gates Iron Works. Over the next few years an intense legal battle followed, finally ending in U. S. District Court when Gates lost the suit. Fraser & Chalmers claimed in their 1901 advertising that the Comet Adjustable crusher was the only gyratory machine on the market which permitted variation in product size without changing the internal liners. In fact, product size could be changed simply by turning the large handwheel seen to the lower left in this photograph. Seven production models of the Adjustable Comet were available for 1901, with the largest of them having a capacity of 100 to 200 tons of rock or ore per hour, depending on the character of the material. The huge Model G crusher required 150 horsepower to operate — it weighed in at 102,000 pounds.

Crushers

The Non-Adjustable Comet crusher was virtually identical to the Adjustable Comet, save for this feature alone. In the machine shown here, adjustment was available only to the extent of compensating for wear of the liner itself, giving no opportunity for changing product size. Fraser & Chalmers highly recommended this machine for ore reduction purposes, while promoting the adjustable machine for quarry work where various sizes and grades of rock were desirable. Advertising literature noted that in the larger sizes, entire car loads of ore could be dumped into the hopper of this machine. Actual capacity hinged on available horsepower, with the company indicating a rule of thumb that required one available horsepower for every ton of material to be reduced per hour.

The original Gates gyratory crusher design was designated as the Style "B" machine, but it had the major disadvantage or requiring the dismantling of the entire upper part of the crusher for repairs to the eccentric and bevel gear in the bottom of the machine. Finally, Gates solved this problem by introducing the Style "D" crusher in 1895. It was virtually identical in design to the Style B, but offered a removable bottom plate which permitted removal of the eccentric without a major teardown of the machine. When Gates Iron Works became a part of Allis-Chalmers Company in 1901, the Style D crusher immediately became part and parcel of the new A-C crusher line.

For laboratory or assaying purposes, Fraser & Chalmers offered the adjustable Comet in a tiny size, noting that it was ideal for the prospector after his work in the field. The same adjusting features were built into this machine as were found in the larger models. Weighing 1,200 pounds, the Laboratory model used a 3 x 7 inch feed opening and required but 2 horsepower for operation. The Allis-Chalmers merger of 1901 brought Fraser & Chalmers together with Gates Iron Works. Merging of these two strong, but competing product lines gave the Allis-Chalmers organization an immediate position of dominance in the crusher business. Despite fierce competition from rival firms, the A-C dominance of the crusher industry endured for decades to come.

In 1901 Gates Iron Works introduced the Style "K" rock and ore breaker, following the gyratory design as before, but borrowing the McCully innovation of a top spider bearing for the crusher's main shaft. In so doing, Gates was able to build a somewhat larger machine than before — the largest Style K had a 21-inch feed opening. Also of interest, the McCully gyratory crusher design was then built by Power & Mining Machinery Co. at Milwaukee. This firm was one of the strongest competitors with which Allis-Chalmers had to contend — that is, until 1924 when A-C purchased this division from the Worthington Corp.

Crushers

Recognizing the need for portability, this Gates gyratory crusher is mounted on steel trucks. In localities where rock was plentiful, moving the crusher from one site to another during a season shortened the hauling distance of finished product. In other areas where rock, especially limestone, was scarce the supply was often quite small at a given site, requiring rather frequent moves of the crushing equipment. With the advent of large steam traction engines in the early 1900s, moving the crusher from one site to the next was relatively simple, even though most of the other work was still done by hand. Allis-Chalmers continued the practice of building portable crushers for many years.

Power & Mining Machinery Co. developed the McCully gyratory crusher in 1910. Ten years later this firm announced the Superior McCully Fine Reduction machine. After acquiring Power & Mining Machinery Co. in 1924, Allis-Chalmers announced that they were currently building this machine in 6 and 10-inch sizes, with an 18-inch model being in development. The smaller model offered two 6 x 40 inch openings, giving it a maximum capacity of 24 tons per hour with ¾ inch product, or 48 tons per hour with 1½ inch product. The larger machine featured two 10 x 52 inch openings for a maximum capacity of 135 tons per hour reduced to 2½ inch product. This machine required approximately 100 horsepower for successful operation. Pressure lubrication was featured, and steel castings were used throughout.

A photograph from a 1913 catalog of Power & Mining Machinery Co. illustrates the No. 1 and No. 10 McCully gyratory crushers. Even the small No. 1 machine is nearly as tall as the workman in this photo, with both the men and the machine being dwarfed by the huge No. 10 model. Weighing 180,000 pounds, the No. 10 had a capacity of 210 tons per hour and required about 150 horsepower. Available, but not illustrated were three additional sizes, larger yet than the No. 10 machine. Of these, the Mammoth with its 42 x 136 inch feed opening is listed — it had a capacity varying from 700 to 1,300 tons per hour and required on the order of 300 horsepower. Total weight of this unit was 425,000 pounds.

In late 1922, Allis-Chalmers announced a new style portable machine using the Style N, No. 4 machine. This was the first such unit mounted as a portable, and it had given satisfactory service over the previous summer at a local gravel pit. For portable installations, an elevator was also furnished to carry product to portable holding bins, but due to the length and size of the elevator, it was furnished as a separate component, rather than being mounted as an integral part of the crusher itself. A single photograph portrays the immensity of these machines rather poorly, although the massive truck wheels convey a vague idea of the huge size and weight involved with this machine.

Crushers

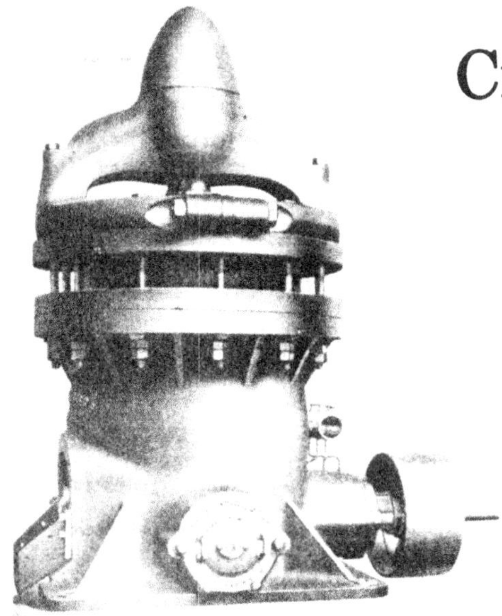

Allis-Chalmers introduced the type "C" crusher in 1926. It followed the Superior McCully design and likewise embodied many features found in the A-C Type "K" crusher. The major difference between these machines was the extra heavy design of the Type C crusher. It was designed for the very worst conditions of east coast operation where extremely hard trap rock was encountered with A-C noting that this machine weighed, size for size, about 20% more than a comparable Type K crusher. Likewise, the main shaft, drive shaft, and gearing were much heavier to accommodate a slightly larger crushing head and somewhat higher operating speed. The Type C was designed to be a primary reduction machine, with further reduction being made in other equipment.

Shown here is a McCully gyratory crusher with the large bevel drive gear exposed, illustrating the relative ease with which this component could be serviced or replaced. Also shown is the access door through which the gear drive could be adjusted or serviced. The McCully design was originated by Robert McCully of Philadelphia, with Power & Mining Machinery Co. acquiring world rights to manufacture this machine about 1910. Eventually, Power & Mining Machinery Co. became a division of the Worthington Corp., but in 1924 Worthington sold this division to Allis-Chalmers and left the ore and mineral processing business to concentrate on their pumps, engines, and other equipment.

With the development of steam shovels and other means of handling large volumes of rock and ores, the size of crushing equipment increased dramatically. By 1907 Allis-Chalmers was building a gyratory unit with a 36-inch radial receiving opening — this increased to 42 inches by 1910, and to 54 inches in 1917. During 1919 Allis-Chalmers reported that their two huge No. 27 crushers installed previously for Utah Copper Co. had been giving excellent service. Introduced in 1917, the No. 27 used the dual 54-inch opening previously noted, and was built with a dual discharge. Manufacturing problems abounded, especially since the top casting was too large to be shipped in a single piece. Thus, a redesign was necessary whereby the top casting could be built in two separate pieces for field assembly. Allis-Chalmers noted that their machines had a distinct advantage over competing units, in that the main shaft was forged under a hydraulic press in the company's own shops, and from this process on to the final finishing, quality control was evident at every step to insure the highest quality and longest possible service. The machine shown here was equipped with manganese steel heads and concaves, but for use in limestone it could be equipped with chilled iron instead. Capacity of the No. 27 ranged from 500 to 900 tons per hour. In this photograph, a workman standing next to the discharge chute is completely dwarfed by the size of the machine.

Agriculturists saw the need for lime in the seventeenth century, and perhaps earlier. Crushed limestone was not a practical matter though until suitable crushing machinery was developed in the mid-19th century. With the advent of successful crushers and the development of small farm tractors, Allis-Chalmers introduced the L.B.H. Hummer in 1916. The initials stood for (L)ow Feed; (B)ig Rock; (H)igh Discharge. In reality, this completely portable outfit was a series of machines consisting of a 12 x 7 inch jaw crusher, an intermediate conveyor to a "Hummer" pulverator — a hammermill that further reduced the size of the material, and finally to a discharge auger. Capacity ranged from 1½ to 3 tons per hour, depending on the size of the finished product and the physical ability of the operator.

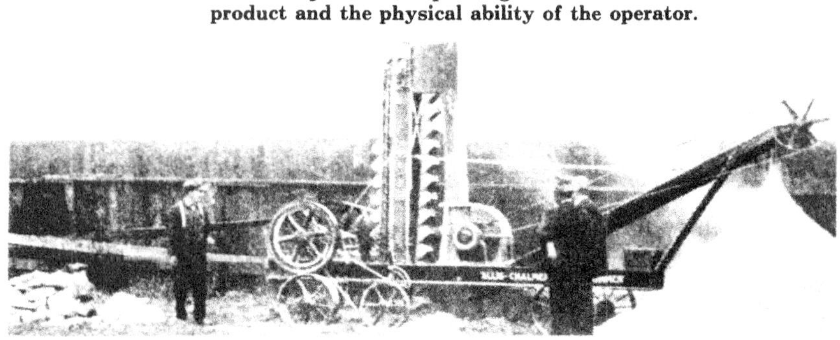

Crushers

Although the original photograph is of rather poor quality, an accompanying article of 1930 indicates that this huge machine is but one of three 60-inch Superior McCully crushers sold during February of that year. Two of these crushers were destined for the Anglo Chilean Nitrate Corp. in Chile. They were of cast steel construction throughout, and arranged for direct-connection to electric motor drives. Another of the 60-inch machines was sold during February, 1930, to Inland Lime & Stone Co. of Inland Junction, Mich. It was to be operated by a 250 horsepower Allis-Chalmers induction motor running at 2,200 volts. The company noted that at this time, February, 1930, they had thus far built six of these giant 60-inch machines.

In April, 1930, Allis-Chalmers announced the No. 248 Fine Reduction Jaw Crusher. Among the many design features were Timken bearings on the eccentric shaft, easy, on-the-go adjustment of product size, and extensive use of manganese steel castings. The outer frame, visible in this illustration was a one-piece steel casting with heavy reinforcing ribs for absolute strength and rigidity. The 24 x 8 inch jaw opening could process 5 tons per hour at a ¾-inch finished size, with this figure ranging up to 30 tons per hour for 2-inch product. Approximately 35 horsepower was required to operate this machine. It weighed 12,500 pounds.

In November, 1926 Allis-Chalmers announced the completion of two new crushers. Built for Chile Exploration Co., a subsidiary of Anaconda Copper Mining Co., these two machines were shipped to Chuquicamata, Chile. At the time, these two crushers were the world's largest, having a capacity of 2,000 to 2,500 tons per hour for EACH machine. The huge feed opening could absorb an entire 70 ton carload of ore at one time! A workman sitting on the rim of the feed opening gives some relative idea of the tremendous size of these crushers. For some years these remained the world's largest crushers, although they eventually gave way to even bigger machines.

Another crusher, the Newhouse design, first appeared in 1929. While the original design held some engineering difficulties, this machine began the "oil adjusted" crusher design. Allis-Chalmers spent the next decade refining and improving the idea until they developed the "Hydroset" control which became standard equipment on all Superior Gyratory and Hydrocone crushers. During the 1930s Allis-Chalmers redesigned the entire machine with a shallower cone angle than was used in earlier gyratories and went to great lengths in redesigning the crushing head on scientific principles. These engineering efforts resulted in vastly improved designs by 1940, but due to the imminent onset of World War II, many of these improvements were not fully implemented until 1947. Newhouse crushers placed the electric motor drive above the top support spider, and with the top-drive all gearing and countershafts were eliminated.

Crushers

During the 1930s Allis-Chalmers offered the Fairmount single-roll crusher. This machine was intended for limestone, dolomite, phosphate rock, and other less tenacious materials. It was not at all suited to granites or other very hard materials. Steel plate and manganese steel castings were featured throughout, and nests of heavy springs were required to absorb excessive shocks and pressures. Single roll machines operated against a curved anvil lined with chilled iron or manganese concaves. Great capacity was a distinct feature of the single-roll machine, with this particular model having a 60 x 84 inch feed opening.

Allis-Chalmers achieved a first with the introduction of the hydraulically controlled gyratory crusher in 1936. The following years saw further developments, although the manufacturing efforts were hampered during the World War Two years. However, in 1953 Allis-Chalmers set another record by building the world's largest gyratory crusher for Reserve Mining Co. for installation at the Babbitt taconite plant on the Mesabi Iron Range. The crusher is capable of crushing taconite at the rate of 3,500 tons per hour from pieces as large as 10 x 60 inches. Requiring over 1.25 million pounds of steel, this crusher has four major steel castings in the shell that weighed from 120,000 to 200,000 pounds each, and has walls up to 8 inches thick. The main shaft of the crusher, a steel forging about 27 feet long with a maximum diameter of 4½ feet, weighs over 100 tons.

Shortly after completing the huge 3,500 ton gyratory crusher for the Mesabi Range, Allis-Chalmers announced the first 84-inch Hydrocone crusher built in the Allis-Chalmers plant in Newcastle, Australia. Shown here in its final test assembly, this machine towers over 15 feet high. This particular unit was destined for a new installation at Jarrahdale, Western Australia, to prepare bauxite ore for processing at an alumina refinery in Kwinana, Australia. Along with this huge crusher, Allis-Chalmers of Australia also provided additional crushing and screening equipment on the same site.

A 1955 photograph illustrates the 60-109 primary crusher spider being shipped on a special depressed well railroad car to the Mesabi Iron Range. This huge machine, capable of crushing over 3,500 tons of ore per hour was announced in 1952, but engineering design, patternmaking, casting, and final machining and assembly processes delayed final delivery for over two years. Considering however, the enormous amount of work involved in producing this huge machine from the drawing board to the finished product, the ensuing time was somewhat less than might have been expected. Even though the Allis-Chalmers shops contained some of the largest machine tools to the world, the huge task of building this gigantic machine must have pressed even these facilities to their ultimate limit.

Cultivators

Paralleling the announcement of the new Model UC, "All-Crop" tractor in 1930, Allis-Chalmers also announced a cultivator built specifically for use with this tractor model. As initially offered, this unit was available in the choice of two or four-row models, with the four-row style being illustrated here. To help distribute the weight of the four-row model, caster-type support wheels were used. They were placed to run mid-way between the individual cultivator gangs. In their 1930 advertising, A-C claimed theirs to be the industry's first quick-detachable unit, stating that only ten minutes was required to mount this cultivator, with another ten minutes being required for its removal. To facilitate raising and lowering of the cultivator a power lift, operated by a foot pedal, was provided.

The 1930s saw a rapid fire series of tractor developments at Allis-Chalmers. Most significant to the industry and to agriculture in general was the introduction of rubber tractor tires by A-C. By 1935, Allis-Chalmers offered cultivators in three distinct groups, the No. 20 series to meet the needs of northern row crop farmers; the No. 40 series for southern territories, and the No. 50 series for California and eastern Texas territories. Pipe beam gangs were featured in the No. 50 series cultivators — this permitted shorter gangs with a total of eight spring-trip shanks. Also available were 2 and 3-row listed crop cultivators in a pull-type design.

For the southern farmer, Allis-Chalmers offered the 250 Bedder and the 450 Listing Plow in 1935. Separate lever controls handled each gang independently, with the gang spacing being fully adjustable on the tool bar. With their 1929 acquisition of LaCrosse Tractor Co. of LaCrosse, Wis., Allis-Chalmers immediately gained the benefit of many years experience in cultivator and lister design. While maintaining many of the earlier LaCrosse designs, Allis-Chalmers built on past experience in offering new and innovative tractor-mounted designs such as the model illustrated here.

Tractor designs changed during the years, as illustrated by this late-model WC tractor. Cultivator designs improved as well, although this illustration is of the 20 and 40 series cultivators introduced some years before. A 1947 price book indicates that the No. 20 series was available as the No. 21-B or the No. 22-B, both equipped for power lift operation, and not available with a hand lift. The difference between the two model numbers consisted in various shovel configurations, with this entire series carrying a price of $156. The mulching attachment was an extra cost item priced at $20. This attachment was placed behind the rear wheels to erase tire marks which otherwise would provide an excellent channel for surface runoff.

Cultivators

Allis-Chalmers offered an extensive line of No. 60 series cultivators for the Model B tractor. The Nos. 61A, 62B, and 64 units were of one-row design spaced from 32 to 44 inches. A No. 65 series permitted two rows of 18 to 28 inch spacing, or a single row spacing of 32 to 44 inches. The No. 66 was a four-row unit with a spacing variable from 18 to 24 inches. A unique feature of these various styles was that they could be field converted into virtually any of the other styles if the situation demanded. An exceptional variety of attachments was available for the No. 60 series cultivators, including the No. 600 fertilizer side dressing attachment, and the special bean cutter and vine turner attachment shown here. This unit carried a cutting blade about 35 inches long, attached to an adjustable standard. A hand lift was standard equipment, but this series could be supplied with a power lift for an additional $5.50. Depending on the size and the equipment supplied, this series ranged in price from $62 to $73.

While the No. 60 series cultivators were built specifically for the Model B tractor, the No. 80 series was designed to fit the Model C tractor. The two-row model with 8 round-shank spring-trip shanks and spare point shovels listed at $96, while the No. 82 with 14 two-ply spring teeth was priced at $104. The largest model, the No. 86 four-row style carried a list price of $90, but did not include the attachments supplied as standard equipment with the two-row models. An exceptional variety of attachments could be supplied. These included various shovel sizes and styles for vegetable cultivation. Lift equipment was not included in the base price, since the No. 60 series could be supplied in either a hand lift or as a power lift unit. Carrying the concept of interchangeability even further, A-C offered a series of front-mounted planters using the No. 80 cultivator frame. Shown here is one such unit, with the cultivator beams in plain view. A typical planter attachment as shown in this photograph carried a price of $160 to $200 over and above the price of the cultivator.

A 1947 Allis-Chalmers price list shows that four different models were available within the No. 50 series cultivators. Built specifically for the A-C WC tractor, these were shipped from the factory with a power lift, with no hand lift being available. The No. 59 Disc Cultivator shown here used a total of twelve 14-inch discs in the front, and three 10-inch full sweeps on the rear of the tractor. This equipment listed for $189. The No. 58, identical in all respects except for the lack of the rear sweeps was priced at $149. For the midwestern farmer who preferred a shovel-type cultivator, Allis-Chalmers offered the No. 55 and No. 56 models with respective prices of $225 and $185.

Bedding is a process similar to ordinary plowing, except that the dirt is thrown both ways from the center instead of being laid to one side as with an ordinary moldboard plow. This leaves series of furrows and ridges which in some cases are again leveled with a process called rebedding. Allis-Chalmers offered various bedder styles over the years, typical of which is the No. 151 bedder built specifically for the Model B tractor. The No. 151 was so designed and so built that it became an integral part of the tractor. This attachment was available as late as 1956, carrying a list price of $40, although this figure did not include lifting equipment or the plowing bottom. To accommodate various demands, a variety of these were available and could be ordered at the same time as the bedder itself.

Cultivators

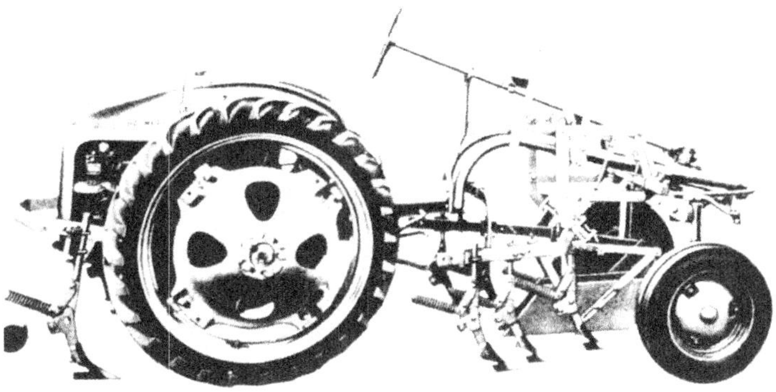

With the introduction of the Allis-Chalmers Model G tractor came a host of attachments to fit virtually any situation, with special attention being given to the needs of truck farmers, nurserymen, and orchardists. The G-1 spring-trip cultivator shown here carried a list price of $52. Pin-break trips could be substituted for the spring-trip design — this gave a price deduct of $14. Lift equipment could be hand or power operated at the purchaser's option. In addition, a wide variety of shovel shapes were available to suit individual requirements. Extra-cost options and attachments included a fertilizer side-dresser for $54.

When Allis-Chalmers introduced the D-10 tractor in 1959, the new 110 Series cultivators appeared concurrently. Of front mounted design, the 110 Series offered round tubular front gangs and a rear furrowing bar. Intended primarily for the truck gardener, tobacco farmer, and similar applications, the 110 Series cultivators could also be supplied with a $42 fertilizer attachment for sidedressing the crop. Depending on the type and number of shovels used the price ranged from $187 to $200. The standard hydraulic lift package added another $12, while the delayed lift system which permitted independent control of the front gangs and the rear furrowing bar added another $19 to the overall cost. Shown here is a D-10 High Clearance model with the one-row Series 110 cultivator.

While the 110 Series cultivators for the D-10 were available only as a one-row style, the 120 Series cultivators for the D-12 were built as a two-row design. Depending on the shovel equipment and other accessories, the price ranged from $159 to $292. This front mounted unit could handle row spacings from 28 to 42 inches, with the two fertilizer units shown on this tractor adding another $84 to the base price. Special attachments included various types of shields to accommodate different crops. This D-12 is cultivating a large field of carrots near San Antonio, Texas.

During the 1950s and early 1960s Allis-Chalmers followed an aggressive pattern in cultivator design, offering new models that were in keeping with contemporary technology and in harmony with current tractor models. This four-row design is mounted on a D-15 Diesel tractor, and is operating at the Allis-Chalmers farm near Racine, Wis. On this large acreage, Allis-Chalmers actively pursued the improvement of their entire farm equipment line through actual field trials and working under normal field conditions, rather than depending on calculated results or laboratory tests.

Cultivators

This D-17 tractor has its 601 Series four-row cultivator in the folded position to illustrate the ease with which it could pass through narrow farm gates. The 600 Series included two different styles, and the 660 Series of similar front-mounted design was also available in several configurations. A major difference in these designs was that while the 600 Series was built for the D-17 tractor, the 660 Series could be fitted to either the D-17 or the D-19 models. While the 601 Series cultivator shown in this photograph carried a 1962 list price of $1,075, the 660 Series cultivators ranged in price from $1,219 to $1,318, depending on the configuration and mounting equipment.

During the mid-1960s Allis-Chalmers offered a series of rear-mounted cultivators, including the 500 Series available in 2, 4, 6, and 8-row designs. Shown here is a No. 2541, 8-row rear mounted unit on an Allis-Chalmers D-21 tractor. Designed for Category II or III 3-point hitch, the 2541 included two 18-inch blade-type stabilizers to minimize side sway. The basic unit carried a 1967 price of $1,852, with a hydraulic fold-up adding another $77. Also required was a tool bar carrier — in the Category III size this listed for an additional $90. Since a large variety of sweeps were available, these two were priced in addition to the base price, as were any shields that might be required. The unit shown here is working in soybeans.

Prior to ending cultivator production in 1983, Allis-Chalmers continued offering a wide range of styles to fit virtually every farming method and every major row crop. The 1983 line included the 90 Series units of rear-mounted design, and available in 4, 6, and 8-row configurations. The Model 95, built for 31 to 40 inch row spacings carried a list price of $4,157 on the 6-row model or $6,906 for the 8-row cultivator — this model included a hydraulic fold-up as standard equipment. Model 10 toolbar cultivators as shown here, were available in 4, 6, and 8-row models as a non-folding version, while the hydraulic fold-up style was built only in 6 or 8 row configurations. The Model 10 series ranged from a low of $2,046 for a 4-row non-folding model up to $4,919 for an 8-row hydraulic fold version.

In keeping with current agricultural trends, Allis-Chalmers introduced a rotary cultivator in 1971, and according to company records discontinued its 100 and 200 Series rotary cultivators the following year. Designed to pulverize crusted soil, the rotary design also had the advantage of leaving uprooted weeds on top of the soil where the summer sun quickly ended their threat of damage to the young crop. Depending on the setting of the rotary wheel units, dirt could be thrown toward the row, away from the row, or could be turned completely around for heavy soil churning action. Each spider gang could be equipped with 3, 4, or 5 wheels, depending on the row spacing. Company advertising claimed that the rotary cultivator design could be operated at field speeds up to 10 mph.

Defoliators

During the 1967-1971 period, Allis-Chalmers offered a defoliator in two basic sizes. The Model 90, later known as the Model 900 gave a 90-inch cutting width, while the Model 128, later designated as the Model 1280 had a 128-inch cutting width. Both were double-drum units, although on special order the Model 90 could be supplied as a triple-drum machine for true one-pass defoliation. These machines were primarily intended for the sugar beet industry, doing a thorough job of cutting and removing the beet tops and depositing them to one side of the machine. By so doing, the next process of actually lifting the beets could be accomplished quickly and efficiently. During 1968 the Model 128 Double Drum Defoliator listed at $2,255, while the smaller Model 90 in a double-drum design listed at $2,020. Adding the optional third drum to the Model 90 raised the list price by another $573.

Dickson Mfg. Co.

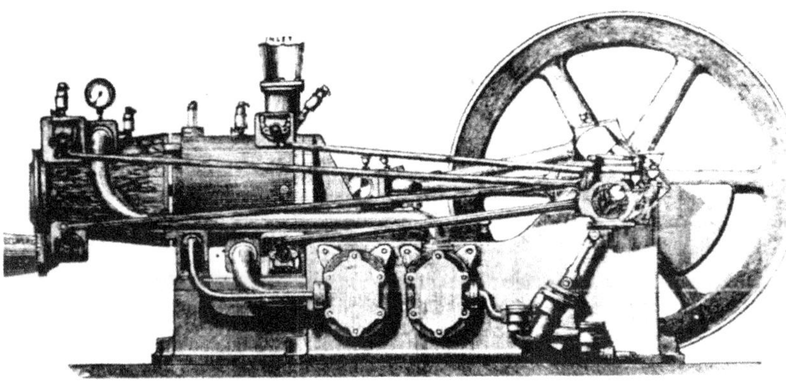

Although historians generally have given more than ample coverage to the participants in the Allis-Chalmers merger of 1901, Dickson Mfg. Co. of Scranton, Penn. usually receives short shrift. An 1895 Dickson catalog illustrates however, that this company was indeed innovative, even offering refrigerating machines as part of a broad general line that included locomotives, blowing engines, hoisting and pumping engines, coal breakers, and mine ventilating fans. In 1895 at least, Dickson manufactured the Stockport gas engine under license from the British owners. Their "Universal Dry-Air Refrigerator" shown here had been around for ten years or more, having been originally developed by T. B. Lightfoot of London, England. This unique machine used no chemical refrigerants of any kind, but instead it used an air compressor system, utilizing the cold air delivered by the release of highly pressurized air down to atmospheric pressure. Dickson offered this machine in nine different sizes having ice equivalents per 24 hours of 960 to 144,000 pounds. With an ambient temperature as high as 90 degrees F., it was capable of temperatures ranging from -30 to -70 degrees F. The smallest machine required about 4 horsepower, while the largest needed 500 horsepower. These machines could be belt-driven, or could be supplied with an integral steam engine.

Disc Bedder

Beginning in the 1930s Allis-Chalmers offered bedders, bedder planters, and similar equipment to meet the needs of southern growers. Typical of this rather extensive line is the rear-mounted disc bedder shown here on a A-C D-19 tractor. This rear-mounted design was available in 1956 for instance, with either a 4-row or a 6-row design, carrying respective prices of $139 and $153. Front-mounted bedders of that year could be furnished for either the D-19 or the 190 tractor models, and offered more options than did the rear-mounted style. Options for the front-mounted series included a cultivator attachment, covering frames, planter attachment, and seed press wheels.

Disc Harrows

The 1929 purchase of LaCrosse Tractor Co. of LaCrosse, Wis., put Allis-Chalmers squarely into the business of building farm tillage implements. LaCrosse Tractor Co., formerly known as LaCrosse Plow Co., was established in 1865, and incorporated in 1893. Under the leadership of A. Hirshheimer, its founder, LaCrosse became a major builder of plows, cultivators, and other implements. When A-C took over in 1929, this factory became their LaCrosse Works, specializing in tillage equipment. Initially, Allis-Chalmers continued the LaCrosse line with little change, including the tandem disc harrow shown here for tractor use. It was available in 7, 8, 9, or 10-foot sizes.

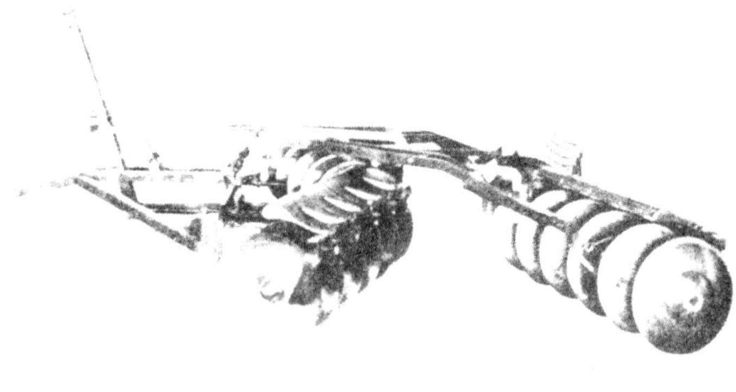

To coincide with its existing tractor line and to match the new "WC" tractors, Allis-Chalmers offered this 8-foot tractor disc harrow in the 1930s. It followed the same general lines as the earlier LaCrosse models. Farming methods of the 1930s included moldboard plows which tended to lay the furrow over in a continuous slab rather than having the pulverizing action found in high-speed, high-lift moldboards of later years. Consequently the field surface after plowing was in need of considerable leveling with the disc harrow — this often required two or more passes, even with the tandem model shown here.

The Allis-Chalmers disc harrow line also included single discs in addition to the tandem line. Many farmers of the 1930s preferred this style, often "lap-discing" the field. In this process, after making a full-width pass across the field, subsequent passes moved over half the width of the disc harrow, thus double-discing the field as a single operation. This 14-foot Allis-Chalmers model of the early 1930s typifies the A-C line at that time — it, like competitive makes, used blades, about 16 inches in diameter. Due its light weight and small blades, among other reasons, these machines had nowhere near the penetration and aggressiveness found in later developments.

A 1930s photograph illustrates an Allis-Chalmers "UC" tractor pulling a No. 8 tractor disc harrow. This is probably the second pass over the field in preparation for planting. The No. 8 Series disc harrows were built in 5, 6, and 7-foot sizes to accommodate varying tractor sizes, with 8 and 10 foot models available to farmers having sufficient power to handle these machines. These models were all available with either 16 or 18-inch blades, with the factory list price ranging from $110 to $179, depending on the width and blade size. No. 8 single acting disc harrows were offered in 8 and 10-foot sizes for 1940 — again, depending on the width and blade size, factory list prices ranged from $73 to $93.

Allis-Chalmers pioneered the snap-coupler hitching method in 1953, first offering it on the CA and WD tractors. With the 1955 introduction of the WD-45 tractor, the snap-coupler was firmly established as an easy and reliable method of attaching various farm implements. Included was the No. 20 disc harrow shown here on a WD-45 LP-gas model. No. 20 mounted disc harrows were available in 5½, 6½, 7, and 8-foot sizes. Front tractor weights were a necessary requirement, although at slight extra cost. Prices for 1957 ranged from $297 to $289, depending on the width and blade size selected. This particular unit is shown operating somewhere in Pacos County, Texas, about 1956.

Disc Harrows

Another mounted disc harrow of the 1950s was the No. 30. single-acting style. Apparently equipped with 20-inch blades, this one is tearing up some old grass land near Athens, Tenn., with an A-C WD-45 tractor providing the power. Note the front wheel weights on this unit. Shortly after World War II, Allis-Chalmers announced the 11-B disc harrows in a pull-type version. These double-acting machines were built in 5½, 7, and 8-foot sizes. The 11-B series remained on the market for several years, carrying a 1957 list price that ranged from $236 to $272.

Of extra heavy construction, the Series 40 mounted disc harrows were suited to the WD, WD-45, and later to the D-14 tractors when equipped with the patented Snap-Coupler design. Built in 6 and 7-foot models, it could be furnished with smooth discs front and rear, cutaway discs front and rear, or front cutaway blades and smooth rear blades. This unit, operating near Fayetteville, N.C., appears however to have smooth front and cutaway rear blades. To get a better idea regarding its heavy construction, A-C literature notes that the Series 40, 6-foot model with a 9-inch blade spacing and sixteen 22-inch blades weighed a total of 1,221 pounds. Pricing for 1957 ranged from $386 to $433. This particular WD-45 does not appear to have any additional front end weight.

The A-C No. 30 Series Disc Harrows were built at the company's plant in Athens, Ga. An example is the No. 36 mounted style shown here on a D-10 tractor. These single-acting units were available in 6 and 7-foot models carrying respective list prices for 1957 of $358 and $362. Additional cost options included adjustable blade scrapers for $30. Smooth or cutaway blades were available, with this unit working near San Antonio, Texas, illustrating the latter style. Cutaway blades offered greater penetration and highly increased aggressiveness, an important necessity when turning over heavy stands of grass.

Originally intended for use with the WD, WD-45, or D-14 tractor equipped with the Snap-Coupler, this No. 109 disc harrow was also an excellent match for the D-15 tractors. In this case a D-15 Series III tractor is shown with a No. 109 disc harrow. This model was built only in an 8½ foot size, but could be equipped with the choice of 16 or 18-inch smooth blades or alternately with 18-inch cutaway on the front and smooth blades on the rear gang. When so equipped the No. 109 carried a factory list price of $366, compared to $338 for the smooth blade version. This model used a 7⅜ inch blade spacing.

Allis-Chalmers No. 200 Series disc harrows were ordinarily supplied with a standard tongue for attachment to a tractor drawbar, but alternatively could be equipped for direct connection to the Snap-Coupler. For the WD-45 and larger A-C models this style was built in a 10-foot size offering the choice of 16 or 18-inch smooth blades, plus the option of front cutaway and rear smooth blades in an 18-inch diameter only. Whether using a standard drawbar hitch or the Snap-Coupler, a remote hydraulic cylinder was required to raise and lower this implement. Prices for the No. 210 ranged from $583 to $614.

Disc Harrows

The No. 213 disc harrow was built in a single width of 12½ feet. With this exception it was identical in design to the No. 210 model with a 10-foot width. Using a standard 7⅞ inch blade spacing, the No. 213 carried a total of 40 blades, either of 16 or 18 inch size, as selected by the purchaser. Pricing for this model with the 16-inch blades was $583, with this figure rising to $606 when the 18-inch blades were used. This photograph, taken near Guthrie, Okla., shows a WD-45 LP-gas model pulling a No. 213 disc harrow equipped with the Snap-Coupler hitch.

A photograph taken at the Allis-Chalmers farm near Racine, Wis., shows a pair of No. 213 disc harrows behind a squadron hitch. In an attempt to provide better utilization of existing equipment against dramatic increases in tractor horsepower, the squardron hitch was devised. While the concept was used to some extent in certain areas, farmers generally preferred integral implement designs without the extra bother of changing machine position for road transport or storage. Thus came disk harrows with widths of 25 feet or more in a single unit. By the use of a hydraulic folding system, these machines could be reduced to half their operating width for road transport.

This unusual photograph was taken at the Allis-Chalmers Farm near Racine. Shown is the Allis-Chalmers T-16 tractor pulling a pair of No. 214 disc harrows. They are attached to an A-C No. 2 Squadron hitch. Apparently the No. 2 hitch was designed so that one disc harrow followed the other when traveling on the road. As the model designation implies, the No. 214 was part of the larger 200 Series disc harrows. At the small end of the line was the No. 209 with a width of 8½ feet. It was priced at approximately $550. The No. 215 was 15 feet wide — it sold in the range of $780 to $1,130. The No. 214 style shown here had an operating width of 13½ feet, and could be supplied with 16, 18, or 20-inch blades. It was available in either a solid frame or a fold-up style, with pricing having a range of $722 to $1,087. Although these units were ordinarily supplied with a standard tongue for drawbar hitch, they could be easily converted for the A-C patented Snap-Coupler system if desired.

A-C No. 215 disc harrows for 1962 could be supplied with a regular solid frame or in a hydraulic fold-up version. Also available was the choice of 16, 18, or 20-inch blades, these then could be furnished as smooth blades, or optionally with cutaway front and smooth rear blades. Depending on blade size and other options the price ranged from $781 to $1,131. Also available as an extra-cost option was the Snap-Coupler hitch in lieu of the standard drawbar for an additional $65. A special Traction Booster drawbar and hitch for use with the Snap Coupler added yet another $110. The Series 200 disc harrows were built in the 1957-1966 period. This No. 215 disc harrow is behind an A-C D-17 tractor.

Yet another example of the Series 200 disc harrow is the No. 213 model shown here behind an A-C WD-45 Diesel tractor. Like other models of the 200 Series, the No. 213 could be furnished with 16, 18, or 20-inch blades at the option of the purchaser. Depending on blade size, this unit weighed from 1,800 to 2,000 pounds. This placed considerably more downward pressure on the individual blades, giving them far greater penetration than had been possible in the two-section disc harrows of the 1930s. Increased tractor horsepower permitted higher field speeds, and this promoted the churning action necessary for reduction of rough ground into a smooth seedbed.

Disc Harrows

Series 200 offset disc harrows were classified as a grove-type harrow and were designed especially for orchards and nurseries. The No. 308 had an operating width of 7½ feet and was furnished with 16-inch grove-type blades with depth pans. It listed at $455 in 1962. The No. 308 could also be furnished from the factory in a 10-foot size, using the same blade equipment as previously noted. The 10-foot model could however, be field converted to a 14-foot model at an additional cost of $271 over the base price of $875. Shown here is a Series 300 disc harrow behind a special D-14 Allis-Chalmers Orchard tractor. It is operating in an orchard near Bangor, Mich.

Allis-Chalmers successfully promoted the Snap-Coupler system as an easy means of attaching and detaching various implements. Likewise, the patented A-C Traction Booster system effectively transferred drawbar pull into greater traction. The combination worked very effectively to make farming operations easier and more economical — easier because of less time and labor, and more economical due to the fuel savings derived from the improved traction provided by the Traction Booster system. This 1960s photograph illustrates an 8½ foot A-C mounted disc harrow behind an A-C D-15 Series III tractor.

Built in the 1961-1969 period, the 2000 Series disc harrows were offered in 13½, 18½, and 20½ foot sizes. The No. 2020 shown here was built only in a fold-up style, with available options including the choice of 16, 18, or 20-inch blades. Depending on the blade size, weight varied from 3,260 to 3,530 pounds. Pricing also varied with the type of blade used — a full set of 16-inch smooth blades gave a factory list price of $1,416. Equipping the No. 2020 with 20-inch cutaway blades on the front and smooth blades on the rear gang gave a factory list price of $1,617 in 1962. Here an Allis-Chalmers D-21 tractor is pulling a No. 2020 disc harrow working in corn ground near Wharton, Texas.

Like the No. 2020, the No. 2018 disc harrow shown here was available only in a hydraulic fold-up version. Only the No. 2014 model could be furnished with a regular full-length, non-folding frame, although the No. 2014 could be converted to a No. 2020 fold-up unit by purchasing a $550 field conversion kit. All three models could be furnished at extra cost with Snap Coupler and/or Traction Booster conversion kits. Standard equipment on the 2000 Series disc harrows included BAL-PAK anti-friction bearings in the disc gangs, adjustable disc blade scrapers, and hydraulically controlled dual wheels for transport and depth control. This No. 2018 disc harrow is being pulled by an A-C D-19 tractor.

While the majority of A-C disc harrows came from LaCrosse Works, those of the heavy duty variety were built at Oxnard (California) Works. Typical of the Oxnard production is the Model J offset disc harrow shown here. Nine different size were available, ranging in width from 6 to 18 feet. All could be equipped with the choice of 22, 24, or 26-inch blades, and while smooth blades were standard equipment, cutaway blades could be supplied at extra cost. In 1956, a 9-foot Model J disc harrow equipped with 24-inch smooth blades carried a factory list price of $704. Substituting cutaway blades for smooth blades added another $43 to the list price. The price did not include scraper equipment, nor did it include lifting equipment — this could be hand or hydraulically operated at the choice of the purchaser. Squadron hitches were available for coupling two equal sizes units up to a 12-foot width per unit or a maximum operating width of 24 feet.

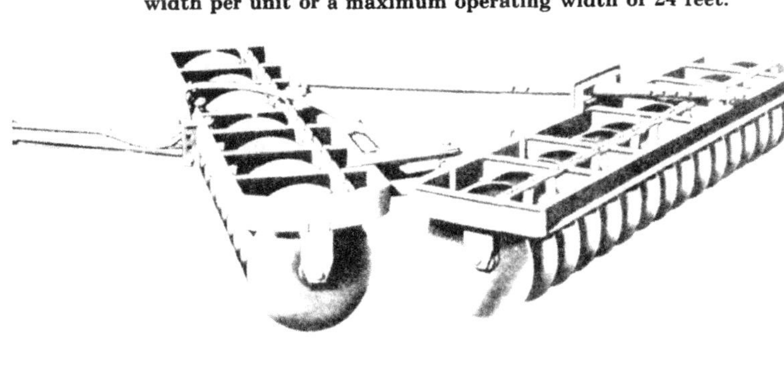

Disc Harrows

Model K disc harrows were built between 1968 and 1973. Initially they were offered in seven models ranging from 6 to 10½ feet in width, but after Serial No. 1400 the 12 foot and 13½ foot models were added. As this photograph demonstrates, the offset gangs provided aggressive action, even in very heavy plant growth or surface residue. This particular model is 10 feet wide, and is being pulled by an A-C WD-45 tractor. The model shown here carried a list price of approximately $827 in 1962, with the small 6 foot model being priced at $564. A 22-inch diameter blade was standard equipment on this series, although 24-inch blades could be supplied at extra cost.

Offered from 1968 to 1982 the Style L disc harrows were available in seven different sizes ranging from 9 to 21 feet in width. Although 22-inch blades were standard, 24 or 26 inch blades could be furnished either as an original order option or for later conversion. Pricing varied from $1,000 to $2,250 depending on size. Specifically, the 18 foot model was priced at $1,860, but this did not include the main pole priced at $100, nor was the hydraulic control equipment supplied — this added yet another $292, and a mechanical angle lock was another option priced at $89. A-C also offered a squadron hitch for the Style L disc harrow, hydraulically controlled, and furnished for either two or three individual units. With proper sizing, 40½ feet could be covered in a single pass! This Style L disc harrow is equipped with front and rear cutaway blades, and is being pulled by an A-C D-21 tractor.

Large size offset disc harrows require considerable power, as is illustrated by an A-C D-21 tractor pulling a 13½ foot Style K disc harrow. This particular model is equipped with notched or cutaway blades on both the front and rear sections, giving the machine an exceptionally aggressive action in tearing up, aerating, and pulverizing the soil. Blade scrapers were not part of the $1,365 base price for this 13½ foot model — they added another $82 to the cost. The hydraulic control cylinder was also a $77 extra. Other derivatives of the "K" series was the "KT" built from 1969 to 1973 and the "WKT" built from 1969 to 1974. Space considerations preclude a showing of each individual model of the rather extensive offering of Allis-Chalmers disc harrows.

As previously noted in this section, Allis-Chalmers offered an extensive line of disc harrows beginning with the LaCrosse designs of the early 1930s and continuing to the sophisticated, hydraulically-controlled units of the 1980s. Simply stated, to illustrate each of the various models would restrict the number of illustrations available for other implements, so this section must be considered as an overview of the total output. The 3400 Series disc harrow shown here was offered from 1977 to 1982, with all disc harrow production apparently being suspended the following year. Typical of later offerings was the 2600 Series disc harrow, built in sizes ranging from 25 to 32 feet, and priced from $18,580 to $20,260. The collapse of the farm economy beginning in the 1980s dramatically reduced demand for new farm equipment, forcing Allis-Chalmers and many other companies to partially or totally withdraw from a vanishing market.

Dryers

Allis-Chalmers has made many pioneering developments in the ore and mineral processing industry, including the design and manufacturing of large kilns. With this technological advantage the company was able to build large rotary dryers for use by the grain milling, paper, and mining industries, to name a few. The Annual Review for 1953 illustrates a recently constructed unit with what was believed to be the largest ever built for a rotary dryer — 12½ feet in diameter and 37 feet long! This was part of what the company classified as a 10 x 80 foot unit. Either gas or fuel oil could be used, with this gigantic dryer consuming 700 gallons of fuel oil per hour. The combustion chamber design was such that the entire flame remained within the chamber, allowing only the hot gases, approximating 2000 degrees F. to pass into the dryer itself. Further information will be found under the "Kilns" section elsewhere in this volume.

Drying Fan

Although innovative in design, the Allis-Chalmers drying fan never went far beyond the experimental stage. This example from the late 1940s was probably intended for use in drying corn and other grains — if nothing else the B-15 Allis-Chalmers engine probably cranked out a major draft for anything in its direct path. Note that the headlights are turned back toward the fan to permit nighttime inspection. As often happens with new developments, this one was several years ahead of its time — grain dryers were not yet widely used in the late 1940s. Twenty years later when grain dryers came into widespread use, untiring electric motors operated the fan and required less attention, no refueling, and less maintenance.

Electrical Equipment

Following the earlier developments of Thomas Edison, Elihu Thompson, and others, the electrical equipment industry became very active during the 1890s. Like many new developments, once the basic problems were solved the path was open to further improvements, a phenomenon that seemed to feed upon itself, producing new ideas and new electrical equipment almost as a geometrical progression.

Ten years before Allis-Chalmers was organized, Edwin Reynolds had worked out details of mounting a slow speed alternator directly to the crankshaft of a Reynolds-Corliss steam engine. The idea had been tried once before in England but had failed. Reynolds however, was successful in obtaining a contract with the Narragansett Electric Light Co. of Providence, R. I. for the installation of two 500 horsepower Reynolds-Corliss engines with direct-connected alternating current generators.

Specifications provided for a cross compound engine with 19 and 36 inch bores, along with a 48 inch stroke. The engine was to operate at 90 rpm, and was intended for 125 pounds steam pressure with a 26-inch exhaust vacuum.

Since Edward P. Allis Co. had no facilities for building the alternator they contacted Thomson-Houston Co. and Westinghouse Electric Co. Both of these firms refused to build this type of alternator at first, but finally acceded, with each firm building one of the units after Allis Co. built the flywheel spiders on which the revolving poles were mounted.

While it must of course be conceded that these competing firms were instrumental in the development of electrical power, Edward P. Allis Co. held its share of "firsts" too, with these 1892 units being the first successful direct-connected alternators ever built and placed into successful service. They remained in operation for some years, finally being replaced by steam turbines.

Due to a certain amount of cooperation within the industry, Edward P. Allis Co. seemed to make no effort to enter the electrical equipment field. However, the rapidly changing market was facing the introduction of steam and hydraulic turbines, and even the Reynolds-Corliss, now doomed by the turbine was seeing less and less application for electric generating service. Thus, during 1904 the recently formed Allis-Chalmers Co. made a lease agreement with Bullock Electric Mfg. Co. of Cincinnati — a lease that finally terminated in an outright purchase of the company. Norwood, a Cincinnati suburb was the site of the factory, and from

Electrical Equipment

Norwood Works came countless numbers of motors and generators over the coming decades.

Allis-Chalmers had jumped into the electrical equipment industry with both feet by 1905, obviously intending to capture their share of the new market. Their acquisition of steam and hydraulic turbine designs was obviously beneficial, since the majority of either style was used for electric generating service. In addition, Allis-Chalmers had recently captured American manufacturing rights for the huge Nuremberg gas engines, quickly adapting them to electric service. The strategy paid off, for in 1907 Allis-Chalmers installed nine 2,000 kw direct-connected gas engine-generators at U. S. Steel Corp., at Gary, Ind. This was the world's largest gas engine-generator installation, and was also notable for the fact that these units could operate in parallel with each other and with other systems.

During 1909 Allis-Chalmers began manufacturing their own switchboards, reporting a year later that they had built some 350 different boards for a total of 850 separate panels. At this point in time, switchboards were of marble, so mounting the individual components was a difficult task.

By 1915 the company had built transformers as large as 6200 kva, and rated as high as 125 kilovolts, making these the largest transformers ever built at that time.

Another 1915 achievement was the contract for three 4,600 kw vertical generators for St. Lawrence River Power Co. of Messena, New York. These units operated at 115 rpm, delivering 575 volts of DC power. At the time these were the largest vertical generators in the world, and except for a single installation in France, were the largest direct-current generators ever built.

Achievements piled up through the years, highlighted by the 1920 introduction of the Style SAR single phase motors. During 1922 the company acquired the contract for the huge Mt. Shasta generators of Pacific Gas & Electric Company. The Mt. Shasta generators had a capacity of 35,000 kva with an output of 11,000 volts at 60 cycles. At the time these were likely to have been the largest hydraulic-electric units ever built.

A 15,000 kva, 132 kilovolt single phase transformer was built for Public Service Gas & Electric Company of New Jersey in 1925, making this the largest of its kind ever constructed. A year later Allis-Chalmers announced their first totally enclosed induction motors, with these motors using Timken tapered roller bearings.

During 1925 Allis-Chalmers gained licensing rights to the Reyrolle metal-clad switchgear of Alphonse Reyrolle & Co. Ltd. in England. In 1926 this became the basis of A-C Armor-Clad switch gear, and was the first metal-clad gear in North America. The first installation was in the Waukegan station of Commonwealth Edison Company. Also during 1926 Allis-Chalmers built their largest direct current motor, a huge 7,000 horsepower machine operating at 700 volts and capable of 50-120 rpm.

Another record came in 1928 with the construction of the largest horizontal shaft type stator yoke of welded

With the 1904 acquisition of Bullock Electric Mfg. Co., Allis-Chalmers planted both feet in the rather new electric power industry. The Reynolds-Corliss steam engine pioneered by Edwin Reynolds and built by Allis-Chalmers was not yet dead from obsolescence, but was certainly suffering from the competitive effect of steam and hydraulic turbines. For a time at least, Allis-Chalmers offered small engine-generator units of the style shown here, even though the company built a far greater reputation on huge generating plants capable of powering entire cities. This particular unit of about 1906 uses an ordinary Allis-Chalmers vertical slide-valve engine direct-connected to an Allis-Chalmers direct-current generator.

Edwin Reynolds pioneered the direct-connected engine and generator, despite concerted opposition from the manufacturing and engineering fraternities who considered the idea ill-advised and unworkable. Reynolds prevailed however, building the first such units in 1892. Typical of the period is this 1909 installation in the power station of Walter Baker & Co. of Dorchester, Mass. In this Boston suburb, Baker manufactured chocolate and cocoa products. The two 1,000 kw units shown here are direct-connected to vertical steam engines of the cross-compound type, having 22 and 48 inch cylinders with a 48 inch stroke. The generators are of the 3 phase style, delivering 600 volts at 60 cycles.

steel construction. This one had a capacity of 25,000 kva and was connected to an A-C 32,200 horsepower double overhung, impulse type water wheel operating on an 800 foot head. This installation was at the San Francisquito No. 1 Power Plant of the City of Los Angeles, Cal.

Taxing their manufacturing facility to the limit, Allis-Chalmers made the world's largest steel forging in 1929. Approximately 42 feet long, and weighing 269,000 pounds, this forging comprised the rotor for a 115,000 kw turbo-generator at the Waukegan Plant of Public Service Co. of Northern Illinois.

Intense competition in the electrical industry kept net profits very low, leaving very little new capital for research and development of new designs. The 1931 purchase of American Brown Boveri Corp. plus the acquisition of the capital stock of Condit Electrical Corp. gave a needed boost to the Allis-Chalmers electrical line. The electrical Department was now able to offer mercury arc power rectifiers, Brown-Boveri voltage regulators, railway electrification equipment, and high voltage (110 kv and higher) circuit breakers. From these developments, plus the 1928 acquisition of Pittsburgh Transformer Co. stabilized the Allis-Chalmers position in the industry.

Allis-Chalmers had actively promoted the electrification of machine tools and factory equipment for some years eliminating the lineshaft, pulleys, and leather belts that had been accepted practice. Walter Geist of A-C was awarded Patent No. 1,662,511 in 1928 for the multiple v-belt drive, termed by A-C as the Texrope drive. This was probably one of the most significant of all the Allis-Chalmers patents, and probably did more for the Electrical Department in terms of new motors, new motor contactors, and new electrical devices than any single item of the 1920s and 1930s. Manufacturers clamored for the new Texrope drives, anxious to dispense with the flat leather belts of the past. First year sales of Texrope drives were so significant that Texrope achieved divisional status within the company by August, 1929.

The effects of the Great Depression were not so quickly felt in the Electrical Division or in the Hydraulics Division because of backlogged orders written before the 1929 stockmarket crash. Eventually however, these departments experienced the same financial crunch that was being felt throughout the company, and in fact, the Sawmill Division never recovered from the effects and was finally discontinued. Hoover Dam, originally known as Boulder Dam, gave Allis-Chalmers new impetus during the 1930s. By getting the contract for the first four hydroelectric units, Allis-Chalmers was at least able to maintain the status quo during the lean years of the late 1930s, even though intense competition kept profits very small, or even nil. The stator yokes for the first four Boulder Dam units were of welded steel construction with radial openings for the water coolers. They were designed for an output of 40,000 kva at 13,800 volts. Allis-Chalmers also secured the contract for the high-voltage metal-clad switchgear — these huge circuit breakers were rated at 4,000 amperes and 23,000 volts, with a maximum

Typical of the 1909 installation was this power house of McKell Coal & Coke Co. of Kilsyth, W. Va. The unit illustrated here was driven by a Reynolds-Corliss engine having a 25 x 42 inch bore and stroke. Operating at 125 rpm, it handled this 500 kw, 3-phase generator which delivered 6,600 volts at 25 cycles. In the foreground is a separate exciter unit driven by its own steam engine. Eventually the 25 cycle frequency was dropped in favor of 60 cycles per second — the human eye is capable of discerning the flickering of incandescent lamps operating on 25 cycles, and it is a most annoying phenomenon. Although the Reynolds-Corliss engine shown here was fairly economical, steam turbines offered distinct manufacturing and operating advantages.

Always ready to meet a challenge, Allis-Chalmers built this unusual flywheel equalizing generator in 1910. It was installed on the ore loading docks of the Toledo & Ohio Central Railroad at Toledo. Three electrically operated ore hoists on a pier of the Maumee River had a capacity of 10,000 tons in 10 hours. When the hoists were operating simultaneously, they pulled a load of 1,600 to 1,800 amperes at 550 volts D.C. Independently, each hoist required on the order of 400 amperes. The intermittent and sudden peak loads raised havoc at the generating station, so Allis-Chalmers was awarded a contract to build and install this equalizing generator. The huge flywheel of this machine was designed so that at peak loads of 10-15 seconds the 1800 ampere load would be held to about 600 amperes by delivering its energy to the generator to which it was connected.

Electrical Equipment

interrupting capacity of 2,500,000 kva. By 1940 Allis-Chalmers had installed or had on order 7 of the 11 hydroelectric units in the Hoover Dam project on the Colorado River.

Most of the 1940s decade was involved with war production, followed by reconversion to peacetime requirements. Some of the technologies developed because of World War Two led to major improvements in electrical equipment, and the 1950s saw a major upswing in power requirements in every sector of society from the home to the factory. Seeing the need for higher transmission voltages, Allis-Chalmers developed a special test transformer in 1952. Rated at 600 kv, it was built to determine whether fundamental methods of transformer design could be relied on at such high voltages. Successful completion of these tests led to a September, 1953 contract with a West Coast utility for six 345 kv, 200,000 kva transformers. These six units, comprising two individual transformer banks were at the time, the largest in the world. These units were shipped in 1955 and 1956.

The 1953 purchase of Buda Co. of Harvey, Ill., put Allis-Chalmers squarely into the engine business, and also had the advantage of acquiring Buda's extensive background with engine-generator sets. This was especially important, since most A-C experience to date had been with extremely large prime movers, and the company had developed little or no market for the increasingly important portable or semi-portable engine-generators. Sales of these units achieved significant importance with a military contract for engine-generator units to help power the Distant Early Warning radar system.

Despite increasing power requirements the Electrical Division had problems making ends meet during the 1960s. In fact, the company had moved production of hydraulic turbines to York, Penn., in 1959 after acquiring S. Morgan Smith Co. Steam turbine construction was suspended in the early 1960s, but resumed for a time after the formation of Allis-Chalmers Power Systems in 1970. This organization operated as a subsidiary of Allis-Chalmers and Kraftwerk Union A.G. of Germany.

Allis-Chalmers Power Systems seemed to get considerable attention and several large contracts shortly after it was organized. Eventually though, this firm found itself in the doldrums, and on January 1, 1978, the Electrical Products Group merged with Siemens A.G. of West Germany to form Siemens-Allis Inc. Curiously, Allis-Chalmers followed this action by another agreement with Kraftwerk Union A.G. to form the Utility Power Corp. to manufacture and market steam turbine-generators in the United States. Allis-Chalmers was a 15% partner, with Kraftwerk Union, a subsidiary of Siemens A.G. owning the remaining 85%. In order to consolidate its competitive position, Siemens-Allis announced plans to move its large motor and hydro-generator operation to brand new facilities adjacent to the Utility Power Corporation site near Bradenton, Fla. However, in late 1986, Siemens-Allis announced plans for the complete closing of the Florida facility in late 1987, citing lack of orders world-wide as the reason for abandoning the huge new facilities.

Shortly after being organized in 1901, Allis-Chalmers set about acquiring the manufacturing rights to large gas engines, finally being successful in obtaining a shop license for the Nuremberg engines of German origin. Allis-Chalmers "Americanized" the design, and in 1909 sold this huge twin tandem double-acting gas engine with an integral 2,500 kw generator to a steel company. The engine frames alone weighed some 90 tons each. Due to the rather low operating speed of under 100 rpm, speed regulation became a critical factor, especially with alternating current generators. This required huge flywheels to assist in speed regulation, and this of course added still more cast iron to the total weight of the machine.

The 1909-1910 period saw immense growth of the Electrical Division at Allis-Chalmers. Included were transformers of immense size for this particular time frame, with this one being of the water cooled type, and rated at 2,000 kw and 66000/13200 volts transformation. Water-cooled transformers simply used coils immersed in the oil within the transformer tank. Water circulated through the coils, keeping the operating temperature within safe limits. These were some of the largest transformers A-C had built, and were installed for Milwaukee Electric Railway & Light Co. at its West Allis substation, terminating at another substation in Kilbourn Falls, Wis. Over 2,000 gallons of oil were required for each transformer, and when filled, each unit weighed approximately 55,000 pounds.

Electrical Equipment

Norwood Works, the A-C plant formerly operated by Bullock, was a thriving factory, building motors and generators by the thousands. This 1910 scene shows workers, including some women, winding coils at Norwood Works. Standardization of various components meant that motor, generator, or transformer coils for small size units could be built assembly-line style, cutting the manufacturing cost and permitting the company to remain competitive in a continuing tug-of-war for market position. In this 1910 photograph, note the almost complete lack of electric lighting, compensated for by huge window areas on both sides of the work bay.

Allis-Chalmers was a leader in the electrification of factory equipment and machine tools. This 1910 Allis-Chalmers file photo illustrates an Ingersoll slab milling machine converted to electric motor drive. Lathes, punch presses, saws, blowers, and sanders — these were all eligible candidates for installation of electric motor drives. The effort was very successful, with Allis-Chalmers playing a major role in American industrial development, simply by helping eliminate the troublesome and eminently dangerous lineshafts and belt drives so typical of the past. The introduction of variable speed motors further enhanced this concept, adding a new dimension to machinery operation.

In 1909 Allis-Chalmers began manufacturing their own switchboards, thus developing the electrical product line still further. This board was in the power house of Isaac G. Johnson & Co. of Spuyten Duyvil, N.Y. The Johnson firm was organized in 1853 to make iron castings, and by 1909 had moved into such exotic materials as Monel metal. This board is of natural marble, although Allis-Chalmers recommended to the sales force that marble with a black finish was preferable, since matching the natural product for appearance required great time and care in the shops. Due to the nature of the marble slabs, all instruments were surface mounted, this requiring only the drilling of holes to pass the necessary mounting bolts and current-carrying lugs. The disconnect switches are entirely open, having insulated handles.

An early photograph looking down one gallery of Norwood Works shows an endless row of lathes and other machine tools, with dozens of generator spiders stacked like cordwood down the center of the area. These are awaiting their turn to be machined for use in converters, generators, and certain styles of motors. After being machined, a spider for an alternating current generator was fitted with its field poles. The heavy spider casting also provided a flywheel effect — this was imperative to assist in providing the precise speed regulation essential to proper operation of the finished machine. Rather substantial electrical machines were built at Norwood Works, although it appears that some of the very largest generators might have been at least partially built in the West Allis Works of Allis-Chalmers.

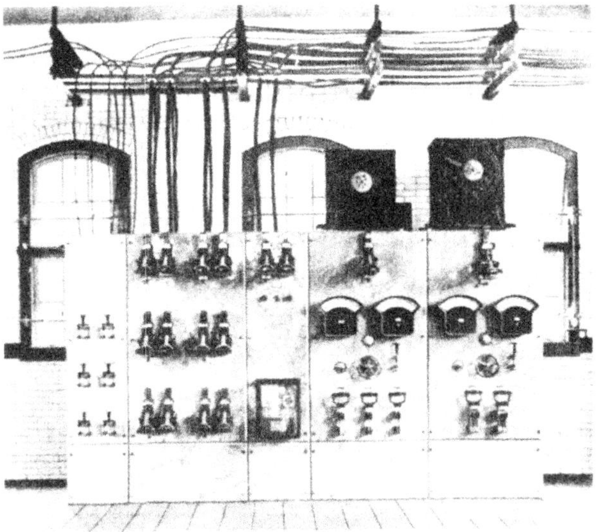

The old saying of "What goes up must come down" applies as well to high voltage transmission lines. Raising the voltage by the use of transformers is imperative for long distance movement of electrical power, but requires a transformer at the other end to bring the voltage down to a useable level. For 1910 Allis-Chalmers offered their lighting transformers with 5% and 10% taps built in. When required, these taps could be used to obtain the same secondary voltage at the far end of the line as at the generating station. A-C advertising noted that "this makes it possible to use lamps of the same voltage either near the generating station or at distant points."

Electric railways required immense sources of DC power, with the rotary converter being a logical choice in most situations. Converters provided direct current with a rather high ripple which was not objectionable for many applications, and was more cost effective than an engine-generator set or a motor-generator using an alternating current motor coupled to a direct current generator. Combining the AC and DC windings on a common shaft was a far better approach, and thousands of converters were placed into service prior to the advent of commercially saleable rectifiers, and in recent years the diode. Shown here in a photo of about 1915 are two 500 kw, 60 cycle, 240 volt rotary converters installed in the plant of the Merchants Heat & Light Co. of Indianapolis.

Electrical switchboards of the 1920s had become far more sophisticated than those of even ten years earlier. Plant engineers were more conscious of machine operation and maintenance and demanded better instrumentation to have a running assessment at any time. More and better metering was required to determine power usage by the customer, and in the case of feeder lines, constant monitoring became a way of life. Accurate voltage regulation at the generator was probably the most tedious problem, with most companies relying on the Tirrell regulator or other designs. This board however, displays a considerable amount of open copper, but within a few more years the electrical industry would see so-called dead front switchboards with no live components whatever exposed to the attendant.

In 1919, America's post-war industry required ever greater power to operate new and larger manufacturing equipment. This huge reversing type, plate mill motor was rated at 800 horsepower and capable of delivering 2500 horsepower. It was also capable of exerting a maximum repeating torque of 400,000 foot-pounds from 0 to 37 rpm, and could deliver an emergency torque of 425,000 foot-pounds. From full speed of 37 rpm, back to 0, and up to 37 rpm full speed in the opposite direction required but 3 seconds. Under these circumstances the commutator load could vary almost instantaneously from 0 to 4000 amperes. Due to its well proportioned design, this motor was said to be capable of handling this immense load with "black" commutation, that is to say, without any sparking whatever at the brushes.

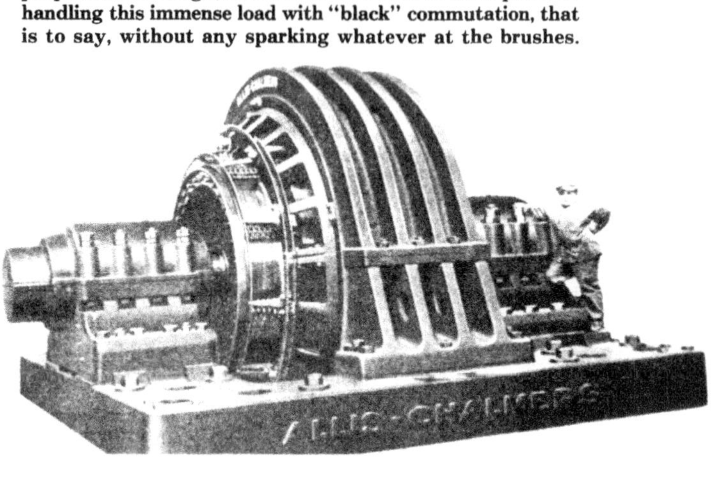

During World War One, Allis-Chalmers supplied a large number of 15 kw 110 volt self-contained engine-generator sets to the U.S. Government. During 1919, the company announced that these units, plus a small 5 kw model were available to the general public, pointing to their advantage as an independent, portable source of power. Despite this promotion by the 1919 sales force, current research has turned up little evidence to show that any substantial number were sold for private use, nor has any data come to light indicating how long these units remained in production after the initial announcement. The size and simplicity of these units seem to have made them an excellent participant in the post-war market for home generating units, but Allis-Chalmers appears not to have pursued this market.

Electrical Equipment

While of no immense significance perhaps, 1920 marked A-C's entrance into the building of single phase motors. Recognizing that many small factories and nearly all homes were furnished with single phase power, this Type "SAR" motor helped maintain a portion of the market that A-C had apparently done little to promote in their frenzy to maintain top rank position among the heavy equipment builders. Rated at ¼ horsepower, 60 cycles, and 110/220 volts, the Type SAR motor was highly recommended for use on domestic refrigerators then coming into general use. The general tone of A-C's promotion seems to lean toward the sale of these motors to refrigerator manufacturers, probably in large-size orders, rather than to orders of perhaps ten at a time to a local hardware store.

Allis-Chalmers first demonstrated their version of the Reyrolle armorclad switchgear in 1926. This was the first metal-clad switchgear built in North America, and it would soon revolutionize American concepts of high voltage switchgear design. The design was built under license from A. Reyrolle & Co. of Hebburn-on-Tyne, England. Reyrolle had been building this equipment for several years previously, with the result that by 1926, most switchgear equipment in Great Britain was already metal-clad. Although this 1926 model seems exceptionally crude and heavy, it was capable of handling 300 amperes at 15 kv and was rated with a 75,000 kva interrupting capacity. From this design, modern American metal-clad switchgear gained its pedigree.

In June, 1922 Allis-Chalmers announced that work was nearly completed on two 35,000 kva, 11,000 volt generators. These two units, of 60 cycle design, and operating at 257 rpm, were intended for the Mt. Shasta Development of Pacific Gas & Electric Co. The complete hydroelectric units were built at West Allis, and were at the time the largest 60 cycle waterwheel generators ever erected. Total weight of the generator itself was 550,000 pounds, of which the rotor and shaft constituted 265,000 pounds. The stator frame was built in four separate sections and assembled on-site, while the rotor poles were built up of laminations stacked on large bolts to ensure accurate alignment. The poles were six feet in length, with the poles being dovetailed and keyed to the flywheel spider.

In 1922 Allis-Chalmers built this Corliss engine of drop-piston valve design for Corn Products Refining Co. of Kansas City, MO. Of twin tandem compound design, it used two 27 inch high pressure cylinders, two 50 inch low pressure cylinders and a common stroke of 54 inches. Corn Products Refining Co. chose this unit because they believed it to be the most economical of all prime movers. Allis-Chalmers noted however, that this engine was no more economical than the standard Corliss design, although the drop-piston engine permitted higher rotative speed. The alternator has a capacity of 2,500 kw and is center-mounted between the two engine sections. A huge flywheel helps maintain accurate speed regulation.

Electrical Equipment

One of the largest of its kind ever built, this huge 10,000 kva furnace transformer was installed for U.S. Ferro-Alloys Division, Vanadium Corp. of America, at Niagara Falls, N.Y. in 1929. The primary side was rated at 12,000 volts, and was provided with taps to maintain full capacity on the secondary. When operated at 80 volts, this transformer could deliver 72,000 amperes. The immense size of the secondary windings and their terminations at the top of the transformer dwarf the workman standing beside this unit. Due to the engineering experience gained from this particular unit, Allis-Chalmers announced plans to move forward with other styles of this unit by the following year.

For years, electrical engineers attempted to devise easy and economical ways of obtaining DC power from AC lines. Rotary converters were widely used, as were motor-generator sets. After the 1931 purchase of American Brown-Boveri Inc., Allis-Chalmers had the advantage of providing industry with the mercury arc power rectifier which Brown-Boveri had developed. It was an outgrowth of the old-style glass bulb rectifier developed some 25 years earlier. Without getting into a detailed engineering explanation, the mercury arc rectifier functioned under the valve action of a metallic arc burning between two electrodes in a highly evacuated space. Current can only pass if the cathode is excited and if the anode is maintained at a temperature at which no electrons are liberated. Mercury is used because it is easily condensed and returned to the cathode. Mercury arc power rectifiers were widely used before being outdated by advancing technolgy. This unit is capable of delivering 3000 amperes of direct current.

Another major acquisition of the American Brown-Boveri purchase was the Rocking Contact voltage regulator for alternating current generators. Voltage regulation was a never-ending problem, especially for plants with sudden load changes. This regulator could be used for an individual generator or for parallel operation of alternators. It could also be adapted to voltage and power factor regulation of synchronous motors, speed regulation of DC motors, voltage and current regulation of DC generators and battery chargers, power regulation of turbo-alternators, electric furnaces, wood pulp grinders, and similar machines. In addition to this great flexibility, the Brown-Boveri design permitted automatic paralleling of synchronous machines. Many other regulator designs depended on vibrating contacts, switching contacts, and other methods, all of which were a constant source of annoyance. The Brown-Boveri design emerged as one of the best designs of the 1930s and remained as the regulator of choice for some years to come.

A 1935 photograph illustrates machine work on the stator yoke for a 40,000 kva, 13,800 volt generator to be installed to Boulder Dam. This work is taking place at West Allis Works on the huge 40 foot vertical boring mill. The mill itself had been installed several years earlier, and was at the time, the largest vertical boring mill in the United States, and perhaps in the world. While it seems hard to believe, the huge table on which the workmen are standing, and to which the stator yoke is secured, turns beneath the cutting tools. Because of the dimensions of the boring mill, note that a heavy duty stairway permits workmen to reach the platform next to the actual tool bars. Huge equipment such as this mill gave Allis-Chalmers at least a tiny advantage in securing large and heavy work, since A-C engineers had the engineering experience to do the work with confidence.

By 1937, transformer design reflected improved technology and manufacturing techniques. New features included an oil sealed, inert gas protective system to compensate for pressure changes within the transformer tank due to temperature changes. The large unit on the right hand side of this photograph was equipped with blowers beneath each set of radiators. While no radical changes were made on the 1937 power transformer line, the A-C regulating transformers were improved by the use of a 33 position step mechanism with each step providing a tap change of ⅝%. During 1937 the company built several of these units in 33, 46, and 69 kv styles.

Allis-Chalmers reported that for 1937 a definite trend toward the use of metal-clad switchgear had been established. During this particular year the company had developed a complete line of vertical lift type, metal enclosed switchgear in capacities of 25,000, 50,000, and 75,000 kva. In the switchgear shown here, the oil circuit breakers are raised into or lowered out of position by lifting screws. In the lowered position, the circuit breaker could be completely removed from its cubicle. This design followed many features found in the original Reyrolle design introduced several years earlier.

Destined for a huge hydroelectric installation, this rotating field is undergoing final inspection at the Allis-Chalmers shops in West Allis. A unique feature is the welded steel construction seen in the spider — this design gave greater strength than cast iron, and with a dramatic reduction in total weight. The individual poles are secured to the spider by means of steel plates and huge bolts, with the final assembly giving a unitized structure of immense strength. Centrifugal forces become enormous in huge rotors such as this one. As an example, a 10 foot rotor operating at only 100 rpm attains a rim speed of 31,40 feet per minute.

Transformer design for 1938 was to see little change during the next decade. First of all, the designs had been reasonably defined and exhaustive engineering work had determined the methods which would give the best results and minimum losses. With the onset of World War Two, many new designs were tabled for the duration, leaving new experiments for another time. This huge transformer of 1938 is of overwhelming size, making the technician in the foreground almost invisible. This technician is apparently testing components of the voltage regulator system integrally built into the transformer. Note the huge expansion tank above the transformer.

Building huge transformers was one thing, but shipping them was quite another. Due to the great height, a special underslung railway car was used, and due to the immense weight, two sets of trucks were at each end of the car for a total of 16 wheels on the rails. Before setting the transformer on the car all oil was drained and the tank was filled with an inert gas, usually nitrogen. The bushings were removed and shipped separately to prevent damage to the fragile porcelain. Then the transformer was placed on the rail car and several workmen toiled long and hard to secure it for the journey. This was no small task, requiring blocking and bracing almost beyond imagination. Note the heavy bracing attached to this unit.

Norwood Works turned out A-C Ampac welders by the thousands during the Second World War. This scene at Norwood Works illustrates a line stretching as far as the eye can see, and completely loaded with Ampac welders in various stages of construction. Allis-Chalmers was able in this instance, to convert some of their vast transformer engineering experience into this compact machine, one that was able to get the job done with a minimum of steel and copper — materials that became very scarce during the war. For reasons which the current research has not uncovered, the end of hostilities seems to have signaled the end of Ampac and Weld-O-Tron construction, since A-C advertising of the early 1950s seems to make no mention of it, nor do Annual Reports appear to show that this product line was sold to another firm.

Introduced in 1940, the A-C Weld-O-Tron was an A-C welder utilizing a common core for both the reactor and the transformer. During World War Two this machine was improved and in 1942 a new line of A-C welders was offered to industry, these machines having a capacity of 200 to 600 amperes. This machine required substantially less copper and steel for its production than did a comparable motor-generator welder, so the War Production Board (WPB) strongly encouraged its use wherever possible. Small repair shops across the country welcomed the small Ampac welder of 1943, shown here. The Ampac was ideal for repair of farm and industrial equipment.

Allis-Chalmers contributed mightily to the war effort, and was able to do so without refitting their factories completely to production of war materials. As an example, the 1941 Annual Review notes that several U.S. Navy tenders recently launched, including this one, were completely equipped with Allis-Chalmers motors and generators. Typical of the generators might have been their 814 kw model. Operating at 750 rpm, this close-coupled diesel-electric generating unit was designed for marine service even before hostilities began, and already was compact, lightweight, and shock-proof.

A 1944 scene at Norwood Works illustrates thousands of diesel engine starting motors waiting for final packing and shipment. Allis-Chalmers was able to provide this equipment to the military on rather short notice, since Norwood Works was already equipped to produce these motors in quantity. While a great many of these starting motors probably were used on the company's own engines, it seems entirely likely that wartime demands saw considerable quantities built to government specifications for use on engines built by other companies.

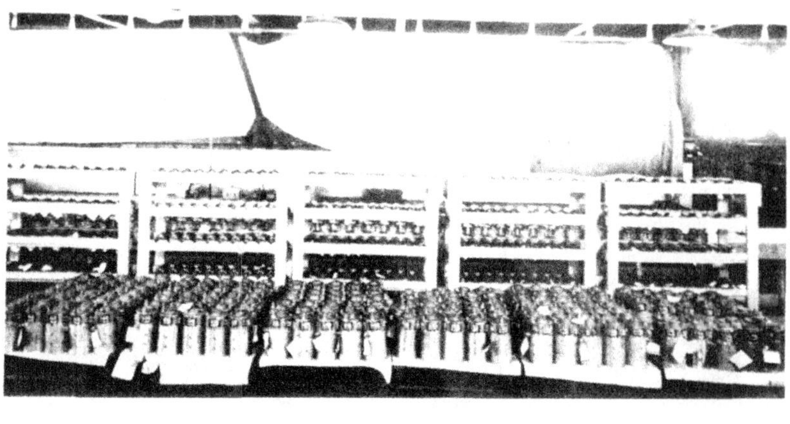

While the mercury arc rectifier was the rage of the 1930s, the 1950s saw the Excitron-type rectifiers for heavy direct-current requirements. This photograph from a large chlorine plant shows twelve 6000 ampere, 600 volt, 12-tube Excitron-type rectifiers in operation. At left are shown the anode breakers; at right the cathode breakers with the disconnect switches; and in the center is the substation containing the excitation, phase control, vacuum measuring, and protective equipment modules. This unnamed plant of 1952 was completely equipped with Allis-Chalmers equipment from the incoming 154 kv line down to the low-voltage substations.

Electrical Equipment

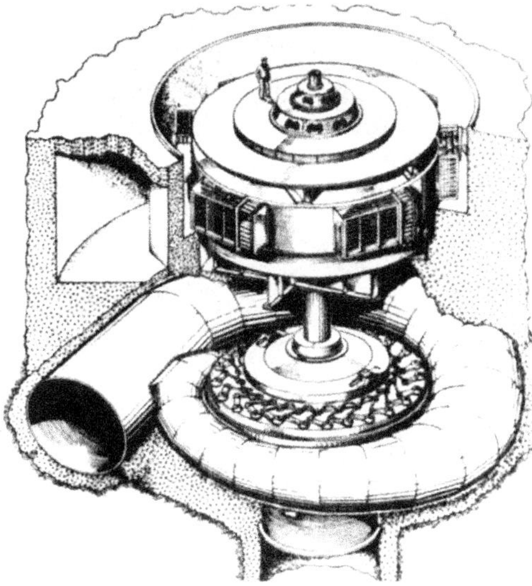

During 1952 Allis-Chalmers announced a contract with the Tennessee Valley Authority (TVA) to build the world's largest electric motor. Completion was scheduled for 1955 for installation in the Hiwassee Dam in southwestern North Carolina. The reversible pump-turbine will operate in one direction as a turbine, and the other direction as a pump. Functioning as a turbine, the huge motor will serve as a generator, and a rather large one at that, since the turbine has a rating of 120,000 horsepower. On the other hand, when operating as a pump, this huge unit had a capacity of 3.3 billion gallons per day. At the time, Allis-Chalmers was the only company in the world building complete hydraulic and electric units of this type. Allis-Chalmers pioneered the reversible pump-turbine concept.

To completely normalize and stablilize huge rotor shafts, Allis-Chalmers built a special "hot balance" pit at West Allis Works. This area closely approximated the operating temperatures encountered in actual service, and permitted the manufacturer to balance the shaft under actual working conditions. This gigantic 98-ton rotor was for the world's first 300,000 kilowatt turbine-generator, and was built for Detroit-Edison. This huge turbine-generator unit went on the line in 1958. This same year, Allis-Chalmers received its first order for 345 kv power circuit breakers.

During the 1950s, Allis-Chalmers was instrumental in developing weatherproof metal-clad switchgear, with this 1958 Shelter-Clad switchgear being an example. Prior to marketing, A-C conducted exhaustive tests on their equipment, and in this photograph a section of Shelter-Clad gear is undergoing a water test far more severe than would be encountered even under hurricane force winds. This unit was tested while live and under full load to determine whether heavy storms might have any ill effects on its operation. Achieving weatherproof equipment of this caliber required extensive engineering.

A 1950s photograph shows a 75 ton generator rotor with the slots being machined. This huge rotor is 37½ feet long, and will be part of a 357,000 kw steam turbine-generator unit for Consolidated Edison's Astoria Station Unit No. 5 in New York City. Actually, this rotor was installed in a 171,000 kw generator operating cross-compound with a 3600 rpm, 186,000 kw unit. Machining a shaft of this type required highly experienced machinists. Despite the size of the shaft, machining dimensions were extremely critical. A major machining error on this huge forged shaft might require that it be scrapped, costing the company thousands of dollars.

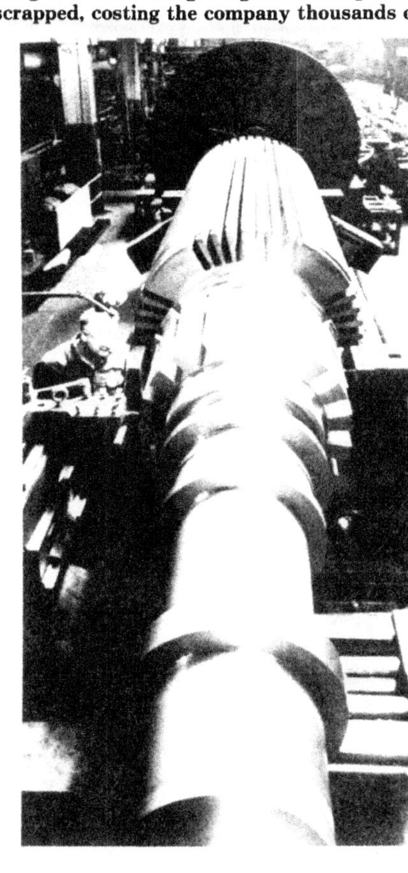

Electrical Equipment

This 1964 photograph was taken at the Big Eddy Station of Bonneville Power Administration at The Dalles, Oregon. Shown are six Class OA, single-phase rectifier transformers. These ultra-high voltage units were developed for the United States Department of Interior. Also during 1964, Allis-Chalmers built four 500,000 volt reactors, the largest in the United States. Under construction at the end of 1964 were three 500,000 volt power transformers — these when completed would constitute the world's largest transformer bank.

During 1966 Allis-Chalmers reported that their extra high voltage (EHV) installations included virtually every major transmission inter-tie across the United States. This equipment included large power transformers, air blast circuit breakers, disconnect switches, and high voltage reactors. Also of note for 1966 was the vacuum capacitor switches shown in this photograph. This switching method provided utilities with a safe, reliable and economical means to switch capacitor units for power factor correction and voltage regulation on and off of transmission lines, even though these capacitors might be many miles away from the generating station.

The decade of the 1960s saw immense growth of the electrical power industry and for Allis-Chalmers this decade was rather profitable. A-C provided some of its best electrical engineering achievements in this decade, not only in the area of large power equipment, but also in equipment like this electrolytic tinning line at Granite City Steel Company. Shown here are 68 Allis-Chalmers gear motors to power the plating rolls, with additional Allis-Chalmers motors and motor-generator sets being located throughout the remander of this production line.

During 1963 Allis-Chalmers completed this country's first Cyclotron magnet. This huge device was built for Washington University Medical School and Associated Hospitals Medical Center in St. Louis. It was designed for use in the research and treatment of cancer. This six million volt unit was built with hollow, water-cooled, copper conductor coil windings with 128 turns, requiring 590 amperes. This is supplied by a solid state silicon rectifier rated at 650 amperes and 75 volts. Allis-Chalmers had been building the Betatron unit for several years, with much of the initial research being derived from the company's role in the Manhattan Project of World War Two. Further information on this subject may be found under the "Betatron" section of this volume.

Recognizing the need for mobile transformer units, this 1961 outfit consists of a 2,000 kva, three phase unit used by a West Coast utility. It includes manual, no-load tap changers on both the high and low voltage windings. The immense weight of this unit required a special truck mounting in order to prevent roadway damage. Portable units such as this one remain an important part of large power maintenance, especially in case of difficulty with a stationary transformer unit. Although a certain amount of down time might be expected while the changeover is made to a temporary unit, the delay is far less than would be required if it were necessary to move in another stationary transformer unit.

Electrical Equipment

Allis-Chalmers announced in 1971 that they were building a huge hydroelectric generator to be installed in the Dworshak Dam at Orofino, Idaho. The massive size of this unit required the generator housing to by constructed in two halves for shipment, as shown in this photograph. Presumably this 1971 equipment was built by Allis-Chalmers Power Systems, the joint venture firm formed in early 1970 by Kraftwerk Union AG of Germany and Allis-Chalmers. From all appearances, Kraftwerk Union was subsequently acquired by Siemens AG, also of Germany, and from this acquisition came the eventual formation of Siemens-Allis.

The late 1960s and early 1970s seem to have been the peak years for the production of large power equipment, with the total number of units declining each year after, or at best showing only a minimal increase. A 1967 A-C file photo illustrates one of the company's largest transformers. In fact, it is the largest regulating transformer ever built, and is being readied at West Allis for shipment to Consolidated Edison Company of New York. This gigantic unit measures 70 feet long, 37 feet wide, and stands 29 feet high. On July 11, 1977 Allis-Chalmers issued a news release announcing that the company had agreed with Siemens AG of Berlin and Munich, West Germany, to form a new company to be headquartered in the United States and engaged in the electric products business. The new company, Siemens-Allis Inc. was expected to begin operations in January, 1978. Under the terms of this agreement, Allis-Chalmers placed all of the businesses and facilities of Allis-Chalmers Electrical Products Group.

Elevators

Between 1960 and 1966 Allis-Chalmers produced the Flight-Line portable elevator. Available in lengths from 26 to 50 feet, its 20½ inch width gave great capacity. Ostensibly built for ear corn and small grain, it was also well suited for baled hay, silage, bagged material, or other products being moved to storage. An extra long 11-foot spring-balanced feeder was standard equipment, as was the shielded PTO drive. Where required, the Flight-Line could also be equipped for electric motor or gas engine drive. A shovel-type hopper was available when required, as was an adjustable bale chute. When elevating bales or bags, removing the end plate and hood at the top end of the elevator was all that was required. Early models of the Flight-Line were of 17 inch width, later moving to the 20½ inch width.

Farm Loaders

Always looking for new and innovative methods, A-C engineers developed this rear-mounted loader in 1950. Apparently it was intended to operate with the A-C power lift system, and seems to have been designed as an easy-on, easy-off unit. Due to its simplicity the manufacturing cost would be quite low. The advantages of rear mounting were at least partially offset by the continual need for the operator to look backward. Present information is sketchy, but if indeed this unit was placed into regular production, the total number built was rather small. Note that this machine is unloading into yet another A-C innovation, the front-unloading manure spreader.

This rather curious loader design has an unknown history. Either it was designed by A-C engineers for possible production, or it was built by an outside vendor for use on the Allis-Chalmers tractors. Despite a huge engineering and manufacturing organization, Allis-Chalmers appears to have purchased their farm loaders from an outside firm for a number of years, even though the required engineering and manufacturing allotment to build their own unit would have been rather small. In fact, the rather simple construction of this particular unit embodied structural steel members, not even requiring formed and welded sections.

A 1950's photograph shows the D-15 tractor with an A-C farm loader. At this point in time, loaders were using tubular steel wherever possible. This lightened the machine considerably, while offering great strength. Farm loaders began appearing with increasing regularity in the late 1940s, and within another decade, almost every farm had a loader. There is no question that for the livestock farm, hydraulic loaders were one of the greatest labor savers ever brought to the farmstead. Although a few short line companies offered loaders with adapter kits for various tractors, many farmers preferred to buy a loader engineered to fit their own particular tractor. Thus, the market for loaders was there, with a great many A-C tractor buyers being potential buyers of a loader as well.

Beginning in 1965 and ending in late 1983, the 400 and 500 Series farm loaders were available. Shown here is a model 417 loader on an Allis-Chalmers D-17 Series IV tractor. This particular unit had a maximum lift of 9 feet and could be equipped with a standard manure bucket or with various size material buckets up to 80 inches wide. An 80 inch dozer blade was also available, and was easily mounted to the lifting arms. Only the 417 loader used the tractor hydraulic system; all other loaders of this series required a separate front-mounted hydraulic pump.

Field Cultivators

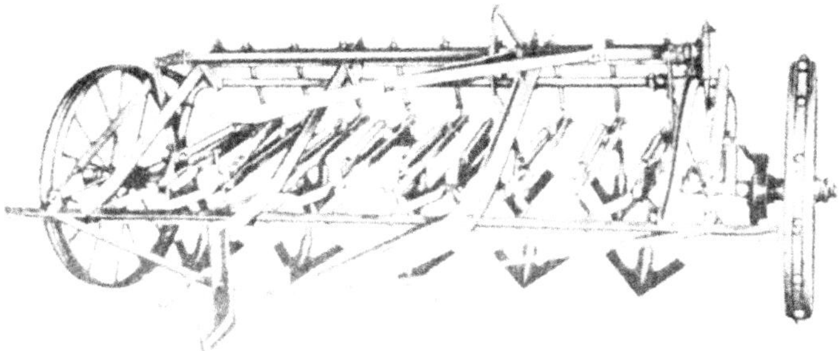

Possibly as a result of the 1929 purchase of LaCrosse Plow Co., Allis-Chalmers announced in late 1932 that it had introduced an entirely new line of field cultivators. This machine of 1930s vintage displays state-of-the-art design, at least so far as Allis-Chalmers was concerned. Farm tractors were just beginning to use rubber tires, but farm implements had yet a while to wait for this feature. Hydraulic lift systems were embryonic yet, so this unit featured a ground-driven mechanical lift. A rope leading from the lift unit to the back of the tractor seat permitted the operator to raise or lower the implement at will. Despite its crudity of design against the refinements of the 1980s, this field cultivator was indeed compact and simple.

Beginning in 1940, Allis-Chalmers introduced a new series of field cultivators — this basic design remained in production until 1966, going through numerous changes during that time. With the advent of the A-C Snap-Coupler system, every effort was made to provide implements matched to this hitch method. This WD-45 tractor is equipped with a 10-foot mounted field cultivator. For 1957, it was priced at $307, while a 12-foot model carried a list price of $333. Both units carried pneumatic gauge wheels for maximum depth control, and a stabilizer unit was available as an $18 extra cost option. This unit is operating near Lancaster, Wisconsin.

A factory photo illustrates the 1000 Series field cultivator, offered between October, 1968 and December, 1971. An exclusive feature was the elastomeric cushions at each shank mounting. This type of mounting gave quiet operation and permitted the shank to vibrate in every direction for effective soil shattering action. For this reason, A-C dubbed this machine the *Silent Weed Killer*, The 1000 Series was built in 10½, 13½, and 15½ foot sizes, with 3 and 4-foot wing attachments available for all but the largest unit. Truss rods were included with this attachment to give extra strength to the main frame. Although the 1000 Series field cultivators were available in a conventional pull-type design, they could also be purchased in a standard 3-point hitch configuration, using either Category 1 or Category 2 hitches.

Large farming operations require large equipment, so Allis-Chalmers met this challenge with their Squadron Hitch system. Hydraulic lines run to each individual field cultivator, and hydraulic control of the hitch frame permits one unit to follow the other for on-highway transport. Although it seems certain that Allis-Chalmers sold a substantial number of the Squadron Hitch units, and likewise sold a substantial number of field cultivators, the unrelenting competitiveness of the farm implement market became virtually unbearable by the early 1980s. This, coupled with cash flow problems on the farm, virtually put the farm equipment industry into limbo, forcing production lines to stand idle. Thus, Allis-Chalmers like most other equipment builders, withdrew its least profitable lines in order to hedge against further losses.

Allis-Chalmers introduced the 1200 Series field cultivators in 1970, followed by the 1300 Series in 1973. This large pull-type machine could be purchased in a primary width of 10½ feet; this then was expandable all the way to 21½ feet. As in the 1000 Series field cultivators, the high torque build-up in the elastomer clamp gave superior soil-shattering action. A non-folding design was used in models up to 13½ feet, with a manual-fold being featured up to 16½ feet in width. Hydraulic folding for transport was featured in all models 16½ feet and larger. In 1983 the 10½ foot basic unit was priced at $2,100, with this figure climbing to $3,953 for the 21½ foot hydraulic-fold model.

Fertilizer Spreaders

Between 1968 and 1970 Allis-Chalmers attempted an entry into the fertilizer spreader market. Farmers were using ever-increasing amounts of dry fertilizers, creating a need for equipment that would distribute it over the fields quickly and accurately. The 402 and 402-F models were intended to achieve that goal. The 60 cubic foot box had a 4,000 pound capacity, and was equipped with a stainless steel drag chain. Also featured was chlorinated rubber paint over a zinc chromate undercoat to minimize corrosion damage. Distribution was achieved with a 23-inch single dish-type spinner. Although no specific production figures have been located, presumably the total number of units produced was not sufficient to maintain regular production.

Firefighting Equipment

A 1942 photograph illustrates what Allis-Chalmers called a "Mobile Pumping Unit." Presumably this was intended as an aid in fighting fires, although it might have had numerous other uses as well. The 1942 license plate indicates that the unit was roadworthy, and probably capable of average road speeds. At this point in time, little has been learned of the actual reason for building this unit — perhaps the company intended it as an in-plant unit for fire duty. Possibly it was hoped that a market for a mobile pumping unit could be developed, and since 1942 was the beginning of the defense effort, there is a chance that some military use was planned.

It seems quite likely this nifty little unit was originally intended for fighting fires, especially since another photograph shows a hand-operated extinguisher mounted to the running board. Because of the war effort, and the limitations on chrome, nickel, and other important metals, it seems curious that this machine is equipped with fancy chrome hub caps. The hood was hinged below the windshield and could be lifted from the front for engine servicing. While this unit is perhaps best classified as a curiosity, it demonstrates once again the innovation of design on which Allis-Chalmers built an international reputation.

Allis-Chalmers quickly adapted the Buda forklift line as their own, doing so immediately after acquiring Buda in 1953. It is not now known whether Buda originated this localized fire fighting apparatus or whether it was a brainstorm of A-C engineering, but the fact remains that this adaptation of a factory truck to fire duty brought an even greater measure of safety to personnel, facilities, and equipment. By suitably modifying the original equipment and adding several pieces of fire fighting apparatus, this small maneuverable unit could be on-site and fighting a fire long before fire companies could respond.

Flour Mills

Webster's Biographical Dictionary notes that Oliver Evans (1755-1819) has been called the "Watt of America," obviously drawing a parallel with the famous James Watt, inventor of the steam engine. Evans constructed the first high pressure steam engine in America sometime prior to 1802. Thirteen years earlier, in 1789, Evans designed a flour mill equipped with elevators, conveyors, and other machinery that permitted one man to do the work formerly required four workers. Initially the Evans system was greeted with the skepticism so characteristic of the eighteenth century, but the early doubts were soon replaced with enthusiastic acceptance. After gaining acceptance of his method, Evans compiled a technical manual, *The Young Mill-Wright and Miller's Guide*. From its initial release in 1795, this immensely popular book went through fifteen editions by 1860, and was the earliest foundation of American milling practice. The *American Heritage History of American Business & Industry* notes that "Evans's flour mill liberated the American economy from subsistence agriculture and paved the way for numerous other advances that brought technology and science to the service of the farmer."

Revolving millstones were known as early as 300 B.C. According to Varro (232-147 B.C.) the first revolving millstones were invented by the Volsinians, the inhabitants of a city in Etruria, which city was destroyed by the Romans in 280 B.C. Despite some refinements, twenty centuries passed with no major changes in the milling of flour. Not until the 1770s did any activity occur in gradual reduction milling, with these efforts coming primarily from Austria, Italy, Switzerland, and Hungary.

Ignaz Parr (1778-1842) of Austria is usually spoken of as the father of high milling. Parr experimented with a process of re-grinding the separated grits, making from it an "Extract Flour." It was of such high quality that demand far exceeded the supply. Following Parr's experiments, others attempted refinements to high milling using stone buhrs. Simply stated, high milling attempted only a partial reduction of the grain each time through the machine, rather than attempting to reduce the grain in a single pass and casting off whatever would not pass through the sifting process. The first account of absolute high grinding in America was at Northfield, Minn., in 1866. A farmer asked that some of his wheat be ground that way, and the results were so encouraging that many American millers attempted high milling using their stone mills.

In 1868, a shop at Faribault, Minn., built a machine described as "a box with a sieve on top and a blast of air under the sieve." This is the earliest known account of a combination of a sieve and air blast for purification of middlings, and led directly to the flour purifier. In 1871 the resulting "Patent Flour" was introduced to New York and Boston. The demand almost immediately overtook available supplies, with Patent Flour advancing in price by $3 a barrel over the regular grades.

During the 1820s, flour mills fitted with rolls instead of stones were tried in France, Italy, and Switzerland. Unsuccessful at first, the concept of gradual reduction by the use of rolls continued its development, with the Pesth Cylinder Mill of 1867 being the first working roller mill. During 1874, Friederich, Wegmann of Wegmann, Bodmer & Co. in Zurich introduced the porcelain roll of some of the Pesth mills in what is now Budapest, Hungary.

Edward P. Allis Company exhibited cast iron mill rolls at the Centennial Exposition at Philadelphia in 1876. The following year, Oscar Oexle representing the Wegmann interests was invited to Minneapolis by Cadwallader C. Washburn. In 1878 he designed America's first all roller mill in Washburn Mill "C." This *experimental* mill contained sharp corrugated rolls, smooth iron rolls, and porcelain rolls, making three regular grades of flour in five reductions. A careful study of various publications and correspondence at the time, and some years later, indicates that the porcelain roll system was not adopted at this time, and further indicates that this mill was not envisioned then or later as being a merchant mill.

William Dixon Gray came to work for Edward P. Allis Co. in June, 1876. The firm was a widely recognized mill manufacturer, and in fact, its earliest activities date back to the firm of Decker & Seville who based the majority of their business on fitting millstones. Gray closely followed the early developments of high milling, concluding that roller mills "would do better work and make better flour than millstones." Gray also objected to the train of gear wheels on the Wegmann machine and invented a belt system that would soon revolutionize this relatively new manufacturing art.

The first Gray belt roller mills were installed in June, 1878, in the Eagle Mill of J. B. A. Kern in Milwaukee. Immediately afterward, Gray supervised the installation of eight roller mills for Schoellkopf & Matthews at Niagara Falls. However, both of these units used millstones in combination with the roller mills.

Early in 1878, C. C. Washburn was building a new "B" mill, and while it was under construction the existing "A" mill exploded. Since the new foundation for "A" mill was longer than necessary, Gray convinced Washburn to set up a small experimental roller mill using Allis equipment. Within a year Washburn was convinced of the roller mill system, so Gray approached Charles A. Pillsbury with the same idea. Pillsbury finally consented to the conversion of his Excelsior mill into a roller mill. With Gray's roller mill now firmly in hand, Edward P. Allis Co. exhibited it for the first time at a grain miller's convention in Cincinnati during June, 1880. Later that year E. P. Allis bought the old Kilbourn Mill in Milwaukee, completely remodeled it, and equipped it as a model roller mill. Thus, the Daisy Mill was America's first flour mill completely equipped with rollers. In less than a year it was open to prospective buyers and others interested in the process.

Flour Mills

Proof of Gray's success is shown by the fact that over 25,000 pairs of Allis rolls were in use by 1888, and Walter Peterson notes in *An Industrial Heritage* that by 1894, eighty-five percent of the combined capacity of Minneapolis, Duluth, and Milwaukee mills were equipped with Gray rolls.

It would appear that with the introduction of the Gray Roller Mill, E. P. Allis Co. almost immediately ceased production of buhr stone mills, a previous cornerstone of the company. Not until late 1926 when Allis-Chalmers purchased Nordyke & Marmon Company did it once again resume the buhr stone business. Curiously, neither Edward P. Allis Co., or in later years, Allis-Chalmers, built much of the auxiliary equipment required for a complete mill. Apparently they were content to furnish the roller mill, plus the necessary equipment to follow the flour through to the finished product. Allis-Chalmers seems to have had an aversion, at least for some years, to build the necessary cleaning and separating, and tempering equipment needed before the wheat entered the mill. In fact, a 1910 A-C *Sales Bulletin* notes that "receiving separators, scourers, suction fans, and dust collectors are not made by Allis-Chalmers Co." Prior to the 1926 acquisition of Nordyke & Marmon, A-C also built no equipment for corn milling or rice milling.

Ellis Nordyke set up shop building flour mills in 1851. Soon after, A. H. Nordyke, his son, joined in the business under the title of Nordyke & Son. Sometime in the 1850s D. W. Marmon joined the firm, and after Ellis Nordyke's death in 1871 it was incorporated as Nordyke & Marmon Co., Richmond, Ind. From this little building at Richmond the company grew immensely during its first half century, relocating to Indianapolis in 1902. At this point in time, Howard Marmon invented a V-4 air-cooled engine with pressure lubrication — certainly a stroke of genius compared to state-of-the-art engine building in 1902. For the next two decades, Nordyke & Marmon operated a highly successful business in flour mills and related machinery, while Marmon Motor Car Co. attempted making its own mark on the automotive scene. In 1927 Allis-Chalmers purchased the flour mill division, ostensibly because Nordyke & Marmon wished to further pursue activities in automobile development. The Marmon Sixteen, a huge and expensive car with a V-16 engine was the very epitome of Marmon design. Introduced in 1930, it was the end of the Marmon line.

While the Flour Mill Dept. functioned as a separate company division, a 1908 company *Sales Bulletin* indicates that friction clutch pulleys and couplings were also built by this department. In fact, Allis-Chalmers offered as many or more varieties than anyone following this line of endeavor. Sizes ranged from a small 10-inch size having a rating of 9 hp at 100 rpm. This size sold for $85. At the top end was a 66-inch model listing at $960. This one had a capacity of 550 horsepower with a shaft speed of 100 rpm. It could be furnished with a maximum shaft bore of 12 inches.

During 1909 the Flour Mill and Saw Mill Departments were combined to form the Milling Machinery Dept.

Few changes were made to the roller mills in the following years, but Allis-Chalmers continued to offer new styles of bolters, purifiers, and other flour mill machinery. Also introduced in 1923 were oat crushing rolls, intended not primarily for the milling industry, but rather for the livestock feed industry, and for horses in particular. University tests had shown conclusively that crushed oats gave far better feed efficiency, with the company noting to their salesmen that "conditions have changed and the making of feed is now more of a local proposition and accordingly increases the market for such machines as our oat crushing rolls."

With the 1926 acquisition of Nordyke & Marmon, Allis-Chalmers bolstered its sagging sales of flour mill equipment, at least temporarily. Buying N & M did nothing to enhance sales of existing A-C equipment, but at least it broadened the line to some extent. The Depression years of the 1930s, and intense competition were only part of the problem — the roller mills wouldn't wear out! Once the market had been filled with sufficient mills to produce the nation's flour supply, sales dwindled quickly. By 1930, most of the earlier stone buhr mills had been converted, so there was little market to pursue. The onset of World War Two saw no major increase in sales either, so the Flour Mill Dept. languored. With little new production in sight, the last large mill was built about 1960. As late as 1981 however, Allis-Chalmers Corporation is still listed in the *Thomas Register* as a manufacturer of roller mills.

Further information may be found on pages 9, 10 and 11 of this volume under the *William Dixon Gray* heading. To assist the reader, we have chosen to include portions of the Nordyke & Marmon line built prior to this company's acquisition by Allis-Chalmers. The latter continued to manufacture many of the N & M machines, at least for a few years, but it remains unknown at present when N & M parts were discontinued. In order to provide a sense of chronological continuity while still maintaining a separation of the individual product lines, selected Nordyke & Marmon captions constitute the beginning part of this section, followed by early E. P. Allis mills, and finally with the Allis-Chalmers line.

Flour Mills

Typical of the 1880s mill line was this Nordyke & Marmon under-runner geared mill. These were the same as the pulley mills except for the bevel gearing. Although the pinion gear on the vertical shaft was of cast iron, the large bevel gear was filled with choice rock maple cogs for quiet operation. When worn, the wood cogs could be easily replaced with a new set. For 1882 Nordyke & Marmon offered these mills in sizes from 26 to 48 inches, referring to the diameter of the stones. The all-wood framework was in itself a piece of engineering and required skilled workers in its construction. Nordyke & Marmon sold these mills by the thousands long after Edward P. Allis & Co. had opted to build the roller mill exclusively.

Like Decker & Seville, and later, like Edward P. Allis, the Nordyke & Marmon fortunes were literally built on millstones. Like Allis, the Nordyke & Marmon enterprise used only the imported French buhr stones, and these were "selected by our expert at the quarries in France." Most millers considered the French buhrstones to be the very best, looking at various American granites as being distinctly inferior. An 1882 Nordyke & Marmon catalog notes that "we usually keep from two to four thousand blocks, and fifty to one hundred pairs of stones on hand." These were shaped by skilled craftsmen into the form shown here, finally being given the proper "dress' by other highly skilled artisans.

In our age of electric motors and v-belt drives, gear motors, and other conveniences, the engineering challenges facing the mill builder seem immense indeed. As an example, this Nordyke & Marmon mill of 1882 is set up with a unique belting arrangement whereby power could be transmitted to the vertical mill shaft from a horizontal countershaft. This quarter-turn conversion required considerable engineering, and indeed, some of this is evident in the unusual alignment apparatus of the rack-and-pinion style which carries the idler pulley in its standard.

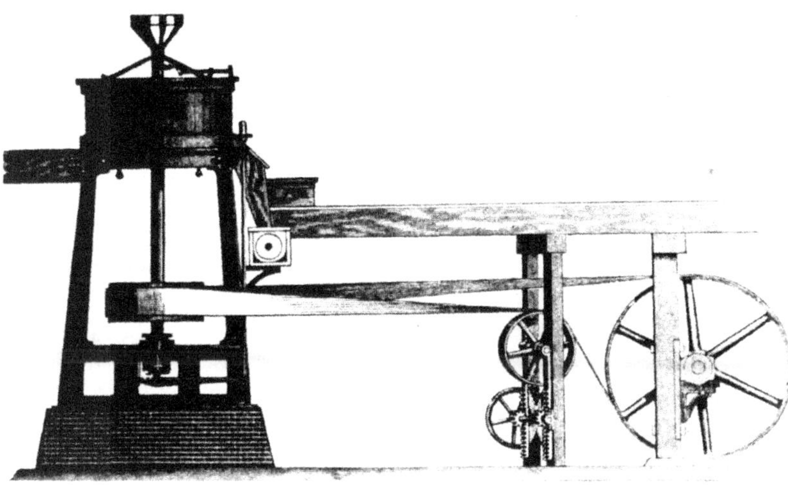

One of the best known Nordyke & Marmon mills was their Plantation Mill introduced in the late 1870s. The company continued building this design at least into the 1920s, although it would appear that after A-C assumed control in 1927, production finally ended. A 1919 Nordyke & Marmon catalog lists this unit in 14, 18, and 22-inch sizes carrying respective prices of $155, $195, and $280. Designed specifically for the use of farmers and others not familiar with the idiosyncracies of merchant mills, this little mill offered features that included a steel relief spring which automatically backed the runner stone away from the bed stone when the adjusting handwheel was released. With this mill, farmers could grind crushed ear corn for their livestock and when required could grind wheat or rye into flour for the kitchen table.

An 1882 catalog cut illustrates a Nordyke & Marmon Plantation Mill at work, being powered by a small portable steam engine. Although this catalog cut carries the Nordyke & Marmon title on the engine, it appears that this company did not actually build steam engines, but purchased them from other firms, particularly Atlas Engine Works of Indianapolis. Even though the 14-inch mill required only from 3 to 5 horsepower for successful operation, this went up to a range of 7 to 12 horsepower for the 22-inch size. The Plantation Mill was generally set up to discharge directly into a waiting basket, but could, at extra cost, be equipped with a bolting attachment that sifted the flour to some degree.

Flour Mills

Large merchant mills used a series of stone buhr mills, often driving them with water power. During the 1880s, neither Nordyke & Marmon nor the Edward P. Allis Co. built their own water wheels. Allis apparently sold and installed the Burnham wheel, and perhaps others, while Nordyke & Marmon settled on the famous Leffel turbine which was said to be the best known of all water wheels. The vertical shaft design of the Leffel wheel nicely matched the vertical shaft design of the upper-runner and under-runner flour mills, leading to some unique engineering designs.

Grinding ear corn required that it first pass through a crushing device, and Nordyke & Marmon responded with a heavy duty earn corn crusher. Livestock men liked this machine because ground feed was more palatable to their herd. In this setup the corn crusher is driven from the back side of the mill, with a small elevator carrying the crushed corn directly to the hopper of the mill. Curiously, Edward P. Allis Co. seemed to be interested only in the merchant mills, apparently having little or no impetus to build or market the small farmer-owned styles. Although Allis began business with the manufacture of buhr stones as a primary source of income, Gray's roller mill seems to have brought a hasty end to this line of work, while in contrast, Nordyke & Marmon appears to have been successful on both fronts.

Nordyke & Marmon offered a rather substantial line of elevator machinery and auxiliary equipment, while Allis concentrated primarily on roller mills and flour processing machinery. Elevator boots were a Nordyke & Marmon specialty, with the company ready to furnish a cast iron design as shown here, or elevator boots of wood construction. The Nordyke & Marmon engineering department was also able to fabricate special elevator equipment for specific applications.

Shortly after Edward P. Allis Co. introduced Gray's roller mill, Nordyke & Marmon responded with their own version. During the 1880s Nordyke & Marmon recommended the gradual reduction system of milling, suggesting that from one to three breaks be made on roller mills, with the final reduction being made on existing buhr stones. Their 1884 price list shows a figure of $470 for a four-roller mill equipped with corrugated rolls and in a 6 x 12 inch size. The largest model, using 9 x 30 corrugated rolls carried a price of $850. Beyond the obvious milling advantages, roller mills also required about one-third as much power as a comparable stone mill.

Like Edward P. Allis Co., Nordyke & Marmon offered little in the way of precleaning equipment. Instead the equipment of various specialist companies was used, with the Invincible scouring, polishing and separating machine being a case in point. Wheat and other grains had to be thoroughly cleaned before grinding. Many mills first put the grain through a receiving separator which eliminated most of the weed seeds, straw, and other foreign material. Also available was a magnetic separator with an 1882 catalog noting that "The introduction of wire-binding attachments to the reaper made this apparatus absolutely necessary." These machines used air currents, vibratory sieves, beater rods, and brushes, or combinations thereof in their work of cleaning the grain for milling.

Flour Mills

Quoting from an 1884 Nordyke & Marmon catalog, "We claim wheat cannot be made too clean so long as it is not injured in the operation . . ." With this in mind the company presented the California Brush Finisher, a machine that combined the action of the Brush Smutter and Combined Separator into a single unit. No riddles or shakers were used; this to prevent damage to the grain. Brush finishers used a revolving brush working against a stationary one, with the grain passing between them and literally brushing each kernel. The smallest of the California Brush machines had a capacity of 15 bushels per hour and carried a price of $125. The largest model was priced at $300 and could clean 125 bushels per hour.

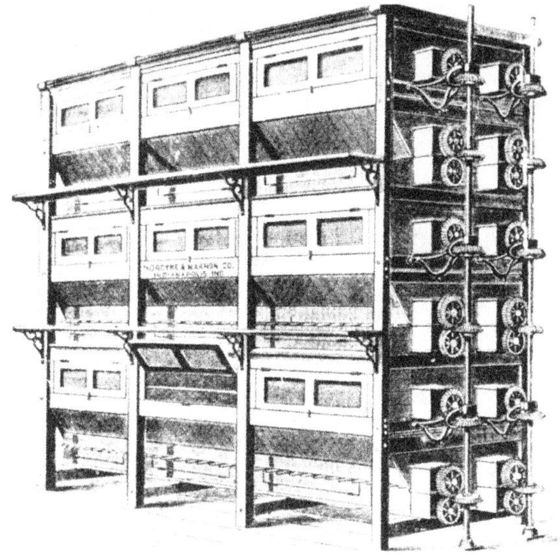

Early forms of English describe *boult* as a machine for sifting flour. Eventually the word was shortened to *bolt*, referring to a machine of this category, and specifically, *boult* referred to a screen or a fine transparent cloth used for sifting. From these terms come the bolting machines or bolting chests so familiar to flour milling. Nordyke & Marmon began business with Ellis Nordyke's 1851 patent on a bolting machine, and from it came a profusion of improvments, leading to this six-reel bolting chest of 1880. This particular machine took the First Premium at the First International Millers' Exhibition held at Cincinnati that year. A small mill might use but one of the six reels shown here, while a large mill might employ a substantial number of these huge six-reel bolters.

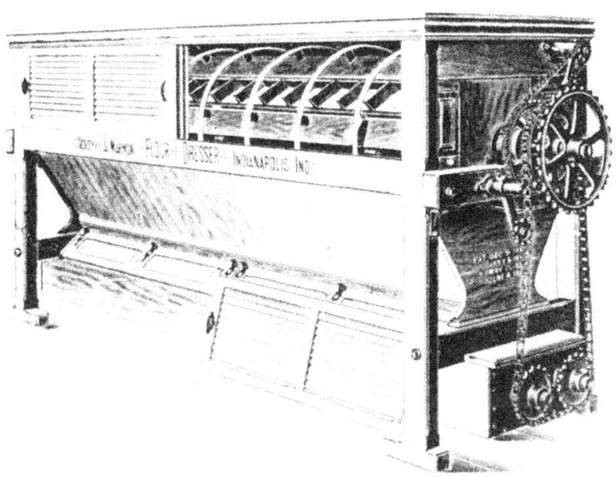

Unlike the bolter which generally used silk fabric for sifting the flour, the round reel flour dresser was designed for separating and bolting medium ground cereals and similar products. Typical applications included hominy, grits, corn meal, and similar items. These machines were constructed almost entirely of wood, expertly fitted, and with the exterior beautifully finished. All interior parts were of wood, with no metal touching the flour. Even the conveyors and spouts were made of wood, remaining substantially this way until the introduction of stainless steels. Despite a substantial amount of lumber and considerable labor, this particular machine was priced at about $360.

Those who have raised buckwheat know it is not part of the wheat family at all, deriving its name from the resemblance of its triangular fruit to beechnuts. The word has several ancient origins, among them being the German term, *buchweizen*. The early settlers often used buckwheat as the first crop because its heavy leaf canopy effectively choked the remaining weeds. Buckwheat flour was quite popular at the table and for livestock. Nordyke & Marmon offered this buckwheat shucker in 1882. It used iron rolls and a shaker sieve to remove, insofar as possible, the outer shucks and the black specks so characteristic of buckwheat flour. Only a few mills specialized in buckwheat flour, and its importance as a crop has diminished to the extent that "genuine buckwheat pancakes" remain largely a memory of days gone by. This little machine was priced at $150 for 1882.

An advertising cut from a Nordyke & Marmon catalog typifies the custom or merchant mill of the 1880s. To the left is the boiler and steam engine in a separate building — it is belted directly to a large stone buhr mill. The grain is received at the first floor level, and is conveyed to the second story where it drops to a cleaning separator directly over the mill. A second elevator, not shown in this engraving, carries the ground flour to the bolter, with the finished flour being received below into a flour packer — a mechanized device for putting the flour into sacks or barrels. Nordyke & Marmon, like Edward P. Allis Co., was ready, willing, and able to install custom mills from start to finish, and in fact, built a great many merchant mills.

Flour Mills

A1920s photograph illustrates the Nordyke & Marmon roller mill equipment at the Eckhart & Swan Milling Co. of Chicago. This paticular mill had a capacity of 1,200 barrels of flour per day. The *Farmer's Encyclopedia* of 1891 notes that a barrel of flour weighs 196 pounds. This tome also indicates that on the average, 2½ bushels of wheat were required to make 100 pounds of flour. Thus, a 1,200 barrel mill would require nearly 6,000 bushels of wheat per day to produce over 235,000 pounds of flour. The open flat-belt drives would be a horror to today's safety engineers, but to the credit of this installation, the lineshafting is mounted on the floor below.

To the credit of Nordyke & Marmon, this company offered an extensive line of corn milling machinery in addition to the roller mills designed more particularly to wheat, rye, and similar grains. This three-roll, two-break corn mill typifies the line, having a capacity of 10 to 15 bushels of meal per hour. Spreading the rolls for a coarser grind, as for livestock feed, gave a capacity of 30 bushels per hour. Nearly a century after Edward P. Allis Co. began building roller mills, it is now, and probably will remain a mystery as to why they failed to enter auxiliary markets using modifications of the Gray design.

Another example of the Nordyke & Marmon aggressiveness was this Silver Creek corn meal bolt and refiner. Knowing that corn meal for human consumption was a sizeable market, Nordyke & Marmon offered virtually a complete line of equipment, beginning with the receiving elevators to refiners like the one shown here. As its name implies, this unit was intended to screen the corn mill and sift out the fines, thus leaving a highly desirable product. The corn starch industry was yet another example of Nordyke & Marmon endeavor, with some of the equipment lending itself admirably to the tasks of screening and purifying this product.

Nordyke & Marmon was one of the few companies to build a samp mill. *Samp* was a term picked up by the early settlers from the Massachusetts Indians and is a corruption of their term, *saupac*. The term is nearly synonomous with *hominy*, another Anglicized version of an American Indian term. This machine was intended to simplify the making of hominy, and combined the breaking and hulling apparatus with a bolting reel which delivered the finished samp, completely separated from the offal. Nordyke & Marmon also supplied rice grading reels, rice glazing drums, and many other machines for the milling industry. It appears however that the company's preponderance for automobiles, coupled with Howard Marmon's determination to compete with General Motors might have been a factor in the 1926 sale of the flour mill machinery division to Allis-Chalmers. Tragically however, Marmon's dreams of becoming another automotive mogul vaporized with the Great Depression.

Now seldom seen, the mill sheller shown here was once rather popular. Farmers wishing to sell ear corn directly to a grain elevator could do so, with the elevator doing the shelling as the corn was unloaded from the wagon. Nordyke & Marmon offered small spring shellers in one and two-hole designs — these Eagle shellers were probably built by Eagle Mfg. co. of Lancaster, Ohio. The large commercial sheller shown here was small by later comparison — the largest one, called Size No. 2 had only a 75 bushel per hour capacity. On entering the shelling cylinder, the corn dropped directly onto a vibrating screen, while a built-in fan discharged the shucks and dust to a remote location. The cobs emerged from the cylinder and dropped through a chute to the floor below. List price for this machine was $120.

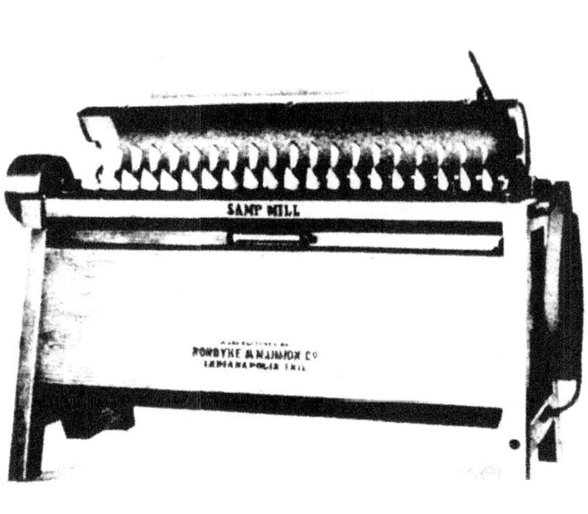

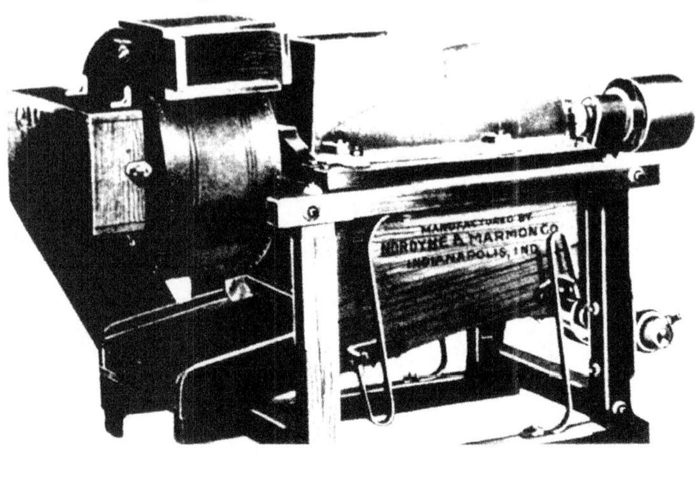

Flour Mills

A curious application of the stone buhr mill was this specially designed coffee mill. A 1919 Nordyke & Marmon catalog notes that "the mill is neatly finished in red with heavily nickeled trimmings which, with the polished brass hopper, gives it a very handsome appearance." Nordyke & Marmon built this little ten inch mill for grinding coffee, spices, and similar materials. Because of its specialized design, it would be likely that sales were rather low, making it even more of a delight for today's sophisticated collectors wanting something a cut above the usual. In 1919 this machine sold for $200, with part of the expensive price probably due to rather low production, and highly finished parts also contributing mightily to the final price tag.

Reliance Works of Edward P. Allis & Co. advertised "Fr. Wegmann's Patent Porcelain Roller Mill' in an 1880 book entitled, *Practical Hints on Mill Building*. Allis had secured manufacturing and sales rights to the Wegmann design, offering it initially, but soon modified it to the designs of William Dixon Gray, Allis' superintendent of this department. The Wegmann roller mill, developed in Europe, had attracted considerable attention because of the high-quality flour it could produce. Despite the substitution of v-belt drives for flat belts, the use of steel instead of cast iron for the frame, and the change to roller bearings from the original babbitt style, roller mill design changed but little from the concept manifested in this machine of 1880.

Edward P. Allis Company began business dressing and fitting millstones, among other things. This remained a major part of the operation until William Dixon Gray introduced his roller mill through the company in 1878. At that point the company seems to have lost no time in eliminating this line of endeavor, preferring instead to concentrate solely on the roller mill. Allis used imported French buhr stones exclusively — millers had learned that these were superior to all others. The careful selected blocks of stone were then shaped by hand. The bed stone remained stationary, and was usually set into the frame with cement. The runner stone was reinforced with an iron band shrunk over its periphery. Both stones were dressed, that is, furrows of specific dimensions and of a specific design were cut into the stones. The millstone dress shown here is fairly typical, although records of the United States Patent Office indicate that well over sixty different designs of millstone dress had been patented by 1873.

Gray's Patented Noiseless Roller Mill was an overnight sensation! Despite some early doubts by merchant mills, these were quickly banished when the high quality flour from these mills was traced to the kitchen. Pillsbury Mills did in fact, follow some trial batches of the flour produced by these mills all the way from Minneapolis to New York and quickly found that the new "Patent" flour received enthusiastic welcome. Gray's first major change over the Wegmann design was to eliminate the noisy gearing which the latter used on the roll drives, replacing the gears with leather belting. Shown here is a small package roller mill system, complete with a conveyor to carry the whole grain to the rollers.

Flour Mills

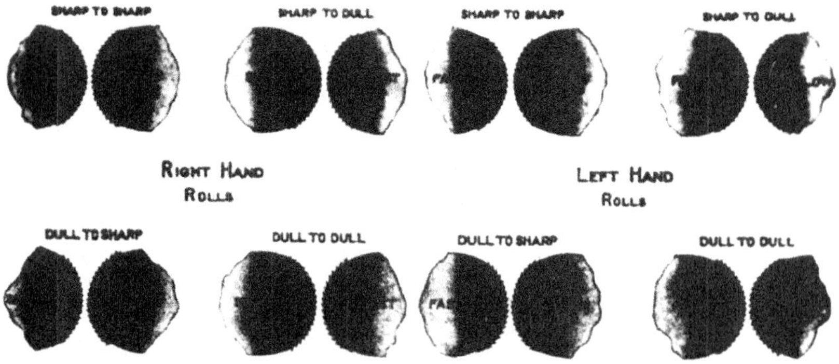

Various types of grain and various milling methods required different arrangement of the rolls. The corrugated rollers could be set: 1) sharp to sharp; 2) sharp to dull; 3) dull to sharp; and 4) dull to dull. These four possible configurations for right hand rolls were also possible with left hand rolls, making eight different configurations possible. The grinding operation uses corrugated rolls termed "breaks" for reducing wheat to middlings and cleaning the bran, while smooth rolls are used for middlings reductions. Early practice at least, dictated 12 corrugations per inch for the first break, and by the fifth break this was up to 24 corrugations per inch of roll periphery. In all cases, one roll runs faster than the other, with a typical first break set running the fast roll in a 2½:1 ratio, with a typical middlings break using a speed differential of 1⅓:1.

A very early, but undated price list of Edward P. Allis & Co. lists Gray's Patent Noiseless Belt Roller Mills in sizes from 9 x 14 to 9 x 30 inches. The first figure denotes the diameter of the roll, with the second figure indicating the length. Within these varying sizes were many different configurations, each designed for a specific part of the milling process or for the capacity requirements of the individual mill. A typical instance would be a divided four-roller machine having one set of corrugated rolls and one set of smooth rolls. A machine of this arrangement, and using 9 x 14 rollers carried a list price of $500. In most mills, emphasis was placed on *gradual* reduction of the grain to flour, along with *gradual* purification of the flour. This regimen required bolting and purifying equipment after each break in order to separate the bran, the middlings, and the flour.

For their own in-house use, as well as to satisfy the experimental needs of merchant mills, Allis-Chalmers offered this experimental reduction machine. It was intended for handling only small amounts of grain, and in fact, the reduced material was deposited in a small drawer beneath the machine. The experimental mill was so designed that various rolls could be used interchangeably, and speed ratios between rolls could also be altered with relative ease. Having this type of equipment available was essential to merchant mills, especially when encountering new varieties of wheat with different milling characteristics. Speed differential changes when required on the production mills could then be made with a certainty of being correct, and prototype rolls could be made for special applications at far less cost than producing full size models.

For checking the milling characteristics of the grain prior to a run, Allis-Chalmers offered the Milwaukee Granulator, also known to the trade as a "post roll" since it was usually mounted on a supporting post of the mill building. This simple little machine was belt driven for the convenience of the operator. Simplistic though it is, it demonstrates once again the propensity of Allis-Chalmers to supply virtually any needed equipment for reduction milling, including the necessary "test" equipment.

Gradual reduction milling using the roller mill requires tempering of the grain before the first break. Tempering prevents the total destruction of the bran to powder during the first break. This has historically been achieved using either water or steam. Even though this required fairly simple equipment, Allis-Chalmers as late as 1909 noted in a company *Sales Bulletin* that "when desired the Kirk Water Tempering Machine is supplied, but not made by us." In fact, a careful study of the A-C flour mill machinery line indicates that the company built virtually no machinery that was used prior to the roller mill, although they offered a most extensive line of bolting and purifying equipment. Even after acquiring Nordyke & Marmon Co. in 1927, Allis-Chalmers seems to have opted for other vendors when it came to grain cleaning and tempering equipment.

Flour Mills

The Allis-Chalmers Reliance Purifier was sold for many years, finally being replaced in 1921 with the Allis Ball Bearing Purifier. Except for the change to ball bearings on the fan and eccentrics, this machine remained virtually unchanged from its original design. As previously noted, gradual reduction milling also included gradual purification of flour, that is to say, the gradual elimination of bran and middlings, as the other half of the equation. Although this equipment seems simple enough from the exterior, there was in fact, a great deal of engineering involved, along with almost endless trials and experiments, all aimed at getting the absolute maximum yield from every bushel of wheat.

Early in 1920 Allis-Chalmers announced a new line of package packers. These were designed initially for the flour milling industry, but the company noted that the same machines had, or had the potential at least, for use in many applications, including cereals, clays, cotton seed hulls, graphite, pigments, soaps, starch, sugar, and many other products. According to various Allis-Chalmers literature, the company had worked for several years on a successful design before this machine came to the market. As its name implies, the packer was intended to at least partially automate the packaging process, doing so with remarkable efficiency compared to previous methods.

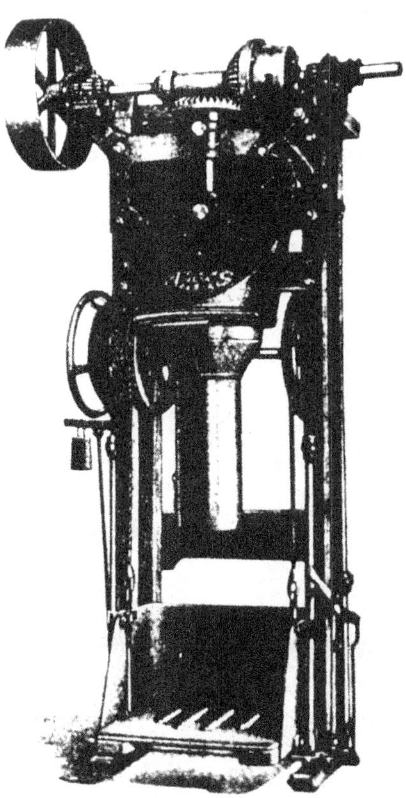

By 1909, and perhaps earlier, Allis-Chalmers had introduced their Universal Bolter with an overhead Vibromotor Drive. As milling technology developed, bolting or sifting, came to be a generic term which included three distinct stages. Scalping separates the coarse particles of broken wheat from the flour and middlings made on the break rolls. Grading consists of separating or grading the different sizes of middlings made by the corrugated rolls so that each size may be treated on separate purifiers. Flouring, or actual bolting, is the final stage in the production of flour, originally using a very fine grade of silk cloth. State of the art flour bolting for 1910 saw the best grades of bolting cloth coming from Switzerland. This material ranged from No. 10 silk with 11,236 meshes per inch to No. 14 silk with 19321 meshes per inch.

"A Study in Symmetry" might be a good title for this photograph. Typical flour mills of the 1920s (and much later) used flat leather belts operated from a central lineshaft. Joints in the belts generate an unmistakeable clickety-clack sound as they pass over the pulleys, and with dozens of belts in evidence, each with its own individual sound the performance was nearly symphonic. Many old-timers have related to the Author that listening to the lineshafting was one of the most relaxing things ever invented. Compared to the work they accomplished, the leather belts caused little problem, with injuries from the open drives being a greater cause for concern. Also to be reckoned with, certain conditions could generate static electricity in the bolting room. This, in combination with the flour dust, was probably responsible for many flour mill explosions, some of them tragic indeed.

Flour Mills

Recognizing a continuing market for crushed oats as livestock feed, Allis-Chalmers offered the Oat Crushing Rolls in 1923. The company noted, that although motor trucks were coming into use, a great many horses were still being used for "delivery of ice" and other purposes. The oat crushing rolls were, as the name implies, smooth rolls which flattened the kernels, making them more palatable, easier to eat, and far easier to digest. For a company which seems to have specialized in the merchant mill, it is indeed a change of direction in approaching a relatively small market, and one that embraced virtually no aftermarket. However, during the 1920s commercially produced livestock feeds began to be produced on a more or less localized basis, whereas in earlier days commercial feed production was almost completely under the control of large milling companies.

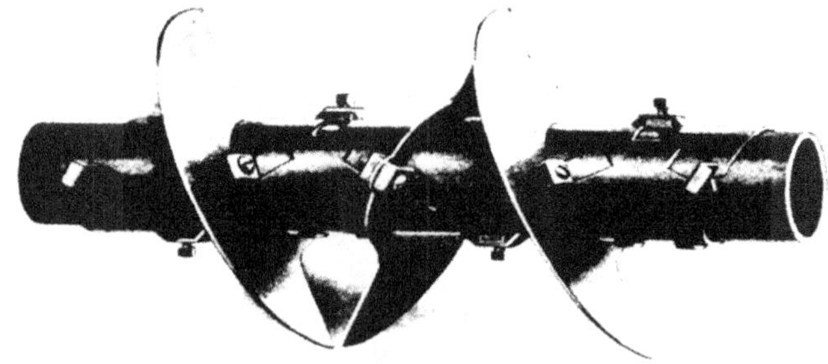

While perhaps insignificant in the total perspective, the Allis-Chalmers reversible auger of the 1920s was indeed a unique addition to the line. This auger was designed so that the flow could be divided from the center to each end, from each end to the center, or whatever configuration was required for a specific job. In the 1920s, divided flow augers were not easy to build, not at least without having a factory at the ready. Changing circumstances in a mill also required that the product flow be changed with a minimum of bother, and for these purposes the reversible auger was an answer. Using a special clamp arrangement for each individual section of the auger flight, these were so mounted on the tubular shaft that they could be removed or turned to suit the operation at hand.

During the 1920s Allis-Chalmers introduced the Vertical Grinder. Intended primarily for the livestock feed industry, it was an attrition mill with a 3600 rpm roller bearing motor mounted vertically within the base. This gave a very compact machine. A 1927 letter from a user indicates that the 25 horsepower model could grind about 4½ tons per hour of ordinary livestock feed, although for very fine grinding the capacity dropped to 3,500 pounds per hour. During the 1920s decade, Allis-Chalmers seems to have been somewhat more interested in developing other product lines for the Flour Mill Division rather than concentrating solely on roller mills and their attendant equipment. Some 40 years of roller mill building on the part of A-C and its competitors had brought the market nearly to saturation.

Allis-Chalmers experimental mills found their way into the linseed oil industry by 1925. Set up with an A-C Texrope drive, this roller mill is equipped to crush flaxseed. Flaxseed is used in the manufacture of linseed oil, and it is advisable to learn the oil-bearing qualities of the seed before purchasing it. In a note to their salesmen, A-C indicated that some of these mills had already been sold to linseed oil manufacturers, as well as to the U. S. Government. During the 1920s Allis-Chalmers also experimented with equipment required for the manufacture of soybean oil as extracted from the seed. At this point in time, soybean oil had assumed little of the importance as a cash grain crop that would be attached to it in coming decades.

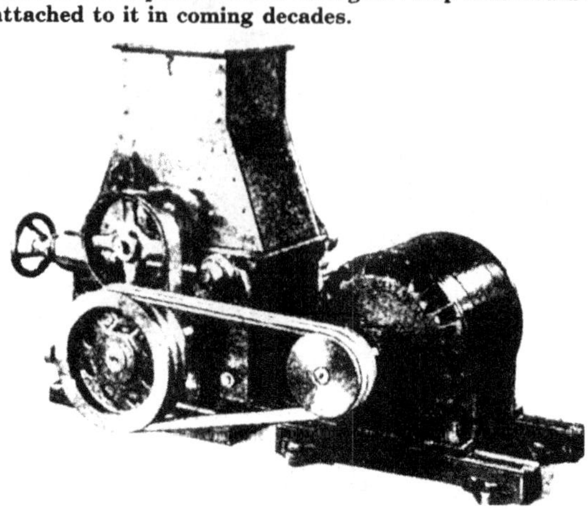

In 1930, Allis-Chalmers announced the new Universal Purifier. This machine featured an aspirator using a feed roll to deliver the middlings in an even sheet to the action of the air current. Opposed ball bearing eccentrics insured a positively balanced machine. The main suction pipe was located at the extreme tail end of the machine where the strongest air current was required. Although this machine obviously uses ball or roller bearings, it strangely enough, shows no v-belt drives which by 1930 Allis-Chalmers was promoting with all possible vigor.

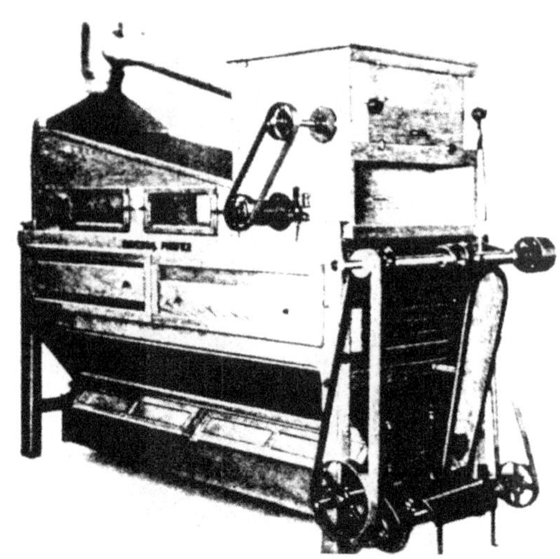

Flour Mills

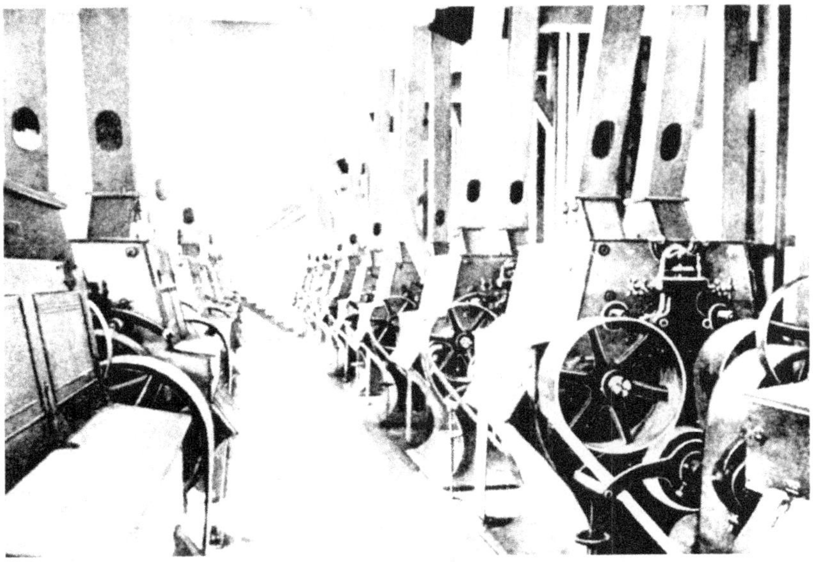

A 1920s scene from the Atlanta Milling Co. of Atlanta, Ga., illustrates two of the four lines of roller mills in this establishment. Allis-Chalmers specialized in these installations, and to their credit, the company had made, supplied, and installed some of the world's largest flour mills. Despite care in manufacturing the machinery, and care in handling the reduced grain, dust was often a problem, with the extremely fine flour leaking through places thought to be virtually airtight. As noted earlier in this section, flour dust in the right proportions with air can be, and is, highly explosive. Even with today's high technology which includes sophisticated methods of detecting potential explosions of flour dust, even before they happen, flour mill explosions still occur, often with a great cost of human life.

Nordyke & Marmon Co., acquired in 1927, brought several new products into the Allis-Chalmers line. These included this flaking mill. During 1931 A-C announced their entry into the breakfast food field with their flaking roller mill. A typical design used smooth steel rollers, 20 inches in diameter and 24 inches long. The rolls were designed so that they could be heated or cooled on their interior as required. Corn flakes for instance, required proper preparation and cooking of the grits, along with the addition of syrups and other substances required by a specific formula. After cooking, the grits are dried to about 15% moisture and then placed in tempering tanks. After leaving the tempering tanks the material is sent to the flaking rolls which flake the product to the desired thickness. From the rolls the flake material passes through rotary toasting ovens, and from there to the packaging department.

Although Allis-Chalmers had been building a barley crushing machine at least since the 1920s, the 1927 acquisition of Nordyke & Marmon brought with it an extensive line of malt mills designed especially for the brewing industries. Effective January 16, 1920, the Eighteenth Amendment prohibited alcoholic beverages in the United States, so for the next 13 years, until December 5, 1933, there was little market for distillery equipment. An exception however was the Act of Congress allowing for 3.2% beer. After a 13-year hiatus the brewing and spirits distilling industries were anxious to resume business, requiring a great deal of new equipment in the process. For these requirements Allis-Chalmers offered malt cleaners, malt reduction mills, and various other machines unique to this industry.

Designed especially for producing commercial livestock feeds, the Allis-Chalmers attrition mill of the 1930s was indeed a demon. Two large electric motors drove the grinding disks in opposite directions, literally disintegrating the grain. Recognizing the problem of dust and its attendant dangers, Allis-Chalmers used totally enclosed self-ventilating motors. The intake and exhaust connections for each motor are obvious in this illustration — metal piping carried air to and from each motor. Although attrition mills were quite popular, hammer mills were rapidly coming into use for commercial livestock feeds. Due to their simplicity of design the hammermill soon overtook the attrition mill in popularity.

With the introduction of the Tex-Rope Drive patented by Allis-Chalmers the company was quick to adapt their own machinery to this unique system. In so doing the flat leather belts of the past were eliminated, and were replaced with drives subject to little of the slippage formerly associated with flat belts. Power was also transmitted more efficiently and maintenance of the v-belt drive was minimal. This Style A roller mill of 1933 is little different in principle than Gray's Noiseless Mill of 1880, although by this time the roller mill was equipped with ball bearings, improved feed devices, and other niceties.

Flour Mills

A 1937 photograph illustrates the Allis-Chalmers five-sieve low-head gyratory sifter. Of far smaller dimensions than the huge bolting reels of earlier times, the gyratory sifter had its own self-contained motor drive, requiring none of the laborious work of setting up lineshafts and aligning a series of pulleys. This self-contained machine was secured to rubber mountings that required no special modification of the building for installation. Always at the forefront of self-contained drives, Allis-Chalmers was a leader in converting industry away from the time honored lineshaft system. In fact, the Allis-Chalmers Tex-Rope patents for the multiple v-belt drive were probably one of the company's most significant achievements.

Square sifters 27 sieves high saved a great deal of space on the bolting floor. These units of 1937 vintage typified state-of-the-art design at this time. The chests were made of the finest wood. Due to the severe service demanded by the vibrating sieves, remaining examples of these huge sifters are a tribute to the art of wood joinery. After being completed the sifters were given a high quality finish comparable with the best furniture. A series of spouts beneath each sifter is connected through a flexible cloth connection. Various grades of flour are derived from each level and routed within the sieve to the appropriate discharge, finally ending up on the floor below. The multiple supports at each corner of the sifters are actually wooden rods.

Another 1937 photograph illustrates a 10 x 42 inch three-reduction mill. Designed especially for the distillery industry, this sieve-type mill embodies a substantial degree of safety engineering. Although some shafts and belt drives are still unprotected, much of the mechanism is completely enclosed. Enclosing the working parts was not entirely aimed at safety however. The development of ball bearing drives required shielding of the high precision surfaces from dust and contaminants. Furthermore, lubrication was a problem. From these needs came the development of enclosed gear boxes with all parts, bearings included, running in a constant bath of oil. This forward step in mechanical design, while more expensive initially, lowered overall maintenance costs immensely.

Already in the 1920s Allis-Chalmers was experimenting with various means of extracting oil from soybeans. As soybean oil became more important so did the need for specialized machinery to achieve this goal. Shown here is a 1937 model of a 32 x 40 inch flaking mill designed specifically for a soybean solvent extraction plant. While no specific data has been located on this unit, one's imagination quickly conjures an image of a huge machine indeed, using the 32 inch roll diameter as a yardstick. During the 1930s Allis-Chalmers stood ready to design virtually any machine for the milling industries, and did in fact, develop some unique machines during this period.

Flour Mills

During the 1930s the Flour Mill Dept. learned firsthand about hard times. Sales were poor, partially because the aftermarket was very small, partially because of market saturation, and partially because of the great depression. These and other factors led to a decline from which this department never recovered. Meanwhile however, new developments emerged, with this 1937 photograph illustrating a self-contained battery mill. It used two direct-drives and a base integral with the mills. The design was unique and represented some excellent engineering, being in fact, a sign of things to come. Sales were however, disappointing at their best.

A 1947 photograph illustrates the interior of a large Kansas mill driven by Nordyke straight-line double roller mills. They were group-driven from lineshafts mounted on the ceiling of the floor below. Even with a limited amount of advertising data available for this research project it is obvious that after the Nordyke & Marmon acquisition, Allis-Chalmers selected what they considered to be the best machines from both lines, relegating the remainder to obsolescence.

A 1956 photograph illustrates nine of 30 new Allis-Chalmers double roller mills destined for a new flour mill in Haiti. At this point in time all moving parts are shielded or enclosed. Ball bearings have now been standard equipment for some time, with glass inspection doors preventing foreign material or human hands from contaminating the grain. Even with these changes, Allis-Chalmers roller mills were still of the same basic design as envisioned by William Dixon Gray some 75 years earlier. This in fact was one of the major problems with the roller mill — its initial design was of such caliber that subsequent improvements were inconsequential, and certainly not of enough importance to warrant the installation of all-new equipment.

During the 1950s Allis-Chalmers installed a substantial number of flaking mills for the Kellogg Co. This illustration shows 15 of the latest Allis-Chalmers design at Kellogg's Battle Creek, Mich., plant. Each of the mills shown here uses 28 inch rolls, with other Kellogg facilities using another 31 mills of this size. In addition, Kellogg was at that time operating another 135 Allis-Chalmers flaking mills using 24 inch rolls. As is obvious from this photograph, the milling industry of 1956 looked far different than the wooden building with a steam engine and lineshafting shown earlier in this section. Wood conveyors have been replaced with those of stainless steel. All moving parts are completely enclosed, and lighting comes not from rows of windows but from modern fluorescent fixtures. Modern technology permitted the development of improved roll designs, with the operator being able to regulate roll temperature almost at will.

A note accompanying this 1966 photograph states, "More than 80% of all ready-to-eat breakfast foods produced in this country are made with A-C flaking mills, shredding mills, and rotary dryers." This particular illustration shows one of eight flaking mills installed the previous year for a leading breakfast food manufacturer. Allis-Chalmers noted that these eight machines were capable of producing 4,800 boxes of flaked cereal per hour! Since, as noted, 80% of all flaked foods were being produced on A-C machinery, it seems obvious that the market saturation had been reached. Stated another way — Allis-Chalmers now had far more production capacity for roller mills, flaking mills, and related equipment than it had prospective buyers.

Forage Equipment

A 1940s factory photograph illustrates an early example of the Allis-Chalmers forage blower. The stiff hopper used on this machine required backing the forage wagon up to the hopper for each load, and for inexperienced tractor operators, backing up a wagon can be a tremendous problem! This particular unit is equipped with an A-C engine complete with variable speed pulleys — most likely from the already well-known "60" grain harvester. Allis-Chalmers was an industry leader in the use of auxiliary engines on a wide variety of farm equipment. While sufficient power could be delivered with a drive belt, aligning the tractor to the blower was always a problem, especially since anchoring the blower was oftentimes difficult. Tractor PTO drives of the 1940s were in many cases not of the caliber required for heavy loads, so drive belts or integral engines as shown here came as the only alternatives.

After a great deal of experimental work, Allis-Chalmers announced a new forage blower in 1949. This unit was regularly supplied for belt drive, although it could be supplied with an engine drive at extra cost. The 9-foot feeder permitted easy unloading — it was equipped with two heavy coil springs to balance the weight for easy raising and lowering. A conveniently placed throwout lever controlled the feeder; it could be operated from either side of the machine. The first series of Allis-Chalmers forage blowers were marketed between 1949 and 1962 with over 31,000 units being produced during this period. Shown here is the usual long-hopper version, together with an A-C auxiliary engine. Optional equipment included a short hopper designed especially for trucks and wagons equipped with a built-in unloading mechanism. Once installed, this $155 attachment became an integral part of the blower. The attachment included the necessary gear reduction case and telescoping driveshaft whereby the power required for actual wagon unloading was derived from the blower itself. Allis-Chalmers forage blowers were unique in that they used a 9-inch discharge pipe which offered 65% greater carrying capacity than the 7-inch pipe used by most of the competition.

The 30-L forage blower introduced in 1949 was complemented with the 30-S unit from 1962 to 1964. The suffix codes translated to machines having either a long feeder or a short hopper. Although actual production of the 30 Series seems to have ended by 1964, this series is listed in Allis-Chalmers Dealer Price Books as late as 1967. Meanwhile it was being replaced with the 480 Series blowers. shown here is an experimental version of the 480-L blower. A comparison of this machine to the earlier 30 Series indicates that at this point at least, many of the parts were actually from the earlier models, but were undoubtedly modified at a later stage of development, and certainly were upgraded to full-fledged 480 Series status before actual production.

Shown here is another 480 Series blower, this one having the short hopper for direct unloading from a power feed box. In fact, the box itself was an A-C design of the early 1960s, a time when Allis-Chalmers was determined to capture a larger share of the forage equipment market. By the early 1970s this determination had all but vanished. It is easy to speculate on the reasons for this change of direction, but some of the possible factors include a change in farmer demand, the company's sales approach or faults thereof, and changes in the corporate approach. Regardless of the factors or combinations thereof, Allis-Chalmers forage equipment probably achieved but a small percentage of its real market potential.

The 480 Series forage blowers were offered as 480H, 480S, and 480L models. The suffix codes designated the "H" as having a short hopper with an auger conveyor; the "S" as using a short hopper with a raddle chain conveyor; and "L" was a unit equipped with an 8½ foot hopper conveyor using an endless rubber belt. All models of the 480 Series featured a 6-blade fan with formed cups. The 48-inch diameter housing was 9 inches wide. All three units weighed approximately 1,200 pounds each. A 1967 Price Book indicates that the 480H was listed at $588, with a figure of $738 for the 480S model. Topping the line was the 490L blower with a list price of $965. Most 480S models could be converted to the 480L blower as an extra cost option of $295. Production of the 480 Series ran from 1964 to 1971.

In the 1972-1974 period Allis-Chalmers offered the 660 forage blower. It was somewhat larger than the 480 Series, having a 60-inch diameter housing. Apparently the long hopper was discontinued, even as an extra-cost option, although the short hopper, chain raddle design of this illustrated model could be substituted for an auger conveyor. Like previous A-C forage blowers a throwout lever was situated directly over the conveyor as a safety precaution. This unit, like its predecessors was furnished with one telescoping pipe section and a curved delivery hood. Extra equipment included delivery pipe in various lengths, flexible elbows, and other attachments required for specific applications.

During 1940 Allis-Chalmers announced their new forage harvester. This simple little machine used the same type of header as was found on the then-current "40" All-Crop grain harvester. A close look at the design also reveals that additional parts of the design were modified from the same machine. The company claimed that their Model B tractor would handle the new forage harvester — this was due in part to the use of an endless rubber conveyor to fill the wagon, rather than the usual blower found on most machines. This design was in itself a unique feature, illustrating that Allis-Chalmers engineers recognized the abundance of small tractors on farms and designed a machine to match the available horsepower. Only 25 of these units were built in 1940, with another 150 coming out the following year.

A very rare picture coming from somewhere within Allis-Chalmers departments illustrates some unusual equipment. Powering this A-C forage harvester is the company's Model F tractor, an experimental unit of the 1940s which went approximately to the prototype stage and from there to the shelf. The belt conveyor system of this forage harvester is still in evidence, and in this case is delivering cut corn ensilage to a GMC truck of late 1930s vintage. Allis-Chalmers promoted the forage equipment as constituting a one-man system, compared to the large work force required for cutting corn with a corn binder, loading it on hayracks, and feeding it into an ensilage cutter at the silo.

A 1943 photograph from the Allis-Chalmers Archives illustrates the company's forage harvester at this point in time. Due to World War Two, production of the forage harvester was suspended in 1942 and did not resume until 1947. Demands for military equipment probably kept design work to a minimum during this period, although this particular unit illustrates an improved pickup attachment, and is equipped with an A-C auxiliary engine. As previously indicated, much of this early design was actually borrowed from the "40" All-Crop harvester.

About 1948 Allis-Chalmers substituted a blower for the earlier conveyor equipment on the forage harvesters. The sales effort included reference to their "2-in-One" design that allowed the forage harvester to be used for standing grasses, using a direct-cut attachment; a pick-up forage harvester utilizing a raddle-type pickup, or a row-crop machine when equipped with the special row-crop attachment. An early 1950s photo shows an Allis-Chalmers WD tractor pulling the A-C forage harvester with a pickup attachment. Since this particular unit is PTO operated, it can be safely assumed that the WD had all the load it needed most of the time, and perhaps more than needed at other times. The A-C forage harvesters were equipped with a knife grinding attachment to maintain them as razor-keen cutters — allowing them to dull to any extent greatly increased the horsepower requirement.

Since the A-C forage harvester could be set up in numerous configurations, pricing began first with the base machine which carried a 1957 base price of $1,030. Optional lift equipment, either hand or hydraulic was furnished at no extra charge, but the PTO attachment added $130 to the base price, while an auxiliary engine raised the base by $551. The row-crop attachment for standing corn listed at $427, compared to $383 for the standing grass attachment. A pickup attachment for windrowed crops added by $97, although it first required use of the standing grass attachment, being mounted to this device. Shown in this field scene is the A-C forage harvester with an auxiliary engine and the pickup attachment. It is being pulled with an A-C WD tractor, and cut forage is being delivered to an A-C forage box.

Introduced in 1959, the Model 50 forage harvester replaced A-C's earlier style. Built somewhat heavier than its predecessor, the Model 50 was available with either four-knife or six-knife cutting heads. A knife sharpener was built in as standard equipment. Also included was a rear swinging drawbar for trailed wagons. Allis-Chalmers recommended a three-plow tractor or larger to operate this machine, but could also supply the Model 50 with the D-14 auxiliary engine at extra cost. Row crop and grass units were available, as was a pickup unit to be used in conjunction with the grass attachment. Although some A-C records indicate that production ended in 1962, others show that these machines were available as late as 1965. The base machine was priced at approximately $1,228 with field attachments adding somewhat to this figure.

Introduced in 1963 the Model 782 forage harvester was available until 1975. This high-production machine carried a base price of about $3,000. Several attachments were available, including a 6½ foot pickup at $800; 1-row attachment, $946; 2-row attachment, $1,200; 3-row attachment, $1,800; and a direct-cut mower attachment with a list price of $1,200. The cutting mechanism consisted of a 300-pound cylinder mounted in two large roller bearings. Six 36-inch spiraled and cupped knives were used, and this could provide cutting lengths as close as ¼-inch. By using three or two knives, plus interchanging gears in the main gear box, a total of twelve cutting lengths were available all the way up to 2¼ inches. Standard equipment included an automatic clutch stone stopper for the feed rolls, an adjustable pitch delivery chute, and a concave bottom.

Offered between 1961 and 1966 the Model 56-F flail harvester was yet another approach to forage harvesting. This machine was equipped with a 72-inch rotor. Its 36 knives were mounted to a 24-inch diameter rotor. Operating at its rated speed of 1,456 rpm this rotor with a top speed in excess of 100 mph made short work of cutting and pulverizing the standing crop. Allis-Chalmers recommended a "full three plow tractor or larger" to operate this machine, so it can be imagined that the D-17 Allis-Chalmers tractor shown here had its work well planned. During its five year production run, only about 1,500 machines were built, probably a disappointingly low figure for a machine that listed at under $1,400. This photograph also illustrates an Allis-Chalmers forage box.

The 720 Forager Field Harvester was yet another Allis-Chalmers style. Offered from 1972 to 1975 this machine could be equipped for either a 540 or a 1000 rpm PTO drive. A huge 108 inch square feed opening gave a huge capacity. Another important feature was a reversing mechanism, controllable from the tractor seat, that permitted reversing rotation of the feed rolls and row-crop gathering unit for fast clean out. Shear pins protected the cylinder drive and drive line. Eight tungsten carbide knives on the cylinder were designed to cut and throw the crop in a single operation. A built-in sharpener permitted fast, easy honing of the knives without ever removing them from the machine. In addition to the pickup attachment shown here, the 720 was available with one-row or two-row crop attachments.

Having a capacity of 60 tons or more per hour, the 780 forage harvester required 50 PTO horsepower at the bare minimum, and likely had a requirement nearly double that figure when operating at its maximum capacity. Offered between 1967 and 1975 the 780 could be furnished with one, two, or three row attachments, a 6½-foot pickup attachment, or a 6½-foot mower attachment for cutting standing forages. With a base weight of nearly two tons, this big machine carried a price of about $3,000, a figure that might be substantially higher after purchasing the desired cutting attachments. This field scene illustrates a 780 forage harvester behind an Allis-Chalmers D-17 Series III tractor. Bringing up the rear is an Allis-Chalmers forage box.

Two models of the two-row standing crop attachment were available. One was built to accommodate row widths of 36 to 42 inches, while the narrow row version handled row widths of 28 to 32 inches. Cutting two rows at a time required considerable power, although by the late 1960s many farmers had upgraded their tractor fleet to include at least one large workhorse for jobs like forage harvesting. Five feed rolls were used on the 782 forage harvester to provide an even flow of material to the cutting cylinder. Allis-Chalmers also offered a 3-row standing crop attachment. It was built to work on 20-inch rows. As a practical matter, cutting three rows of standing corn at a time dictated a tractor at least in the 90 horsepower class, with many operators topping the 100 horsepower mark for powering this machine.

By the late 1960s Allis-Chalmers once again offered a Power Feed Box that utilized state-of-the-art design. Of the same dimensions as the 1962 version, this design used beaters to provide uniform delivery to the cross conveyor. As in the previous model, two separate unloading raddle chains moved material forward to the beaters. Complete shielding of all moving parts is evident. The unloading mechanism could be operated by any tractor having a standard PTO shaft, and by using the four different settings on the ratchet drive and by varying the PTO speed, the rate of unloading could be precisely matched to the job at hand. For a time at least, Allis-Chalmers appears to have offered several styles of running gears matched to the Power Feed Box, although the Feed Box was designed for mounting on any standard running gear.

In 1962 Allis-Chalmers offered this Power Feed Box to complement an already extensive line of forage harvesting equipment. Apparently some previous efforts had been made to market forage boxes, but these seem to have been quite limited. This 1962 version had an inside width of 76 inches and an inside length of 13 feet 9 inches. The standard box was 32 inches high. Extensions were available of 22 inches so that the double extension model shown here had a height of 76 inches. With this configuration the wagon had a capacity of 546 cubic feet. Two separate steel chain raddles were used, and a quick detachable integrally shielded PTO shaft was provided. Company records tend to indicate that this model was built only in 1962.

Forklifts

Industrial equipment was but one of many areas of endeavor within Allis-Chalmers. Through the acquisition of the Buda Company at Harvey, Ill., in 1953, Allis-Chalmers gained the benefit of Buda's prior work with fork lift trucks. Buda had been a pioneer of fork lift development and was by 1950 offering a substantial number of models. Forklift trucks offered tremendous savings in labor over conventional hand methods and permitted the handling and loading of heavy loads quickly, efficiently and safetly. With the advent of forklift trucks a silent revolution began in warehouses and factories with the transition taking place in only a few years. This Buda warehouse style forklift is moving two heavy pallets of material which in the earlier days of manual handling would have required several men.

During the 1950s Allis-Chalmers built and sold thousands of F-Series forklift trucks, including this newly announced FT-100 model of 1957. This particular forklift had a capacity of 10,000 pounds and was available with diesel, gasoline, or LP-fuel engines, with the propane model shown here. Optional equipment included the overhead guard and the headlights mounted on the forklift mast. Gasoline and propane versions used the Allis-Chalmers Power Crater engines ostensibly developed for A-C farm tractors, but admirably suited to this task as well. Company advertising noted that in forklift service, many of their engines were running 10,000 hours or more before a major overhaul was required.

The Model F lift trucks were painted in the traditional "safety yellow." Other safety features included two braking systems, a remote mounted fuel tank, and familiar automotive-type controls. Electric wires were all on one side of the unit and the fuel line was mounted on the opposite side. The Power-Shift models automatically disengaged when the driver left his seat, and optional engine protective controls would automatically shut down the engine if oil pressure dropped or water temperature rose above preselected points. The precision built masts had been developed from years of experience, and by the 1950s were ranked among the best in the industry.

Due in large part to their design, F-Series lift trucks had excellent grade climbing ability. Allis-Chalmers did not overpower the machines, yet enough reserve power was built in for the heaviest jobs. The excellent stability came from properly engineered weight distribution incorporated into a design that lowered the center of gravity, placing the weight near the floor. Another important feature was the patented self-adjusting, pressure-compensated packing on the lift piston. Hydraulic systems under constant and heavy use tend to leak, and oil spots on warehouse floors present a hazard to foot traffic. The lowering valve was of pressure compensated design to assure safe and controlled lowering of the carriage.

Forklifts

As part of a carefully conceived plan, many components of the F-Series lift trucks were interchangeable from one size to another. This represented a considerable savings to companies operating a substantial fleet of these machines. Beyond the basic FT models, Allis-Chalmers offered the FS series in 6,000, 8,000, and 10,000 pound models. These lift trucks differed in that a quick removeable counterweight was used, as was a high underclearance for special applications. The GS safety trucks were equipped with special safety features that permitted operation in certain hazardous areas. For the ultimate in safety the FDX Safety Diesel model was recommended, since it had no electrical system with its potential for a spark that might set off an explosion.

Allis-Chalmers FT-Series lift trucks were available in fourteen different styles with lift capacities ranging from 2,000 to 10,000 pounds. Although every model could be furnished with the standard cushion tires shown here, the 2000, 2500, 3000, 4000, and 6000 pound models could alternatively be equipped with pneumatic tires. Laminated and wire-shielded tires were also available. Four different mast configurations were available, depending on specific requirements. In addition to the standard constant mesh transmission, FT-Series lift trucks could also be furnished with a Power Shift torque converter drive. Servicing the hydraulic pump was simple, requiring only the removal of the grille and panel — removal of the counterweight was unnecessary.

A 1965 photograph illustrates an Allis-Chalmers F50 lift truck suitably modified for a fork lift gangplank conveyor system. Apparently this unit was developed to aid in ship loading, yet another application of the lift truck. As in other areas of endeavor, Allis-Chalmers was not satisfied with only conventional applications of their lift trucks but remained perpetually involved with their own engineering department and the suggestions of other users so that new and potentially useful designs could be produced. These machines were built under the highest engineering standards, using for instance, a constant mesh transmission that eliminated gear clashing.

In 1965 Allis-Chalmers announced a new retractable overhead guard for its lift trucks. The guard was available as an optional feature on new lift trucks, or it could be retrofitted to existing models. With this feature a lift truck could be operated in high ceiling areas where an overhead guard was imperative, but could also work in areas or pass through doors where low clearances would otherwise prevent. This illustration shows the retractable overhead guard in its raised position, while a ghost image shows the same unit when retracted. Due to the weight of the guard, and also for operator convenience, this device could be advanced or retracted hydraulically with the operator never leaving the seat.

Allis-Chalmers also supplied a forklift attachment for their utility tractors. On construction jobs and other applications where the small cushion tires of a conventional lift truck would be impractical, this tractor-mounted unit would get the job done. The greater width of the utility tractor also permitted a higher lift or uneven terrain, with this unit having a maximum lift of 21 feet. Fork lift attachments of various sizes and having various features for specific applications were built specifically for several of the Allis-Chalmers utility tractors, with this D-15 Utility and its 21-foot forklift being a typical example. This particular outfit is loaded with tobacco wrappers in a processing plant at Red Lion, Penn. The photograph was taken in 1962.

Forklifts

Allis-Chalmers lift trucks quickly made their way into virtually all phases of commerce and industry. From large factories to relatively small lumber yards, the lift truck was about to revolutionize material handling methods. With the advent of lift trucks came the later fulfillment of the palletized concept whereby large quantities of otherwise unmanageable materials such as this huge stack of lumber could be handled in minutes rather than hours, as was required by hand methods. A-C lift trucks could be equipped with a host of attachments. Included were such things as cotton bale clamps, carton clamps, paper roll clamps, and slip-on arms. Other optional features included a side shift, revolving carriage, and rotating clamp, to name only a few.

A-C 500 Series forklifts were designed for conventional purposes but also had a specific application in handling of sod, orchard, and field crops. Using industrial tractor designs, these machines could negotiate any location accessible to an ordinary farm tractor. Now the orchardist for instance, could eliminate considerable labor from the harvest, replacing it instead with the untiring output of the forklift. The Model 510 was equipped with a 10 foot mast and had a maximum capacity of 4,000 pounds. A shuttle clutch provided instant forward, reverse, or neutral — it was supplemented with a hand lever to be used as necessary. A number of special attachments were available, and other masts could be provided with lifts ranging up to 21 feet.

From the previously built I-60 forklift came the I-600 model in 1971 and 1972. The I-60 was equipped with an A-C Model 160 gasoline engine or with the A-C D-175 diesel. While the standard mast had a 12-foot height, a special 21-foot style was also available. Optional equipment included cast iron counterweights, swing tines, a dozer blade, snow blade, and log grapple. Sod farms in particular could use the optional flotation tires, and an enclosed cab was available for opertor comfort and efficiency. Allis-Chalmers also offered the 500 and 600 Series as rough-terrain machines using an extra high clearance drive axle. For 1971, a Model 600 forklift using a gasoline engine and the standard 10-foot mast sold for slightly over $10,000. The 630, of the same dimensions, but equipped with a 30-foot mast listed at approximately $11,900.

The I-60 and I-600 forklifts were closely related, and in fact the latter was a direct descendant of the former. In a gasoline version it listed at approximately $6,500, while the diesel style carried a factory list price of slightly over $7,000. Much of the total cost was used in building the mast equipment and in the hydraulic system. Forklift hydraulics are often subjected to the most severe duty, so in recognition of this fact, a substantial over-design is dictated if long service is to be expected. Although Allis-Chalmers enjoyed a substantial portion of the forklift market during the 1960s, the following decade saw the entry of virtually every tractor manufacturer into this lucrative sideline. The result was that by 1972, almost every American tractor builder also offered forklifts, either as an attachment or as integral units.

Forklifts

Although this photograph illustrates the 510 forklift with conventional fork tines, an immense number of attachments were available that minimized hand work — work that was sometimes hazardous with manual methods. As an example, the tine rotator attachment was designed to work with a palletized waste bucket to do the housekeeping chores of two men in a fraction of the time formerly required. In operation, the standing waste bucket was picked up by the fork lift, and after getting positioned over a waiting truck the rotator simply dumped it, ready to go after another. The pulpwood grapple rotated and opened to catch its load. A front-mounted bucket attachment with a one-yard capacity converted the forklift into a high performance loose material loader.

The A-C 706 forklift was built in several styles and offered masts of 14, 21, and 30 feet. The basic model was followed by the 706-B - similar to the 706 but including a ROPS cab. The 706-C forklift of 1980 and 1981 could be furnished with a gasoline engine — its factory list price was approximately $33,000, with the diesel model carrying the same price. Another of the series was the 706-D Rough Terrain model, designed for the worst operating conditions. One of the most innovative models was the 766 Tow-Eze Rough Terrain model. It was factory designed with a hitch tongue on the rear of the machine that permitted towing down a highway from one job to another. This unique machine listed at approximately $38,000.

Although current research is indeterminate on the point, it appears that the 840 forklift was built only in 1975. This articulated model was furnished only with a diesel engine. The 14-foot version listed at approximately $27,000, with the 840 having a 30-foot mast being only about $700 more expensive. The articulated frame permitted excellent maneuverability, even in tight areas as often found on construction sites. Due to its heavy design this unit had a huge lifting capacity, as is evidenced by four huge blocks of concrete on board this particular machine. Production ended quickly, presumably because of poor sales.

In a 1977 news release from A-C's Matteson, Ill., facility the company announced a family of sideloading industrial trucks designed and built in North America. Sideloaders were designed for a broad range of users with loads that might be long or awkward to handle by conventional means. The sideloaders could carry these loads parallel to the direction of travel, enabling them to move in areas only slightly larger than the sideloader itself. The Allis-Chalmers ACS 100 and ACS 120 had respective capacities of 10,000 and 12,000 pounds. Although a 12-foot lift height was standard, other masts were optionally available. Standard features included a year-around operator cab, hydrostatic power steering, and all-roller, I-beam, two-stage Allis-Chalmers masts.

Forklifts

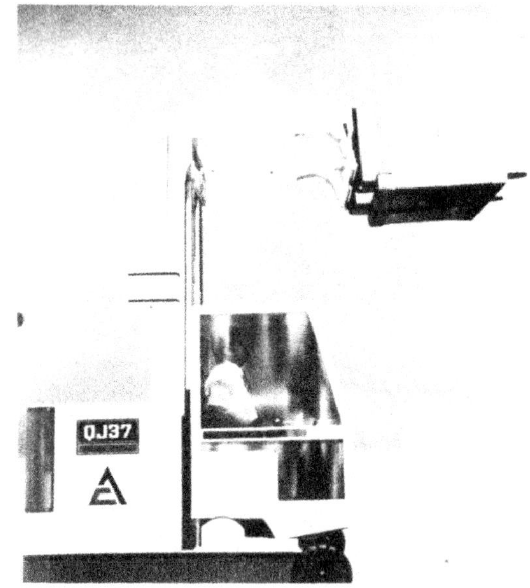

A 1974 Allis-Chalmers news release announced an advanced concept of electric lift truck design. Called the Quantum Jump-37, this 2000 pound capacity, demonstration prototype used the force of the lowering mast to put new power into the battery. The ventilated, all-enclosed cab goes up with the operator, or the forks can go up without the cab, depending on how much precision the operator needs for placement of the load. All operations of this unique prototype were made by pushbutton or joystick controls. Current research has failed to determine whether Allis-Chalmers followed this design with anything approaching a production model, but there can be no doubt that the QJ-37 was indeed a step into the future!

Electric hand trucks are frequently found in all types of manufacturing facilities, enabling workers to handle palletized items with relative ease, and practically eliminating the risk of back injury caused by improper manual lifting. These simple, relatively inexpensive units carry a storage battery unit. All operator controls are mounted in the steering tongue and permit the operator to raise and lower the forks hydraulically, likewise permitting forward or backward operation at will. Despite an excellent line of lift trucks and related items, the economic climate of the early 1980s prompted Allis-Chalmers to withdraw from this market — a market highly competitive, not only from other American builders, but also from overseas manufacturers as well.

Fraser & Chalmers

Various historical accounts of Allis-Chalmers and its predecessor companies have, for the most part, concentrated on Edward P. Allis Co. and its beginnings, leaving but a slight percentage of the historical account to Fraser & Chalmers, another of the participants in the Allis-Chalmers merger of 1901. While Allis was busy making flour mills and sawmills, Fraser & Chalmers was manufacturing a host of equipment for ore and mineral processing. Additionally, this huge concern was manufacturing large steam engines, pumping engines, boilers, and other items.

The *Crushers* section of this volume sets forth some of the early history concerning Fraser & Chalmers and also presents a brief sketch of the interrelationship of Fraser & Chalmers with Gates Iron Works, yet another partner of the Allis-Chalmers merger. To avoid redundancy the reader is directed to this section for an early history of the company.

History has it that merger talk came about as a result of a hotel lobby discussion between Edwin Reynolds of Allis and William J. Chalmers of Fraser & Chalmers. Add William Marburg, a New York promoter, and presto — within a few months the great combine is formed. From all appearances, Edwin Reynolds had a grand vision of a few huge manufacturing plants that would built the Reynolds-Corliss steam engine, ostensibly to power the generators soon to be demanded by the public as they got their first sampling of electrical power. Chalmers for his part seems to have had an entrepreneurial bent — this coupled with an uncanny business savvy had already made him very wealthy. In retrospect though, the Reynolds-Corliss steam engine of 1900 was not yet dead, but it had certainly acquired terminal symptoms. America's vast number of untamed rivers were ripe for hydroelectric power, and within a few years these would become highly developed. The steam turbines of Curtis, Parsons, and DeLaval to name a few were making immense progress as efficient prime movers that required but a fraction of the weight per delivered horsepower that was evidenced in the Reynolds-Corliss engine. Thus, the Reynolds dream of an America literally covered with his steam engines was never to be realized. In fact, there are some indications that the 1900 financial status of Edward P. Allis Co. was not all that it might have been, so talk of a merger might have been more a matter of practicality than of euphoric hope.

Despite the fact that Fraser & Chalmers at the time of the merger was only nineteen years old (1872), the company had grown to an immense manufacturing operation in less than two decades. During 1884 the company built their first pumps in accordance with Prof. Riedler's designs. They were an overnight success, and by 1888 the Riedler pumps had been well adapted to city water systems, sewage systems, and many other applications. Although Fraser & Chalmers was the licensed builder for the United States, and perhaps for North America, the Riedler pumps were built under license in other countries by other manufacturers.

Fraser & Chalmers

However, from 1884 to November, 1897, over 1,200 Riedler pumps had been built. An 1897 Fraser & Chalmers catalog notes that the company had recently installed one of the huge pumping engines at the mines of Chapin Mining Co. at Iron Mountain, Mich. This pump had a capacity of 2,200 gallons per minute agains a head of 1,700 feet. Another interesting installation was that made for Pope Mfg. Co. of Hartford, Conn. in their tube works. This installation consisted of triplex differential Riedler pumps driven by a Fraser & Chalmers vertical triple-expansion condensing engine. The pumps for Pope Mfg. Co. were required to operate in conjunction with an accumulator against a pressure of 1,250 psi.

The principal feature of the Riedler pump was its mechanically operated valve. The valve, circular in form, is designed for a lift of 1 or 2 inches, and the area is so designed as to reduce the speed of the water flowing through it to only a few feet per second. As a Fraser & Chalmers catalog notes, "In other words, the valve area is figured entirely on the basis of plunger displacement." Although the valve opened automatically at the beginning of the suction stroke, closing was precisely timed by a controller at the end of the stroke. The bottom line was that despite the great engineering achievements of Edwin Reynolds in the areas of steam engine and pump engineering, the Riedler design was of the caliber that it could not be ignored as a strong competitor.

Fraser & Chalmers also built a strong reputation on their patented Sederholm boiler. This design retained the advantages of the standard horizontal tubular boiler, but added a series of cylindrical drums placed underneath the main boiler drum. The small drums and the bottom of the main boiler were placed directly in the fire, with flue gases returning through the usual boiler flues in the main shell. In short, the Sederholm boiler was found to be an extremely efficient steam generator compared to most units on the market, and efficiency in steam generation meant dollars saved in fuel costs. Thus this boiler became a major item in the Fraser & Chalmers shops.

Beyond the crushing equipment already noted in a previous section, Fraser & Chalmers was by 1900 building an extraordinary variety of equipment for roasting, smelting, and refining of various metals and minerals. Sampling plants, copper converters, blast furnaces for lead ores, and reverberatory furnaces — these were just a few of the many Fraser & Chalmers catalog items. In addition, the company noted their readiness to design, engineer, manufacture, and install special equipment for required applications.

Against this backdrop then, Fraser & Chalmers looms large indeed as a partner in the 1901 Allis-Chalmers merger. It appears to be a logical assumption then, that the merger tended to strengthen the holdings of each individual partner. Allis, while not in immediate trouble, might well have been had the manufacturing of Reynolds-Corliss engines continued with no look at steam turbines or hydraulic power. Fraser & Chalmers might soon have saturated the flash-in-a-pan economy of the mining industry. Although the question of whether the general public benefited from the merger might still be the subject of a protracted philosophical discussion, a pragmatic view might hold that the merger was a benefit since by combining their engineering strengths and manufacturing skills, these companies were able to make technological advances in far less time than if they had acted singly. In fact, had each of these previously competing firms remained on their own, it seems entirely possible that failure would have been imminent. With failure and the closing of the factories, workers and the local economies might have been dramatically affected.

Fraser & Chalmers is more often associated with rock crushers than with its other equipment. Rightly so perhaps, since their Comet crusher was one of the best known machines of the gyratory style. By the time of the 1901 Allis-Chalmers merger, Comet crushers were in use all over the world. As indicated in the *Crushers* section of this volume, Comet crushers were of many sizes, ranging from a small 3 x 7 inch laboratory model up to a huge 18 x 42 size having a capacity of 200 tons per hour. The largest model required something like 150 horsepower for successful operation — the largest model weighed well over fifty tons! This early photograph depicts the Comet in its earliest form, with the original photograph probably originating at the Fraser & Chalmers factory at Chicago.

Minerals processing machinery was big business for Fraser & Chalmers, with this company building virtually every machine required for mechanical mining and ore processing. This huge ore dryer of the 1890s had a diameter of 36 inches and a length of 18 feet. Built specifically for a mining operation, it embodied state-of-the-art drives, at least for the 1890s. A series of open-gear reductions followed the flat belt drive, and it was equipped with the usual tight-loose pulley arrangement. The lineshaft pulley was at least as wide as the two pulleys shown here. It was usually of wood and had no crown, while of the two pulleys on the countershaft illustrated here, one was bushed and ran loose on the shaft, while the other was keyed to the shaft in the usual manner. When running idle, the belt was shifted with a shifting device, or perhaps with a wooden pole to the tight pulley, starting the dryer drum into motion. To stop it, the belt was again shifted to the loose pulley.

Fraser & Chalmers

Fraser & Chalmers could not of course build the Reynolds-Corliss engines of Edward P. Allis Co., but they could, and did build a Corliss design early in their manufacturing career — their earliest one being the Bates engine. Ironically, reciprocating steam engines saw little use until after the 1876 Centennial Exposition at Philadelphia. At this time Corliss Steam Engine Co. displayed the famous Centennial engine, but it was of beam design, similar to Watt's first engines. Edwin Reynolds, the factory superintendent for Corliss, soon afterward came to Allis' Reliance Works at Milwaukee, and it was Reynolds who pioneered the girder frame design similar to the Bates-Corliss shown here. Regardless of the various detail changes made in subsequent years the Reynolds-Corliss remained the hallmark, the touchstone, by which Corliss steam engines were judged.

Fraser & Chalmers, unlike Edward P. Allis Co., made a strong showing as boilermakers. This company either developed or built under license, some of the industry's best and most efficient designs of the era. Curiously however, Reliance Works built engines, but no boilers, at least not to any extent. A handwritten note on the original photograph indicates these to be examples of the Strong boiler, with these specific units being 90 inches in diameter and 18 feet long. The two large flues would seem to be the main flues, with a multitude of return flues below them. At first glance this arrangement would lead one to believe that the boilers are simply setting upside down, but the large mounting lugs indicate otherwise. Possibly these boilers used a downdraft arrangement or some induced draft system. Fraser & Chalmers also built the Sederholm boiler, one of the most efficient boilers of its time.

An 1890s photograph illustrates a huge triple expansion steam engine built for the DeBeers mining interests in Africa. This engine was a combination tandem and cross compound design with the first stage of 11½-inch diameter at the very top, and the second stage, 19-inch cylinder directly below. To the left is the third stage cylinder with a diameter of 30 inches. This engine used a 36-inch stroke. Access to the engine was obvious with the cast iron stairway and numerous walkways. Since these engines often operated for months at a time without ever shutting down it was imperative that provision be made for the engineer to inspect and oil various working parts at regular intervals. The large flywheel is grooved not for v-belt drive but for rope or cable drive.

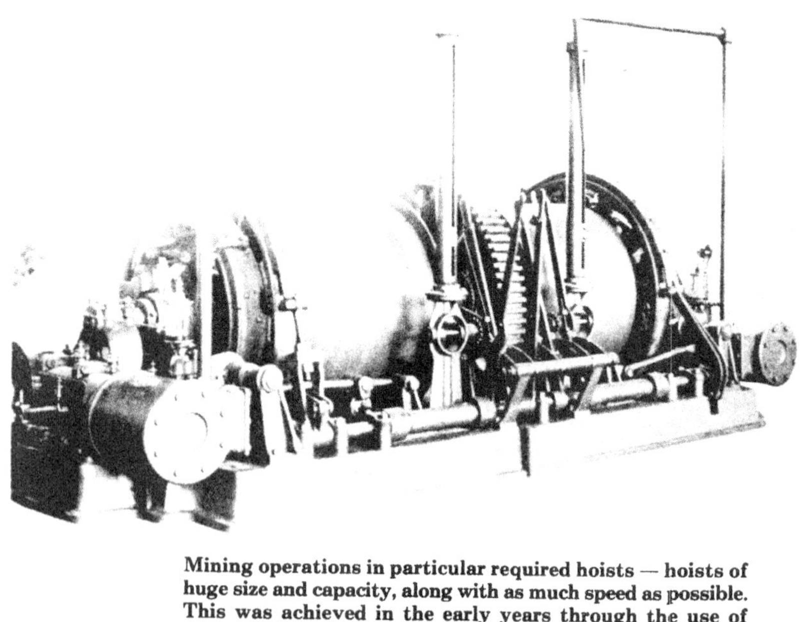

Mining operations in particular required hoists — hoists of huge size and capacity, along with as much speed as possible. This was achieved in the early years through the use of steam hoists such as this large Fraser & Chalmers model. These large drum hoists could be started, stopped, or reversed almost in an instant. Doing so however caused great strains on the equipment, so extra iron was necessary to carry not only the rated load but the strain load imposed intentionally by emergency conditions or unintentionally by unskilled operators. The large steam cylinders are evident at each end of the bed casting.

Fraser & Chalmers

Large steam engines were usually supplied as condensing engines, that is to say, the exhaust steam was condensed and reused. Condensing the exhaust steam lowered the back pressure on the engine, making it more efficient. This required a vacuum pump in many instances. Certain Corliss engine designs also required a vacuum pump for the dashpot valves. To these ends, Fraser & Chalmers supplied this combination pump and condenser unit, with this particular one designated for an installation at Pope Mfg. Co. in Connecticut. This small unit was apparently of slide valve or piston valve design and used a conventional flyball governor system for speed regulation.

Fraser & Chalmers offered an extensive line of hoisting engines, and in so doing probably became a sizeable thorn in flesh for Edward P. Allis Company. No doubt Edwin Reynolds, the chief architect of the Reynolds-Corliss design took considerable notice when Fraser & Chalmers installed a hoisting engine of this size. Unfortunately, current research does not reveal the destiny of this particular unit, although it is obviously of considerable size, and also is obviously of the Corliss design. The average reader of today is often totally unfamiliar with the reasons why an engine as mechanically complicated as the Corliss was so popular among steam users. The answer boils down very simply to its great economy in the use of steam. This meant great economy in the use of fuel, and that converted to an immense doller savings each year.

Yet another example of Fraser & Chalmers hoisting machinery is this huge 22 x 60 inch reel hoist. Built for a large mining company sometime during the 1890s, it typified the trend to mining equipment of ever-increasing dimensions. From the engineering viewpoint this hoist is unique, since the brake shoes were made, not of the usual gray iron castings, but instead were built up with riveted steel sections. Electric welding had not yet come into vogue, and it can be imagined that today's technology would have constructed this hoist using numerous built-up steel sections. A gigantic engine powers the hoist; in fact engine and hoist were erected as an integral unit.

Fraser & Chalmers apparently began building pumps and compressors using the Riedler valve about 1884. An undated photograph illustrates a huge duplex Riedler air compressor built for the Horn Silver Mine at an undisclosed location. The compressor units are in the foreground of this illustration — the compressor pistons were connected directly to the engine pistons with reach rods. At the other end of the steam cylinders were the usual piston rods and crossheads which attached to the crankshaft and flywheel. The flywheel was used only as a means of regulating speed — no belted load was taken from it. This huge compressor used cylinders of 14-inch bore and 24-inch stroke, and was of double-acting design.

An undated photograph illustrates one of two 15,000 horsepower pumping engines installed for Bethlehem Iron Co. An 1899 Fraser & Chalmers catalog points with pride to some of the companies installations around the world. In fact, Fraser & Chalmers installations were occasionally so large as to require the chartering of an entire ship for delivery of the equipment. Some areas had few trees and little or no lumber, so the job included perhaps another shipload of lumber shipped direct from a Gulf port. In addition to supplying pumping engines for city water systems, Fraser & Chalmers also sold a considerable number of these engines for irrigation pumping, and not a few were sold to mining companies for hydraulic mining operations.

A more or less typical installation was this cross compound Corliss steam engine with an integral air compressor. Manufacturing plants were by 1890, using considerable amounts of compressed air for daily factory operations — a typical example was the introduction of pneumatic boiler tools which replaced earlier hand methods. Although these units were restricted to a low operating speed, primarily because of the Corliss valve mechanism, their huge physical size still permitted them to generate an immense amount of compressed air. Despite the disadvantages of size and weight, these units did however have the advantage of running for many years with virtually no money spent on repairs.

Recognizing the need for large prime movers in remote and almost inaccessible areas, Fraser & Chalmers offered this sectionalized Corliss steam engine in the 1890s. After the Allis-Chalmers merger of 1901 the new company continued to offer this engine for several years. The entire engine was designed so that no single part weighed more than 300 pounds. Even the flywheel was built in sections. Each individual piece was then bolted together at the job site, a task that surely must have been tedious at its best. Today's high technology manufacturing methods would probably experience some difficulties in building an engine of this design, and in this light it seems almost incredible that it was built a century ago on machines now considered to be archaic.

Stated very briefly, copper converters did their work by bessemerizing the copper matte consisting primarily of copper, iron, and sulphur. The converter burned off the sulphur, the iron contents were slagged off, and the converted copper remained. The so-called parrot converters were the style used by Anaconda Co. and others for this process, at least during the 1890s. This process was introduced into the United States in 1884. Almost immediately afterward, Fraser & Chalmers had shipped converting equipment to all parts of the world. These three units are about to leave the Chicago factory of Fraser & Chalmers in 1899.

Power for an electric railway system was derived from four Fraser & Chalmers Corliss engines built in the late 1890s. Although Edwin Reynolds and others at Edward P. Allis Co. had pioneered the mounting of the generator rotor direct to the crankshaft, this idea soon spread to other companies. Fraser & Chalmers seems to have directed most of their energies to mining equipment, pumps, and compressors until the late 1890s when they apparently saw yet another market for their engines in the form of prime movers for electric generators. By so doing, Fraser & Chalmers walked into the land of the giants — not only was Edward P. Allis Co. a major competitor, the newly formed General Electric Co. was becoming an ominous competitor, and Westinghouse Electric Co. was yet another dominant force.

Fraser & Chalmers

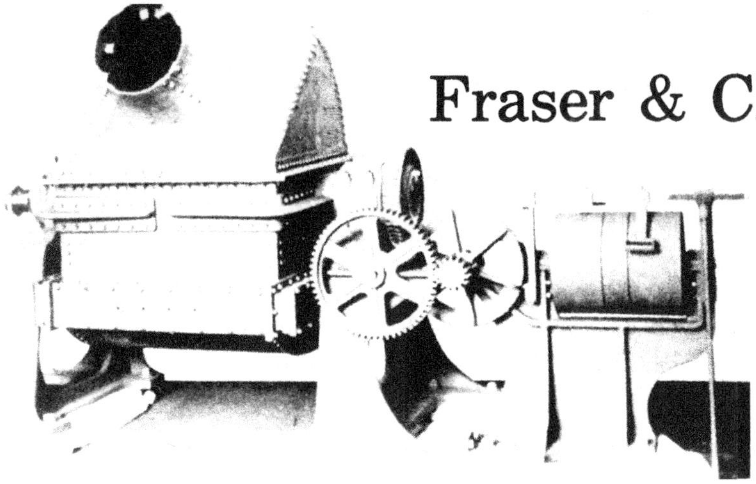

Shown here is a 3-ton converter of the "Stalmann" type. This style was used by the Mt. Lyell Mining & Railway Co. of Tasmania, among others. The bessemerizing process required first that the converter lining be made of crushed quartz mixed with sufficient clay to hold it together. Molten copper matte was then charged into the converter, and air from a blowing engine at 6 to 15 psi was forced through the molten mass. The sulphur combined with the oxygen of the air to form sulphurous acid gas — it passed off. Meanwhile the oxidized iron united with the quartz to form a slag which was poured or skimmed off. After this the molten copper was poured into molds. No extra fuel was required — the heat was supplied by the combustion of the sulphur in the matte.

Stamp mills were a major fixture of virtually every valuable metals mine, with this illustrated style being but one example. Steam stamps like this model were very large and powerful, and could pulverize an enormous amount of ore — something on the order of 60 to 100 times the amount that would be pulverized by an ordinary mechanical stamp mill. As the name implies, stamp mills simply stated, pulverized the crushed ore to powder, thus allowing the concentration stages to begin. Various publications tend to indicate that many of the design improvements were actually suggested by superintendents or employees of various mining operations. After proving their worth, many of these designs were adapted as standard equipment. The tremendous hammering force of the stamp mills required the ultimate in design to keep them from disintegrating through their own activity. A study of this illustration shows the extremely heavy design of this unit. After the 1901 Allis-Chalmers merger, many of the Fraser & Chalmers machines remained in production, some of them for a number of years.

Another Fraser & Chalmers product was the copper furnace, one of which is illustrated here. This one is larger than it might appear — the chamber is 36 inches wide and 90 inches long. Fraser & Chalmers specialized in this equipment, and offered smelting furnaces designed for specific applications such as smelting to silver-lead bullion, copper matte when copper oxide was present, or black copper if the ores were oxides and carbonates. Also built were furnaces for compromise pyritic melting and iron matter smelting. The smelting furnaces were designed to concentrate the ore by separating the worthless gangue from the valuable portion of the ore. Depending on the assay level in the ore, different processes were used to extract the desired metals.

Fraser & Chalmers Co. is usually credited with the development of the frue vanner, a machine designed for concentrating high grade fine material. In the early 1900s a competing company noted in their catalog data that "no device used in the mining industry has gained such a world-wide reputation and become so universally used as the vanner..." The machine used an endless rubber belt supported by rollers so as to form an inclined rubber surface about 6 feet wide. Rubber flanges were built into the belt. As the belt travels around the rollers, ore is slowly fed on the belt in a stream of water. The belt having a steady shaking motion as it travels, settles the deposited material on the belt, and water jets wash back the lighter material, letting only the heavy material pass into a receiving tank below. With this very useful machine Fraser & Chalmers Company, as its original builders, probably built their fortune.

Fuel Cells

Throughout its long history Allis-Chalmers poured immense sums of money into research and development projects. A-C research engineers did not limit their efforts to the ordinary areas of research but often took bold steps in developing new technologies. Probably one of the most significant developments was the fuel cell.

Briefly stated, fuel cells are a direct-current power source utilizing a chemical reaction to obtain useful power. Fuel cells are not batteries, rather, they depend on a chemical fuel being supplied while the cell is in operation. Hydrogen-oxygen fuel cells are the most commonly known types, and with these, Allis-Chalmers announced its first successful experiments.

Fortunately for the historical perspective, a thoughtful A-C research engineer compiled a chronological list of fuel cell developments some years ago. This listing begins with the October, 1959, announcement that Allis-Chalmers had developed and demonstrated the world's first fuel cell-powered vehicle. This was an experimental farm tractor equipped with enough hydrogen-oxygen fuel cells to generate 15 kilowatts of power.

Following the announcement of the experimental fuel cell tractor came a 10 kilowatt ammonia-oxygen power fork lift truck in December, 1960. This was followed in March, 1963, with an improved 7 kilowatt forklift with a lifting capacity of 2,000 pounds. Exactly one year later the company designed and built a 2,000 watt hydrogen-oxygen fuel cell power pack for the U.S. Army's Harry Diamond Laboratory in Washington, D.C. Concurrently with this system, A-C announced that they were, in cooperation with the Atomic Energy Commission and the U.S. Army, studying the development of mobile energy depot systems using fuel cell technology. These research projects led to the March, 1962, testing of hydrogen-oxygen fuel cells in zero-gravity situations.

Not limiting these developments to military uses, A-C built an experimental golf cart in October, 1962. This 3 kilowatt vehicle used hydrazine-oxygen fuel cells.

America was looking to outer space exploration in 1962, and in November of that year A-C built a self-contained 50 watt hydrogen-oxygen fuel cell system for aerospace use by the U.S. Air Force. May, 1963, saw the demonstration of a 1,000 watt hydrogen-oxygen fuel system oriented to aerospace applications. This system used a unique static moisture removal concept and was an important development in aerospace technology. Later that year a 5 kilowatt hydrocarbon reforming fuel cell system was announced for the Engineering Research & Development Laboratory of the U.S. Army at Fort Belvoir, Va. This development constituted the first major air-breathing fuel cell system. During September of 1963 Allis-Chalmers announced the discovery of nickel boride to replace platinum as the anode catalyst in the hydrogen-oxygen and hydrazine-oxygen systems. This development helped lower the production cost of the individual fuel cells.

The many and varied developments of the previous five years led to a January, 1964, contract with the U.S. Air Force to supply a quantity of 750 watt hydrogen-oxygen fuel cells incorporating the static moisture removal system. A month later NASA ordered some 2,000 watt hydrogen-oxygen units.

One of the most unique developments associated with fuel cells was the July, 1964, announcement of STAR-1, the world's first submarine to be powered by fuel cells.

The Allis-Chalmers experimental fuel cell tractor of 1959 was the world's first fuel cell-powered vehicle. At the time of its introduction the fuel cell concept was virtually unknown to the general public, except perhaps in the satirical sense provided by science fiction comic books. To obviate this problem, early Allis-Chalmers announcements also illustrated in a very basic way how the fuel cell worked. Although this was the first announced application of the fuel cell concept, A-C engineers had apparently been working on it for some time, probably as early as 1951. Converting a chemical reaction into electricity is a phenomenon discovered long ago, but as this illustration shows, the fuel cell is not a battery. Various scientific texts and encyclopedias discuss fuel cells in greater technical detail.

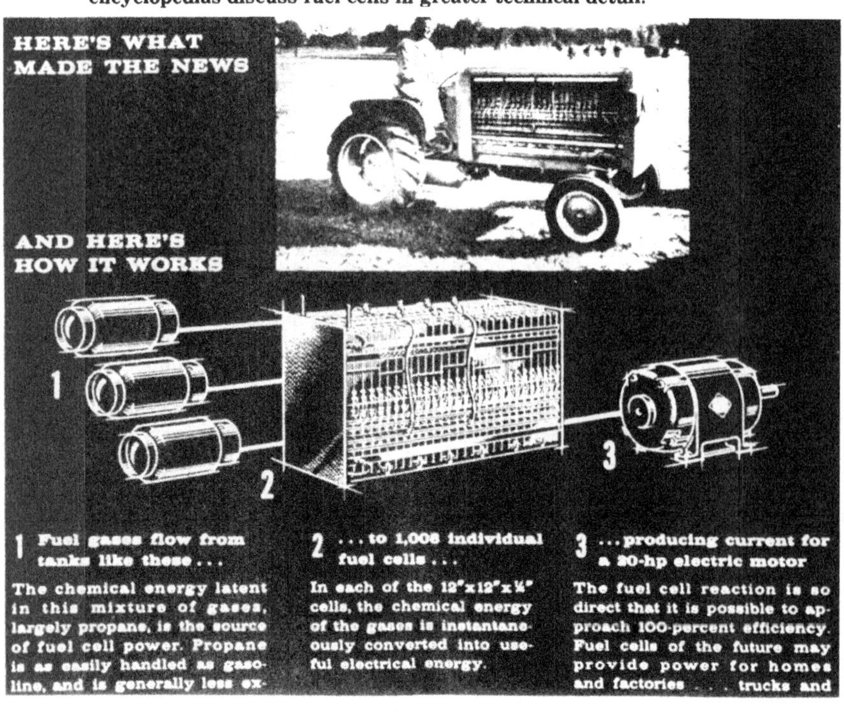

In October, 1959, representatives of leading trade magazines saw an actual field demonstration of the new Allis-Chalmers fuel cell tractor. While obviously in the experimental stages, this unique application of the fuel cell caught the attention of scientists all over the world. Fuel cells were first described by Sir William Grove in England during 1839. During the 1880s this concept gained new impetus, but not until the 1920s were any substantial improvements made. Scientists in many countries attempted to build practical fuel cell systems during the 1930s and probably made extensive experiments on this idea during World War Two. Already in 1959, A-C engineers and other research scientists saw the fuel cell as an answer to power requirements for a newly emerging aerospace technology. Shown here is the Allis-Chalmers fuel cell tractor pulling a two-bottom plow on its maiden voyage.

Fuel Cells

Built by General Dynamics/Electric Boat Division, the STAR-1 successfully went through all its tests, both on the surface and submerged. Allis-Chalmers supplied the hydrazine-oxygen fuel cells for this one-man vessel. A multi-million dollar contract was awarded to Allis-Chalmers in September of 1964 for building 2,000 watt, 28 volt hydrogen-oxygen fuell cells to be delivered to NASA (National Aeronautics and Space Administration). The design included drinking water as a useable by-product. This success was buoyed still further by obtaining another contract in October, 1964, to study fuel cell power supply requirements for the Manned Orbiting Laboratory (MOL) Program of the U.S. Air Force.

The next year, 1965, was filled with great achievements in fuel cell engineering, and a year in which A-C acquired several important contracts. First came a January contract with Lockheed Missile & Space Company at Sunnyvale, Cal., for development of a 1,500 watt fuel cell system to be included in Lockheed's proposal to the Air Force for the MOL system. As a direct result of Allis-Chalmers fuel cell developments came the April announcement of an oxygen concentrator. Only the size of a brief case, this unit could extract pure oxygen from the air when plugged into an ordinary electrical outlet. A-C noted that this system demonstrated a new oxygen purification process known as "electrowinning." In May, 1965, the company began marketing the first fuel cell systems for research, development, and education laboratories. Seven different models were available with outputs ranging from 100 to 1,800 watts, and using either the hydrogen-oxygen or hydrazine-oxygen concepts.

Coinciding with the Annual Stockholder's Meeting for 1963, Allis-Chalmers announced this new fork lift truck powered by fuel cells. Although this was an experimental unit, it was the first fork lift truck ever powered by fuel cells. Technician Gene Behne is in the foreground, and shown standing in the background is Peter A. Antony who engineered this development. While the fuel cell tractor used hydrogen-oxygen cells, this unit was slightly different in that it used hydrazine-oxygen fuel cells. The fuel cell system was capable of developing 7 kilowatts of electricity and the fork lift truck had a capacity of 2,000 pounds. Despite favorable test results, it does not appear that the fork lift truck or the fuel cell tractor ever were sufficiently developed as production units.

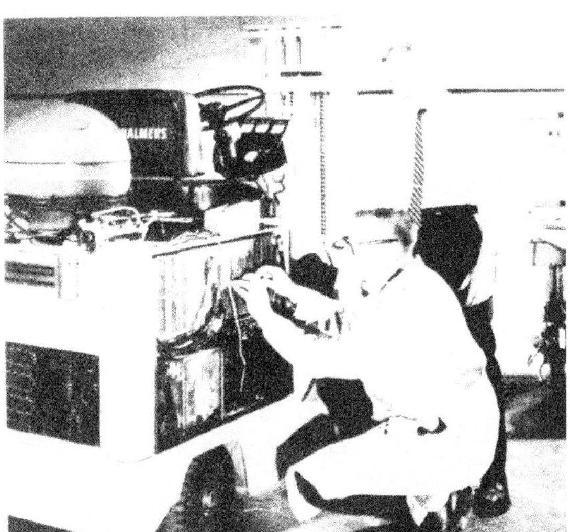

As these activities continued, 1965 saw even greater achievements in the development of fuel cells for aerospace projects. During June of that year an advanced 2,000 watt fuel cell system was announced for NASA. This unit was capable of producing both electricity and drinking water from hydrogen and oxygen for 1,500 hours, more than ample for current space projects. Also in June of that year, the U.S. Army ordered ten fuel cell power supply systems that featured an entirely new thermal control concept. According to A-C engineers, shutter systems were used, similar to Venetian blinds. The shutters were actuated by heat-sensitive units which allowed the cell to operate at its optimum temperature. This radically new concept eliminated the pumps, fans, and coolants previously required.

Apparently the considerable success of A-C fuel cell development prompted the company to build larger quarters, so in January, 1966, an entirely new fuel cell laboratory was announced. This facility was to be located in the Milwaukee suburb of Greendale. Late in 1966 A-C announced successful completion and acceptance tests for an ammonia dissociator built for the U.S. Army. The prototype model provided sufficient oxygen for a 5 kilowatt fuel cell power plant. Then in November, 1966, came the successful orbit of a 200 watt fuel cell boosted into space by a Titan III-C rocket. This was the first time in history that an electric power source for an unmanned satellite had been flown in an orbiting vehicle and controlled from the ground by radio signals.

As fuel cell technology improved, fuel cell systems became more sophisticated. Research in this field also brought about other developments and led to the

Designed as a golf cart, this 1962 unit from A-C research scientists also doubled as a convenient in-plant vehicle. Originally powered by a 200-pound bank of storage batteries, a small fuel cell system offered a simple, lightweight, and very efficient replacement. This 3 kilowatt unit remained strictly experimental despite its potential possibilities. Current research has not provided cost details of the early fuel cell systems, although it might be supposed that production costs were rather high at this point in time. Had the technology been developed sufficiently at that time to permit mass production techniques, then total units costs would probably have been considerably less. Rather low petroleum costs during the 1960s probably generated minimal enthusiasm from the fuel conservation standpoint.

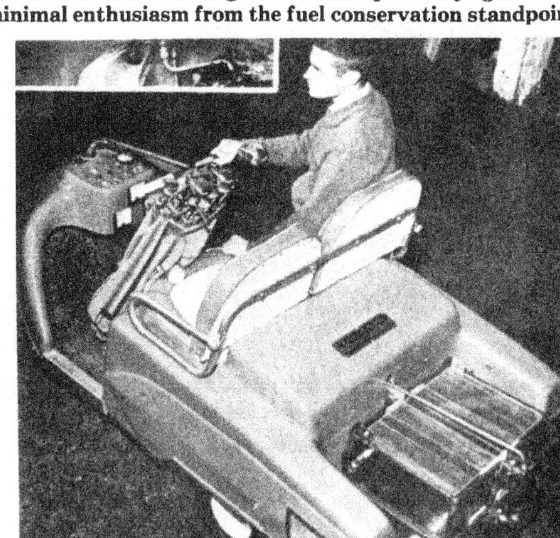

Fuel Cells

exploration of still others. As an example, Allis-Chalmers announced in early 1967 that researchers were exploring the possibilities of improved hydrogen production equipment with an eye toward the economical production of ammonia fertilizers. While this particular idea seems to have gained little headway, the company reported favorably on a research study of electrolytic hydrogen production in a nuclear desalination complex, using of course, fuel cell technology. Work continued on the concept of fuel cell powered vehicles replacing internal combustion engines and providing a solution to the pollution problems then existing. Another important aerospace study contract was awarded to Allis-Chalmers in 1967 — this one for design analysis of a system that would produce hydrogen and oxygen from water on the moon, as part of a lunar energy depot contract to Chrysler's Space Division and NASA. From this research came a water electrolysis unit of improved characteristics.

In addition to the company's 1967 research contracts, two important manufacturing contracts were secured from NASA. One involved the development and testing of an improved fuel cell system, while another advanced system completed 125 days of automatically producing electricity during a simulated space mission. In mid-1968 yet another NASA contract was secured — this one for production of advanced fuel cell power systems of long duration, some of them up to 2,500 hours.

September, 1968, marked another important milestone in fuel cell research for Allis-Chalmers and for aerospace engineering. The Langley Research Center at Hampton, Va., announced that it had completed a 28 day environmental chamber test in which all oxygen was supplied by a water electrolysis unit designed and built by Allis-Chalmers. This development marked yet another step in a system that would provide oxygen for a four-man crew on space missions. Several other aerospace contracts were awarded to A-C in 1968, among them a contract to build improved fuel cell power systems for the Apollo program of NASA.

Only one major contract is noted for 1969 — this one for a fuel cell power system to be used in the U.S. Navy's first deep submergence search and recovery vehicle (DSSV). Then on December 10, 1970, Allis-Chalmers announced that it was discontinuing its advanced electrochemical products department, citing the cancellation of several major military projects as the major reason. In 1971 this business was acquired by Teledyne Corp. for its Teledyne Isotopes Division.

Allis-Chalmers apparently began its first research into fuel cell development in 1951. Not until 1959 was the first fuel cell-powered vehicle announced, and it remained in the prototype stage, never achieving regular production. Despite this fact, Allis-Chalmers introduced many revolutionary developments in fuel cell technology during its two decades of activity in this area of research. Hindsight always remains as a good teacher, but with the company's many achievements in fuel cell research and the proliferation of manned and unmanned space flight in the 1970s and following — had Allis-Chalmers remained in this business, its fortunes might have been different indeed.

With an eye toward military uses of fuel cell power systems, Allis-Chalmers successfully acquired several major contracts for research, testing, evaluation, and production of specialized systems. A 1963 news release indicates that the company had by then developed this small 1000 watt fuel cell power system destined for an Air Force satellite. After successfully passing the zero-gravity tests, along with numerous other operational procedures, Allis-Chalmers was awarded a contract for further production and research. This 1963 system embodied a unique static moisture removal concept that simplified the problem of removing by-product water and waste heat while in flight.

In July, 1964, came announcement of STAR-1, the first submarine in the world to be powered by a fuel cell. Although the vessel was owned by the Electric Boat Division of General Dynamics, the fuel cell power system was designed and built by Allis-Chalmers. This one-man research vessel eventually led to the 1969 research project for preliminary design of a fuel cell powered deep submergence search and recovery vehicle. Despite the many fuel cell power developments pioneered by Allis-Chalmers, in late 1970 several major research and production contracts with the Armed Services were cancelled due to an across-the-board cutback in the U.S. Aerospace budget. The loss of these contracts left a research team with nothing to do and production facilities stood idle. Thus, on December 10, 1970, Allis-Chalmers announced its withdrawal from electrochemical research, completing closing these departments. In 1971 this entire department of the company was purchased by Teledyne Corp. for its Teledyne Isotopes Division.

Gas Turbines

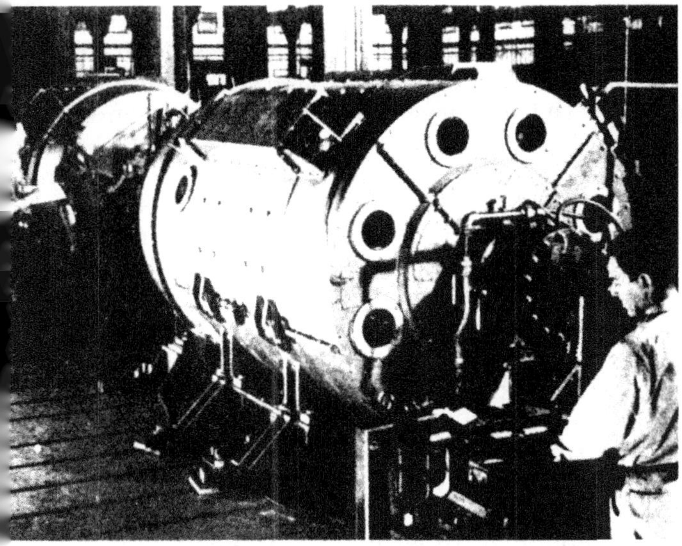

The gas turbine is among the earliest forms of combustion motors — its origins are often attributed to Leonardo da Vinci. Despite its early beginnings however, the gas turbine assumed no major importance until early in the twentieth century. Even at this point in time gas turbine development was very slow despite intensive efforts by several eminent engineers. During World War Two, Allis-Chalmers designed and built the direct-drive propulsion motor shown here. It was part of a diesel-electric marine drive. During this research project it has been impossible to determine the extent of Allis-Chalmers involvement in gas turbine design. Data is also sketchy at best regarding the number of these units actually installed for marine propulsion purposes.

In 1944 Allis-Chalmers noted that over 140,000 horsepower of gas turbine power had been installed in the United States, with the majority of this equipment being used in oil refineries. This 1944 photograph illustrates one of 29 Allis-Chalmers Houdry gas turbines in oil refinery service, with this particular unit generating over 10,000 horsepower in a 100-octane gasoline plant somewhere in Ohio. Allis-Chalmers also pointed out that despite numerous other gas turbine designs, the Houdry unit was the only land turbine application actually operating. The company further noted the potential uses for gas turbines, including such applications as railroad locomotives, power plants, and other jobs requiring a large prime mover.

During 1943 Allis-Chalmers noted that the refining industry was at the time using gas turbines more widely than any other manufacturing industry. This was probably due in part to the ready availability of fuel for the turbine, and due to the small physical size and weight per delivered horsepower compared to contemporary styles of internal combustion engines. Although Allis-Chalmers appears to have been quite active in the building of gas turbines, at least during World War Two, it does not appear that the company had explored this area sufficiently during the 1930s to enable it to offer its own designs. Rather, it appears that the designs were developed by others, with A-C assuming a manufacturing role as its primary responsibility. It also seems likely that this primary role included testing and evaluation work as well. Many of the gas turbine designs created out of necessity during World War Two led directly to development of todays modern jet aircraft engines.

Gates Iron Works

To avoid redundancy the Gates gyratory crusher is not illustrated in this section, having already been described under *Crushers*. Reference to Gates Iron Works is also made in *The Formative Years* section and under the *Fraser & Chalmers* heading.

As one of the four companies forming Allis-Chalmers in 1901, Gates was given a book value of $410,000, making it the smallest of the group. Edward P. Allis Co. had a book value of slightly over $5.1 million, while Fraser & Chalmers was booked at slightly over $3.2 million. Dickson Mfg. Co. carried a value of $1.2 million. Despite its size, Gates Iron Works brought some valuable designs into the merger.

Fraser & Chalmers was noted for its mining machinery and crushing equipment. Allis was of course well known for its flour mills, saw mills, and steam engines. Dickson Mfg. Company contributed boilers, compressors, blowing engines, and mining equipment. Gates Iron Works, like the interrelated Fraser & Chalmers firm built rock crushers, but went a step further by developing an extensive line of cement making machinery. For this development if for no other, Gates Iron Works was a valuable asset to the new company.

Curiously, Gates Iron Works antedated all the other

Gates Iron Works

partners in the merger — its origins dated back to 1842. Philetus W. Gates I and H. H. Scoville first organized the firm, but in 1846 the partnership dissolved with Gates retaining the company. David Fraser was working for Gates at the time, but left in 1848 for the Gold rush in California, returning in 1850. The following year he married Scoville's daughter and went to work for his father-in-law. During 1854 Fraser went to work for Gates once again — his working partner was Thomas Chalmers. Within a short time David Fraser was a foreman and Thomas Chalmers was the Gates plant superintendent.

For reasons not now apparent, the company was reorganized in 1862 and at this point Gates, Fraser, and Chalmers were partners. Ten years later, in 1872 the firm of Fraser & Chalmers was formed, and Thomas Chalmers' son, William J. Chalmers was company bookkeeper. He is the "Chalmers" of the Allis-Chalmers equation.

Shortly after being organized, Fraser & Chalmers began building a crusher design which Gates challenged in the courts as a patent infringement. After several years of litigation Gates finally lost the case, making what previously been a very competitive situation into one of intense dislike. This may possibly have been due in part to the activities of William J. Chalmers. In his book, *An Industrial Heritage*, Walter J. Peterson notes that after the formation of Allis-Chalmers Co. in 1901, "William J. Chalmers almost immediately incurred a virtually unparalleled degree of dislike and distrust in the new organization." Much of this intense rivalry and distrust may have grown from the strong emotions existing prior to the merger between Chalmers and the Gates people. In fact, there are indications that Gates Iron Works, despite its small financial base in comparison to Fraser & Chalmers, had nevertheless become so competitive that the latter firm found itself on the brink of bankruptcy just prior to the 1901 merger.

The atmosphere of rivalry, mutual distrust, and at times, intense hatred continued through 1911. Coupled with these internal problems was the determination on the part of Milwaukee based investors to wrest control from the New York financiers who had become entrenched in the company hierarchy. After ten long years of in-fighting, company executives finally recognized by late 1911 that the entire merger was on the verge of collapse. In early 1912 the company was reorganized to avoid outright foreclosure. One of the appointed receivers was Otto H. Falk. Finally the company could look forward to a new page in its history. By the time of the reorganization effort, many of the original partners had left the company, and when Falk got done cleaning house, still more quietly resigned. Through all these tribulations the company managed to continue building Gates crushers. If battles have winners, perhaps the Gates crusher and cement machinery lines remained as mute testimony of the winning team.

Gleaner Harvester Co.

Of all the inventors associated with the harvester business, few could match the activities of Curtis Baldwin. At various times Baldwin has been titled as a maverick, and among some of his associates it appears that he was frequently titled somewhat below this status. From all accounts it would seem that Curtis Baldwin was an inventor's inventor — one of those unique individuals so caught up in new inventions that all else in life assumed a minor role.

Curtis Baldwin was born in a sod house on the prairies of western Kansas during 1888. Part of his birthright was the dubious honor of pitching wheat bundles each summer. This hot and tiring task germinated the seeds of invention as young Baldwin decided to invent an easier way to harvest the grain.

At the young age of 23, Curtis Baldwin received his first patent (1,004,134) for a "Standing Grain Thresher." First tested in 1910, it led to the formation of The Baldwin Co. at Nickerson, Kan. the following year. The Standing Grain Thresher was a stripper machine ostensibly built along the same line as McKay's early stripper designs widely used in Australia. Despite some moderate success, extensive modifications, and additional patents, the Standard Grain Thresher Co. folded in 1918 and was reorganized as Baldwin Brothers. Two of Curtis' brothers were in this firm. Also included was Clarence A. Stevens.

Whether from a simple case of wanderlust or from disagreements with his brothers, Curtis Baldwin left the firm shortly after it was reorganized, ending up at Savage Harvester Co. in Denver, Col. The *Encyclopedia of American Farm Tractors, (Crestline, 1979)* illustrates the 1921 Savage tractor. Probably using the same tractor design, Savage offered a side-mounted harvester that same year — the harvester design was that of Curtis Baldwin. Despite its advantages when compared to its contemporaries, Savage is said to have built less than 30 of these units before going out of business. At this point in time Advance-Rumely Thresher Co., yet another Allis-Chalmers acquisition, stepped in and purchased manufacturing rights to the design. Indications are that the subsequent Advance-Rumely combines were built at least in part, under license of the Curtis Baldwin patents.

Gleaner Harvester Co.

While Curtis Baldwin was devoting his inventive efforts to the Savage harvester, Cal Stevens and the other two Baldwin brothers were designing a combine around the Fordson tractor. "Around" is an appropriate term, since this combine was actually mounted on the Fordson. To protect their work they were issued Patent No. 1,702,323. After some successful field demonstrations in 1923, Gleaner Mfg. Co. was organized in 1924. Phenomenal sales led to establishment of a new factory at Independence, Mo. in 1925.

With all these successes coming to Gleaner Mfg. Co., and the concurrent hard times for Savage Harvester Co., Curtis Baldwin decided in 1926 to rejoin Gleaner Mfg. Co. In their book, *The Grain Harvesters, (ASAE, 1978)*, Quick and Buchele note that "his reception is not a matter of public record!" The 1926 season was not a happy one. Stephen H. Hale who had advanced much of the capital for the organization was dissatisfied with the operation and demanded a reorganization in 1926. At this point came Gleaner Combine Harvester Corp., assuming the assets of Gleaner Mfg. Co.

After rejoining the company, Curtis Baldwin designed a new pull-type combine, calling it the Baldwin Gleaner. Curtis expected the new corporation to build this machine. They refused, Curtis Baldwin resigned, and in 1927 formed his own Baldwin Harvester Co. A temporary truce enabled Gleaner to assume manufacturing and sales of the Baldwin design, along with the Gleaner models. The uneasy peace fell apart in 1928 when Curtis Baldwin and other investors formed Baldwin Harvester Corp. and began building the New Baldwin combine. Gleaner hastily filed for an injunction that would restrain use of the Baldwin or New Baldwin titles, so Curtis Baldwin switched to the "Curtis" trademark, and in 1929, Curtis Harvesters Inc. was organized.

Curtis Harvesters went broke in the 1930s, but Curtis Baldwin was in no way dissuaded from his inventive bent. With the organization of Curtis Harvesters, Curtis Baldwin set about the development of a "Rotary Reaper" design which embodied many of the principles used in current-production axial flow combines. Not limiting himself to harvester developments, Baldwin then applied these principles to a centrifugal gold separator device. The rotary design led to formation of the Rotary Reaper Corp. at Ottawa, Kansas in 1936. It, like Curtis Baldwin's other companies, brought him fame but was sadly deficient in bringing in a fortune.

Elmer J. Baker, Jr., the Editor of *Farm Implement News*, urged Curtis Baldwin to record his memoirs for future generations. Although he agreed to do so, Baldwin was then totally engrossed with the development of an alfalfa wafering machine, and thus recording his life's work was delayed. Curtis Baldwin never did write the story and died April 19, 1960.

After Curtis Baldwin left the company of his brothers, Gleaner Combine Harvester Corp. continued on its own path to destiny. Almost 5,000 Gleaner combines were built in 1929, and in 1930 plans were made for a corn harvesting attachment. A 1930 advertising circular fully describes this unit. The Gleaner Baldwin Corn Combine was the first such machine built and sold in America. It did not pick the corn through snapping rolls, but instead it cut the entire stalk, carrying grain and fodder through the threshing cylinder. Gleaner noted that this machine "cuts, shells, and cleans standing corn, clearing the field in one operation." Power for this combine was supplied by a Ford Model A engine.

For virtually every manufacturer the Great Depression was a critical illness that often resulted in a terminal disease. Many companies went broke almost immediately, while others were so crippled that they never recovered, even with frantic attempts to delay the inevitable. With Gleaner Combine Harvester Corp. the end came in February, 1931. Farmers could not pay the dealers, dealers could not pay Gleaner, and Gleaner could not pay its suppliers. A newly formed firm, Gleaner Harvester Corp., was formed in April, 1932 and the next two years were spent in liquidating existing inventories. Not until early 1934 did production again resume, with the new company taking a conservative approach by limiting its efforts to sales in wheat growing areas plus a limited export business.

The conservative business approach paid off. Beginning in 1935, and continuing each year through 1954, Gleaner Harvester Corp. showed a net profit after taxes, and paid a dividend on outstanding shares each and every year. The 12-foot pull-type model went on the market in 1934, and with some modifications through the years, stayed on the market through 1952. A small 6-foot size came out in 1937 — it remained in the line through 1953. Several other sizes were built during this period, but all were eclipsed by the above mentioned models.

Probably due in part to the success of the Massey-Harris self-propelled combines, Gleaner began limited production of the Model A and Model R self-propelled combines in 1951. While this was happening, Allis-Chalmers was developing its Model 100 All-Crop combine, a self-propelled model built on the same general pattern as their concurrent and widely sold pull-type machines. Perhaps realizing that the Model 100 All-Crop had been developed to its full potential, Allis-Chalmers began negotiations with Gleaner Harvester Corp. in 1954 for purchase of the company. Various news releases were issued in late 1954 to this effect, but not until January 31, 1955 did Gleaner Harvester Corp. issue a letter to its dealers stating that, "Beginning February first, what has been Gleaner Harvester Corp. of Independence, Mo., will function as: Allis-Chalmers Mfg. Co., Independence Works."

Gleaner

Barely into manhood, Curtis Baldwin manifested his inventive bent with the Standing Grain Thresher, first tested in 1910. Somewhat resembling a giant vacuum cleaner, the large hooded section in the foreground actually sucked the heads of grain back to the separating section. This machine was built on the stripper design, eliminating the need for a sickle. An uneven or weedy stand of grain presented major difficulties however, so construction of this unit was abandoned after several years. The Standing Grain Thresher was a push-harvester, that is, the horses which moved the machine were behind it — one of them is barely visible in this 1910 photograph.

During 1911 Curtis Baldwin was awarded Patent No. 1,004,134 for the Standing Grain Thresher. This 1911 scene was photographed near Wichita, Kansas. It illustrates the back side of the machine and shows how the horses worked from the rear. A gasoline engine powered the thresher, so the horses were only required to provide the ground travel. Despite early and continuing problems with the stripper design, Baldwin persisted in his attempts to remedy these difficulties, finally abandoning the large suction fan system with a slatted canvas conveyor. Although this design was somewhat more efficient and required substantially less power, sales of the Standing Grain Harvester were somewhat less than was hoped for. Apparently, farmers were not yet convinced of this harvesting method.

The 1917 Standing Grain Thresher featured a Waukesha engine, roller bearings, and galvanized sheet metal parts. At this point it was still built as a push harvester with the horses working behind the machine. Curiously, this photograph from a 1917 catalog shows that the machine was manufactured at Sparta, Mich., even though Baldwin had his beginnings in Kansas and appears to have sold most of the machines in Kansas, Nebraska, and Texas. Various accounts indicate that Curtis Baldwin's two brothers were also involved with the company, but an early prospectus of about 1917 lists Curtis C. Baldwin as General Manager — no other Baldwins are listed as officers or directors of the company! C. A. Stevens is listed as secretary. It was Stevens who advanced at least part of the operating capital, and it was likewise Stevens, who with two other Baldwin brothers, continued building Gleaners after Curtis left for other pastures.

During 1917 the Standing Grain Thresher Co. folded and Baldwin Mfg. Co. took over in its stead. Curtis Baldwin seems to have left the firm about this time, or shortly afterward, showing up later in denver, Col. with the Savage Harvester Co. Meanwhile back in Kansas, Gleaner Mfg. Co., following up with some of Curtis Baldwin's earlier designs, came up with a most unique machine. In 1917 or 1918, Gleaner came out with a wraparound design which embodied a Gleaner combine and a Fordson tractor. The Fordson took the country by storm, and because of its low price was often dubbed "the poor man's tractor." Its eight foot cutter bar made it large enough for the average farm, yet narrow enough to pass through a 12-foot gate, a point which Gleaner's advertising literature stressed through this photograph. The auger conveyor which moved the cut grain off the platform and toward the threshing cylinder was unique in its time.

Keenly aware of the severe duty to which the Fordson tractor would be subjected in operating the Gleaner, the company offered a number of accessory items. Apparently these were optional, although they were highly recommended. Included was the Camel Clenair attachment, an air cleaner system highly sophisticated for the 1918 market. An auxiliary water tank gave additional radiator capacity — apparently the Gleaner people knew that the Fordson would get mighty hot on this kind of load! Steering post and seat extensions got the operator out of the major dust zone and gave better visibility. An optional exhaust elbow helped minimize fire hazards by running the unmuffled exhaust straight upward instead of the usual Fordson practice of heading it toward the ground. A study of the specially equipped Fordson shown here will reveal all of the above mentioned accessory items.

Gleaner Harvester Co.

While Gleaner Mfg. Co. was happily building the wrap-around design, Curtis Baldwin was in Denver, Col., making further improvements on a somewhat different machine. This 1921 scene illustrates a Savage harvester at work — it was primarily the design of Curtis Baldwin. From all appearances, this machine duplicated the concurrent Gleaner machine, at least to a degree. Despite what might have been a reasonably good design compared to its competitors, some historians have indicated that less than 30 of these machines were built. This unfavorable turn of events led to the end for Savage Harvester Co. in the mid-1920s with Advance-Rumely Thresher Co. of LaPorte, Ind., taking over the designs. Using portions of this design plus some of their own, Rumely offered a reasonably good combine for several years.

Gleaner Mfg. Co. ceased to exist on December 30, 1926. Through reorganization it was immediately replaced with Gleaner Combine Harvester Corp. Curtis Baldwin was now back with the company, and while there he designed a new 10-foot pull-type combine. Gleaner would not build it, Baldwin resigned, and almost immediately formed Baldwin Harvester Co. Injunctive relief on the part of the other Baldwin firm prompted Curtis Baldwin to rename the firm Curtis Harvester Co. Through an agreement now clouded by time, Standard Steel Co., which Curtis Baldwin contracted to build his machines, the Gleaner Combine Harvester Corp., and Curtis Harvester Co. all agreed to let Gleaner do the manufacturing and marketing activities. From all this came Gleaner's first pull-type machine, the Model P. First offerd in 1928, it was built in 10 and 12-foot sizes.

A late 1920s catalog illustration shows the Gleaner-Baldwin 10-foot machine ready for field work. With this design horses could pull the machine, although the majority were pulled with tractor power. One man of course was required to operate the tractor, with another being needed to operate the combine. The extreme dust conditions, coupled with the heat from the nearby engine were only partially compensated by giving the combine tender an elaborate cushioned seat. From this vantage point the combine tender kept tabs on machine operation and had the additional task of raising and lowering the header as needed. The grain tank dumped by gravity into a waiting wagon. Some Gleaner combines of the early 1930s were powered with Ford Model A engines, while this one appears to be using a Ford Model T engine and radiator.

Although production of the Fordson tractor ended in the United States during 1927, the huge number of these still on farms probably prompted Gleaner to use one for this photograph which also includes a 9-foot Gleaner-Baldwin pull-type combine. Gleaner combines never exceeded a header width of 12 feet until their self-propelled Model A combine emerged in 1951. In spite of, and perhaps because of, excellent sales after 1929, the company found itself in trouble again by 1930. Due to the Great Depression, farmers couldn't buy *used* machinery, much less buy new equipment. Those who had bought new combines prior to the Crash were unable to pay the dealers, the dealers couldn't pay Gleaner, and so the buck stopped with Gleaner. Once again reorganization was in the works.

Although Gleaner seems to have concentrated its sales efforts on the wheat growing regions, the company also made some exports, particularly to South American countries. Apparently this effort was sufficiently proportioned so as to require the printing of Spanish language catalogs — this one illustrating a 12-foot Gleaner-Baldwin model. Although the standard grain tank is shown here, some harvesting methods demanded the immediate bagging of grain, so bagging platforms could be secured as a replacement for the grain tank. The company's conservative approach to design and manufacturing seems to have precluded any hillside combines. Rather, Gleaner offered only level land or "prairie" combine models.

Gleaner Harvester Co.

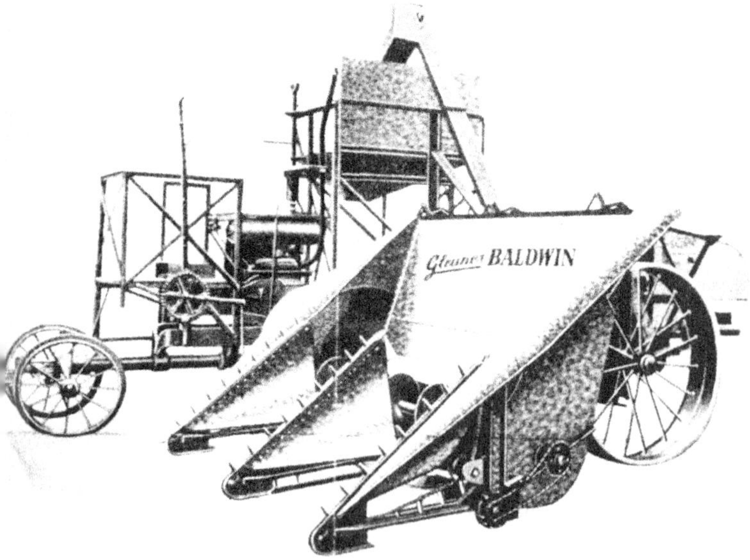

A "first" for Gleaner was the introduction of America's first standing corn combine in 1930. This was by no means a single-use machine — the corn head could be removed and replaced with the standard grain table. The Gleaner used a rasp-bar cylinder, and with this design, all the corn went through the machine, stalks and leaves included. A Ford Model A engine was standard equipment, delivering 32 horsepower. A unique feature was that the Ford engine included electric starter and generator as standard equipment, making hand cranking unnecessary. Only the battery was extra, and electric lights were also available for night operation at a slight extra cost.

The Gleaner Baldwin Corn Combine of 1930 used a no-nonsense approach to cutting the corn. Two large saws were mounted at the bottom of the gathering units, while top-mounted gathering chains brought in any material that did not fall into the auger of its own free will. The auger conveyor was in itself a hallmark of simplicity and completely eliminated the troublesome canvas conveyors used by many competitors. Despite its unique design, the Corn Combine could not have come at a worse time for Gleaner, as the Great Depression was beginning to take its toll on farmers and merchants alike. After the 1931 receivership of Gleaner Combine Harvester Corp. came the 1932 organization of Gleaner Harvester Corp., a new company that spent the next two years selling off excess inventory and tightening the corporate reins in order to weather hard times. The strategy worked and Gleaner continued to show a profit every year after 1934.

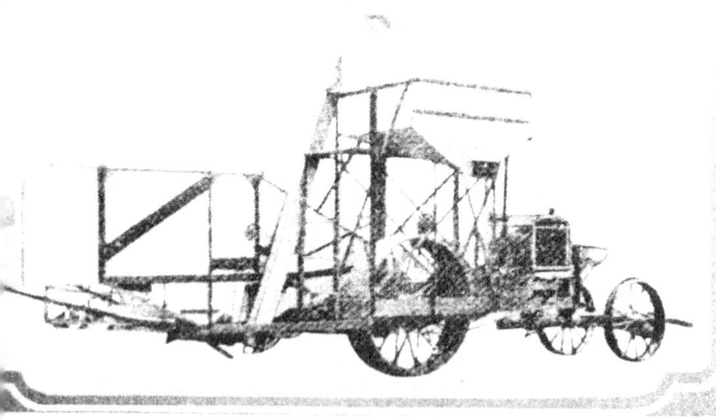

A Gleaner-Baldwin "A" combine, built from 1929 to 1935 was one of its most successful early models. It was built with a 12-foot cut. During the World War Two period Gleaner continued to offer new models, some with rubber tires, and some reverting back to steel wheels. This was necessitated by the rationing of rubber due to the war effort. To better accommodate various grain crops, Gleaner built straw walker or raddle machines simultaneously. The Gleaner Baldwin Model E combine appears to have been the only one produced continuously during World War Two. Various other models seem to have had production suspended for at least part of the duration. In 1953, Gleaner introduced a new and modern pull-type machine, the P-80. It was to be the last new pull-type combine introduced prior to the 1955 merger with Allis-Chalmers.

Gleaner introduced the Model A, self-propelled combine in 1951. This machine used the familiar straw walker separator design. After acquiring Gleaner in 1955, Allis-Chalmers continued building the Model A until 1963. Concurrently, Gleaner introduced the Model R, identical in all respects except that it used raddle separation, preferable for certain seed crops. Production of the Model likewise ended in 1963. Additional data on these models may be found in the *Combines* section of this volume.

Grain Drills

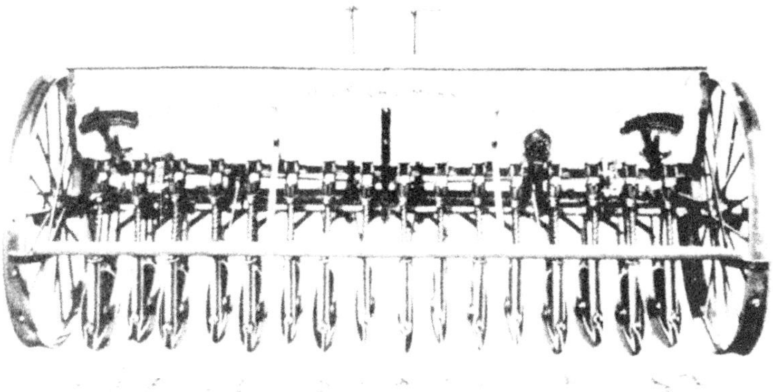

Allis-Chalmers grain drills appear to have followed the 1929 acquisition of LaCrosse Tractor Co. LaCrosse had been building grain drills for some time, following more or less conventional patterns. Although specific data has not been available, it appears that the early Allis-Chalmers drills closely followed the LaCrosse designs. For example, the 1938 version shown here was available for horse power or tractor use as desired. The steel wheels shown on this unit were actually a substitution for the wood wheels offered as standard equipment. Perhaps sales of these drills were below expectations, or perhaps World War Two offered a convenient intervention. Regardless of the reasons, production of these drills was apparently suspended by 1940 or shortly thereafter.

In 1954 Allis-Chalmers announced an entirely new All-Crop grain drill. This unit once again added grain drills to the A-C farm equipment line after the apparent suspension of drill production about 1940. The new All-Crop drill presented a unique approach with it fully mounted design, and in fact, A-C advertising of the period noted that "not since the introduction of the force-feed grain drill in 1851 has any development occurred which deviated so greatly from the original conception of mechanical seeding and fertilizing..." When originally announced in 1954, the All-Crop grain drill was built specifically for the Allis-Chalmers WD, WD-45 and CA tractors.

Allis-Chalmers pointed with pride to their new All-Crop grain drill, noting among other things its easy transport to and from the field. Allis-Chalmers advertising literature of 1954 stated that "you travel in road gear on the highway, just like a handy pick-up truck." All that was required was to raise the drill to its maximum height and hang the covering chains on convenient hooks to prevent unnecessary wear. In addition to its ease of handling on the road, the mounted design afforded easy maneuverability, permitting the farmer to work close to fences. This Wisconsin country scene shows a WD-45 and All-Crop drill on the road — it has just met and passed a 1949 DeSoto with plenty of room to spare.

The waist-high grain hopper on the All-Crop grain drill permitted easy filling. Behind it is the optional grass seed attachment — its flexible drop tubes are well behind the disc openers to prevent too-deep seed placement and a resulting poor stand. A-C claimed that the All-Crop drill offered precision seed placement due to the excellent flotation offered by the rear tractor tires. The mounted design gave better traction, and the process of hitching or unhitching was as simple as opening the jaws on the lift arms and pulling a lever to detach the Snap-Coupler. An accurate, easy-to-read acre meter came as standard equipment. This eliminated guesswork and assured the farmer that he was planting and fertilizing at the desired rates.

Grain Drills

All-Crop grain drills were built from 1954 to 1962. Two sizes were available — the 11 x 7 model and a smaller 9 x 7 size. This translated into an eleven-run drill on 7 inch spacing to be used with the WD or WD-45 and a nine-run drill on 7 inch spacing for the CA tractor. In 1957 the larger model listed at $454 with single-disc openers or at $518 with double-disc openers. A grass seed attachment added $67 to the list price. By 1957, and perhaps earlier, Allis-Chalmers offered a substitute hitch kit which permitted use of the 9 x 7 All-Crop drill with Ford, Ferguson, or Ford-Ferguson tractors and their three-point system. The 11 x 7 unit weighed nearly 900 pounds. This may have been a reason that the grain hopper only held 5 bushels and the fertilizer hopper topped out with a capacity of 400 pounds. Additional weight might have required front weights. No data has been located that might even approximate the total production run of this unique grain drill.

Development of the All-Crop grain drill series certainly was marked with enthusiasm by those assigned to engineer this line, as witnessed by some of their innovative designs. In a broader sense this view does not seem to have been shared by the entire company. Despite some excellent designs, Allis-Chalmers seems to have given the grain drill line a relatively minor position of importance when balanced against the full agricultural equipment line. Perhaps the grain drill market was saturated, leaving room only for a relatively small number of replacement units each year. Due to its rather specialized and short-term use, coupled with little need for replacement parts, no substantive aftermarket could be expected. Despite these adversities, Allis-Chalmers marketed a somewhat larger grain drill line during the 1960s. Matched to increased tractor sizes, these drills were ordinarily sold in the wheel-type version shown here, with the same drill suitably modified and available with the Snap-Coupler hitch. Shown here is a D-17 tractor with a 16-run All-Crop drill.

For 1967 this 18 x 7 All-Crop drill with rubber press wheels carried a factory list price of approximately $900. Since these drills were available with a standard ASAE hitch, or with the patented Snap-Coupler design, a rather complicated pricing structure resulted. Prices were shown for base units with separate prices for other components. For example, various types of openers carried different pricing, as did many other attachments. With this selective pricing structure the farmer could order a grain drill to his specific needs. This field scene was photographed near LaCrosse, Wis.

A rear view of the All-Crop grain drill illustrates its great simplicity. Three levers controlled all feeds — one for the grain hopper, one for the application of fertilizer, and a third lever to regulate seed metering from the grass seed hopper. A single crank adjusted the tension of all torsion springs on the individual gangs, assuring uniform depth of seed placement. The screw-type crank also provided positive penetration pressure control to within a fraction of a pound. The Micro-Feed design incorporated the non-clogging features of a double-run feed with the sliding adjustment of the flute-feed system. This gave incredible accuracy. The fertilizer system featured a Force-Flo design using a spiral spring feed for uniform and positive feeding. The fertilizer attachment could be set up to place fertilizer in the ground, on top of the ground, or broadcast, as desired.

Already in the 1960s, Allis-Chalmers agricultural engineers were experimenting with "minimum tillage," the latest buzz word for soil conservation practices. As happened so many times in the company's farm equipment line, this idea was ahead of its time — not by much, but still sufficiently unique in its concept to get short shrift from the average farmer. Perhaps as a throwback to our agrarian past, farmers remain stubbornly resistant to change, preferring for the most part to stay with proven methods rather than increase the odds for a successful crop by trying unproven schemes. To its credit, Allis-Chalmers seems to have had little hesitation about trying new methods in crop production. Even though this 1960s scene of tractor, disc, and grain drill provided one-trip planting may have seemed a bit unusual at the time, this principle has become a widely used practice in the 1980s.

Harrows

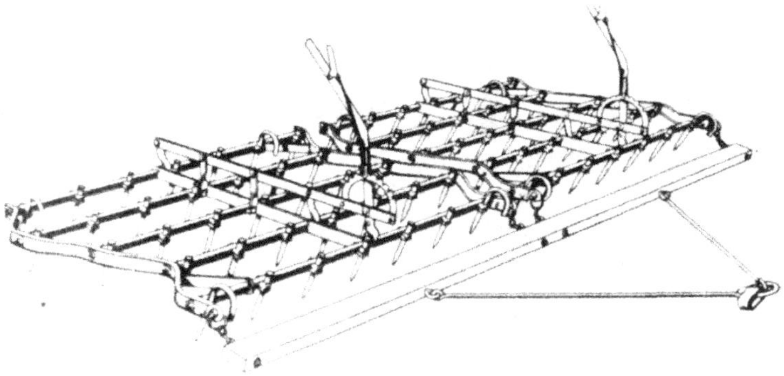

The Allis-Chalmers tillage line descended primarily from LaCrosse Plow Co., purchased by A-C in 1929. LaCrosse was an old established line, thus establishing some credibility among its peers. LaCrosse however had a rather weak marketing organization, a problem that was solved with the 1931 purchase of Advance-Rumely. During the 1930s Allis-Chalmers offered the 100 Series harrows. The U-bar design was a relatively new feature, as were the offset ends. Beyond these features however, the A-C spike-tooth harrows varied little from others in the industry — harrow preferences among farmers often came down to color preferences, with some farmers liking A-C Persian Orange better than red or green lines.

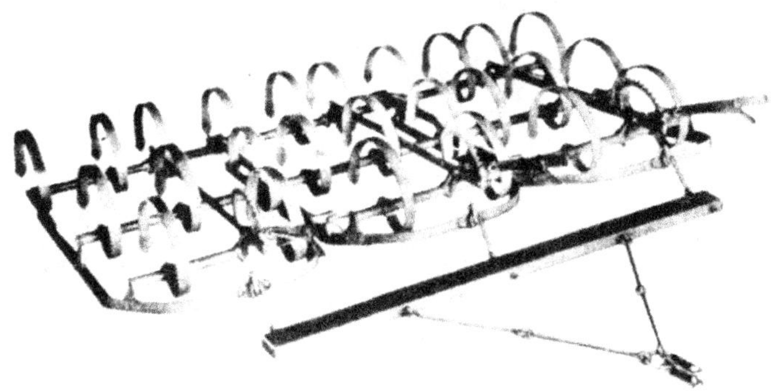

Like the spike tooth harrow, A-C spring tooth harrows descended directly from LaCrosse ancestry. The spring tooth harrow dates back to an 1869 patent of David L. Garver of Hart, Mich. Garver's attempts to market this harrow were largely unsuccessful for some years. By the 1890s however, several hundred different spring tooth harrow patents had been issued, with patent litigation and interferences in such disarray that several large manufacturers merged their patents into a single corporation known as National Harrow Co. National then issued shop licenses to each of the participating companies. In principle the A-C spring tooth harrow of 1935 as illustrated here, differs little from the original Garver machine of 1869; its competitors were also of the same general appearance. Stated another way — spring tooth harrows of the 1930s were by and large a generic item, differing primarily in minor details and the color of paint.

With the advent of modern hydraulic systems and quick-hitch designs came improved manufacturing techniques which permitted the building of improved harrows. Shown here is an Allis-Chalmers D-17 Series 3 tractor with a spring tine harrow. This unit has the proper tooth spacing for corn cultivation. The harrow teeth can deflect in any direction. This feature, plus the higher ground speeds available in modern tractors, provide the ground-shattering action necessary to dislodge weed growth, bringing the roots to the soil surface where the sun effectively reduces them to organic matter. High carbon steel permitted the use of light-weight frame sections, and the fully mounted design eliminated remote cylinders and other accessories.

During the 1960s Allis-Chalmers offered an extensive series of spring tooth harrows — they ranged from 8 to 20 feet in width. Spike tooth harrows were offered in sizes from 11 to 33 feet, and coil tine harrows were available in a similar size configuration. All 1960s models could be purchased in a pull-type version, and certain models could be equipped for either a conventional 3-point hitch or the Allis-Chalmers Snap-Coupler system. With the trend toward minimum tillage systems plus a very competitive market setting, Allis-Chalmers gradually withdrew from secondary tillage implements, and instead concentrated their efforts on primary tillage machinery. This strategy likewise failed, with the company eventually phasing out virtually its entire tillage equipment line.

Hay Crushers

A rear view of the 1949 Allis-Chalmers experimental hay crusher illustrates its rather simple design. Freshly cut hay was carried up the canvas conveyor to the crushing rolls, and from these to a cross apron which carried the crushed material to the side, depositing it in a windrow. This particular machine appears to have used two solid rolls which actually crushed the crop, squeezing out the excess moisture which hastened drying. Smooth rolls however presented horrific field problems in that they tended to wrap, despite all attempts to prevent it. Later research would conclude that crimping or bruising the freshly cut stems hastened curing and drying of the crop as quickly as crushing, but without the major problem of material wrapping on the rollers, plus the additional advantage of harvesting a palatable and nutritious crop. During the 1960s, Allis-Chalmers offered a mower-conditioner system which offered one-pass mowing and conditioning of the hay crop.

During the postwar years of the 1940s, Allis-Chalmers engineers set about developing various harvesting equipment based partially at least on the chassis of the All-Crop combine. This concept doubtless had its beginnings prior to World War Two, but wartime demands precluded any major developments. This photograph, dated December 15, 1949, illustrates an experimental hay crusher using the header and reel from an All-Crop "60" combine, along with numerous other combine parts in the chassis and drive train. Perhaps this experimental model used combine parts to eliminate a lot of expensive prototype work and hand-built parts, or perhaps the concept was much larger, embodying a vision of several different machines with interchangeable parts. Had this concept been carried to the ultimate, the matter of parts interchangeability might have been a major selling point.

Hay Rakes

In a 1950 advertisement Allis-Chalmers noted that their "farm machines have been distinguished by the excellence and originality of their design." The A-C side delivery rake and tedder of 1947 was no exception. It announced two new features to the industry — controlled steering and a power-driven, selective-speed reel. While other hay rakes castered the rear wheels, this new A-C design used a stiff mounting for the rear wheel while castering the front wheel. With this arrangement the rake precisely followed the tractor, regardless of countours or sidehills. When turning the rake pivoted on the rear wheel, leaving the end of the windrow straight, and without the usual large loops in the windrows found at each corner. The model shown here was equipped with optional tandem wheel mountings for use on terraced land or rough, rolling fields.

Allis-Chalmers built the Side Delivery Rake & Tedder from 1947 to 1957. Production during that period totaled some 52,000 units. The PTO drive system was soon emulated by other companies, as were some of the other features, including the rubber-tire wheel mounting. The floating reel was protected by a slip clutch, and some 33 anti-friction bearings were used. All gears were enclosed and run in oil. Two forward raking speed were provided in the reel gearbox, along with a single reverse speed for tedding — the process occasionally used for turning over previously raked windrows. Each pair of heavy duty rake teeth could be replaced by removing a single bolt. In 1958 this model was replaced with the No. 7 rake, similar in design but carrying a new model designation. The No. 7 remained in production until 1962, with its total production only totaling about 2,400 copies.

Hay Rakes

With the introduction of the Allis-Chalmers Model G tractor came a host of ideas for machine attachments. These included an experimental arrangement whereby the production model hay rake was attached to the front of the Model G tractor. This arrangement probably gave a very versatile and maneuverable machine, but for various reasons it never made it past the experimental stages. This rear view photograph was taken at the LaPorte Works on December 16, 1949. Chances are it had just been completed in the experimental shops.

An actual field scene illustrating the Model G tractor with an integral hay rake shows that this scheme worked, and in fact, is said to have worked very nicely. With this arrangement the operator was placed in an excellent vantage point, but this point itself seems to have spelled doom for the plan. Those who experimented with the machine soon found that the clouds of dust rising right behind the rake teeth made it virtually impossible to operate in this location. A study of this photograph makes it apparent that driving into a brisk breeze would catapult the entire load of dust and debris onto the operator. Primarily for this reason, the entire project was abandoned, probably by 1950.

Allis-Chalmers 77 Series rakes were built in the mounted style shown here as the 77-M, or a conventional ground-hugging model known as the 77-G. With this exception, both were virtually identical in design. Originally introduced in 1958, the 77-M remained in production through 1960, with production totaling 2,250 units. During 1961 and 1962 however, 750 of the 77-M rakes were dismantled for building 77-G rakes from common parts. The specific reasons for an early abandonment of the 77-M mounted design is now unknown. The 77 Series rakes embodied the latest in parallel bar design for fast and efficient raking. This Allis-Chalmers photograph shows a D-14 tractor with a 77-M rake complete with the Snap-Coupler hitch design.

Another view of the side rake-tractor combination is shown in this December, 1949, photograph. This mounting method gave the same pivoting action featured in the conventional pull-type model then in production. From all appearances Allis-Chalmers assumed a very aggressive stance toward development of new and original farm equipment designs during the late 1940s, the 1950s, and into the 1960s. Whether from marketing conditions, farmer preferences, or perhaps from limiting factors imposed within the company infrastructure, development of new and innovative A-C farm equipment designs appears to have slowed dramatically by the late 1960s, diminishing to a crawl during the 1970s, and ending entirely of course with the 1985 merger forming Deutz-Allis.

The Allis-Chalmers 77-G hay rake was a conventional pull-type machine with a ground-driven reel instead of the PTO drive. Situations not requiring the features of a PTO drive, such as variable reel speed, were served just as well with the conventional ground-drive design, but without the added expense and maintenance. Allis-Chalmers also promoted the side-mounted mower, with one of these units shown in place on a D-15 Series II tractor. A great many of these 82-S mowers were sold for highway mowing and similar situations far away from the farm.

Hay Rakes

Between 1965 and 1970 Allis-Chalmers offered the 78 Series rake using a conventional ground-drive design. Advantages included a conventional drawbar hitch, permitting the use of any farm tractor, and without the limitations imposed by the Snap-Coupler hitch design. Also featured were sealed tooth bar bearings that required no greasing, yet could be replaced with a minimum of work. Offset rear wheels gave excellent tracking ability and permitted the rake to follow the exact contour of the ground. Heavy duty stripper rods prevented damage to the teeth, and for special circumstances, rubber-mounted teeth could be supplied. Ground driven through a single v-belt, the 78 rake was the acme of simplicity despite its many features.

Hoar Shovel Co.

In 1926 Allis-Chalmers purchased manufacturing rights for the Hoar shovel, an ultra-compact machine that had been designed 10 years earlier. During the 1920s, Allis-Chalmers accounted for as high as 50% of all mining machinery sales in the United States. Adding this air-operated machine would serve to broaden the line still further. Since it was anticipated that sales might not be in any great number, Allis-Chalmers struck a deal with Hoar Shovel Co. of Duluth, Minn., for a 5% royalty on all machines sold. That was the extent of the A-C investment — if the machine sold well, so much the better, but if it sold poorly there was no large monetary loss. Hoar Shovel Co. had built less than 150 of these machines by 1926, and for their part, Allis-Chalmers built substantially less than this figure. This Model S-2 Hoar shovel is shown loading rock in a railroad tunnel.

The Hoar shovel could work in an area only 7 feet wide and 7 feet high. Within these boundaries it was able to revolve a full 360 degrees. Since it was operated with air motors it could work in coal mines and other hazardous locations with a minimum of risk. Despite its crudity in comparision to modern designs, this little shovel was apparently quite powerful. In this illustration it is shown picking up another load of rock. Although some of these shovels were used for tunnel building, the majority of those built seem to have been destined for mining operations. In the case of coal mining for instance, this little machine probably saved a lot of work formerly done by hand.

Some writers attribute the short life of the Hoar shovel not so much to competing makes as to the development of the engine-powered front-end loader. These machines began to appear during the late 1920s, and although heavy and crude, the front-end loader was soon to dominate the field of heavy materials handling. Allis-Chalmers heavily promoted the Hoar shovel in their *Sales Bulletin*, an internal organ published for A-C salesmen. After the late 1920s however, the Hoar shovel disappeared, and was undoubtedly given up as a lost cause. Under the terms of the A-C agreement with Hoar Shovel Co. however, the company stood to lose but little from the endeavor. Shown here is a Model 2 Hoar shovel mucking in Great Notch Tunnel, at Great Notch, N.J.

Hoists

Almost immediately after introducing the Reynolds-Corliss engine came an entire series of hoisting engines — steam-powered machines especially designed and built for hoisting purposes. After the 1901 merger which formed Allis-Chalmers, the latter continued building steam hoisting engines, some of which assumed mammoth proportions. This particular example was built for Quincy Mining Co. It used steam cylinders 48 inches in diameter and a stroke of 84 inches. The drum shown in this illustration is 26 feet in diameter, 12 feet wide, and could carry up to 7,500 feet of 1⅜ inch rope. Various mining industries were the major market for these giant hoisting engines, although others were doubtless used in different applications. An ornate winding staircase leads to the operating platform, with the operator being barely visible in this illustration. Other steam hoisting engines are illustrated in the *Steam Engine* section of this volume.

The advent of reliable electrical equipment and dependable power supplies relegated the steam hoisting engines to obsolescence, and in their place came a variety of electric hoists. Allis-Chalmers was an industry leader in the development of electric hoists, having by 1937 progressed to this huge unit designed for a pull of 30,000 pounds at a speed of 1,300 feet per minute. Allis-Chalmers built ten hoists of various sizes in 1937, but the one shown here was the largest. It was built for Cary Mine, Odanah Iron Co., Hurley, Wis., and operated by Pickands Mather & Co. The drums were fabricated of steel plate, having a diameter of 8 feet and a face of 7 feet. Each drum was designed to carry 4,750 feet of 1¼ inch rope in three layers. Power was secured from a 600 hp induction motor.

Expertise in hoist design was probably a major factor in Allis-Chalmers securing a war contract to build cargo winch parts for shipboard duty. During 1943, Allis-Chalmers set up facilities for quantity production of winch cases. This contract permitted maximum utilization of existing facilities including the foundry and machine shops for the war effort. Although some American manufacturers secured major contracts for assembling war materials, Allis-Chalmers contributed mightily by providing a great many different component parts. Due to the company's highly diversified product lines the A-C contribution to the war effort included such far-ranging equipment as superchargers for aircraft engines. Further data may be found in the section entitled, *Military Production*.

Home Heating Systems

About 1935 or 1936 Allis-Chalmers cautiously entered the home heating market. Marketed through the Home Appliance Division these units could be supplied for burning coal, or in the oil fired design shown here. Company records indicate that by 1937 some 165 units had been installed in Milwaukee alone, with others as far away as New York, Ohio, and North Carolina. The Universal Home Heater employed a more-or-less conventional hot water boiler to which was connected a conventional automotive radiator. The latter served as the heat exchanger, and a fan forced air through the radiator core, delivering it through ductwork to the rooms. The OB-120 unit shown here would deliver 100,000 Btu, but when equipped with a circulating pump and a larger circulating blower it was capable of over 120,000 Btu.

Home Heating Systems

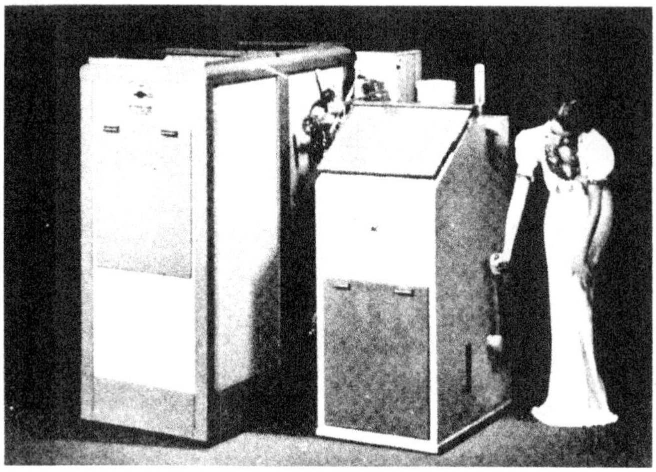

The Allis-Chalmers Universal Home Heater was also offered in two solid fuel models of 80,000 and 100,000 Btu output. This unique system used a very small grate area, and its hopper could carry a 24 hour coal supply. Shoveling of ashes was eliminated by providing an ash receptacle that was simply replaced by an empty unit when full. The company claimed this furnace to be more than 80% efficient over its entire range. To demonstrate the absolute cleanliness of this new system, it is being maintained by a lady in a white evening gown . . . Sure!

By 1940 the Allis-Chalmers Universal Home Heater had also taken on the aspect of summer air conditioning. For those who can recollect the 1940s era, air conditioning was something to be found in a few drug stores or department stores, not in the average home. This particular system circulated water through the same heat exchanger used for heating in the cold weather months — the water supply was run through the coils and from there to the nearest drain. Although considered wasteful in the light of present day energy and resource conservation, the single-pass approach was used to some extent in the early days of air conditioning systems. Probably because of the onset of World War Two, production of the A-C domestic heating and cooling systems appears to have been suspended for the duration, and did not resume after hostilities ended.

Hydraulics Division

The *Century Dictionary* of 1889 defines *hydraulics* as that branch of engineering which pertains or relates to fluids in motion. The Latin term *hydraulicus* actually relates to an ancient water organ invented by an Egyptian named Ctesibius. Under the lengthy title of *A Descriptive and Historical Account of the Methods of Raising Water by Mechanical & Other Means*, Thomas Ewbank sought to set down for history the development of all known methods. His 1848 book is now a rarity.

From the dawn of time, man has sought ways of bringing water to his use. From the ancients, through the age of Leonardo da Vinci, and down through history have come a steady progression in ways of pumping water from one place to another. Beyond this and numerous other uses of water is the subject of harnessing its power — a subject that occupied the best minds and talents of Allis-Chalmers engineers for some years.

A very early advertisement of Edward P. Allis Co., indicates that by the 1870s and perhaps somewhat earlier, Mr. Allis had taken on the agency for N. F. Burnham's patent water turbine. This coincided nicely with the existing line of Allis flour mills and other machinery. Burnham's design was far superior to the ordinary water wheel, and as time and technology permitted, Allis continued to offer the best available designs.

Although the harnessing of water power had occupied European engineers for centuries, progress was slow indeed. Not until the 19th century did the harnessing of water power make substantial progress despite the 1643 discovery by Toricelli that the velocity of a fluid flowing through an orfice is approximately equal to that which a solid body would acquire in falling a height corresponding to the distance between the level of the fluid in that vessel and the center of the orfice.

A dearth of historical data exists regarding early American developments, and in fact, Wilhelm Mueller records in the German title, *Die Francis Turbinen* of 1905 that U. A. Boyden of Massachusetts was instrumental in developing the Fourneyron turbine in the United States. He began in 1840. From this, James B. Francis, the originator of the famous Holyoke Testing Flume, built an improved design in 1849. To translate Mueller, "This turbine is the prototype of all American design." After these initial obstacles were overcome, more new developments followed in the next sixty years than had been witnessed through all previous history. Mueller further comments that while European engineers were engrossed in reducing hydraulic turbines to practice using empirical formulas, American engineers were content to make a design, test it, modify it, and re-test, continuing the process until they had achieved the desired goals. To reiterate, European engineers followed definite procedural rules in making

Hydraulics Division

their designs while in the 1800s at least, American engineers were too impatient, preferring the cut-and-try method instead, and with excellent success too!

Allis-Chalmers made early use of a trademark noting, "Ours the four powers: steam, gas, water, electricity." To substantiate the claim, the company announced in February, 1904, their entry into the hydraulic turbine business. Adding Clemens Herschel to the team gave the new project a distinct touch of class. Herschel was at the time regarded as one of the world's finest hydraulic engineers. Beyond this, Allis-Chalmers made arrangements to work with the world-famous firm of Escher-Wyss in Switzerland on hydraulic turbine designs. In less than a year after the initial announcement, Allis-Chalmers was building their first hydraulic turbine. Of the Francis type, it was rated at 1,250 hp under an 88 foot head and with a speed of 360 rpm. This unit, installed for Rochester (New York) Railway & Light Co., operated until 1952. Its immediate success brought many new orders, and in fact, by July, 1906 the company had over forty hydraulic turbines in various stages of construction.

By 1915 it was obvious that continued hydraulic turbine work would require a test facility. Soon after a temporary outdoor flume was built at West Allis Works, followed in 1920 by the necessary excavating at No. 2 erection shop for a permanent test pit. Ten years later the test pit was completely rebuilt and modernized, resulting in improved turbine design and greater working efficiency.

The great French explorer, Jacques Cartier was the first to historically record the existence of the Great Lakes, but not until 1604 did Samuel de Champlain make mention of the falls at Niagara. By 1757 a tiny portion of Niagara had been harnessed for power, but not until 1881 was the first hydroelectric station built. Subsequently came hydraulic turbines by Leffel, and then by Westinghouse. During 1919 however, Allis-Chalmers completed installation of its first unit at Niagara Falls. This unit was so eminently successful that in 1923 the company installed what was then the largest hydraulic prime mower in the world. This combined hydroelectric unit was designed and built by Allis-Chalmers. Rated at 70,000 hp, it could deliver up to 65,000 kva at 12,000 volts, using a speed of 107 rpm. Its design came from the famous hydraulic engineer, William Monroe White who then stated that units of 100,000 hp or more were entirely feasible. Eventually this prediction came true.

Special tribute must be given to several A-C hydraulic engineers. Arnold Pfau came to Allis-Chalmers from Escher-Wyss in Switzerland. Pfau, who died in 1953, was granted many patents covering hydraulic turbines and was a recognized expert on butterfly valves and governor design. Forrest Nagler (d. 1952) came to Allis-Chalmers in 1906, making the outstanding contribution of developing the fixed blade propeller runner, now known as the Nagler runner. The Kaplan runner, similar to the Nagler, was built by Allis-Chalmers, even though its inventor, Dr. Victor Kaplan of Austria does not appear to have been in the employ of A-C. The personalities noted here barely scratch the surface so far as eminent engineering talent is concerned — a complete listing of all the hydraulic engineers once employed by Allis-Chalmers reads like a "Who's Who" of engineering talent!

Of all A-C hydraulic turbine installations, one of the best known is that at Hoover Dam. Built under appropriations of the Public Works Administration, the first four hydraulic turbines were entirely of A-C design and manufacture, as were the generators and switchgear. Each turbine used a three-piece forged steel shaft with a diameter of 38 inches and a total length of 75 feet. Water from Lake Mead is admitted to the spiral casing — it alone weighs some 200 tons. A-C records show that these first four units weighed some 3,000 tons and required over 150 rail cars for transportation. By 1940 Allis-Chalmers had installed 7 of the 11 units at Hoover Dam. While the Niagara Falls unit of 1923 set world record proportions in size and output, Allis-Chalmers built an 82,500 kva unit for Hoover Dam in 1952.

Beyond this 1934 PWA project, Allis-Chalmers also secured major contracts for the Tennessee Valley Authority (TVA), along with numerous smaller contracts. To accommodate virtually any situation, any amount of fall, or any required volume, Allis-Chalmers remained for many years as the world's only hydraulic turbine builder offering all three major types of turbines — the Francis, propeller, and impulse types. Keen competition and market saturation combined with other factors to end the heyday of hydraulic turbine production. By the 1970s the hydraulic turbine market was but a fraction of that in earlier years.

An early A-C Sales Bulletin illustrates a very early hydraulic turbine at an undisclosed location. The vertical wooden shaft is evident, with yet another of these units in the background. Prior to the establishment of manufacturing concerns specializing in this field, hydraulic turbines took on many forms, often depending on the whim of the millwright in charge. If the water supply was ample, operating efficiency was not a major factor, since the water was free.

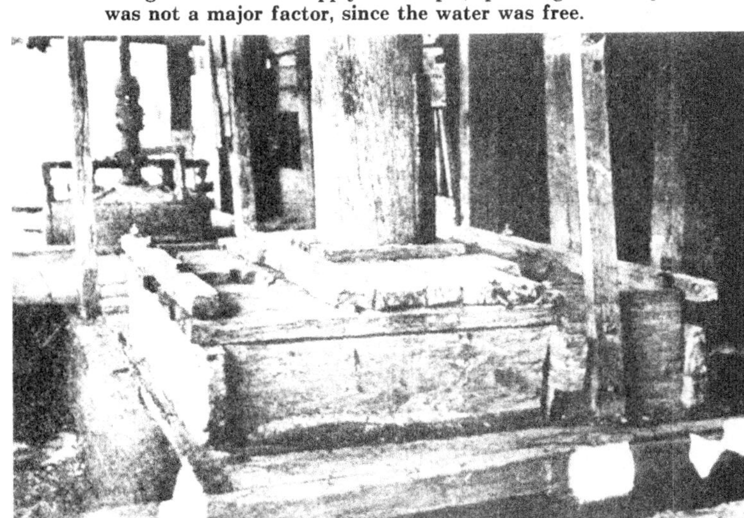

Hydraulics Division

Since the dawn of time man has sought for effective methods of harnessing the power of water. An 1895 catalog of Pelton Water Wheel Company attempted to show the progress of the past fifty years by illustrating the great overshot wheel at Laxey, Isle of Man, England built in the 1850s. This huge wheel, 72½ feet in diameter, was capable of about 150 horsepower. Its efficiency was at best, about 65%, while the much smaller Pelton of similar horsepower had an efficiency of about 85%. Even though no cost figures are available for the Laxey wheel, it is estimated that a Pelton wheel of comparable horsepower would cost no more than one fiftieth as much. Although Allis-Chalmers initially entered the hydraulic turbine business with the Francis turbine, eventually it was A-C who became the only company to offer turbines of the Francis, propeller, and Pelton styles. Ironically, the Laxey wheel still exists as a tourist attraction, while most Pelton wheels are long forgotten.

As hydraulic turbines became more refined, more efficient, and less costly, sales increased dramatically. This 1880s illustration shows a hydraulic turbine belted directly to a sawmill, only one of the many applications for water power. At this point in time the prime mover was of necessity at the site of operations — electrical power lines were still unheard of. Edward P. Allis Co. apparently acted as agents in selling small hydraulic turbines in connection with their flour mill business. This enabled the company to offer prospective customers a complete package installation.

While Allis-Chalmers seemed to concentrate on the Francis-type turbine in the early years of production, Fraser & Chalmers had been offering Pelton water wheels for several years prior to the merger. Details are now clouded by time, but it would seem reasonable to assume that A-C continued to offer the Pelton wheel as an alternative to the Francis turbine, especially since these two hydraulic units had their own unique features which made one style more suitable than the other under certain conditions. This partially dismantled unit shows the Pelton wheel with its buckets attached. Power is transmitted through the connecting rods to the pump cylinders — these are tied together to a common discharge through a massive cast iron manifold above the wheel.

Allis-Chalmers announced their entry into the hydraulic turbine business in early 1904. By securing the services of Clemens Herschel, an eminent hydraulic engineer, Allis-Chalmers was able to gain instant recognition in this field. Their first hydraulic turbine was installed during 1905 at Rochester (N.Y.) Railway & Light Co. It remained in service until the 1950s, finally being abandoned after the generating unit failed. The instant success of this 1,250 hp unit led to the sale of at least 43 hydraulic turbines by July, 1906.

Hydraulics Division

Allis-Chalmers turbines were never carried in stock. Each unit was carefully designed and built to existing conditions and requirements. Shown in this shop photo is a 32,500 kva generator built for Niagara Falls Power Co. Since A-C could build the complete turbine and generator unit, each could be designed specifically for the other for a totally unitized installation. William Monroe White, one of America's most prominent hydraulic engineers was on the A-C engineering staff. Under his supervision and direction Allis-Chalmers would build some of the world's largest hydraulic turbines.

During 1919 Allis-Chalmers installed what was then the world's largest combined hydroelectric unit at Niagara Falls. Under a 214 foot head, this unit was originally rated at 70,000 horsepower, and later re-rated to 75,000 horsepower. In its time this huge hydroelectric unit was the talk of the engineering world, considering the immense task of designing and building components of these huge dimensions. It was precisely on this principle of engineering and building to suit specific requirements that Allis-Chalmers acquired a worldwide reputation. Absolute reliability was a prerequisite and the company steadfastly refused to build to a price. Pick out the two individuals in this photograph and compare their size to the overall dimensions of this unit.

In 1927 the Niagara Falls Power Co. commissioned a privately printed book delineating the history of the company. Regarding this world's record turbine William Monroe White, chief engineer of the Hydraulic Department publicly stated his belief that units of 100,000 horsepower and larger were feasible. This particular machine operated at 107 rpm, generating 12,000 volts of 3 phase power at 25 cycles. Rotating weight totaled some 700,000 pounds. This cutaway view illustrates the general scheme of installation.

An illustration from some Allis-Chalmers advertising material shows the immense size of the Francis turbine runner for the Niagara Falls Power Company unit. Built of cast steel, it featured the vanes, discharge flange and runner crown cast in a single piece! This gigantic 15½ foot runner weighed 100,000 pounds — 50 tons! Few companies had the engineering, foundry, or machine shop facilities to build such immense components, and therein was much of the secret to A-C success in the hydraulic turbine business. Note how the man standing by the runner is completely dwarfed by its dimensions.

Hydraulics Division

Allis-Chalmers was an industry leader in developing plate steel for the spiral casing on hydraulic turbines. Shown here at the erecting shop, this spiral casing was intended for the huge 70,000 horsepower Niagara Falls unit. The casing inlet diameter is 15 feet, while the overall diameter is 50 feet. Plate thickness varies from $\frac{7}{8}$ to $1\frac{1}{4}$ inches. Even the automobile in the lower right hand of this corner is dwarfed by the huge dimensions of the casing.

A 1920s photograph shows a workman seated within a bucket on a huge Pelton wheel, leaving room to spare. The individual buckets were securely bolted to the huge cast steel wheel, and when natural erosion of the metal required replacement, this could be achieved by bolting on an entirely new set of buckets. Extensive research went into proper design of the buckets so as to obtain the greatest horsepower output for a given volume of water.

Propeller turbines achieved commercial success after the Francis and Pelton types were already well established. Due to its own idiosyncracies the propeller turbine required special care in its design. In the Francis turbine, the water enters almost radially inward and is turned downward within the runner. The propeller turbine is an axial discharge runner within which the water is not turned around as in the Francis. Two major types were popularized during the 1920s, the Kaplan propeller shown here, and the similar Nagler style. This particular shop-assembled unit was capable of 40,000 hp at 120 rpm under a head of 70 feet.

Pelton or impulse turbines were built in sizes ranging from small units under 50 hp, up to huge installations of several thousand horsepower. The larger units required massive bearings to carry the radial and axial load. In addition, the bearings were required to carry this load without interruption, sometimes for years on end. Thus it was necessary to design the bearings so that oil could be forcibly circulated, returned, filtered, and sent once again to the bearing. The ever present problem of heat was taken care of by suitable water cooling ports within the bearing shell. This huge Pelton wheel bearing of the 1920s was designed and built by Allis-Chalmers.

Hydraulics Division

The huge Allis-Chalmers erecting shops at West Allis, Wis., illustrate two partially completed spiral casings and turbine units, with additional components in the background. Very few shops anywhere in the United States were capable of assembling such huge components, much less do the necessary foundry and machine work. Although large overhead cranes were used for the heavy lifting, a great amount of manual labor was required as well. Allis-Chalmers built and tested each of these units before shipment, using their own test facility.

Destined for the TVA Power Project, this huge 95,000 pound cast steel runner would eventually operate at 100 rpm. A workman is shown polishing the turbine blades as part of the manufacturing process. Long before this however, came extensive engineering work, followed by the fabrication of extremely complicated foundry patterns. Once the mold was in place, nearly 50 tons of molten steel was required to finish the casting. Extreme care was necessary at each step, since even a minor error which ultimately might result in loss of the casting would represent an enormous loss of time and money for the company.

By the late 1940s Allis-Chalmers had installed four Kaplan-type turbines for the Pickwick Landing Station of the Tennessee Valley Authority. Shown here is one of the first two units ready for installation. Rated at 55,000 horsepower, these were at the time, the largest Kaplan-type turbines in the United States. In his ***Hydraulic Turbine Handbook***, (1948) Arnold Pfau, consulting engineer for the A-C Hydraulic Turbine Dept. notes that Kaplan and/or Nagler turbines were subject to many special considerations in order for them to be successful. With these various engineering requirements in mind, the propeller turbine is eminently successful in operation, but as Pfau notes, "Many failures are on record which are entirely chargeable to disregard, or lack of knowledge, of these requirements."

Propeller turbines were especially suited to low operating heads. As shown here, the propeller vanes could be shifted from within the main housing to provide speed control. This propeller has been machined on the huge 40-foot boring mill installed in the A-C shops, and a workman is checking it for balance. Although the complete propeller weighs approximately 40 tons, less than 70 pounds is required to shift the balance at any point.

Hydraulics Division

Another scene at the Pickwick Landing Station of TVA illustrates the huge components going into final assembly at the job site. The engineering and design of each unit to specific requirements was but one part of the job. Then came the work of the patternmaker and foundryman, following by a trip to the machine shop. The erecting shop likewise had its own special set of requirements and problems, and after successfully passing all the tests there, then came the task of dismantling the entire machine for shipment to the job site. Provided nothing was damaged in shipment, and provided further that other construction work was done per the drawings, all went well. Even so, at its very best, final erection of these huge and infinitely complex machines was a job which might at times require the wisdom of Solomon and the patience of Job.

A 1931 photograph shows one of four large propeller runners installed at the Rock Island, Wash., station on the Columbia River. This particular unit was hand adjustable, normally operating under a head of 32 feet. At 100 rpm, this propeller could develop 21,000 horsepower. This particular unit was designed so that the head could eventually be changed to 48 feet without redesigning the propeller. Note the massive bearing support at the bottom, and compare the size of the workman against one of the vanes.

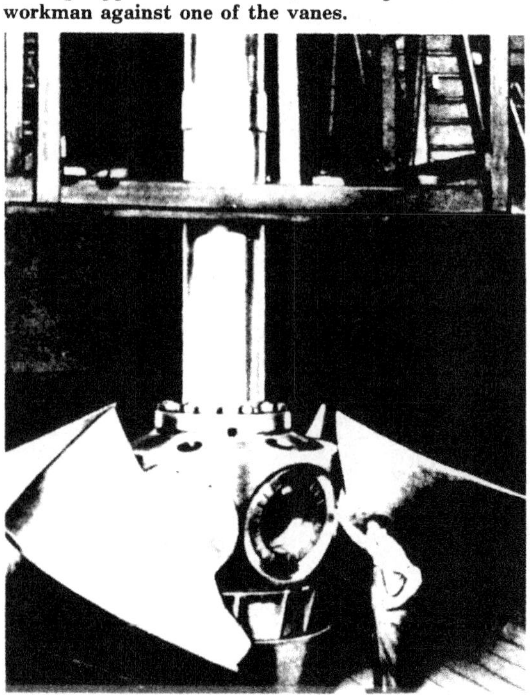

The finest hydraulic turbine built would be useless without an efficient, safe, and accurate speed regulator. The unit shown here was capable of governing 92,000 horsepower, even though some 2,000 cubic feet of water passed through every second! Allis-Chalmers had an excellent reputation for its speed regulators, with much of the credit going to the work of Arnold Pfau of the Hydraulics Department. Pfau specialized in this area of hydraulics, and during his career was awarded a large number of patents. Due to the kinetic energy of such large amounts of water as could be handled by this valve, special design considerations are necessary to avoid disastrous results. As an example, when the pressure regulator stalls the turbine control mechanism, the connecting rod at the top of the regulator is subjected to a tension force of over 600,000 pounds.

An Allis-Chalmers advertising bulletin illustrates the three major types of hydraulic turbines and explains their use. The propeller type at the left is regularly chosen for low heads, up to 100 feet, and in some cases higher. Its higher speed effects appreciable savings in generator costs. The Francis type is used for heads up to 1,000 feet, and occasionally higher, depending on the output capacity of the unit, the quality of the operating water, and the local operating characteristics. Pelton or impulse turbines are used for heads as high as pressure pipelines can be built, but seldom below 200 feet unless water conditions prevent use of the Francis type because of excessive wear.

3 TYPES of Hydraulic Turbines

Hydraulics Division

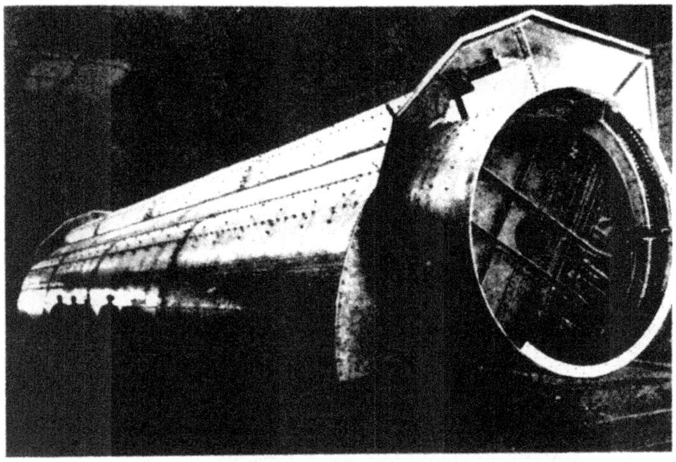

Hydroelectric installations require a large amount of auxiliary equipment. Roller gates are used to control the headwater level in the lake above the turbine, and as the name implies, roller gates achieve this goal. Depending on the size of the installation, the size and number of roller gates may vary. This particular example is 110 feet long and is designed to dam 20 feet of height — a rather formidable quantity of water! As part of the overall installation, Allis-Chalmers furnished roller gates in addition to the primary operating equipment. Once again, A-C was able to handle this work because of its huge facilities at West Allis and elsewhere.

Impulse or Pelton-type wheels are used against very high heads. This particular machine of perhaps a few hundred kilowatts is shown in its assembled position. The wheel cover is removed, illustrating the buckets bolted to the wheel. Due to natural erosion the buckets were subject to wear, but could be replaced when necessary. A large flywheel was used to normalize the alternator speed, with the alternator and exciter tied to the wheel through a common shaft. The governor mechanism is shown in the foreground.

The 1920s and 1930s saw the installation of numerous hydroelectric plants in response to America's electrification projects. Included were a great many impulse turbines such as this unit. Assembled in the A-C shops prior to being dismantled for shipping, it totally dwarfs the workman standing in front of the butterfly valve. There is also a workman standing to the left of the wheel housing, with the huge shaft and bearings being fully his height. Allis-Chalmers was able to capitalize on the sudden need for large prime movers, with their Hydraulic Department working at full capacity for several decades. However, these machines experienced relatively few breakdowns and few were replaced except perhaps when a larger unit was installed. Eventually the market became saturated, and this, together with unrelenting competition made the hydraulic turbine business less than desirable from the profit-and-loss standpoint.

While having great pride in setting world records for its large turbines, Allis-Chalmers also could supply rather small impulse units, including this 350 hp, 312 kva outfit for the hydroelectric plant of Champion Sillimanite Corp. It used a speed of 1,200 rpm and operated under a head of 1,170 feet. A-C also built Francis-type turbines in sizes as small as 32 horsepower, with a typical unit of this size operating at 1,800 rpm. Being able to offer an extremely broad range of sizes and styles certainly must have given Allis-Chalmers a competitive edge in a market that was competitive to the extreme.

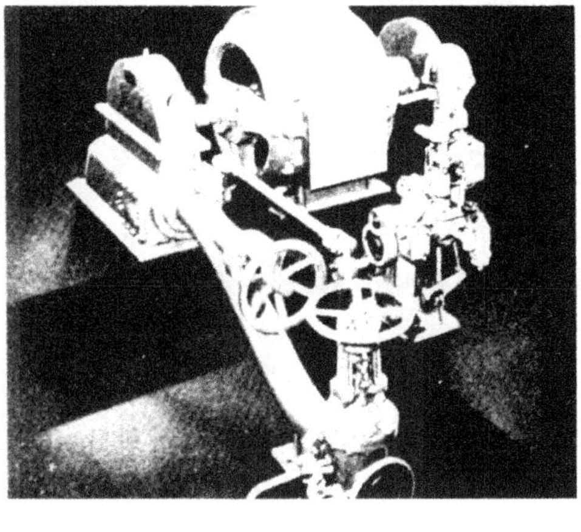

In 1928 Allis-Chalmers installed what was then the world's largest impulse turbine in terms of physical size. Built for the No. 1 San Francisquito plant of the City of Los Angeles, this impulse turbine was rated at 32,000 horsepower under a head of 800 feet and with a speed of 143 rpm. It had an overload capacity in excess of 40,000 horsepower. The 143 rpm speed gave a frequency of 50 cycles, but raising the speed to 171.6 rpm gave a 60 cycle frequency. This unit is of the double overhung type.

Hydraulics Division

The Public Works Administation (PWA) was organized on June 16, 1933. Shortly thereafter, Allis-Chalmers received a $1.1 million contract to build four large turbines at Boulder Dam, now known as Hoover Dam. In 1934 this, the first of several cast steel spiral casings was shipped to the Boulder power house. Each of the four turbines used a cast steel spiral casing weighing over 200 tons, and having an inlet diameter of 10½ feet.

A company photograph illustrates a 45-inch Type F standard runner with steel plate vanes. This unit had been in service continuously at the Wausau Street Railway plant at Wausau, Wis., for one year before being removed, and had been under a heavily fluctuating load. The original 8 x 10 photograph better illustrates the erosion of the vanes that had taken place over this period of time, with some of the wear probably due to impurities in the water, and a portion of the wear due to the fluctuating load. Due to these factors and many others, turbine runner design was a subject which occupied extensive research time using various wheel configurations and various metals for the blades.

An A-C file photo shows work in progress on the Boulder Dam project. Each of the four turbines was rated at 115,000 horsepower, a world's record for size and power. The forged steel shafts of these units were made in three separate pieces, 38 inches in diameter, and totaling some 75 feet in length from runner to the top of the generator. At maximum power each turbine discharges nearly 3,000 cubic feet of water *each second*! About 150 carloads of parts and equipment were required for the first four units, and by 1940, Allis-Chalmers had installed or had on order 7 of the 11 units in this project.

Allis-Chalmers by no means confined its hydraulic engineering efforts to the continental United States. Another large project of the 1930s was the Sho Development Plant of Asano Co. in northern Japan. The initial project used three units of 120,000 hp operating under a head of 120 feet. The A-C sales effort was aimed not only at U.S. and foreign government agencies — municipal power plants as well as large industrial operations were also courted in an attempt to secure their projects. This photograph was made from a model built in the A-C Engineering Department.

Allis-Chalmers was eminently capable of building very large machinery by virtue of its huge machine shop and other equally large manufacturing facilities. The company's gigantic vertical boring mill with a 40-foot table was the largest such machine in the world. Because of its immense size, this machine could greatly simplify machining projects such as this turbine speed ring for the Pickwick Landing Station on the Tennessee River, another PWA project. Even though the table of this gigantic machine was 40 feet in diameter, the speed ring is somewhat larger, as evidenced by this photograph.

Hydraulics Division

New turbine installations were almost at a standstill during World War Two, but after hostilities ended, America quickly retooled to industrial production. Other countries likewise began to once again look for additional hydroelectric power. This 1953 photo shows the nozzle pipe assembly of a 150,000 hp turbine in British Columbia. Built by Canadian Allis-Chalmers Ltd., this unit was at the time, designed for the highest head in the Western Hemisphere, 2,850 feet. The water was secured from a 10 mile tunnel through a mountain to drive this immense unit.

During World War Two, Allis-Chalmers built considerable equipment for the Armed Forces. Subsequently, the advanced technologies acquired during this period were coupled with the company's extensive experience in building variable pitch Kaplan runners to provide yet another market — reversible pitch marine propellers. This 1953 photograph illustrates a tugboat, the *Nancy Moran* in New York Harbor. It had been in operation for two years and featured a single lever control of blade pitch and engine speed from the pilot house. Pitch control was achieved by a hydraulic servo-mechanism operating a cam in the propeller hub. Since the mechanism was located inboard, the diameter of the hub was not increased. With this device, inching or standing against a current was possible.

During 1954 Allis-Chalmers continued fabrication of parts for a huge pump-turbine installation for the TVA's Hiwassee Dam project. Rated at 80,000 horsepower under a 190 foot head, this unit also discharges 3,900 cubic feet per second when working as a pump. Because of the immense size of the various parts, careful design was necessary to keep individual components within railroad shipping limits.

Another 1953 achievement was the installation of a large unit at the San Bartolo (Mexico) power plant. Here some Mexican workmen are shown riding along as one of the nozzle pipes is being lowered into position. This particular installation used a six-jet vertical impulse turbine operating under a head of 1,233 feet. Rated at 39,000 horsepower, it uses a speed of 428.6 rpm. Its generator is rated at 28,000 kva and operates at 13,200 volts. Note the unique cable trolley system at the top of this photograph — an arrangement unique to hydroelectric installations.

Developed for the Hiwassee Dam Project, this pump-turbine is, as its name implies, either a pump or a turbine as was required. An artists conception of this gigantic installation gives the reader some perspective when compared to the man standing to the left of the photograph. A-C hydraulic turbine operations eventually ended up at the York (Pennsylvania) Works, and in 1978 the company announced a standardized line of hydro-generators. The new compact units were equipped with everything needed to convert the energy of falling water into 1,000 kilowatts of electrical power. A-C called this new unit a "tube" turbine because it used a propeller turbine enclosed in a fabricated steel tube.

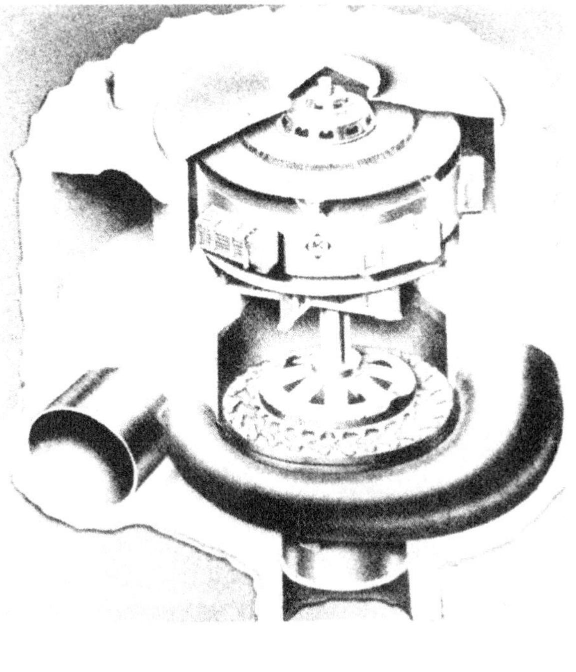

Industrial Loaders

The modern industrial loader was a concept far ahead of technological abilities. With the coming of the tractor came the need for materials handling equipment more versatile than the steam shovel. Early attempts to build loaders were only partially successful. At the very best, these machines were extremely heavy and cumbersome — their great weight alone made clumsiness an inherent feature.

The 1930's was a decade of great technological advancement, with Allis-Chalmers pioneering the use of pneumatic tires on tractors. Other major innovations of the period include the development of diesel engines suitable for tractor use, along with refinement of hydraulic power as a replacement for the winches and hoists of an earlier time. Each of these components was insufficient to permit the introduction of successful industrial loaders, but collectively these many and varied developments led to the first efforts.

Allis-Chalmers moved into the crawler tractor business with its 1928 acquisition of Monarch Tractor Co. During the 1930's several related lines were purchased, including Ryan Mfg. Co. in 1930, and Hy-Way Machinery Co. in 1931. After the attack on Pearl Harbor, non-military production slowed dramatically, and in some cases ceased entirely. For Allis-Chalmers a similar situation held true — a major portion of the manufacturing facilities were turned to war production. During this five year period however came many technological advances that might have been delayed for decades had it not been for the immediate and imperative need of the war effort. Benefiting immensely was the area of hydraulics, especially as applied to high pressure designs for cylinders and actuators.

Walter A. Geist served as A-C President from 1942 to 1951. Mr. Geist had the abilities to carry A-C through the difficult and trying times of World War Two, but in his place came W. A. (Bill) Roberts in the president's chair from 1951 to 1955. Roberts had a vision of the future and proceeded to acquire several companies, one of which was LaPlante-Choate Co. in Cedar Rapids, Iowa. LaPlante-Choate was well known for its industrial equipment, particularly its large dirt scrapers. The stage was now set for the industrial loader.

Historical data regarding the introduction of the A-C industrial loaders is sketchy, but from all appearances, it followed shortly after the 1952 purchase of LaPlante-Choate Co. By 1960 the line included a number of "tractor loaders" in various sizes, but sharing in common the features of a torque converter transmission and planetary axles. With the 1974 merger of the industrial equipment line into Fiat-Allis Construction Machinery Inc., production of these machines took the Fiat-Allis name.

By 1960 Allis-Chalmers was offering an extensive line of tractor loaders designed specifically for the purpose. The TL-12 model shown here was a slightly small version of the popular TL-14 loader. The latter size could be supplied with the choice of a popular A-C built G-262 gasoline engine, or an A-C 6000 diesel. Both developed about 100 horsepower. The TL-14 had a carrying capacity of 5,500 pounds, offered a maximum lifting capacity of 13,000 pounds, and had a maximum breakout force of 18,800 pounds. Standard equipment included a bucket level indicator, Hi-Traction differentials, and power steering. Accessories might include special buckets, backfill blades, or a crane hook.

Shown here ready for shipment, the TL-16 tractor loader gave a carrying capacity of 7,000 pounds. Like other models in the series it offered four-wheel-drive with rear wheel power steering. The Hi-Traction differential transferred power to the wheels with the best footing, and an exclusive single-lever speed and direction control gave any of six speed and direction combinations. An Allison power shift transmission was standard equipment — it was used in conjunction with a single-stage torque converter. The TL-16 was offered with the choice of an A-C 4DI-344 diesel engine or a Hercules 6G0-339 gasoline model.

A 1959 publicity photo illustrates the big TL-40 tractor loader working on a road building project. Huge machines like the TL-40 enabled the construction of roads and other projects that previously would have been impossible based on the high cost of manual labor. Advanced hydraulic systems gave the operator the ability to pry out and load rocks and boulders that otherwise would have been difficult to handle at best, and in a worst case scenario, totally impossible. With today's casual acceptance of the industrial loader as a way of life in the construction industry it is difficult indeed to imagine a return to this phase of American life prior to its development.

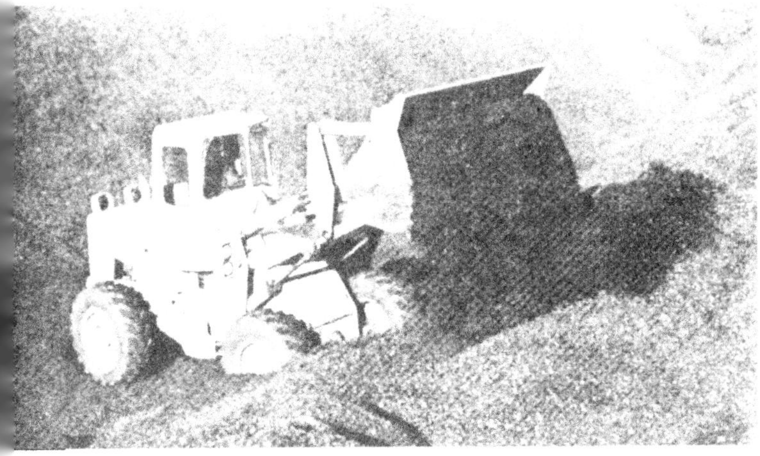

In 1965 Allis-Chalmers announced its new articulated loaders. This TL-545 is shown working at the steam plant of the University of Wisconsin at Madison, Wis. The Sof-Shift transmission provided full power forward and reverse shifting, and a new air-over-hydraulic dual axle brake system was featured. The A-C 2900 Mark II diesel engine was standard equipment. Its turbocharged design delivered 102 flywheel horsepower. Weighing 17,700 pounds, the TL-545 offered a maximum lifting capacity of 17,000 pounds. The 2900 Mark II engine was of six cylinder design and used a $3\frac{7}{8}$ x $4\frac{1}{4}$ inch bore and stroke.

Ever interested in acquiring military contracts, Allis-Chalmers developed this "break-apart" front-end loader in the early 1960s, and built 40 of these units in 1964 for delivery to the U. S. Marine Corps. They were designed for disassembly and easy helicopter transportation into remote and hard-to-reach construction areas. Weighing $9\frac{1}{2}$ tons when assembled like the unit in the background, the four sections include the bucket, rear wheels and hydraulic linkage, front axle and wheels, driver's compartment and transmission, and the diesel engine. Whether A-C built additional units for military or domestic purposes is presently unknown. The innovative design is however, typical of A-C engineering developments over the years.

Introduced in 1964, the TL-645 was another of the A-C articulated tractor loaders. This machine could turn in a $15\frac{1}{4}$ foot circle and had a bucket capacity of $3\frac{1}{2}$ yards. Placing the lift and dump cylinders on separate hydraulic circuits gave the TL-645 a combination of rolling and prying action for maximum breakout force. Features included an Allis-Chalmers Model 3500 diesel engine with 192 horsepower. Its six-cylinder turbocharged design included a $4\frac{1}{4}$ x 5 inch bore and stroke, yielding a 426 CID. The Model 3400 engine was of the same design but was naturally aspirated. TL-645 loaders were very popular, especially for rock quarries and similar materials handling applications.

Allis-Chalmers claimed their TL (Tractor Loader) machines to have the highest degree of maneuverability in the industry. These machines were designed to swing through a full 90 degree arc for extremely fast loading and unloading cycles. This particular TL-645 machine is equipped with special log forks and is working at the Bemis Hardwood Lumber Co. near Robbinsville, N.C. in 1968. Additional attachments for the A-C modular design included a fork lift, dozer blade, back hoe, crane arm, and road roller. Additional modifications were available for the loader bucket. Beyond all these possibilities, Allis-Chalmers was prepared to engineer and build special attachments for virtually any application.

This 1967 photograph shows the combination of articulation and an all-hydraulic boom action to allow precise handling of loads with complete control. For transport, the sideboom unit tucks up against the tractor loader, and does not interfere with normal operation. "Hitch-King" became a registered A-C trademark for the self-aligning and quick adjusting hinge pin design used in these articulated loaders. The special design of the big $2\frac{1}{2}$ inch diameter hinge pins provided an extra large bearing surface and permitted easy adjustment by simply tightening an adjusting nut. This design practically precluded the need to replace pins, bearings, and bushings — a job that was difficult and very expensive, to say nothing of the down time.

Internal Combustion Engines

One of the most interesting areas of A-C product development involves internal combustion engines. In April, 1902, the Board of Directors authorized company officials and engineers to study the current styles, with Edwin Reynolds apparently playing a major role in this endeavor. About a year later Reynolds announced that Allis-Chalmers had acquired manufacturing rights from Nuremberg Machine Company of Germany and was their sole licensee for the United States, Canada, and Mexico. Half a decade later Allis-Chalmers emerged as the largest gas engine builder in the United States.

Working from an original Nuremberg engine, A-C engineers "Americanized" it. That is, they modified it to better suit current American practice. In the process the engines received numerous design improvements. Thus in less than two years time, Allis-Chalmers began searching for the best design, secured manufacturing rights, set up a test engine, modified it to their own liking, and began offering engines ranging from 250 to 3,000 horsepower. All of these engines were of horizontal, four-cycle, double acting design — that is, the engine worked on both sides of the piston. Placing two of these double acting cylinders in tandem gave an output ranging up to 3,000 horsepower, and when built as a twin-tandem unit the output went up to a rated 6,000 horsepower.

These gigantic engines were aimed primarily at industries using waste gases from blast furnaces, but despite this rather specialized application, 81 units were built between 1905 and 1930, with all but thirty of them being built between 1905 and 1912. Company records indicate that gas engines quickly became a major part of the company's business, since by June, 1906, orders for gas engines constituted over 20% of the unfilled orders.

The first engine, a 1,000 horsepower model went into service during 1905 for the Milwaukee plant of Illinois Steel Co. Then in 1906 came what was probably the company's most impressive installation. A single powerhouse, 105 feet wide and 966 feet long housed seventeen Allis-Chalmers gas engines, each rated at 4,000 horsepower. The great success of the initial installation prompted the installation of another seventeen units between 1910 and 1918. Using waste gas from the blast furnaces, these engines powered huge generators which supplied the company's entire electrical needs, and in fact furnished power for the city of Gary, Indiana, until the beginning of World War Two. The last large gas engine, built in 1930, was also the largest size built — it was rated at 10,000 horsepower but had a reserve capacity approaching 15,000 horsepower.

Coinciding with the industry trend toward self-igniting oil engines, Allis-Chalmers announced a series of their own models during 1915. During the early 1920s oil engines were gradually phased out in favor of diesel engines — all of these styles were of the heavy stationary type. The horizontal diesels could be equipped with specially built pumps for oil pipelines — a great many of these were built for this service.

In 1932 Allis-Chalmers announced the introduction of package power units for the first time, using engines from the Model E, U, and L tractors then in current production. All that was necessary was to slightly modify the engine and add the extra accessories for this purpose. This move apparently met with some success, and while sales of the power units increased, orders for the large stationary engines of earlier years dwindled to nothing.

Paralleling current tractor production, Allis-Chalmers continued offering package power units using the same engines. Finally in 1953 Allis-Chalmers purchased Buda Company of Harvey, Ill., and this acquisition put A-C well into the engine business. Using then-current designs and following them with new and improved engine styles, Allis-Chalmers continued to offer an exceptionally wide range of engines — See *Buda* section for further information.

Already in 1902 Allis-Chalmers began searching North America and Europe for a suitable gas engine design. Although most engine builders of the period opted for the small farm engines, A-C chose to follow its general pattern of building exceptionally large units, far past the capabilities of almost any other U.S. manufacturer. Their first engine of 1904 retained the "Nurnberg Gas Engine" title, even though A-C engineers had already begun the process of "Americanizing" the design to better fit the peculiarities of American engineering practice. The first engine of 1,000 horsepower was installed in the Milwaukee, Wis., plant of Illinois Steel co. It was of four cycle, double acting design, with a simple connecting rod for its four cylinder bores.

The large horsepower output of the A-C gas engines precluded the use of belting to drive machinery, but by virtue of the immense power involved, required a direct-connected design in nearly every instance. This photo from the A-C Archives shows a twin tandem, double acting engine set up with compressors on the opposite end. These could be set up for air or gas compression, depending on the particular situation. This photograph gives no indication of the location of this particular engine. The A-C gas engines ranged from and 18 x 24 inch bore and stroke up to a 52 x 60 inch size.

Internal Combustion Engines

A 1910 photograph illustrates a typical electrical generator installation. In this case, two 4-cycle, double-acting, twin-tandem engines are installed, each having a rated capacity of 2,000 horsepower. This particular size would have used cylinders of a 32 inch bore with a 42 inch stroke. Since Allis-Chalmers was also in the electrical generator business, the company was in a position to build a complete turnkey installation including everything from the engine to the switchboards.

The overall size of Allis-Chalmers gas engines, together with their immense output required equally large castings. Each component had to be capable of carrying its rated load plus a considerable overload without any difficulty whatever. Thus the gas engine main frame assumed considerable proportions, as is evidenced by the workman standing to the rear of this main frame casting. When installed, the main frame was actually embedded in concrete approximately up to the horizontal line below the crankshaft opening. The rigidity of the frame was further augmented by carrying the foundation bolts entirely through the body of the frame.

Completely dwarfed by one of the engine cranks, the workman in this photograph gives some perspective regarding the immense size of these engines. The crankshaft carried not only the entire power load delivered by the double-acting pistons, it also carried an immense flywheel ranging from a mere 12 foot diameter for the 260 horsepower model, up to a diameter of 24 feet in the 5,900 horsepower size. As has been pointed out in various sections of this book, machining such huge components was work that could be done in very few American shops and this fact alone gave Allis-Chalmers a distinct advantage.

One of the largest A-C gas engine installations was an initial construction of seventeen units for the Gary plant of Indiana Steel Co. Each unit was rated at 4,000 horsepower. Of four-cycle, double-acting, twin-tandem design, these engines used a 44 inch bore with a 54 inch stroke. Operating speed was approximately 86 rpm. Floor space required included a width of 38 feet and a length of 60 feet. This initial installation of 1906 was eminently successful and resulted in the installation of another seventeen units between 1910 and 1918. These engines operated off the waste gases from the blast furnaces, harnessing energy that would otherwise have been simply burned off.

In 1929 Allis-Chalmers built its largest gas engine for the South Chicago plant of Illinois Steel Co. Two of these units were installed and each was direct-connected to a 6,000 kilowatt electric generator. A photograph of this installation shows the immense foundation required for these engines, as well as the large access area beneath the engine for fuel lines, exhaust lines, water pipes, and other necessities. A spare cylinder is shown in the foreground — it had a 60-inch bore, with these engines using a 64-inch stroke.

Internal Combustion

A main frame for one of the A-C 10,000 horsepower engines is shown on the rails, ready for delivery to Illinois Steel Co. in 1929. Each of these enormous engines weighed nearly 2,000 tons, requiring forty rail cars. Fuel consumption totaled 1 million cubic feet of gas per hour. Other dimensions of this monster were: Crankpins: 30" diameter x 27" long; Main Bearings: 40" diameter x 64" long; Diameter of shaft in flywheel: 43" inches; Piston Rods: 20" diameter; Flywheel: Cast steel, segmental, 26 feet in diameter; Cooling water required: 100,000 gallons per hour. Other dimensions and weights include: Cylinder, 168,000 pounds; Connecting rod, 30,500 pounds; Crankshaft with cranks and wheel hub, 265,000 pounds; piston, 9,000 pounds, and flywheel, 185,000 pounds. Rated speed of these engines was 83⅓ rpm. Special design features included water cooled main bearings. Starting was achieved by air pressure of 200 psi. These engines could be brought from a standstill to synchronous speed in less than two minutes.

At this point in time it remains uncertain as to exactly when A-C began building "diesel oil" engines. Although these engines were enormously heavy compared to the delivered horsepower, they were nonetheless somewhat lighter than the earlier Nuremberg design. There are indications that production of these engines may have started by 1915 and perhaps even earlier. However, a 1920 price list shows that the three-cylinder "triplex" engine shown here was rated at 270 horsepower and used a 16 x 24 inch bore and stroke. The engine alone carried a 1920 list price of $34,990. It is coupled to an A-C heavy duty pump suitable either for water or oil pipeline service.

A 1920 factory price list indicates that A-C oil engines were built in 14 x 21; 16 x 24, and 18 x 27 inch cylinder configurations. The smallest size was available as a single cylinder, duplex, or twin duplex engine with ratings of 70, 140, or 280 horsepower. The 16 x 24 model was rated at 90 hp per cylinder, and in addition to the above styles could also be furnished either as a triplex or a twin triplex engine. The largest size was built only in duplex or twin duplex sizes rated respectively at 220 and 440 hp. The 16 x 24 twin triplex model weighed 165,000 pounds and carried a factory price of slightly over $69,000. This particular unit is shown direct-connected to an Allis-Chalmers 60-cycle alternator.

During the 1920s and 1930s Allis-Chalmers made a special effort to install these diesel engines in oil pipelines. When equipped with an A-C pumping unit, the company could provide a complete package installation. A typical single cylinder, 14 x 21 engine was priced at $14,820 — it developed 70 horsepower. Weighing 37,000 pounds, it used an 8 foot flywheel which alone weighed 13,000 pounds. These engines used air injection systems, with a 1920 catalog noting that "development of solid injection for diesel engines is still in the experimental stage, and in our opinion is not as reliable as the air injection which we use." Whether any of these interesting engines still survive is presently unknown.

A 1920s photograph shows an Allis-Chalmers 20-35 tractor engine connected to an electrical testing machine. This procedure was intended to "break in" the engine and to determine whether any problems might exist. After the test run the engine was ready for installation in the tractor chassis. As an offshoot of this work, Allis-Chalmers began offering their tractor engines, suitably modified, as a package power unit. This particular scene shows additional test stations in the background.

Allis-Chalmers announced in late 1932 that their model E, U, and L engines could be supplied as package power units. Despite this early announcement, company records are sketchy on this point. This 1939 photograph illustrates an L-90 power unit operating a sawmill at Mayhill, New Mexico, for the Mershon Lumber Co. This huge gasoline engine used six cylinders of 5¼ x 6½ inch bore and stroke for an 844 cubic inch displacement. Normal operating speed was 1,050 rpm, with maximum output rated at 110 horsepower. About 1950 this engine became the L-844 model, apparently with little change in design.

The E-60 power unit, first announced in 1932, was built until the late 1940s when it was replaced with the E-563 engine. The bore and stroke dimensions of the E-60 were identical with those of the L-90, except that this engine used only four cylinders for a total displacement of 563 cubic inches. Like the L-90, this unit was rated at 1,050 rpm. Maximum brake horsepower was 74, but for continuous service, 60 horsepower was available. A gasoline carburetor was standard equipment, with proper equipment for kerosene or distillate being optionally available. At extra cost, Ensign fuel equipment could be supplied for natural or manufactured gas.

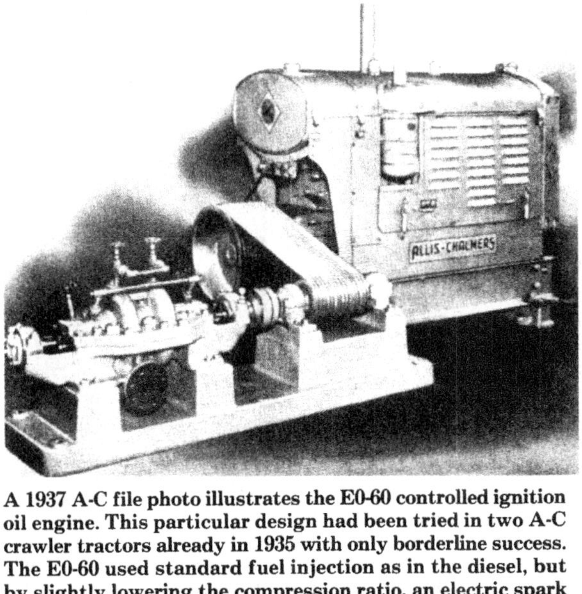

A 1937 A-C file photo illustrates the E0-60 controlled ignition oil engine. This particular design had been tried in two A-C crawler tractors already in 1935 with only borderline success. The E0-60 used standard fuel injection as in the diesel, but by slightly lowering the compression ratio, an electric spark was used to detonate the fuel charge. The slight advantages offered by the "controlled ignition" system were outdistanced by the increased manufacturing cost, plus additional maintenance expenses. This E0-60 is belted to an A-C "MM-2" pump and connected with an A-C Texrope multiple v-belt drive.

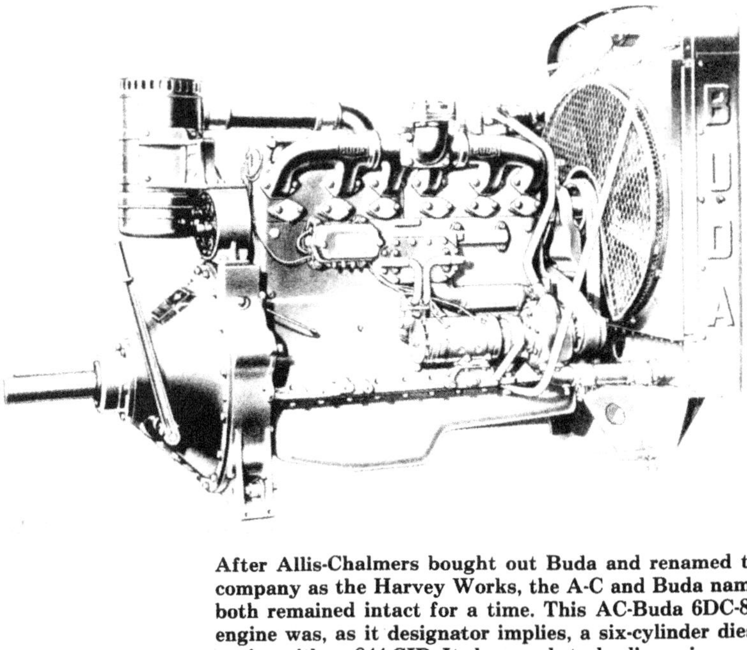

After Allis-Chalmers bought out Buda and renamed the company as the Harvey Works, the A-C and Buda names both remained intact for a time. This AC-Buda 6DC-844 engine was, as it designator implies, a six-cylinder diesel engine with an 844 CID. Its bore and stroke dimensions, and its displacement were identical to the L-90 engine that A-C had introduced two decades before. This particular model had a maximum output of well over 200 horsepower, and proved to be very popular wherever diesel power was required.

Internal Combustion Engines

With the advent of the new tractor line in 1938 came a new series of power units using the newly developed engine line introduced at that time. The B-15 power unit shown here was identical to the engine used in the Model C tractor. Rated at 1,500 rpm, this little four cylinder engine carried a $3\frac{3}{8}$ x $3\frac{1}{2}$ inch bore and stroke for a displacement of 125 cubic inches. Due to the intervention of World War Two, very few B-15 units were built until 1945, and by 1950 the same power unit was designated as the B-125, although it retained most of the features found on its immediate ancestor.

Beginning in 1948 the W-226 power unit emerged as a very popular style and size. Its four cylinder design boasted a 4 x $4\frac{1}{2}$ inch bore and stroke yielding a 226 CID. Rated at 1,400 rpm, it was weighed about 900 pounds as a complete package unit. Production continued until about 1957 when this engine was apparently replaced with a nearly identical model, the W-201. Indications are that once again, as in many other cases, the new model designator was the major change, with other specifications remaining virtually the same as before. Revamping the tin work did however provide a cosmetic change.

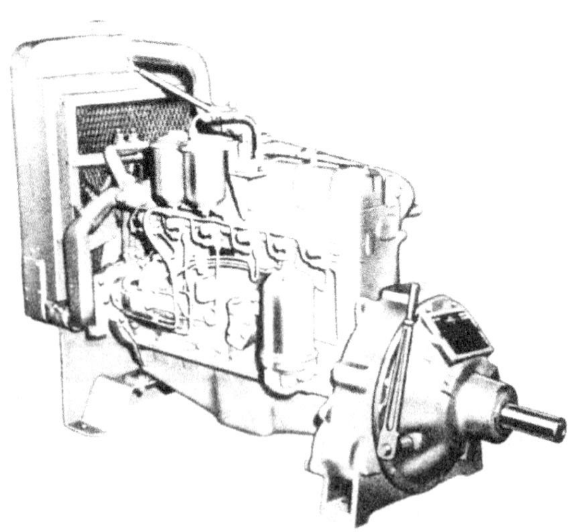

The A-C D-262 engine was an extremely popular part of the diesel engine line for a number of years. Its six-cylinder design included a $3\frac{9}{16}$ x $4\frac{3}{8}$ inch bore and stroke. At 1530 rpm it delivered 50 horsepower. For some years the G-226 was offered as a gasoline version delivering approximately the same horsepower, but using a four-cylinder design with a 4 x $4\frac{1}{2}$ inch bore and stroke. Only a very few of the many A-C power units are illustrated in this section — in many instances various engine styles are described together with their specific applications elsewhere in this book.

The 1950s saw the introduction of the D-516 diesel engine, along with its companion the D-344 engine. Both used a $4\frac{7}{16}$ x $5\frac{9}{16}$ inch bore and stroke, with the D-344 having four cylinders and the D-516 carrying six cylinders. The designator numbers indicate the cubic inch displacement for each engine. Rated at 88 and 131 horsepower, the D-344 and D-516 models used a controlled turbulence fuel injection system featuring an energy cell diametrically opposite the injector. These engines, like others of the A-C line could be furnished as open or enclosed power units, fan-to-flywheel units, or for various special applications.

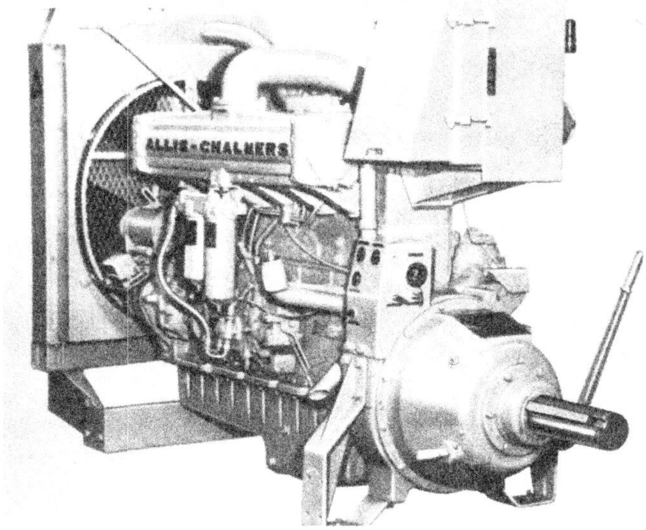

The Allis-Chalmers industrial engine line was very extensive, and at the top fo this series stood the Model 25000 engine. Its sister, the Model 21000, was naturally aspirated, while the engine shown here was turbocharged and intercooled. Both carried six cylinders of 5¼ x 6½ inch bore and stroke for a displacement of 844 cubic inches. At a rated speed of 1800 rpm the 21000 delivered 375 horsepower, compared to 435 horsepower for the 25000. This engine was also built as a "Cross-Continent" truck diesel. The most recent version, the Model 25000, Mark II engine carried a rating of 450 horsepower.

About 1960 Allis-Chalmers further expanded its off-highway diesel engine line to include the Model 10000 and Model 11000 series. These engines carried the same bore and stroke as the earlier D-516 engine, and were in fact an evolutionary outgrowth of this model. A noticeable change however was a rated speed of 2200 rpm for this new series, giving respective ratings of 145 and 220 horsepower. The difference in horsepower was due to the naturally aspirated design of the Model 10000, while the Model 11000 was turbocharged. Oil fields, construction, quarries, and irrigation were typical applications. Additionally, these engines could be specially equipped for marine propulsion or auxiliary duty when required. This particular Model 10000 is shown with a direct-connected emergency standby generator for a hospital.

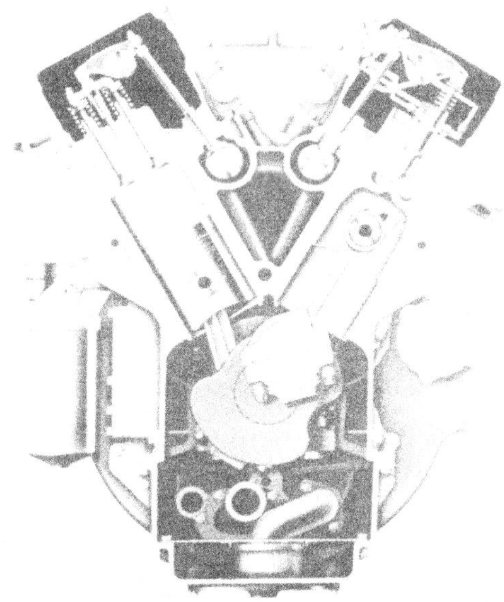

For the largest package unit applications Allis-Chalmers offered the huge V-12 diesel engine during the 1960s. Two models were available — the Model 61000, rated at 800 horsepower, and capable of 2100 foot-pounds of torque at its rated 2200 rpm, and the Model 65000 which offered 900 maximum horsepower. Both engines used a square 6 inch bore and stroke, with the basic difference in these engines being that while the Model 61000 was turbocharged, the Model 65000 was also intercooled. Adding the intercooling package alone gave another 100 horsepower! The pistons were oil-spray cooled and dual camshafts were featured. The rear gear train provided virtual freedom from torsional vibration for engine components. A power-take-off unit came as standard equipment.

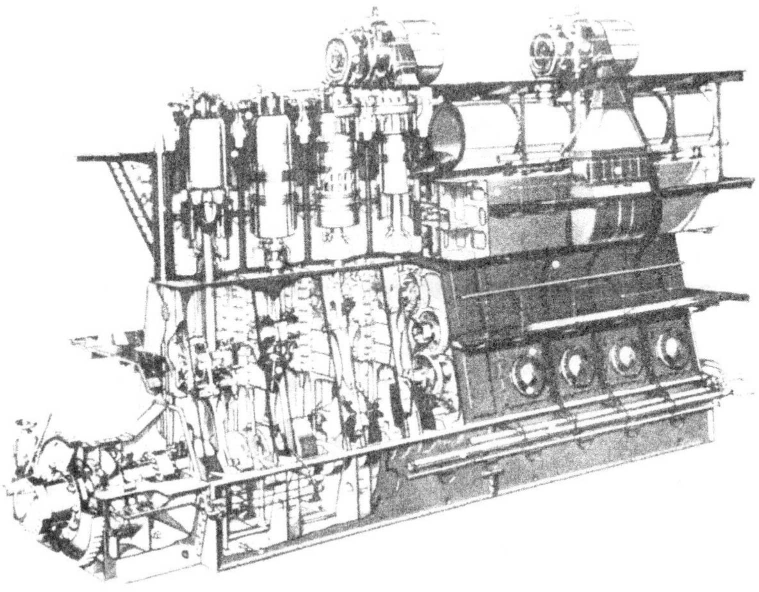

During 1980 Allis-Chalmers commenced the construction of three huge engines for Sulzer Bros. Ltd. The engines were destined for container ships ordered by American President Lines from the Avondale Shipyard. Shown here is one of the 43,200 horsepower engines for this project. This cross-sectional view scarcely does justice to the immense size of the engine, although major portions of the internal mechanism is illustrated. Despite this flurry of activity in building very large engines, a rapidly changing economic scene in the early 1980s brought further production to a halt. A 1981 communication by David C. Scott, Chairman of the Board for A-C and directed to John Lehman, then Secretary of the Navy, requested aid in assuring consideration of the "slow-speed crosshead diesel engine capability recently established at Allis-Chalmers." Further dialogue in this regard apparently met with no success. Another four years later, and Allis-Chalmers would be out of the engine business entirely.

Kilns

The 1901 Allis-Chalmers merger brought with it the expertise of Fraser & Chalmers as builders of ore and mineral processing machinery. Just when this firm began building rotary kilns and similar equipment has not been definitely established, but there is no question that Allis-Chalmers almost immediately began building this equipment after the corporation was organized. The 1924 acquisition of the Power & Mining Machinery Division of Worthington Pump & Machinery Co. added considerable impetus to an already established A-C engineering and manufacturing ability. This 1924 illustration shows a standard revolving cooler. Rather than heat material for processing, it was built to cool a product previously processed under a considerable temperature.

Allis-Chalmers offered a large number of kiln designs, and with suitable modifications, these rotary units could be used in ore and mineral processing, or could be designed to handle temperature sensitive materials such as processed feed grains. In this 1939 illustration, the primary combustion chamber is to the right, while the gear drive system for actually revolving the huge drum is at the back, farther away from the intense heat of the burners. Depending on the specific process involved the fuel system varied. Heating of the kiln or drier might be derived from fuel oil, gas, or in the case of lower temperature requirements, might even depend on steam heating coils.

By 1940 Allis-Chalmers had built some of the world's largest kilns. This one, loaded on rail cars for shipment, measured seven feet in diameter and had a length of 120 feet. Plainly evident in this photograph are the many riding rings which supported the kiln, while a huge gear flange is obvious in the foreground. Although the destination of this particular kiln is unknown, it possibly might have been designed for a Portland cement plant, especially since the cement industry depends on huge kilns for processing of its materials.

Ever on the lookout for new applications, this unidentified A-C photograph of the 1950s illustrates a rather sophisticated installation, together with a large amount of auxiliary equipment. By the 1970s Allis-Chalmers had fully developed its "Kiln-Gas" system which was designed specifically for the gasification of coal as an approach to highly efficient electric power generation with a minimum of airborne contaminants. While the 1970s looked at a resurgence in coal-fired power plants as the new wave of the future, nuclear power stepped in as another ultimate solution. By the late 1980s, coal is once again under serious investigation as the best and most dependable source of electric power, except perhaps for hydroelectric systems.

Allis-Chalmers pioneered the concept of welded sections for kilns, driers, and coolers. Riveted construction was the only choice previously, but welding provided stronger joints and dramatically lowered manufacturing time. This huge kiln of 1939 is shown under construction, with the agitator plates in plain view. Natural wear of the plates only required their removal and welding in of new plates. Although this phase of industry is usually overlooked, a great deal of engineering talent was required in this rather specialized area of research and development.

LaCrosse Plow Co.

LaCrosse was able to trace its lineage back to 1860. At the helm was Albert Hirsheimer, and from his inventive genius and managerial talent came a diverse line of farm tillage tools. LaCrosse was the first company to market a practical balanced frame cultivator, the first power-lift grain drill, the first power-lift tractor plow, along with a number of lesser achievements.

LaCrosse had an excellent line of tillage implement and enjoyed tremendous prosperity for several decades. Eventual market saturation coupled with intense competition gradually diminished the company's good fortune, so during 1915 a move was made into the farm tractor business. This venture was coupled with the purchase of Sta-Rite Engine Co. of LaCrosse. Initially at least, success seemed to be assured. However, the tremendous competition in the tractor business, together with the U.S. economic recession of 1920 soon found LaCrosse in tight straits again. A proposed 1921 merger with Oshkosh Tractor Co. further aggravated an already bad situation. The bruised and battered remains of LaCrosse continued in existence until 1929. At that point President Falk of A-C sensed an opportunity to purchase a line with fine integrity and name recognition at a bargain price. Initially, the new A-C implement line from LaCrosse Works more or less followed LaCrosse lines, but by the early 1930's it had been revamped to the point that during 1933 Allis-Chalmers introduced a full dozen new tillage tools from LaCrosse Works.

In 1860 Barclay & Bantan Implement Works was organized at LaCrosse, Wisconsin. On May 17, 1865 Albert Hirsheimer bought the plow shop for $1,530. From this small 20 x 50 frame building grew a major farm implement manufacturer. Originally known as A. Hirsheimer Co., the growing firm was incorporated in 1893 as LaCrosse Plow Company, retaining this title until purchased by Allis-Chalmers in 1929. Immediately thereafter came the LaCrosse Works, and from this factory came a great many Allis-Chalmers tillage implements.

Organized in late 1915 as Happy Farmer Tractor Co., the first offering was the 8-16 model. This tractor used the company's own two-cylinder opposed engine having a 5 x 6½ inch bore and stroke. A single forward speed of 2½ mph was provided, and "a Steam Clutch, which is acknowledged as the best" was featured. Other dimensions included an overall length of 13 feet, a width of 6 feet, and a weight of 3,100 pounds, of which 88% was on the traction wheels. The 8-16 Happy Farmer sold for $550.

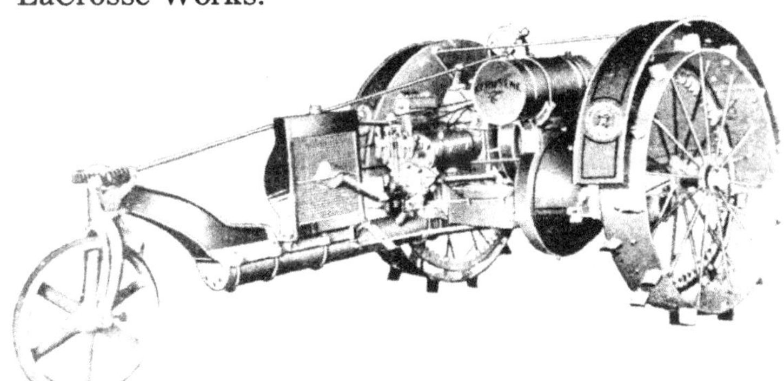

During 1916 Happy Farmer Tractor Co. was re-titled as LaCrosse Tractor Co. This same year saw the introduction of the Model F, 12-24 model — virtually identical in design to the 8-16 except for a larger engine and other minor changes. Priced at $1,075, it weighed 3,800 pounds. A two-cylinder engine of 5¾ x 7 inch bore and stroke was featured, and like the 8-16, one forward and one reverse speed was offered — both of them being 2½ mph.

Did Allis-Chalmers copy the design of the Happy Farmer or did the LaCrosse people plagiarize the A-C design? Although there are obvious differences between the Happy Farmer and the 10-18 Allis-Chalmers models, the similarities are such that it would seem entirely possible that at least some of the design originated with a single source. Although the 10-18 used a one-piece cast frame, the front wheel design from wheel, to wheel fork, to steering mechanism are quite similar. Both used a two-cylinder opposed engine, and both used internal tooth final drives. Current research has not revealed any connection whatever between the Happy Farmer and the A-C 10-18 tractors, if indeed any connection ever existed. At this point in time, details of any such connection may never be known.

LaPlante-Choate Mfg. Co.

In 1952 Allis-Chalmers purchased the LaPlante-Choate operation at Cedar Rapids, Iowa, and promptly renamed it as their Cedar Rapids Works. A-C had, under the direction of President Falk, made some notable acquisitions in the construction field, among them being the Monarch Tractor Co. which became the Springfield (Illinois) Works. This October, 1952, photograph illustrates the first Model TS-300 motor scraper being shipped from Cedar Rapids Works. Adding the LaPlante-Choate line of motor scrapers complemented an existing A-C offering of motor graders, crawler tractors, and other construction equipment. LaPlante-Choate gained fame in the early 1900s with a variety of machines that included horse-powered stump pullers and huge house-moving trucks.

Manure Spreaders

Until the late 1950s, Allis-Chalmers had little to offer in a line of manure spreaders, despite the fact that thousands were purchased every year. Few farm implements are subjected to the physical abuse which is a common denominator for this labor saving machine. Virtually no other equipment is exposed to the deteriorating effects of barnyard acids to the extent of the manure spreader. A-C engineers announced an entirely new spreader design in 1956. This revolutionary machine unloaded from the front, rather than from the back, as was the usual practice. It was truly a unique machine.

A left hand view of the front-unloading spreader is shown in the 1956 file photo. At the very front of the machine can be seen one of the vertical discharge chutes. The discharge beaters were situated in the front wall of the spreader. Allis-Chalmers also equipped some of these spreaders with a special folding apron assembly on the rear of the machine. It actually permitted loading manure directly from the floor, requiring nothing at all in the way of hand or machine loading. This feature is said to have been rather popular on large poultry farms. The Model 110 had a capacity of 110 bushels.

An April, 1956 file photo illustrates a ¾ rear view of the A-C front-unloading spreader. Small vertical openings at the front of the machine spread the manure to each side. The fully enclosed back eliminated much of the wastage that was inevitably inherent in standard designs. Wood sides and a wooden floor were standard features which permitted easy replacement when necessary. This revolutionary design was of course, PTO operated. Extra large tires afforded excellent flotation, even on soft ground.

Looking from the rear toward the front of the machine, the agitating beaters and the discharge beaters are plainly visible. The conventional crossmounted beaters served to loosen the payload. The revolving discharge beaters finished the process of reducing the load to small proportions rather than discharging it in large clumps. Despite the unique features of the front-unloading spreader, the idea did not catch the attention of farmers to a sufficient degree that production came to any large figure. Production of the Model 110 ran from 1956 to 1958. Only 1,500 units were built. The 1956 list price was $745.

Manure Spreaders

In 1960 Allis-Chalmers announced the No. 95-G spreader. Of ground-driven design, it was intended as an economical approach to this machine. Simplicity reigned supreme, and even older tractors without a PTO shaft could be used. Some 2,000 units were built in 1960, none were built in 1961, and another 352 were built in 1962. Some of these were later converted into 95-A PTO spreaders, giving a grand total production figure of only 2,328 units.

Following a rather poor showing for the No. 95 spreaders came a somewhat improved No. 180 PTO drive spreader. Of basically the same design as the No. 95, it differed in that it had a larger capacity and added somewhat larger flotation tires. The framework and tongue were built of heavier material, and the apron chain was of heavier material. Large spreaders require a rather massive apron chain for durability and reliability. A weak or corroded apron chain often breaks while unloading. The only alternative is to unload the remainder of the load by hand — a most unenjoyable task!

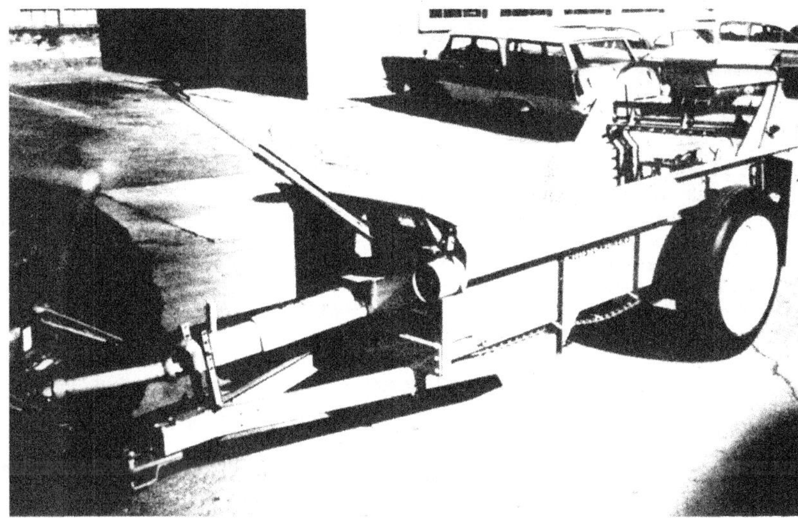

This October, 1960, file photo illustrates the first 95-A manure spreader. It was a virtual twin to the 95-G ground-driven model, except that this unit was powered from the tractor PTO shaft. Several of the 95-A spreaders were re-converted from the original 95-G design in 1961, but production ended that year, with only 525 total units being built. Exactly why the No. 95 spreaders sold so poorly is probably a matter for conjecture, although it must be noted that A-C walked into strong and well established competition for even a tiny share of the market.

Between 1964 and 1969 Allis-Chalmers built Model 180-S manure spreader. Using a single-beater design, it had a huge 180 bushel capacity, extra large tires, and the overall heavy construction required for handling big loads. High speed, heavy duty beaters were required in these PTO spreaders, especially when using hydraulic loaders which tended to leave much of the manure in a compacted state. This situation required the beater to have a formidable shredding ability in addition to giving even distribution of the load.

From 1969 through 1972 Allis-Chalmers offered two PTO spreaders, the Model 281 and the big Model 299. Both were designed specifically as PTO spreaders, using an extra heavy single beater design. Extra heavy apron chains were featured, as were the large flotation tires required for carrying large loads in soft fields. The saw-tooth design of the beater gave the necessary tearing action to achieve the necessary reduction and even spreading. From all appearances, Allis-Chalmers manure spreader production ended with these two models in December, 1972.

281 spreader with optional pan attachmen

Military Production

A photograph of August 26, 1943, serves as a poignant reminder that America was in the midst of World War Two. The billboard requests that "Fathers, 18 to 45" apply immediately at Allis-Chalmers. For Allis-Chalmers, as for most other manufacturers, the turnaround to military production took place very quickly, and to the surprise of the Axis powers, probably happened far sooner than they anticipated.

On September 19, 1942, President Franklin Roosevelt visited the West Allis Works. This was a great honor for the company and its employees. President Roosevelt made a whirlwind two-week tour of private war plants, traveling over 8,700 miles in that period! This photo, taken during a tour of West Allis Works shows President Roosevelt at the extreme left. Seated to his left is Wisconsin's Governor Heil. Standing is Max W. Babb, Chairman of the Allis-Chalmers Board of Directors, and to the far right is Allis-Chalmers' President, Walter Geist.

Wartime requirements included immense quantities of aluminum and other metals. Allis-Chalmers had previously developed its Excitron rectifiers with a primary usage being the direct-current power requirements of the aluminum industry. With the engineering know-how already in hand, Allis-Chalmers played a major role in manufacturing complete banks of Excitron rectifiers for defense plants. This 1943 photograph illustrates 24 banks of 12-tank rectifier units in a large aluminum plant.

A 1945 photograph illustrates a turbocharged P-38 fighter plane. Lockheed introduced this model originally in 1939, and during World War Two the P-38 was involved in nearly every battle theatre at some time or other. Allis-Chalmers specialized in building superchargers for the P-38 and other military aircraft. During World War Two, Allis-Chalmers built over 100,000 superchargers, totaling over 16 million horsepower. The turbo-supercharged P-38 could fly higher and more safely with this A-C built auxiliary.

Allis-Chalmers steam turbines had an enviable reputation going back almost to the beginning of the twentieth century. The company's fine reputation and irrefutable engineering skills proved to be very valuable in the defense effort. The escort carrier shown in this 1943 photograph was driven by Allis-Chalmers turbines. The original U.S. Navy photo notes that the escort carriers "are busy protecting Allied shipping from Axis wolf packs, guarding the flow of material to the fighting fronts."

Military Production

In 1944 Lockheed made a test flight of a prototype XP-80 Shooting Star fighter. Originally powered by a British-built Halford H-1 turbojet, P-80 was too late for World War II, but was nevertheless adopted by the U. S. Air Force. The "P" for pursuit gave way to "F" for fighter in 1948. An A-C file photograph indicates that at least some XP-80 jet propelled aircraft were used by the U. S. Navy to test the A-C DeHavilland jet engines, but details of this work have not surfaced thus far. More importantly from the historical perspective is the fact that Allis-Chalmers was an early pioneer in the design and building of jet aircraft engines.

During the early 1950s Allis-Chalmers bustled with activity in the design and production of aircraft engines. Limited production of the T-34 turboprop aircraft engine after the Pratt & Whitney designs began in 1951. Many of the same engineers who have helped A-C become one of the world's largest producers of aircraft superchargers during World War Two were planning, designing and testing the high temperature gas turbine equipment for this project. Rated at 5,700 horsepower, this new engine had a maximum diameter of 30 inches and a total weight of 2,250 pounds. Its fuel consumption of 0.62 lb/horsepower hour gave it one of the lowest specific fuel consumption ratings of any turbojet in existence. It is shown here on its test vehicle, a converted W.W. II B-17 Bomber. Note that the planes four regular engines are not operating, with all power being supplied solely by the nose-mounted turboprop.

In 1952 Allis-Chalmers did extensive testing of the YT-34 turboprop engine. The model on the test stand was capable of over 15,500 horsepower, with 90% of this power going to the propeller. Testing of these huge engine required a huge amount of test paraphernalia, as is evident from this photograph. Presumably this research and development work, or at least a portion of it, was done under contracts from the U. S. Government. Curiously however, any engineering or manufacturing advantages that A-C might have had over competing firms began to falter by the mid-1950s.

Details of this ejector seat are now sketchy, although it appears to have been designed and built by Allis-Chalmers. The pilot, always wearing a parachute, could automatically eject himself from the aircraft and at least have a chance for survival in case of an emergency situation that would leave no hope at all for either the aircraft or its crew. As in several other instances, it is presumed that this development was at the request of and under contract with the U. S. military.

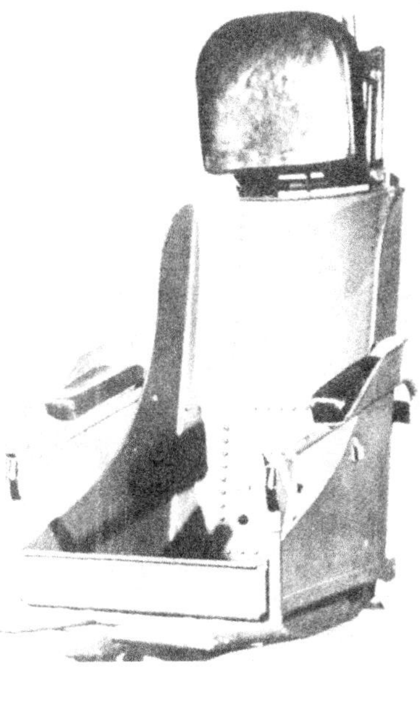

In 1951 Allis-Chalmers announced that a new branch plant at Terre Haute, Indiana would begin building compressors for the Wright J-65 Sapphire jet engine. Intended to power some of the United Nations' best fighter planes, the Sapphire was considered by many to be one of the world's best fighter aircraft. Details of A-C military production remain sketchy at best, but presumably the compressors for these engines were built under government contract, or possibly were part of a sub-lease from another manufacturer.

Military Production

By the time this 1942 photograph was taken, America was well into mobilization for World War Two. This scene, taken at an unspecified site illustrates a small A-C crawler tractor as a ground support unit at a small airstrip. Crawler tractors were an essential part of the war effort — they were able to carve a dirt landing strip in previously impenetrable jungles, doing so in a short time. During World War Two, virtually all of the A-C crawler tractor was directed toward the military effort — an exception being those civilian and agricultural projects which required the use of this equipment.

A 1943 photograph illustrates this sleigh train heading north for the Canol project with supplies and drilling equipment. An Allis-Chalmers HD-10 is pulling this immense train of supplies and equipment overland. Many more Allis-Chalmers were in daily use all through the Aleutian Islands during World War Two. On June 6, 1942, Japanese troops took Kiska and Attu in the Aleutians. Despite continued efforts, it was not until August 15, 1943, that U.S. and Canadian forces regained Kiska.

Allis-Chalmers crawler tractors played a major role in the Lend-Lease program, first proposed by President Franklin Roosevelt on January 6, 1941, and patriotically presented on Capitol Hill with the title of HR-1776. By March, 1941, this bill became law, and from that time until September, 1946, ate up some $50 billion. Over $11 billion of this amount went to the USSR, with this photograph showing Allis-Chalmers crawlers bringing up the big guns of Russian forces.

From the A-C Archives comes this unidentified photograph of January, 1940. Presumably this little all-terrain vehicle was intended for military duty, and it is presupposed that development was an A-C project. Allis-Chalmers was, as has been previously pointed out, very active in the war effort. It is therefore not at all unlikely that this particular project, like many others, came from the Engineering Department. For reasons unknown, this particular vehicle was apparently not selected for military duty, but was instead relegated to obscurity while still in the experimental stages.

The origins of the T-26 Snow Tractor remain unclear, although it seems a likely supposition that it was at least partially developed prior to 1940. This rare photograph of the T-26 model was taken sometime during World War Two, and illustrates the snow tractor complete with top and side curtains for cold weather protection. The radiator shell and louvered side panels appear to be borrowed from standard farm tractor production, but the track mechanism was built entirely from scratch, and was designed specifically for use over snowy terrain.

In their *Illustrated Encyclopedia of Military Vehicles*, Ian V. Hogg and John Weeks (New Burlington Books, 1980) note that "the engine, much of the transmission, and several other components were those of the 4 x 4 Willys Jeep." When adopted by the military in 1943, this unit apparently was given the designation of M7 Snow Tractor, and the ski-mounted M19 trailer was also provided. After being standardized in August, 1943 it was found that demand was less than anticipated, so it was reclassified Limited Standard in November, 1944, and was not retained after hostilities ended. This illustration shows the steering wheels sidemounted on the tractor, and likewise with the trailer. Where conditions permitted the skis could be removed and the rubber tires replaced. The skis were then mounted over the tires in an inverted position, serving dual duty as fenders.

Military

In a 1945 news item, Allis-Chalmers pictured the "new" snow tractor, pointing out that a quantity of these had been gifted to the Army for use in Alaska and the Aleutian Islands. It appears that in the meantime, Allis-Chalmers had substituted a larger 63 horsepower engine for the original Willys engine of only 54 horsepower. Weighing 3,049 pounds, the Snow Tractor had a top speed of about 40 mph (64 km/hr) and a range of 160 miles (257 km). Six forward and two reverse speeds were provided. Allis-Chalmers noted that the many features of this unit, plus its maneuverability suggested possible postwar markets "for the new A-C snowmobile." Had this line of thought been fully exploited, Allis-Chalmers might well have become a formidable competitor in this manufacturing endeavor. This does not appear however, to have happened — with virtually no further comment on the subject after 1945.

Primarily intended for towing artillery, the M4 was built from 1942 to 1960. Rated to pull from 18,000 to 30,000 pounds, it carried a Waukesha 6 cylinder gasoline engine rated at 210 brake horsepower. The M4 had a top speed of 35 mph and a range of 180 miles. It could carry ten men in addition to the driver, and a hatch above the crew compartment gave access to a ring mount which carried a .50 caliber machine gun for air defense. The M4 used the same arrangement of crew compartment, engine room and cargo space, and was in fact, a scaled-down version of the M6 tractor. Allis-Chalmers also built the HD-7W crawler tractor as their version of the M1 prime mover — International Harvester built the TD-14 and Caterpillar built their RD-6 as M1 tractors during the 1941-1946 period.

Development of the M6 tractor began in February, 1942 on two distinct tractor styles — the T22 and the T23. The former featured a fifth wheel connector, while the latter used a cargo area. In June, 1943, the T23 style was adopted and the M6 designation was begun. The M6 used two Waukesha 145GZ engines, each with 191 brake horsepower. These engines operated through a torque converter and a two-speed constant mesh transmission. A built-in winch was provided — it had a capacity of 60,000 pounds.

A rare photograph from the A-C Archives illustrates an M-6 tractor on the production line. Allis-Chalmers built the M-6 from 1942 to 1960. After World War Two ended, the continuing production of the M6 required a considerable working force on the manufacturing end, and in addition, the continuing testing programs required additional personnel. Most, if not all of the M6 production was at the LaPorte Works. Weighing 75,000 pounds, the M6 had a top speed of 20.5 mph and a range of 110 miles.

Bedecked with patriotic banners and streamers, this mighty M6-1235 leaves the production line on July 31, 1945. The largest and most powerful high-speed prime mover in use by American forces during World War Two, the M6 did an amazing job of hauling huge 240 mm howitzers into action. Although largely an undertaking of the LaPorte Works, the entire Allis-Chalmers organization was extremely proud of the M6 and its wartime achievements.

During 1952 and 1953 Allis-Chalmers built the M8, 22-ton cargo tractor. Current research has failed to determine why the A-C production of M8 tractors was of such short duration. This 22 ton cargo tractor is capable of hauling artillery and other weapons, along with ammunition, supplies and gun crews. This view shows the assistant driver behind the .50 caliber machine gun. The assistant driver also has access to dual controls and to the radio equipment. The driver is in the compartment to the right. With a top speed of 40 mph, the M8 was designed to travel in swamps and other difficult terrain and could climb or descend 60% slopes (30 degrees).

In 1951 the Army began work on a new lightweight tank destroyer. Chassis development was done with little problem but armaments remained a dilemma. Meanwhile the Army lost interest in the M50, but the Marine Corps, seeing possibilities for the new design, kept the project alive. In 1963 the M50 106 mm Self-Propelled Multiple Rifle appeared, remaining in production until 1970. Called "Ontos," a Greek word meaning "the thing," this unit was built at the LaPorte Works. Six 106 mm recoilless rifles were used, and two of these could be removed for tripod mounting if required.

The M50 Ontos was used operationally in Viet Nam, usually as a ground support weapon, and rarely as an anti-tank vehicle. A principle objection to the Ontos was its blast, which gave away its position, and the need to reload in the open. The Ontos had a ground speed of 30 mph and a range of 150 miles. Weighing 19,050 pounds, it was powered by a Chrysler V-8 engine with an output of 180 brake horsepower. The armament included six 106 mm RCL rifles M40A1C plus 18 rounds. These vehicles were withdrawn and scrapped in 1970 after 297 were built.

Military Production

An artists rendering of a proposed military vehicle was located in the Allis-Chalmers Archives, and although undated, it probably dates back to the 1960s, a time when Allis-Chalmers was actively pursuing new contracts with the U.S. military. Whether this design ever received serious consideration by the military establishment has not been determined by the present research, but a study of this photograph reveals numerous interesting and innovative features.

A photograph of the mid-1960s shows a fleet of Allis-Chalmers D-12 tractors being loaded for shipment to an unknown Naval Air Station. Allis-Chalmers apparently secured a contract to supply a substantial number of these D-12 Utility models to various military installations. Once on-site, they were destined for front-end loaders, sweepers, winches, and many other attachments, as well a doing a wide range of drawbar work. The sale was consummated through the efforts of the Allis-Chalmers Defense Products Division and the company's Utility Tractor Sales Department.

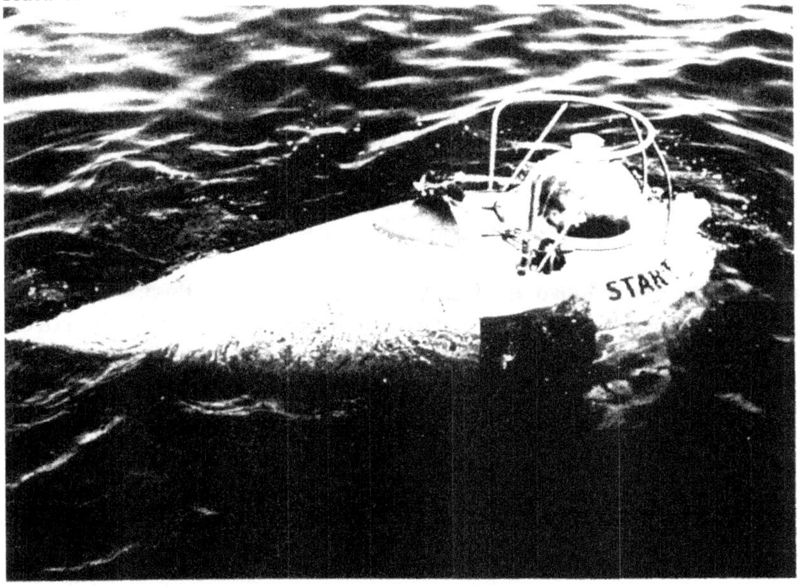

Shown here is STAR I, about to undergo a trial run at the marine research laboratory of the University of Connecticut. The STAR I was a one-man submarine with this research and test vessel owned by the Electric Boat Division of General Dynamics. It is equipped with fuel cell power developed by Allis-Chalmers research scientists at West Allis. The STAR I was the world's first submersible vessel powered by fuel cells, a system whereby the chemical energy of hydrazine and oxygen is converted directly into electricity. This photograph was released to the public on June 29, 1964.

During the 1960s Allis-Chalmers made a concerted effort to gain additional military contracts, even going so far as to organize a division within the company for this specific purpose. One such contract of 1966 called for a new air assault construction vehicle with Allis-Chalmers engines as standard equipment. Called the BEST (ballastable earth-moving sectionalized tractor), it could be flown by helicopter to a remote combat area, reassembled on the ground, and fitted with any of a dozen different attachments to suit the current requirements. These units were equipped with Model 3500M diesel engines built at Harvey Works.

Monarch Tractors

Otto Falk's vision of Allis-Chalmers, as viewed from the president's chair, was one of continuing acquisition and growth. The company's financial position in the late 1920s and early 1930s permitted the acquisition of several choice manufacturing establishments under terms very favorable to Allis-Chalmers. During the early 1920s Allis-Chalmers tractors nearly expired, finally getting a new lease on life due largely to the efforts of Harry C. Merritt.

Seeing immense possibilities for a crawler tractor line, Falk became very interested in the Monarch, one of several lines which had barely weathered out the tough times of the 1920s. Of all crawler tractor lines which might be available, the Monarch had a unique attraction, since this firm had secured manufacturing licenses to certain patents owned or controlled by Caterpillar Tractor Co. Monarch apparently initiated negotiations for sale of the company to Allis-Chalmers, and as Walter F. Peterson notes in *An Industrial Heritage*; "Although Monarch had sales for the previous year (1927) of $1.4 million, Allis-Chalmers was able to purchase the company for $500,000 cash."

Although Monarch was originally organized at Watertown, Wis., in 1913, little information can be located on the company's activities until about 1916. Even at that point, existing data is sketchy at best, that is until an appearance in the 1918 tractor directories of the little 6-10 Lightfoot model. Weighing only 3,000 pounds, the Lightfoot featured a Kermath four-cylinder, L-head engine with a $3\frac{3}{4}$ x 4 inch bore and stroke. A single speed of $1\frac{7}{8}$ mph in forward or reverse was provided, along with a roller chain final drive system. List price was $950.

Another member of the early Monarch crawler line was the Model M, 12-20 Neverslip tractor. Weighing 6,200 pounds, it was only 65 inches wide and 114 inches long. The 12-20 was equipped with an Erd four-cylinder valve-in-head engine having a 4 x 6 inch bore and stroke. Two forward speeds of 2 and 3 mph were provided, along with a single reverse speed of 2 mph. Hyatt roller bearings were featured in the transmission, compared to the bronze sleeve bearings used in the 6-10 Lightfoot model. This tractor carried a 1918 retail price of $1,650.

The company began life in 1913 as a locally owned stock company in Watertown, Wis. During 1919 it was reorganized as Monarch Tractors, Inc. and appears to have been headed toward the Chicago area for a short time under the title of General Tractors Inc. Although details of the corporate structure are now nebulous, there is no doubt that during 1925 Monarch Tractor Corp. was organized at Springfield, Ill. On February 27, 1928, Allis-Chalmers purchased the latter firm, first operating it as the Monarch Tractors Division. Within a couple of years this designation ceased, and production from the Springfield Works became purely Allis-Chalmers.

Early Monarch tractor production at Watertown is largely undefined. Very few references to Monarch tractors can be located prior to about 1916 — the presumption being that between 1913 and 1916, very few, if any, tractors were built. The Monarch Lightfoot 6-10 model appears to have been of the earliest production, along with its larger brother, the 12-20 Neverslip model. Neither of these models were tested at Nebraska, probably because production had ceased in favor of the Monarch 18-30 model — it was tested at Nebraska in August, 1920, under Test No. 56.

Current research leaves it a matter of conjecture as to whether Monarch adopted a marketing strategy of building only the 18-30 and promoting it alone, or whether financial and engineering considerations prevented the company from developing and marketing additional models. Given the fierce competition and the trying financial times of the 1920s, the latter presumption is probably the most likely.

In October, 1924, Test No. 108 was begun at Nebraska on the Monarch Model D, 6-60 tractor. It was followed in April, 1925, with Test No. 113 on the Monarch C, 25-35 model. Two years later, in August, 1927, the 10-Ton, Model F crawler was tested at Nebraska under No. 139. As previously indicated, initial overtures regarding sale of the company appear to have been undertaken by the Monarch people. When the sale was consummated in 1927, Allis-Chalmers instantly acquired a crawler tractor line consisting of two models — the 6-Ton and the 10-Ton tractors. From these models Allis-Chalmers began a nearly half century odyssey in the crawler tractor business, finally combining with Fiat S.p.A. in 1974 to form Fiat-Allis Construction Machinery Inc.

Monarch Tractors

Largest of the early Monarch tractors was the Model N, 18-30 Neverslip crawler. It was substantially larger than its brothers, being recommended for four 14-inch plows. The power plant consisted of a Doman four-cylinder T-head engine with a 4¾ x 6 inch bore and stroke. Rated engine speed was 750 rpm. Although the two smaller models seem to disappear by the early 1920s, the 18-30 with some modifications seems to have remained alive and well until being replaced in 1924 with an entirely new crawler tractor line. When the 18-30 was tested at Nebraska in 1920 (Test No. 56), a Beaver 4¾ x 6 inch engine was used. Due to its favorable performance in Test No. 56, the 18-30 soon gained a 20-30 rating. For 1918 this tractor listed at $2,000, but by 1923 this figure had escalated to $3,500. In 1923 this tractor crawled from Watertown, Wis., to New Orleans, La., a distance of 1,513 miles in 399½ hours for an average speed of 3.79 mph.

Probably introduced for the 1924 season, the Model D, 6-60 tractor was tested at Nebraska late that year (Test No. 108). This model featured a gigantic Beaver six-cylinder engine having a 4¾ x 6 inch bore and stroke for a 1,041 CID! Weighing some 8½ tons, the Monarch D delivered over 51 drawbar horsepower and cranked out over 70 brake horsepower. While earlier tractors were designed with a somewhat lighter chassis and gearing, this "Industrial Tractor" was beefed up at every point to withstand the severe duty encountered in the construction industry. Production apparently ended during 1926.

During 1923 the Monarch Twin Snow Plow was announced. This huge twin rotary machine was intended for the worst conditions encountered in snow belt areas. Apparently a separate engine was used to drive the twin 3-blade fans, one of which is obvious in this illustration. An extremely heavy rudder ahead of the plow served both to steer the machine and divide the snow as it moved toward the fans. Although this photograph is of rather poor quality, it is so far, the only proof this huge machine ever existed.

First marketed in late 1924 or early 1925, the Model C, 25-35 Monarch was tested at Nebraska in April, 1925 (Test No. 113). Presumably it replaced the earlier 20-30 model, but continued to use the same four-cylinder Beaver engine as its immediate predecessor. From all appearances the added horsepower came from raising the engine speed to 1,200 rpm, compared to 950 rpm for the older 20-30 tractor. When tested at Nebraska the Model C delivered about 38 maximum drawbar horsepower. Production probably ended in 1925 or 1926 — precise data in this regard has been impossible to locate.

The Monarch Industrial Model E, 4-40 tractor was in reality a heavy-duty version of the 20-30 model. Both used a Beaver four-cylinder, 4¾ x 6 inch engine for the same flywheel horsepower, but by comparison the 20-30 weighed 8,750 pounds and retailed for $3,500, while the 4-40 weighed 12,000 pounds and sold for $4,200. A K-W high tension magneto was used on the 20-30, but the 4-40 used an American Bosch magneto. Various other modifications were present, but the major difference between these two tractors was larger and heavier components throughout the 4-40 model.

Monarch Tractors

Production of the Model F, 10-Ton tractor began in 1926 and ended in February, 1928, after less than 250 copies had been built. Immediately following the 10-Ton came the Model F, "75" tractor which Allis-Chalmers continued building until September, 1931. Original equipment included a huge Beaver-LeRoi four-cylinder engine having a 6½ x 7 inch bore and stroke. In low gear the "75" pulled over 20,000 pounds in Nebraska Test No. 139. Initially built without a belt pulley, the later copies as built by Allis-Chalmers were modified to include both a belt pulley and a power takeoff shaft. Likewise, the late production of the "75" included a somewhat larger LeRoi four-cylinder engine using a 6¾ x 7 inch bore and stroke. In 1929 this tractor carried a list price of $5,350.

First built in 1927, the Model H, 6-Ton tractor was tested at Nebraska in November of that year under Test No. 147. The test model used a Stearns four-cylinder valve-in-head engine. Rated at 1,000 rpm, it carried a 5⅛ x 6½ inch bore and stroke. Production of this model continued until May, 1928 when it became known as the "50" tractor — Allis-Chalmers then continued its production until 1931. Coincidentally, Allis-Chalmers purchased the Stearns Engine Co. in 1930. Shortly thereafter the "50" was equipped with a slightly larger engine with a 5¼ x 6½ inch bore and stroke. Designated as Allis-Chalmers' own engine, it likely embodied much of the earlier Stearns design. Monarch also introduced a 5-Ton crawler in 1926, but production terminated in February, 1927. No additional data has been located on this model except that only 117 copies were built.

Motor Graders

Subsequent to purchasing Monarch Tractor Co. in 1928, Allis-Chalmers remained acquisitive, hoping to augment the crawler tractor line with compatible road building machinery. Seemingly, Allis-Chalmers gained only the sales rights to equipment of the Ryan Manufacturing Co. in 1930. This firm was building a fairly extensive line of pull-type graders and also offered a self-propelled "motor patrol." The following year, Allis-Chalmers acquired sales rights to the elevating graders of Hy-Way Machinery Co. Company records indicate that these early machines came from LaPorte Works after 1931, with all grader production being centered at LaPorte until being moved to Springfield Works in 1944.

Although several pull-type grader models were offered during the 1930s, by far the greater preference was with the newly developed motor grader. Allis-Chalmers built about 50 of the Model H Motor Patrol during 1932 and 1933, following it with their No. 42 Speed Patrol during the 1933-40 period. This machine was powered by the well-known "E" engine as used in A-C farm tractors. A slightly larger No. 54 motor grader was built between 1934 and 1940. A few #54 "KO" motor graders were built between 1937 and 1940 — these were equipped with the recently developed "KO" Allis-Chalmers diesel engine.

The decade of the 1940s saw several new developments. First came the "AD" series built during 1941 and 1942. Production was suspended at that time due to the war effort, resuming once again at Springfield Works in 1945 and continuing until 1947. In 1948 the new AD-4 motor grader was announced, followed the next year by the BD-3 and "D" models.

Allis-Chalmers also built a specially modified WC tractor with extended frame and equipped with a grader blade. This "W" Patrol was first announced in 1940, with production being suspended in 1942. After World War Two, production once again resumed, continuing during 1945 and 1946. Somewhat similar in design was the "WC Maintainer," first offered in 1938 and built until 1941.

Despite a continuing procession of new grader models, motor grader sales during the 1950s and 1960s remained flat at best. The merger of the Springfield Works heavy equipment line with Fiat in 1974 brought an end to A-C involvement with road building machinery.

Motor Graders

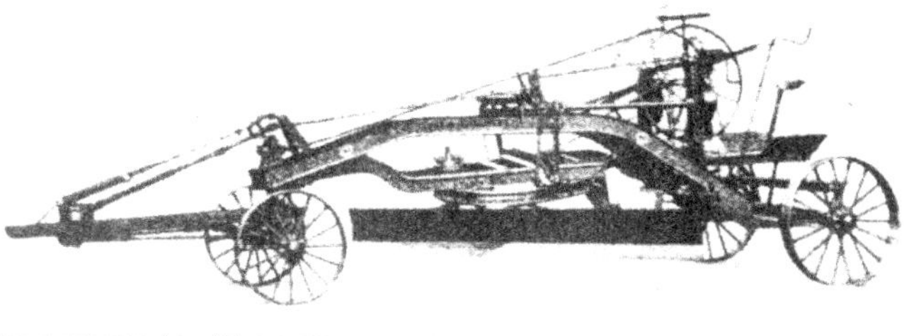

Allis-Chalmers entered the construction machinery business by acquiring sales rights to graders actually built by other companies. By 1932 however, A-C appears to have acquired Ryan Mfg. Co. and Hy-Way Machinery Co. Grader production centered at LaPorte Works, remaining there until it was removed to Springfield Works in 1944. The No. 10-K pull-type grader shown here was built from 1932 to October, 1934. Its immediate successor was an improved No. 10 machine which remained available until 1938.

The Allis-Chalmers elevating grader was built only in a 48-inch size during 1932 and 1934. The elevating grader was widely used in road building because it permitted a relatively easy method of building the grade and at the same time providing the required drainage. The large plow disk can be seen in its raised position. In operation, the disk was lowered and the dirt was carried by conveyor belts to the new road grade. The elevating grader was made obsolete through the advent of tractor-drawn and motorized scrapers. Very few of the Allis-Chalmers elevating graders were built.

The first Allis-Chalmers motorized grader was the Model H Patrol built during 1932 and 1933. This particular model used the "U" tractor engine for power, but operation of the various blade settings was accomplished entirely by hand. The canopy was designed as a sunshade, since this early unit was not equipped, nor probably intended for snow removal duties. Only 50 of these machines were built, and production was ended in March, 1933.

Announced in late 1933, the No. 42 Motor Patrol remained in production until 1940, with over 1,000 of these units being built. This particular machine is of single-axle design, and is obviously equipped with power control of the many blade functions. Adding power control of the blade was a major innovation in motor grader design that dramatically improved productivity and lowered operator fatigue. Note the accessory electric lights added to this unit for nighttime operation.

In 1936 Allis-Chalmers began production of the No. 42 Patrol in a tandem drive wheel configuration. With this exception it was identical to the single-axle version. A 45 horsepower engine was featured, and under favorable footing conditions a maximum blade pressure of 7,000 pounds could be exerted. This basic version of the No. 42 was not equipped with canopy or electric lights, although the blade was power-operated. Note the engine location below and forward of the operator's platform. Over 400 of the No. 42 tandem axle models were built at LaPorte Works.

Motor Graders

Designed as a companion to the No. 42 Patrol, the No. 54 motor grader was announced in late 1934 as a single-axle machine. By the end of production in 1937, over 300 units had been built. During 1937, some 32 of the No. 54 single-axle graders were equipped with the relatively new "K-O" diesel engine which Allis-Chalmers had introduced in certain crawler tractor models. Like the No. 42, the No. 54 shown here carried the engine just ahead of the rear axle.

The No. 54 Motor Patrol was introduced in a tandem drive axle version beginning in 1936. Those equipped with the "E" gasoline engine remained in production until 1940. In addition, a No. 54 tandem machine using the "K-O" engine was announced in 1937 — it remained available until 1939, probably at the same time that this rather unspectacular diesel engine was removed from production in favor of the General Motors two-cycle diesel engine design.

Allis-Chalmers introduced their grader line with pull-type machines, and in fact, continued to offer various models until 1941. The No. 8 grader, a hand-operated model, was built during 1935, 1937, and 1938. The No. 14 power-controlled model shown here was offered between 1933 and 1935. It was redesigned in 1934 to include an A-C Texrope (v-belt) drive from the engine to the control box. A No. 14 hand-controlled model was also built in 1936 and 1937. The No. 12 power-controlled pull-type grader was built in 1936, followed by the No. 110 machine in 1938-39 and the No. 112 grader offered from 1938 to 1941. Production of pull-type graders was not resumed after World War Two.

First built in late 1938, the Speed Maintainer was essentially an Allis-Chalmers WC tractor with an attached grader blade. Current research has not revealed whether the Speed Maintainer was available as a field-installed option, or whether certain modifications required that it be shipped from the factory in its modified form. Hailed by A-C as an ideal low-cost unit, the Speed Maintainer was built from 1938 to 1941 with production totaling 411 units.

First built at LaPorte Works from 1940 to 1942, the "W" Patrol resumed production at Springfield Works in 1945, continuing until 1950. During this split production period, some 3,000 units were built. The hand-controlled grader was mounted beneath a modified Allis-Chalmers WC tractor. A frame extension moved the front wheels ahead for the extra wheel base length required for a motorized grader. Even though this machine weighed but 5,900 pounds, it was capable of a 4,700 pound blade pressure. Standard equipment included a 10-foot blade.

Motor Graders

Announced in early 1941, the completely new "AD" motor grader embodied a three-cylinder, two-cycle General Motors diesel engine. Rated at 1,600 rpm, it was capable of 75 horsepower. Other up-to-date features included chain-driven tandem cases, full-revolving moldboard, six forward, and three reverse speeds. Weighing 21,500 pounds, the AD graders were designed for full-time use, and were intended to be strongly competitive with Caterpillar and other grader builders. Similar in design, but somewhat smaller, the "BD" motor grader was announced in early 1948. A G-M "71" Series two-cycle diesel engine was featured.

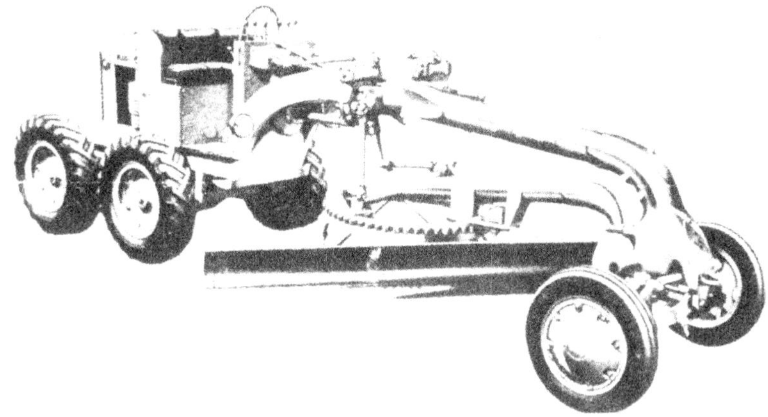

Production of the BD-3 motor grader paralleled that of the larger AD-4 model. Both began production in 1946, and both ended production in 1951. These models were then replaced with the AD-30 and AD-40 models. The BD-3 featured a G-M 2-cycle diesel engine with 78 brake horsepower, making this machine particularly useful for logging road construction and maintenance as well as for conventional road maintenance duties. A great many accessories were available, including a scarifier to break up hard roadbeds.

Announced in 1946, the AD-4 replaced the earlier 40 model. Several new features were evident, foremost of which was the 104 horsepower "71" Series G-M engine. The special tubular frame design had been developed and introduced by Allis-Chalmers. Another unique feature was the "Roll-Away" moldboard. Its involute design was intended to expedite the handling of larger windrows. Six forward speeds were possible, from 2.4 to 16.6 mph, along with three reverse speeds from 2.8 to 6.2 mph. Production of the AD-4 ended in 1951.

New features abounded on the AD-30 and AD-40 motor graders, built from 1952 through 1955. A hydraulically shiftable moldboard permitted instant shifting of the blade a maximum of 20 inches, either to the right or to the left. In addition to this optional feature were other elective items, including a scarifier, fully-enclosed cab, v-type snowplow, and electric lights. Hydraulic steering was apparently furnished as standard equipment. Optionally available was a Hancock side caster elevator to take the fresh windrow and deposit it some distance away from the machine.

Production of the BD-2 and BD-3 Motor Graders ran from 1948 through 1955. Although Allis-Chalmers captured a portion of the usual government-operated road maintenance market, the company appears to have been quite aggressive in specialty areas, such as logging roads. In this endeavor, Allis-Chalmers appears to have been eminently successful. This particular 1952 scene illustrates a BD-2 motor grader at work on the logging operations of the B & B Logging Company, at Hayfork, Cal. The machine is caring for 7½ miles of logging road in steep and mountainous terrain.

Motor Graders

Built from 1949 to 1965, the Model D motor grader proved to be one of the most popular in the entire A-C line. Designed for small to medium-sized jobs, it featured an Allis-Chalmers G-226 engine rated at 58 brake horsepower. The Model D featured completely hydraulic controls, eliminating the cumbersome and complicated gear box of earlier models. Welded pipe and channels made up the frame of the machine. The basic version shown here sans frills of any kind was often purchased for construction work where an enclosed cab was unnecessary.

In addition to the standard Model D gasoline-powered grader, Allis-Chalmers also offered a Model D "gasoline special." It used the same engine and chassis as its smaller counterpart. A major difference however, was the use of substantially larger 8.25-20 rear tires on the "special" compared to 7.50-20 tires on the standard version. Another notable difference was a hydraulically operated blade circle, compared to the unhandy method of dropping one corner of the blade, pulling back on the circle catch block, and driving forward to turn the blade on the circle.

Virtually identical to the Model D gasoline model, the Model D diesel-powered motor grader was offered from 1954 to 1965. Production of this model apparently followed the earlier BD-2 and BD-3 graders. Weighing 9,350 pounds, the Model D Diesel featured a 10-foot blade, although a 12-foot moldboard was optionally available. Special equipment items included a hydraulically-operated power circle turn, wheel leaning device, power shift moldboard, scarifier, and rear-end loader attachment.

As in the standard Model D Diesel motor grader, the Model D Diesel Special featured an Allis-Chalmers D-262 six-cylinder diesel engine. The main difference between this model and the lower-priced, and unenhanced version was in the fact that most of the options offered for the lower priced model came as standard equipment on the Model D Diesel Special. Despite its many innovative features, production of the Model D Diesel only approached a fraction of the output enjoyed by the gasoline version.

Allis-Chalmers offered a large number of optional attachments for their graders, but perhaps none was more unique than the rear-mounted hydraulic loader especially designed for the Model D motor grader by Tractomotive Corp. Mounted on the rear of the grader, the ⅝ cubic yard bucket takes advantage of the grader's built-in hydraulic power system. In addition, the loader bucket was interchangeable with the A-C precision shoulder maintainer — a unit designed to eliminate double passes after the road shoulder is shaped by spreading the windrow.

Motor Graders

An Allis-Chalmers D-516 engine was featured in the Model 45 motor grader. Introduced in 1955, it remained on the line until 1962. The 127 brake horsepower from this machine permitted the use of a standard 12-foot moldboard, although a 14-foot moldboard was optionally available. Weighing 23,800 pounds, the Model 45 featured six forward and three reverse speeds. Power steering was standard equipment, as was power-operated front wheel leaning and needle bearing universal joints.

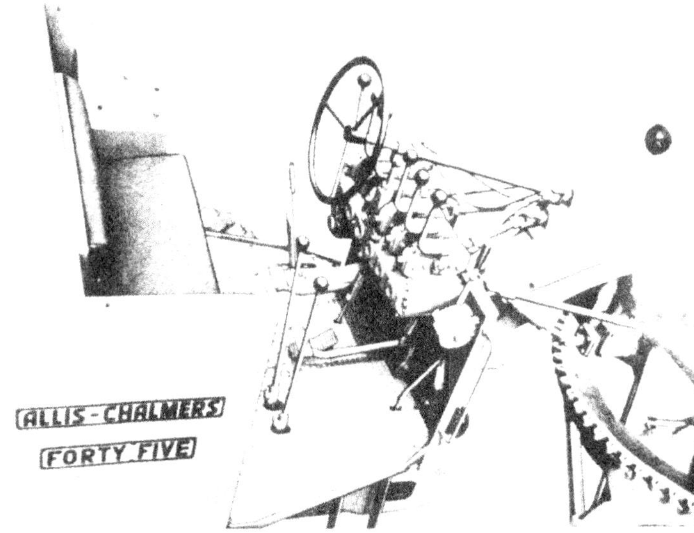

A closeup of the Model 45 operator's platform illustrates a streamlined design that featured a high-visibility gauge console to the left of the operator. Although this illustration does not show an operator's cab, the majority of these units sold for country road maintenance in the snow belt were equipped with a fully enclosed cab, heater, and defroster. The D-516 diesel engine was designed for pushbutton 24-volt starting, although the unpredictable weather of the snow belt made this a questionable proposition without the aid of a heated building or manufactured starting aids.

Snow removal has been and remains one of the most challenging tasks facing road workers. This 1950's photo illustrates a Model 45 equipped with a 9-foot, v-plow, plus a hydraulic wing assembly to further move the snow out of the roadway. Adding this complete assembly of v-plow and wing assembly added over two tons to the base weight of the machine. Although Allis-Chalmers had built v-plows and wing assemblies for some years, the equipment built for the Model 45 and Model 145 graders was actuated entirely with hydraulic power, eliminating troublesome winches, cables, and linkages of the past.

Built from 1959 through 1963, the Model 145 motor grader featured an 80 horsepower Allis-Chalmers D-344 engine. Its 344 CID resulted from a 4⁷⁄₁₆ x 5⁹⁄₁₆ inch bore and stroke in each of four cylinders. Weighing in at 21,540 pounds, the Model 145 used mechanical power operated controls, plus an optional power hydraulic steering booster. A 12-foot standard Roll-Away moldboard was featured — it was capable of over five tons of downward blade pressure. Standard features and options were basically the same as with the Model 45 concurrently produced.

In 1963 Allis-Chalmers ended production of the Model 145 motor graders. Already in 1962 however, production had begun on the M-100 grader, and entirely new design featuring a square tubular frame member instead of the usual round frame of earlier years. For reasons unknown however, production of the M-100 appears to have been sharply curtailed by 1965. Smaller models of this new series were the M-60, of which only about 100 copies were built in 1963, and the M-70, offered from 1963 to 1965.

Motor Scrapers

Like many other technological advances, large dirt scrapers reached practical stages of development largely because of World War Two. New factories, new roads, and new reclamation projects demanded equipment that could move huge quantities of earth in a relatively short time. Elevating graders were the method of choice during the 1930s, but these machines were too cumbersome and too slow for wartime demands. Each new advance of Allied Forces required new roads, new runways, and new staging areas. To this end, the tractor-drawn scrapers played a significant role. Granted, the development of these machines would probably have occurred without the intervention of global conflict, but there can be no doubt that World War Two acted as a catalyst of change. For their part, Allis-Chalmers were unable to offer much in the way of earth scrapers in the wartime effort. The A-C contribution was to reach its epitome in several other areas.

A-C involvement with earthmoving equipment began with the purchase of Monarch in 1928 — this was enhanced with the Model U Industrial tractor and its look-alike counterpart, the United, sold by Equipment Distributor's Assn. in the 1928-1930 period. In the 1935-37 period Allis-Chalmers built the TW Speed Ace, a self-propelled earth hauler that could take its load from an elevating grader or from a power shovel. Only a partial success, the Speed Ace left the scene, and with the imminent onset of World War Two, little further action was taken with this type of equipment.

During 1952 Allis-Chalmers acquired LaPlante-Choate Mfg. Co. of Cedar Rapids, Iowa. This firm had long been identified with heavy earthmoving equipment, and had pioneered the development of tractor-drawn and motor-powered scrapers. This acquisition put Allis-Chalmers securely into road building and construction machinery.

Baker Mfg. Co. of Springfield, Ill. was added to the fold in 1952. Their pioneering work on bulldozers and auxiliary equipment proved to be a significant asset to the construction machinery lines.

Just prior to World War Two, Allis-Chalmers introduced a new crawler tractor line featuring the General Motors "71" Series diesel engines. However, in late 1953, GM purchased Euclid Road Machinery Co. of Cleveland, Ohio. This put GM and A-C on a collision course, and one that undoubtedly was a major factor in the A-C acquisition of Buda Mfg. Co. that same year.

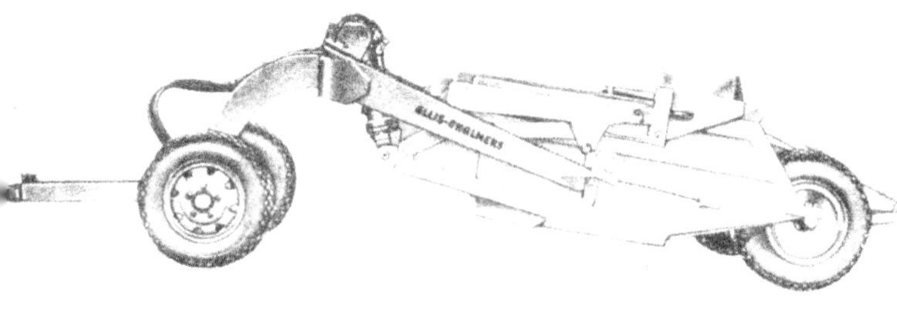

At the small end of the A-C scraper line were the C-20 (1946-53); C40 (1946-53); and C-44 shown here. The C-44 was built in 1946, 1948, and again in the 1955-57 period. Hydraulically operated, it had a heaped capacity of 5.4 cubic yards. A positive forced ejection was used, and an apron opening of 6'2" was provided. The C-44 could make a maximum cut of 7½ inches and could spread up to 12½ inches. Shipping weight of the C-44 was about 6,500 pounds. Four 11.00 x 20, 12-ply tires were standard equipment. Production of the C-44 and other small-end scrapers totaled only a few hundred units.

During the 1947-56 period, Allis-Chalmers built about 500 of the C-106 cable-operated scrapers. Designed for use with a double-drum cable drive unit as was popular in the 1940s and 1950s, the cable system provided operator control of the loading and unloading processes. The Model 106 shown here had a heaped capacity of 8½ cubic yards. Overall, it was 25¼ feet long, 9⅓ feet wide, and 7 feet high. Four 14.00 x 20, 16-ply tires came as standard equipment. The empty weight of 10,300 pounds could easily swell to a loaded gross weight of 35,000 pounds or more. A single copy of the C-84 tractor drawn scraper was built in 1947, plus another in 1948, but no information has been located on this model.

Allis-Chalmers offered the C-108 tractor-drawn scraper from 1946 to 1957, building approximately 1,400 units during this period. Its cutting width of 8½ feet made it an ideal match for large crawler tractors, permitting the tractor to enter the previous cut without crowding the sides. The C-108 had a heaped capacity of 12 cubic yards, and to carry this load plus the units own weight of 15,250 pounds, standard equipment included four 16.00 x 20.00, 16-ply tires. Optionally however, the C-108 could be equipped with rear tires of 18.00 x 25.00 size — these were of 20-ply construction.

Motor Scrapers

With the purchase of Buda, Allis-Chalmers was finally given the luxury of equipping their tractors and industrial machines with their own engines, rather than relying on an outside vendor, or an outright competitor!

The 1950s and 1960s saw rapid industrial expansion throughout the entire United States. The boom years of these two decades translated into countless construction projects, large and small. Under these circumstances, Allis-Chalmers was able to maintain a share of the construction machinery market. Declining sales, heavy competition, and the lack of a well-defined sales organization all began to trouble the A-C construction machinery line by the late 1960s. Finally in 1974, Allis-Chalmers merged their construction machinery lines with Fiat S.p.A. of Italy to form Fiat-Allis Construction Machinery Inc. This marked the end of A-C involvement with the construction machinery business.

At the top of the tractor-drawn scraper line was the Model 315, a huge cable-operated scraper with a heaped capacity of 20 cubic yards. Using the rule-of-thumb figure of 3,000 pounds per cubic yard of earth, this gave a payload capacity of 60,000 pounds in addition to the empty weight of 27,500 pounds! The C-315 scraper featured a cutting width of 9⅔ feet and a maximum cutting depth of 10 inches. By the late 1950s Allis-Chalmers seems to have completed a phaseout of the tractor-drawn scraper line, concentrating their efforts on an extensive line of motor scrapers and motor wagons.

Between 1950 and 1958 Allis-Chalmers produced over 1,400 of the TS-200 motor scrapers. Featuring a Buda 6-DA-779 diesel engine, the TS-200 was capable of over 175 horsepower. Five forward speeds were available, with the highest gear being somewhat over 20 mph for rapid travel from the loading area to the new grade. Air brakes were provided on all four wheels, and positive hydraulic steering was used. The scraper was hydraulically operated. Empty weight of the TS-200 was over 38,000 pounds. Over 1,000 copies of a TS-300 motor scraper were built in the 1947-54 period.

Always looking for new applications, Allis-Chalmers announced this huge 75-ton tractor in 1963. The bulldozer unit, not illustrated in this photograph, was 18 feet wide and 6 feet high. Dual engines, fore and aft, provided a total of 870 horsepower. If desired, this unit could be split into separate components, each of which was capable of handling a large roller, material wagon, or other attachment. Several of these gigantic tractors were said to be headed for the Peace River Dam Project in British Columbia, but little additional information can be located at this time.

The TS-300 motor scraper was the largest piece of equipment manufactured at Cedar Rapids Works in the period through 1954. Allis-Chalmers purchased LaPlante-Choate Co. in 1952, promptly renaming it the Cedar Rapids Works. In addition to normal production for road construction, reclamation, and similar duties, it appears that at least some of the Cedar Rapids production was aimed at military uses. America of the early 1950s had just come through a major war, and was almost immediately involved in the Korean conflict.

Motor Scrapers

Introduced in 1957, the TS-160 motor scraper offered a struck capacity of 7 cubic yards or a heaped capacity of 9½ cubic yards, for a payload of 12 tons! The 155 rated horsepower came from an Allis-Chalmers TDS-516 supercharged diesel engine. The Kon-Tork differential was a term registered by Allis-Chalmers to promenade the special design which automatically controlled the amount of torque delivered to each driving wheel in a quantity dependent on the traction conditions. Five forward speeds up to 25 mph were provided, as was 24 volt direct electric starting for the diesel engine.

Allis-Chalmers probably offered the TR-160 Motor Wagon during the same 1957-61 production period as their TS-160 Motor Scraper. The TS-160 as illustrated here, used the same TDS-516 A-C engine with an output of 155 horsepower. The R-160 wagon had a heaped capacity of 12 cubic yards, and offered a rear dump mechanism with double-acting hydraulic cylinders. The tractor unit could be changed from scraper to wagon. When set up as shown here, total empty weight was 28,100 pounds.

Production of the TS-260 motor scraper began in 1955. This machine had a heaped capacity of 17 cubic yards for a normal payload of over 20 tons. A cutting width of nearly 10 feet was used, and the maximum depth of the cut varied from 9 to 14 inches. The hydraulic design used a pair of 7-inch cylinders to raise and lower the bowl, and the ejector system was also fully hydraulic. The cutting edge for the scraper was made up of three identical sections, each being interchangeable and reversible. This advertising illustration of 1959 shows a 260 with a full load of 17 cubic yards.

Interchangeable with the scraper was the R-260 wagon. Its 15 cubic yard or 20 ton capacity made the TR-260 an ideal machine for transporting large quantities of material for a considerable distance. Like the 260 Motor Scraper, The TR-260 Wagon used an Allis-Chalmers Model 16000 diesel engine — its six cylinders had a displacement of 844 cubic inches. Standard equipment included 26.5 x 25 tires with a 20-ply rating. The complete outfit shown here required slightly over 26 feet in width for a minimum turn.

By 1965 Allis-Chalmers had upgraded the 260 motor scraper and was offering the 260E, an advanced design which utilized an elevating loader. The new loader design, coupled with a larger 355 horsepower engine made the use of a pusher machine less of a requirement than before. Loading was much faster too — this big machine could pick up a full 23 cubic yard load in a minute's time! Note the revised grille design as compared to the earlier models.

Shown here in operation is a TS-260 scraper picking up a load with the assistance of an HD-21 Allis-Chalmers crawler tractor. This particular scene was photographed at U.S. Highway 66 and I-66 near Clinton, Okla. The immense roadbuilding projects of the 1960s, including the building of much of the interstate highway system, required huge numbers of motor scrapers and other heavy equipment. The TS-260 was capable of 230 horsepower, but loading usually required the additional help of a pusher machine.

Although Allis-Chalmers was quick to adapt their equipment to new and unique applications, this innovative bent was often manifested by other companies as well. In this 1965 photograph a T-260 prime mover has been wedded to a 50-ton Browning compactor for use on an earthfill dam. A close look at this photograph illustrates the effect of the compactor — even on new fill, the earth is packed so solidly that even a 50-ton machine barely leaves tread marks!

The blasted walls of a rock quarry form a ghostly backdrop for this 1966 model of the 260 tractor. In this case the tractor has been coupled to an Athey 33-ton rock wagon, and is bringing big rock to the crusher. Even a crusher with a capacity of 100 tons per hour would require only three visits per hour with this combination. Although very large and very expensive, the production from these big machines made them very cost-efficient over smaller equipment.

Built from 1955-62, the TS-360 motor scraper was equipped with a 340 horsepower, A-C 21000 diesel engine. Its turbocharged, six-cylinder design used a 5¼ x 6½ inch bore and stroke for a displacement of 844 cubic inches. The scraper unit had a heaped capacity of 30 cubic yards, or a payload of 34 tons. This, plus the tractor-scraper weight of over 63,000 pounds gave a loaded gross weight of over 131,000 pounds. Available during the same period was the TW-360 wagon — it had a capacity of 24 cubic yards.

Motor Scrapers

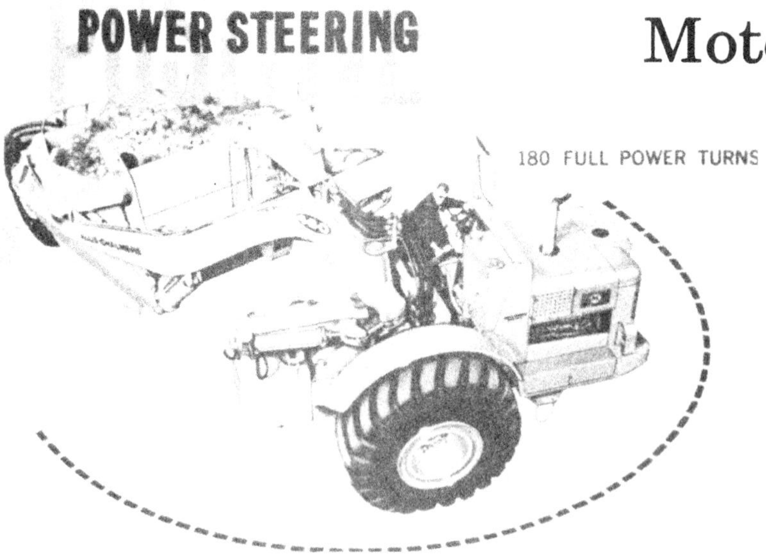

"Full maneuverability under full power!" That's what Allis-Chalmers had to say about their 460 motor scraper in this 1963 advertisement. The constant, double-acting hydraulic steering system in this unit permitted fast, non-stop turns at the cut and fill, along with responsive finger-tip steering even under full loads and at hauling speeds. The double-acting hydraulic jacks exerted a push-pull force over the entire turning range. In addition, this big 435 horsepower, 32 yard unit required only 37 feet to make a 180 degree turn.

A 1962 photographs illustrates a 460 motor scraper scooping up some 30 cubic yards of earth at the Tule Springs, Nevada, archaelogical site where researchers are looking for evidence of early man living there 32,000 years ago. This was the first time an earthmover had been used on an archaelogical exploration. The 460 motor scraper was equipped with an A-C 25000 diesel engine of four-cycle, turbocharged, intercooled and direct injection design. This machine, weighing 72,000 pounds, plus its average payload gave a gross loaded weight of 144,000 pounds — well over 70 tons!

Introduced in 1962, the 562 motor scraper topped the A-C line in terms of size and capacity. While the smaller units were equipped with an engine only in the forward tractor section, this giant earthmover also carried an engine in the rear of the unit. The two 435 horsepower engines thus combined to give an effective output of 870 brake horsepower. The operator maintained full control of both power units, and the air brakes to the front and rear wheels were synchronized for safety and reliability.

A 1964 A-C file photo illustrates a 562 motor scraper utilized by the huge Detroit Edison electric utility in coal handling. Total weight of the unit shown here is 102,000 pounds for the tractor-scraper combination. Detroit Edison purchased this machine with an air conditioned cab for operator comfort in all kinds of weather. A graph contained within an A-C advertising brochure for the 562 indicates a maximum tractive effort of about 55,000 pounds in second gear. Operating in low gear, the tractive effort approached 140,000 pounds.

Lilliputian by comparison, the A-C box scraper for the WD-45 and similar tractor models was an ideal unit for farmers, landscape gardeners, and similar applications. Motor scrapers, graders, and other machines of their size were completely useless for the small repair and finishing jobs, so this tractor mounted scraper was an ideal answer. Although historians and others may at times find fault with certain aspects of Allis-Chalmers' affairs, the engineering departments in almost all branches of the company consistently demonstrated ingenuity and imagination, as is once again demonstrated in this unique tool.

Mowers

By the mid-1930's Allis-Chalmers had developed a side-mounted sickle mower as an attachment to the new WC farm tractor. The "W Safety Mower" was designed to "not only eliminate the dangers of other types, but make hay cutting an easier, faster, and more pleasant job." Mounting this unit ahead of the rear wheels put the mower in full view of the operator. A hand lever was provided to raise and lower the cutter bar. This initial offering was built in five, six, and seven foot sizes.

The No. 7 trail-type mower was built between 1956 and 1958. Prior years had seen the No. 3 (1948-54); No. 4 (1948-52); and No. 5 (1955-56). The No. 7 however, was distinctly different since it included the new and revolutionary TWIN-WHEEL drive. This design completely eliminated the troublesome sickle drivers of the past. In this design the sickle is powered through two short connecting rods in the same plane and is always level with the sickle, regardless of angle or height of cut. The twin flywheel disks, balance each other in opposite directions and also balance the sickle in its movement.

A field scene demonstrates the No. 7 mower behind a D-14 Allis-Chalmers tractor. Despite its several unique advantages, the No. 7 used very small wheels in comparison to its competitors. With the introduction of the TWIN-WHEEL drive, Allis-Chalmers earned a respectable position in the industry, especially since it made most previous forms of sickle driving mechanisms obsolete. The No. 7 was available in rear mounted or trailer styles, and was built with either 6 or 7-foot cutterbars.

Probably due in part to the popularity of the new TWIN-WHEEL drive, Allis-Chalmers built 13,250 of these No. 80-R mowers between 1958 and 1962. Economically priced at $570, the 80-R was built with either a 6-foot or 7-foot cutterbar. The 80-R was designed specifically for Allis-Chalmers tractors equipped with the SNAP-COUPLER hitch, a hydraulic lift system designed by Allis-Chalmers to accommodate their own tractors and implements. Exclusivity was widely practiced in the industry for several years, with the 3-point hitch system finally emerging as a universal hitch that permitted various makes of tractors and implements to work together without difficulty.

Operating in rye grass, this 80-R mower is mounted an Allis-Chalmers D-14 tractor. On the initial order, customers were given a choice of forged mower guards or open face guards, depending on local circumstances. Likewise, smooth or serrated sickles were available. An outer shoe divider was optionally available for about $6, and a pasture clipping attachment added another $18 to the list price. Those wishing to pull a hay conditioner behind the mower for once-over cutting and conditioning needed a $98 hay conditioner hitch that included the necessary extension shaft, bearings, and brackets.

Mowers

A factory photograph of 1960 illustrates the 80-T mower with an A-C hay conditioner in tandem. Although listed as having a production period of 1959-62, a 1967 Price Book continues to list this model at a book price of $485. The letter suffix in the various model designations included "R" for rear-mounted, "T" for trailer-type, and "S" for side-mounted. With the exception of the mounting method, the mowers within a specific model series were virtually identical. Popular for only a few years, the combination of mower and hay conditioner in a tandem arrangement was quickly supplanted by single-application mower-conditioners that did the entire job without requiring two separate implements.

The side-mounted No. 80-S mower was built from 1959 to 1962, but despite this data, the 80-S is still listed in the November, 1967 price book at $312. The side-mounted machines were built specifically for A-C tractors as early as the CA and WD models, and could of course, be attached to most of the later tractor models as well. In addition to the 80-S model, Allis-Chalmers offered an 80-SI mower, similar in design, but built especially for industrial applications such as highway mowing. This model was available only for the D-Series tractors and could not be supplied for the earlier CA and WD models.

Recognizing the universal application of the three-point hitch system, Allis-Chalmers offered the 80-R3 mower beginning in 1962. This machine was of proximate design to the other 80-Series mowers, but was built with the three-point hitch system as an integral part of the design. This mower could therefore not be used with the SNAP COUPLER system developed by Allis-Chalmers. Weighing about 700 pounds, the complete unit carried a 1967 list price of $580.

▲

The No. 82-Series mowers were introduced in 1965. The rear-mounted 82-R and 82-S models were built until 1974, but a No. 82-SI industrial mower which was made specifically for certain D-Series Allis-Chalmers tractors was discontinued in 1968. The 82-R mowers were built for tractors using either Category I or Category II 3-point hitches. The No. 82-S shown here on an A-C 160 tractor was priced at $630 during 1967. This model could be supplied for tractors either with SNAP COUPLER or 3-point hitch.

Last in the A-C sickle-bar mower series was the No. 82-T. Although last, it was nevertheless a high quality machine. Priced at $585, it was produced from 1965 to 1974. Peculiar to this model was the offset wheel design which made square turns a practical reality when mowing hay crops. The outer wheel was set approximately in line with the cutter bar, but the inner wheel was set several inches forward so that the pivot point was moved ahead when making a turn. This change dramatically altered the action of the cutter bar in making a turn and permitted square cuts with relative ease.

Mowers

During 1962 Allis-Chalmers offered the No. 160 and No. 172 rotary mowers for industrial and highway use. These had respective widths of 5 and 6 feet. A-C tractors were ideal for equipment such as center-mounted mowers. The "Power Director" clutch with its high and low range, provided eight forward speeds and the ability to shift-on-the-go for fast mowing. Ease of changing wheel spacing gave correct wheel widths for stability on slopes. For reasons unknown, Allis-Chalmers seems not to have pursued the rotary mower market.

For 1966 and 1967 Allis-Chalmers offered the No. 829 Mower-Conditioner, a combination machine that permitted once-over cutting and conditioning of forage crops. This PTO-driven machine had a 7-foot cutter bar, four-bat reel, and a hydraulic lift system. Due to the heavy requirements of this machine, a 60 horsepower gear box was standard equipment. Weighing 2900 pounds, the No. 829 was priced at $2,430. Like most of the A-C forage equipment line, the 829 was built at LaPorte Works.

Current research has not revealed a precise entry date for the Allis-Chalmers hay conditioners, although A-C engineers appear to have considered several designs already in the 1940's. This 1967 model carried a price of $846. When run in tandem with a sickle mower, an overrunning clutch was desirable, if not imperative, and this option added another $30 to the base price. Features included 6-foot conditioner rolls, with a fluted steel lower roll and a rubber-covered upper roll.

Between 1967 and 1973 Allis-Chalmers listed the 904 Mower-Conditioner, a state-of-the-art design which combined the cutting and crimping job into a single operation. This model used a 9-foot cutting width, requiring a minimum of 35 PTO horsepower on the tractor, and preferably 50 or more horsepower. The upper steel roll was complemented with a rubber-covered lower conditioner roll. Depending on conditions, the 904 could operate at ground speeds ranging from 4½ to 8 mph.

Ending the series of Allis-Chalmers mower-conditioners was the No. 390, built between 1972 and 1974. A 1972 advertising bulletin credits the 390 as a machine that would do the job "once over rightly." The high speed sickle operated at 1,550 strokes per minute, and a steel-fingered reel virtually eliminated cutterbar plugging, that venerable bane of mowing machines over which countless expletives and profanities have been uttered during the past couple of centuries since its invention. By the 1970s, forage equipment was a highly competitive venture, and for reasons not immediately apparent, the company began a withdrawal from lines it deemed unprofitable. Mower-conditioners obviously were doomed to extinction, at least so far as the A-C trademark was concerned.

Although technically at least, the Roto-Shredder line was not a "mower" in the strict sense, at least it qualifies in the generic form. During the 1970s Allis-Chalmers sought to broaden the farm equipment line, adding these versatile shredders as part of the package. Three models were available — No. 82, No. 128, and No. 144. These model numbers also corresponding to the respective cutting width of these machines. When equipped with steel flails these machines were ideal for de-vining potatoes or shredding cornstalks and other crop residues. Rubber flails could be substituted when defoliating sugar beets or other special applications.

Nuclear Power

This undated photograph illustrates a stationary cutting and brazing machine at work in Machine Shop #7 of Stone & Webster Engineering Corp. By itself this photograph has little significance, but this operation was part of the super-secret Manhattan Project which ultimately led to the first atomic bomb. Allis-Chalmers, like many other companies, was involved with the Manhattan Project during World War Two. Like most of the many firms involved in building the equipment necessary for this project, Allis-Chalmers engineers probably had little conception of this project's enormous potential for either war or peace.

Very few photographs have thus far been located concerning A-C involvement in the Manhattan Project, but this particular scene from the #7 Machine Shop of Stone & Webster Corp. shows a huge winding machine, probably built by Allis-Chalmers. This particular device was designed to wind the silver conductors required for part of the project, but even the description accompanying this wartime photograph is so vague that inferences are difficult to establish.

During World War Two, Allis-Chalmers erected the Hawley plant at West Allis specifically to manufacture materials for the Manhattan Project. After August, 1945, further work on the atomic bomb came to an end. Company President Bill Roberts felt very strongly that work should continue, especially regarding peaceful uses of atomic energy. With Roberts' untimely death in 1955 at the age of 57, a great proponent of atomic power was silenced. This particular photograph shows a tensioning machine at use in winding large electric coils with silver conductors.

In October, 1954, Allis-Chalmers secured a contract with the Atomic Energy Commission to build an atomic power plant near Lemont, Ill. Intended primarily as a test facility, this was the first of many such contracts. This photograph from the Argonne National Laboratory illustrates the foundation layout for an atom smasher as part of a 47-acre complex. The identified areas of this photograph are: 1) Powerhouse where Allis-Chalmers motor-generators are in use; 2) Center building for electrical equipment; 3) Pre-accelerator where protons will first be accelerated; 4) Linear accelerator that will shoot protons into; 5) Ring Building. Building (5) is an underground unit housing a 4,800 ton ring of steel magnets from which sub-nuclear "bullets" are deflected into; 6) Mason Building, and; 7) Proton Building. The latter two buildings were equipped with target atoms. This complex was placed into operation during 1963.

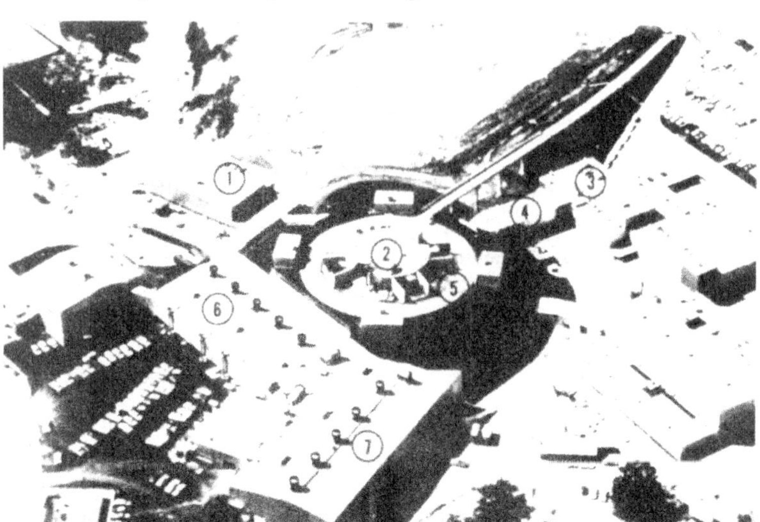

Bill Roberts' vision of nuclear power was continued by his successor, Robert Stevenson. During 1961 Allis-Chalmers completed the Elk River Reactor, a boiling water type in which the company took great pride. Considerable money was devoted to further research and development, but the early 1960s were too early — technological barriers and an apprehensive populace were intangibles that research money could not overcome. Thus it was that rumors began surfacing in the early 1960s to the effect that the company would withdraw from the nuclear power business. Finally in March, 1966, withdrawal was exactly what happened.

Ore & Mineral Processing Equipment

Under the heading of *Crushers*, a great many styles of these machines are illustrated. However, one curious design is this portable breaker — a Size No. 2, Style "D" complete with elevator and dust screen. The unit as shown here weighed 12,600 pounds. Portable or semi-portable crushers were a requirement for small mining operations and rather remote areas accessible by roads and trails, yet still sufficiently inaccessible so as to render the importation of huge stationary machines impractical. Rock breakers, crushers, and rolls were among the first steps in reducing the ore to a powder.

Edward P. Allis Co. marketed a considerable number of stamp mills, owing much of their success to the improvements made by Edwin Reynolds. Curiously, the huge factories of Fairbanks, Morse & Co. at nearby Beloit, Wis., also made a specialty of mining equipment and supplies, with their 1905 catalog totaling nearly 700 pages. Within this volume is illustrated most of the current Allis-Chalmers mining machinery line, including this ten stamp battery. Stamp mills were, as the name implies, intended to stamp rock to a powder, and the machine shown here was equipped with two mortars containing five stamps each. Depending on the character of the ore, each individual stamp might weight from 450 to 1400 pounds.

Several sub-headings of this volume have dealt with the emergence of Allis-Chalmers Mfg. Co. as a major builder of ore and mineral processing machinery. In particular, the reader is directed to the heading of *Crushers*, as well as the *Fraser & Chalmers* heading.

As a business genius, Edward P. Allis had few peers. Yet it was the famous Edwin Reynolds who moved the flour mill and sawmill manufacturer into a highly diversified company. Already in 1887 Edwin Reynolds, had built the first steam stamp with a solid cast iron anvil block to replace the oaken "spring timbers" then in use. Productivity increased dramatically under the Reynolds system, and within a short time virtually every steam stamp with its hardwood timbers was replaced with the Reynolds design. Thanks to this Reynolds invention, plus numerous others pertaining either to steam engines or hoisting machinery, enabled the company to ride the crest of American expansionism prevalent during the latter decades of the nineteenth century and continuing for a time into the next. The 1901 merger which brought four immense companies into a single organization known as Allis-Chalmers served to solidify their individual and collective positions within the ore and mineral processing industries for decades to follow. As with other areas of endeavor, market saturation was bound to occur, leaving Allis-Chalmers with excess manufacturing capacity in markets already filled to overflowing with their own products as well as those of competitors. Fortunately however, the ore and mineral processing lines proved to be the most enduring and dependable of the company's ventures, with Allis-Chalmers Corporation retaining its capabilities for manufacturing conveying, screening, and processing equipment into the 1980s.

A very indistinct photograph remains of an ore train and steam traction engine at the North Star Mines. The original photograph was very poorly made, but a close look reveals a rather unusual steam traction engine which appears to be quite different than the usual American varieties. Since a considerable expense was involved in setting up a crushing and concentrating operation, ore trains were a rather common sight, finally losing out to railroad cars and semi-trailers.

Ore & Mineral Processing Equipment

Tremain stamp mills, as illustrated here, were intended to meet the demands of mines in their first stages of development, enabling the operator to establish a crushing, amalgamating, and concentrating plant in the shortest time, and for the least amount of money. The Tremain mill was entirely self-contained, minimizing erection time. About 10 horsepower was required to operate a single unit, each of which weighed about 3,300 pounds. Tremain stamp mills had a capacity of 8 to 20 tons in 24 hours depending on the drop speed and the type of rock being reduced.

The A-C mining machinery line did not stop with crushers and concentrators — it extended all the way down to the smallest requirement for the mining industry. The elevator unit shown here was but one example of the many styles available on order from the company. In addition, early A-C advertising includes ore feeders, bin gates, frue vanners, ore buckets, and countless other items.

Various types of rock and different ores all require different methods of extracting the valuable from the worthless, and certain applications made the roasting furnace necessary. Allis-Chalmers was prepared to build a complete roasting furnace system as shown in this photograph from a 1905 catalog. Its immensity can be judged better by comparing the workman in the lower left foreground to the roasting furnace in the background. The A-C strategy of offering a complete operation, whether it was a crushing plant, concentrator, or roasting furnace served the company well over the years, gaining for them an excellent reputation and considerable prestige.

As part of the separating process, jigs became essential in certain operations. Almost always used with coarse materials, and built in many different forms, jigs depend on the tendency of a mixture of ore particles with varying specific gravity to arrange themselves into layers according to gravity when given a specific motion by a column of water. This 1905 Hancock jig represents one of many different forms, and it, like most others of the time, used a long wooden tank as the major component. Careful sizing of the raw material plus careful adjustment of the length and speed of the vibrating stroke were all imperative to successful operation. The Hancock jig was invented by H. R. Hancock of South Australia.

Mechanized mining required mechanized hoisting equipment, with the steam hoisting engine being an imperative necessity. Allis-Chalmers offered many different steam hoisting engines in more or less standard configurations and sizes. Specially designed hoisting engines were likewise available. The operating levers can be seen in the left foreground, and from this position the engineer was able to maintain full control of the hoist, even though his position in relation to the mechanism would now be deemed unsafe.

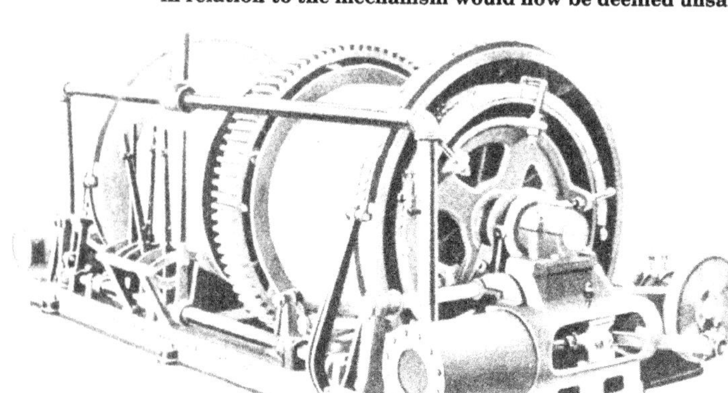

Ore & Mineral Processing Equipment

As early as 1905, Allis-Chalmers announced that they had perfected the necessary machinery for comparatively new method of recovering gold by dredging. In operation, the gravel is excavated by the dredge buckets and carried to the dump hopper; the dredge buckets shown to the left of the illustration carry about 15 per minute to the hopper. In addition to the dredge buckets were centrifugal pumps, separating equipment, and other devices suited to the particular conditions. The unit shown here was built by Allis-Chalmers for the Bonanza Basin Gold Dredging Co.

The tailings sampler was just one of many smaller devices required for a successful concentrating operation. As its name implies, the tailings sampler was needed to obtain accurate samples of the tailings so that a check might be obtained on the operations of the mill. Also required in ore and mineral processing plants were screens of every kind, from placer grizzleys for hydraulic mining to rolled and flat screens for extractors and other applications.

During the 1930s, Allis-Chalmers introduced several new secondary grinding mills, with the "Preliminator" mill shown here being designed specifically for the cement industry. Already in 1915 Allis-Chalmers had introduced the "Compeb" mill which revolutionized the art of grinding. The Preliminator design was a ball-mill arranged for preliminary grinding and applied in non-metallic fields such as the cement industry. It preceded the finish grinding mills, and could be equipped with peripheral screens so that as a unit it could furnish a sized product, or it could be used in a closed circuit with external screens or classifiers.

Shown here is a water-jacketed lead furnace. This device, while not originally built by Allis-Chalmers, became a part of the A-C line with their 1924 acquisition of Power & Mining Machinery Co. The small unit shown here had a capacity of 30 tons per 24 hours, and was intended for small mines, and those in remote areas. This unit, like many other items in the mining machinery line, was so built that it could be dismantled with no piece weighing more than 300 pounds, thus permitting mule back transportation if desired.

New for 1937 was this Peirce Smith Copper Converter built by Allis-Chalmers for the United Verde Plant of Phelps Dodge Corp. The converter is 13 feet in diameter, 30 feet long, and is built with a 1¼ inch steel plate shell. The conncection pipes are designed and arranged for temperatures up to 600 degrees F. Due to the high temperatures, constant duty, and corrosive conditions involved, engineering and designing machinery of this type was and is, extremely challenging.

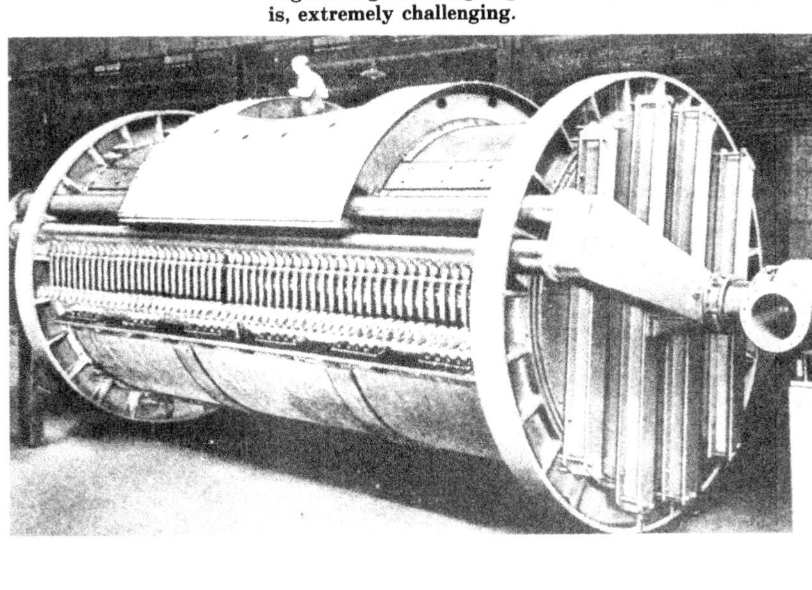

Ore & Mineral Processing Equipment

For certain very soft ores, for coal reduction, and for selected types of mineral processing, the hammer mill was an ideal machine. In terms of first cost, daily capacity, and in maintenance costs, the hammer mill possessed certain advantages. Once again, Allis-Chalmers accommodated the market with their usual heavy duty designs, as illustrated here. The comment has been made that many A-C machines were overdesigned to the point that they never wore out, leaving virtually no aftermarket, and effectively sealing the company's own doom in a certain endeavor. Once the market was saturated, it was time to move to something else. Making the change at the appropriate time marked the difference between success and failure.

Shown here are a few of the 56 roasting furnaces installed at the plant of Anaconda Copper Co., making this one of the world's largest calcining plants. Allis-Chalmers specialized in designing a plant, with the engineering people taking concepts and transferring them to workable drawings. The manufacturing plant then began the task of construction, and whether of cast sections or welded plates, a complete system appeared. After initial testing at West Allis, a single machine or multitude of same was then torn down for shipment and permanent erection at the job site.

Allis-Chalmers made a specialty of screens and screening equipment. By 1937 the company had done considerable research on sizing coal and coke for central heating plants. With the 1930s advent of small automatic strokers for residences came the need for accurate sizing equipment. In came A-C to the rescue, offering low head screen installations similar to the design shown here. Simple at first glance, screening machinery actually represents considerable engineering, and is in reality quite complicated.

A 1942 photograph illustrates one of ten Allis-Chalmers vertical hammermills used to grind corn for use as fuel to alleviate a coal shortage in Argentina. Wartime requirements put commercial shipping into a secondary role. This factor, plus a surplus of grain probably made corn a cost-efficient heating method. Likely as not, this practice was not followed to any great extent or for an extended period of time.

Not content with the development of ore and mineral processing equipment, Allis-Chalmers ventured still further into the fossil fuel processing industries by building huge refinery towers. This 1943 photograph shows one such tower built for depropanizing operations in the petroleum industry as it leaves the Allis-Chalmers shops. Vessel of varying wall thicknesses, lengths, and test pressures were fabricated in the A-C shops, all according to API-ASME construction codes.

Ore & Mineral Processing Equipment

Within this volume, several references are made toward the cement manufacturing machinery of Allis-Chalmers. Without a doubt, A-C played a major role in modernizing this most important industry. A major contribution was the introduction of new, improved, and much larger rotary kilns than had been previously thought possible. This 1946 scene illustrates one such cement plant on the West Coast of the United States. The rotary kiln in the foreground is 11¼ feet in diameter by 360 feet long. These dimensions were made possible through the extensive use of electric welding, yet another area of A-C expertise.

Now looked on with virtual horror, asbestos has been banned from most segments of American society for health reasons. For many years though, the asbestos industry provided an important commodity, with asbestos in various forms being used as an insulator against high temperatures. The A-C Low-Head screens shown here were being used to upgrade asbestos fibers previously considered too low in quality for commercial use. This particular plant was in Quebec, Canada, where the world's most extensive asbestos fields are located, especially in the eastern part of this province. By 1955 Allis-Chalmers had installed over 30 of these units in the region.

Allis-Chalmers boasted an intimate relationship with the metals industries — it was Edwin Reynolds who revolutionized the blowing engines used for melting steel. From these beginnings came one innovation after the other, and in this 1957 photograph, a huge centrifugal compressor casting is being poured at the West Allis Works. Smoke and flames are evident from the vents in the mould. This casting was for the largest centrifugal compressor built by Allis-Chalmers — it could deliver 110,000 cubic feet per minute while operating at 2,650 rpm.

During the 1950s Allis-Chalmers built a great many complete plants for the ore and mineral processing industry. One example is this turnkey operation of 1955 in which huge kilns replaced sintering machines for phosphate nodulizing. The company's initial entry into ore processing came with Edward P. Allis' purchase of some iron ore territory in the Lake Superior region. With the inventive mind of Edwin Reynolds as the catalyst, Allis quickly made major inroads into this industry, and from these beginnings A-C became highly diversified within this field, probably offering more equipment to suit specific conditions and materials than any other manufacturer.

As noted previously, Allis-Chalmers offered a broad and expansive line of screens, not just for the mining and minerals industry, but for grain and flour as well. This 1957 photograph taken near Durango, Mexico, illustrates a series of Allis-Chalmers *Ripl-Flo*, Model SH screens hard at work for an iron mine operated by Cerro de Mercado, S.A. During the late 1950s Allis-Chalmers made an all-out attempt to futher secure its international status by forming Allis-Chalmers International as a major operating division.

Ore & Mineral Processing Equipment

Back in 1905 Allis-Chalmers was offering tram cars and related rolling stock for use in mines. These were operated by cables occasionally, sometimes by mules, and often by hand. As years went by this system gradually improved so that half a century later, in 1958, Allis Chalmers was able to introduce a new 12-man mine car equipped with a semi-automatic control system designed by A-C engineers. Once the speed and direction were set, acceleration and dynamic braking were automatically controlled.

Yet another innovation was the electrically operated car shaker introduced by Allis-Chalmers in 1958. Shown here is one of five 3½ ton car shakers furnished to the U. S. Atomic Energy Commission for unloading coal at their Hanford Works near Richland, Washington. In operation, the shaker unit was lowered by an electric hoist to the coal car where its imbalanced rotor transmitted vibration to the car and of course to the product. Horrendously noisy, these units did however, reduce unloading time dramatically.

A 1958 photograph by Allis-Chalmers International proudly displays a new dolomite processing plant in South Africa. This country boasts large deposits of dolomite, a high magnesia limeston used to line steel-making converters. This particular plant was also equipped with several Allis-Chalmers Low-Head screens and a Model 1051 Hydrocone crusher — it featured a 48 x 42 inch opening. In addition to numerous foreign acquisitions in the late 1950s, Allis-Chalmers also purchased the Tractomotive Corp. of Deerfield, Ill., in 1959 to build tractor loaders, tractor dozers, and backhoes.

During 1958 Allis-Chalmers placed a number of Ripl-Flo screens with A-C crusher drives and associated equipment at this site in the Pyrenees in northwestern Spain. Single and double-deck Ripl-Flo screens were used in this installation. The aggregate from this plant was needed to provide over 1 million cubic yards of concrete for building of the Salto de Aldeadavilla Dam which was intended to supply electrical power to a large portion of northwestern Spain.

Another industry "first" was the stainless steel heat exchanger in this dry process kiln. Built for Northern States Portland Cement Co. of Mason City, Iowa, in 1959 the chain section of this kiln also included stainless steel chain for improved heat exchange. Historically, low carbon chains had been used as heat exchange media in wet process kilns. This photograph taken at West Allis Works shows welders completing fabrication of the stainless steel spirals within the chain section of the kiln.

Ore & Mineral Processing Equipment

A 1959 photograph shows a crushing plant in Massachusetts. Instead of the huge gyratory crushers of earlier years, this installation features a new 36-inch Hydrocone crusher. Hydraulic control of the crushing mechanism offered precise product control with fingertip ease. In addition, the Hydrocone required less headroom than the old design, and had greater capacity and greater strength. With the Hydrocone design came many other improvements, including a new lubrication system, improved methods of changing the oil seals, and a one-piece bottom plate assembly.

In 1960, Iron Ore Co. of Canada ordered a huge dryer for shipment to their vast mining project at Sept Iles in Northern Quebec. Built by Canadian Allis-Chalmers, this huge dryer had a diameter of 13½ feet and a total length of 110 feet, of which a 52 foot section is shown here, loaded and ready for delivery to the job site. This unit was designed for a retention time of 10 minutes in its 110 foot length with a slope of ¾ inch per foot.

Vaguely replicating a cathedral in its loftiness, the plant interior of Miron & Freres at Montreal, Quebec, is almost completely filled with a huge cement kiln and a large Air Quenching dryer. In 1962 this was the largest dry process cement kiln in North America. The kiln is 15 feet in diameter and 550 feet long, while the cooler is 120 feet long. This particular plant was of dramatically different design than conventional cement plants, especially because of its emphasis on compactness, automated operation, and simplified maintenance duties.

A 1963 photograph illustrates an 8 x 16 double-deck Low-Head screen, the largest such screen ever built. This particular unit was destined for scalping and sizing phosphate ore which comes into the unit in sizes as large as 20 x 60 inches. Photographed while under completion at West Allis Works, this screen was destined for the mine of Compagnie Senegalaise Des Phosphates De Taiba near Dakar, Senegal, Africa. Some dimensions of this unit include a width of 10½ feet, length of 18½ feet, and a height of 10⅔ feet.

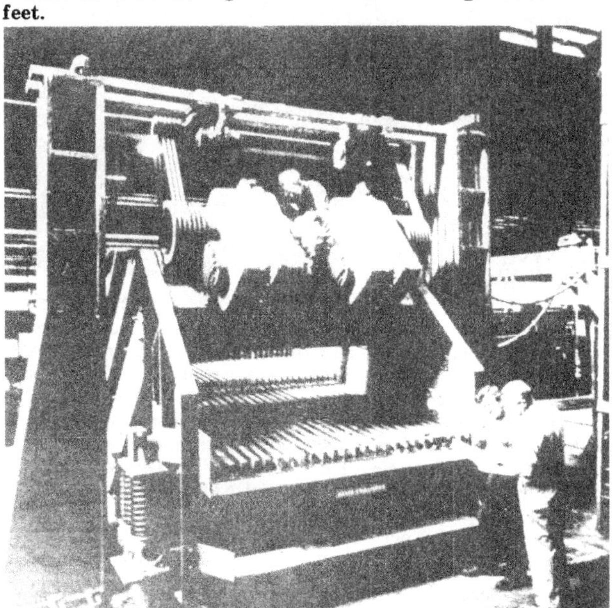

Long identified with mining and mineral processing industries, Allis-Chalmers also attempted to make definite inroads into the petroleum rubber, and chemical industries during the 1960s. Another area of interest was in phosphate processing equipment, with the final product being widely used as a fertilizer for agricultural, orchard, garden, and lawn uses. A 1966 photograph illustrates a phosphate plant at Tarpon Springs, Fla., equipped with Allis-Chalmers machinery.

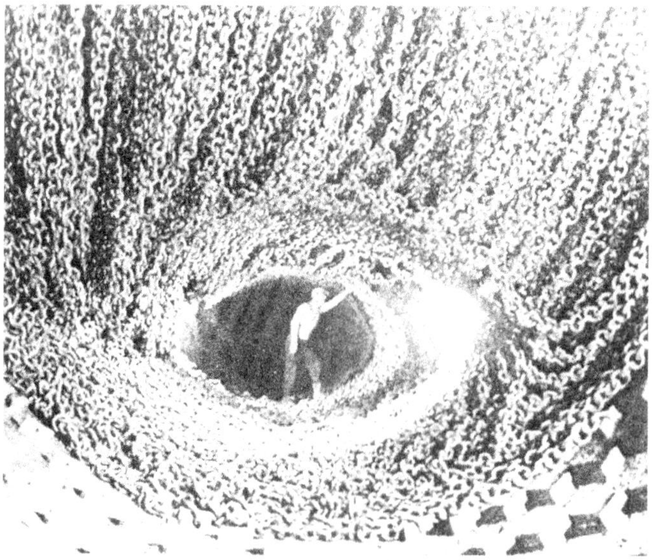

In 1966 Allis-Chalmers completed this huge 622 foot kiln for Universal Atlas Cement Division, a United States Steel Corp. affiliate. Installed at Hannibal, Mo., this kiln shared the title of longest in the Western hemisphere with a twin previously built by Allis-Chalmers for Puerto Rican Cement Co. at Ponce, Puerto Rico. Shown in this photograph is a portion of the 29,000 feet of 1 x 3 inch steel chain which has a total weight of 375,000 pounds, and thus formed the biggest chain system in the world. The chains are used to capture heat from burning pulverized coal gas and transfer it to the wet cement slurry as the huge 18-foot kiln revolves.

Whether it was something relatively simple like the holding bins in the foreground, moderately technical, like the kiln in the center, or highly complicated like the grinding machinery in the background, Allis-Chalmers offered a design and engineering department equal to these tasks. "Reliability" was a key word with A-C engineers and "dependability" was another. Throughout the company's history, few of these large installations were made "off the shelf" but instead, each job was carefully engineered to specific requirements.

In 1964 Allis-Chalmers announced completion of this huge firebrick-lined copper converter, one of three that had been built at West Allis Works for shipment to the Kennecot Copper Corporation's site at Magna, Utah. In operation, these huge revolving furnaces would use injected air to burn sulfur off the molten copper and to oxidize the iron in the copper, carrying it off as slag. Each of these converters was 30 feet long and 13 feet in diameter. Magna, Utah is just west of Salt Lake City.

Typical of a 1960s installation are several conveyors, crushers, holding bins, and many other accessory items. Although the design, manufacture, and installation of mining machinery, conveyors, and crushers remained fairly profitable through the years, this division was unable to absorb the losses of certain other company endeavors whose red ink finally became a hemorrhage. Since this volume is intended to illustrate the progress of Allis-Chalmers as seen through its products, conjectures regarding its managerial adroitness or an occasional lack thereof are left to the statisticians and augurs.

Another 1960s project was the completion of this 70-foot cooler, with a stevedore looking at one section aboard the "Mormacpine" from the Port of Milwaukee. Allis-Chalmers built this cooler as part of a Grate-Kiln cement pellet plant valued at over $3 million. It was part of a new 550 ton per day cement plant to be built near Rio de Janeiro, Brazil by Cimento Portland Branco do Brasil, S. A. Financing for the total project cost of over $8.8 million was provided under an Alliance for Progress loan by the Inter-American Development Bank, Washington, D.C.

Despite the introduction of many new and innovative ore and mineral processing machines over the decades, the 1960s began to see a leveling out of sales at the very best. No matter how good the design of any machine, with no buyers, it is valueless. Market saturation was certainly a factor in the decline of new installations, and due to the high initial costs, most operators preferred to run their current machinery to destruction before buying new equipment. This 1960s photograph illustrates a most impressive Allis-Chalmers setup incorporating crushers, conveyors and vibrating screens into an integrated design.

Steel-frame conveyors and their shadows combine to form an interesting study of angles and perspective in this 1964 company photograph. As previously noted, Allis-Chalmers had long been identified with screen manufacturing, and by the 1960s was one of the world's leaders in this endeavor. In addition to the many Allis-Chalmers conveyors within this photograph are A-C vibrating screens, A-C crushing equipment, and Allis-Chalmers electric motors.

Amid a takeover crisis, oil embargo, and internal difficulties, Allis-Chalmers purchased the Stephens-Adamson division from Borg-Warner Corp. in 1974. This gave Allis-Chalmers the rights to the huge stacker and reclaimer shown in this photograph. A 1976 installation at Vancouver, B.C., Canada permitted the storage of over one million tons of coal for shipment, with the coal being stacked by this machine. When a ship arrives the coal is reclaimed from storage and conveyed to the dock for loading. The Stacker-Reclaimer can process up to 4,000 tons per hour.

Outdoor & Recreational

Earliest of the A-C lawn and garden tractors was the B-10. Built through 1968, it used a 9 hp Briggs & Stratton air-cooled engine. The 12-volt electrical system assured instant starting. Luxury features included a foam rubber seat with back and arm rests. The "Easy-Lift" implement control came as standard equipment and permitted anyone in the family to handle implements with ease, right from the tractor seat. A three-speed sliding gear transmission was standard equipment.

Robert S. Stevenson became President of Allis-Chalmers in 1955. Under his management the company attempted to solidify its existing product lines, discarding those that were unprofitable. Under Stevenson's leadership the company was enjoying improved sales and earnings through a comprehensive plan of reappraisal and reorganization. From 1965-1968 Willis G. Scholl continued the Stevenson plan, despite several takeover attempts. David G. Scott ascended to the helm in 1968 and against this backdrop of hostile takeover attempts and a shift in company fortunes, A-C continued to invest in their Simplicity line as part of a turnaround strategy.

Allis-Chalmers acquired Simplicity Mfg. Co. in 1965. This long-time manufacturer had its factory at Port Washington, Wis. The excellent reputation of the Simplicity line, along with an extensive offering of lawn and garden equipment made Simplicity an ideal acquisition. In fact, the new line of Allis-Chalmers lawn and garden tractors, mowers, and other equipment proved to be one of the finest in the industry. To further bolster the company's position, an additional lawn and garden manufacturing facility was begun at Lexington, N. C. in 1968. The years following 1965 witnessed an ever-growing line of A-C lawn and garden equipment.

Built in 1966, the B-12 lawn and garden model was slightly larger than the concurrent B-10, but of the same general design. It was immediately followed by the B-112, an updated version offered during 1967 and 1968. This B-12 is shown with an E-Z Rake attachment designed to rid lawns of thatch and other debris, depositing this material into a trailer towed behind the tractor. As with their other product lines, Allis-Chalmers was very innovative in presenting new products for their outdoor and recreational equipment lines. A photograph taken at the A-C Sales Farm illustrates a B-10 tractor at work with an attached 32-inch roto-tiller. Also available on this late 1960s model was a 46-inch sidemounted sickle mower. The five-gang trailer mower was ideal for golf courses and large estates with its 8-foot cut. A 42-inch rotary mower was a widely used attachment tailored to average lawn mowing duties. Also included in the attachments was a half-ton dump cart and a center-mounted 42-inch grader blade. The angle-type snowplow and a 36-inch snow blower were available for winter duty.

New for 1967 was the Mow-Bee riding mower from Allis-Chalmers. Designed as a companion to the then-current B-10 and B-12 lawn and garden tractors, the Mow-Bee was powered by a Briggs & Stratton vertical shaft engine. Built with a cutting width of 26 inches, this model offered five different forward speeds. The full-length flexible steel frame permitted the mower deck and front axle to float and follow natural ground contours.

Following industry trends, Allis-Chalmers began offering a hydrostatic drive on the HB112 tractor of 1968, continuing it on the HB212 model of 1969-70. This 12 horsepower model featured electric starting and offered many, many different attachments in addition to the rotary mower deck shown here. While many of the chassis components remained the same as in earlier models, the hood and fenders received new styling, and large chrome hub caps added a touch of class.

Between 1974 and 1979 Allis-Chalmers offered the 405 rear-engine riding mower. Its quality construction included an all-gear, fully enclosed and lubricated transmission with two forward and one reverse speed. A 5 hp Briggs & Stratton engine was featured. A v-belt cluth was used, along with standard wheel-type steering. Accessories included hubcaps and rear-mounted grass catcher.

Introduced in 1972, the 400 Series tractors featured automotive type cam lever steering for precise operator control. The mower attached to the front axle, permitting it to move with the front wheels and follow natural contours. On the safety side, all controls were clearly marked, and interlocks prevented staring while the transmission or PTO might be engaged. The 410 model used a 10 hp Kohler single cylinder, air cooled engine, while the 416 tractor featured a 16 horsepower Kohler single cylinder, air cooled engine with a 3.75 x 3.25 inch bore and stroke.

The 616 and 620 lawn tractors were introduced in 1972, remaining in the line until 1973. Following these models came the 600-Series tractors announced in 1974. The 620 tractor shown here was furnished with an Onan two-cylinder air cooled engine rated at 19½ horsepower. An in-line hydrostatic transmission was standard equipment, along with a 3-speed sliding gear transmission and an 8-pinion limited slip differential. Also standard was a hydraulic lift system and a front-end PTO. A rear PTO shaft was optionally available, as was a large assortment of optional equipment.

The 608 tractor was first announced in 1974, remaining in the line until 1976. It was immediately followed with the 608LT tractor — this style was built into 1979. A 1979 A-C advertising leaflet indicated that a Briggs & Stratton 8 horsepower engine was used. A belt drive was used to the transmission, with the latter featuring five forward speeds plus reverse. A similar but somewhat larger 611LT tractor boasted an 11 horsepower engine but was, for all practical purposes, identical to its smaller cousin except for physical size.

Outdoor & Recreational

Built from 1974 to 1979, the 712 Hydro and 712 Shuttle tractors were identical except for the drive mechanism. That is, as the name implies, one tractor featured a hydrostatic transmission, while the other used a conventional semi-automatic transmission with four forward ranges plus reverse. The design of these models permitted front, center, or rear power take off drives. A Kohler single cylinder, 12 hp engine was featured in both models. As with other models, an extensive range of accessories and attachments was optionally available, including minor items like hub caps to major equipment such as a 42 inch snow blower.

Offered between 1974 and 1979 the 718 Hydro, like other lawn and garden tractors of the Allis-Chalmers line offered one of the industry's most extensive attachment and option lines. Included were items like a 42-inch center-mounted grader blade. It could be set in three different blade positions and attached to the tractor with only two pins. A 42-inch single-stage snow blower was available, as was a 46-inch sickle bar mower. The vacuum collector attachment included a large collector box on a trailer behind the tractor. Center-mounted rotary mowers were available in 42 or 48 inch sizes. Also available were two different rotary tiller models, a 48-inch snow plow blade, a 10-inch moldboard plow, and a lawn sweeper attachment.

The popular 700 Series tractors were available with many options, but one of special importance was the front-end loader designed especially for this tractor series. Its 48-inch width gave it access to many areas far too small for ordinary loaders. Allis-Chalmers also offered a front-end hitch for these models to ease the task of moving and positioning vehicles, boat trailers, aircraft, and the like. A Category "0" 3-point hitch was also furnished, and individual disc brakes were located on the differential output shafts.

Introduced in 1981, the 800 Series tractors included the 808 and 811 models. Both used Briggs & Stratton single-cylinder engines of 8 and 11 horsepower respectively. Except for this difference both tractors were of virtually identical design. Standard equipment included electric starting and lights, plus gear and sector steering. These models could be equipped with either a 36-inch or a 42-inch rotary mower deck, along with a host of attachments from dump carts to spring tooth harrows.

New styling, more power, and additional versatility were all new breakthroughs in the 816GT garden tractor. A new 16 horsepower Briggs & Stratton twin cylinder engine was featured. The Allis-Chalmers exclusive, custom-designed three-speed transmission matched speed to job needs. The base tractor weighed only 430 pounds, but was nevertheless capable of handling a 10-inch moldboard plow, 30-inch rotary tiller, or 48-inch, 3-blade mower.

Outdoor & Recreational

Listed as having a production period of August 1, 1979, through the end of 1980, the 910 tractor featured a six-speed drive as an integral part of the design. The controlled traction differential was an A-C exclusive which gave limited-slip action to automatically transfer some power to the wheel with the best grip. Standard equipment included a Kohler 10 horsepower air-cooled engine. The center PTO was operated with a mechancial cone clutch and belt for center and rear-mounted implements. A front-drive PTO was also available, as was a full line of accessories and allied implements.

Identical except for engine size, the 912 and 916 Hydro tractors featured a hydrostatic drive with a variable displacement axial piston pump and fixed displacement, reversible axial piston motor. This system included a foot operated v-belt clutch between the engine and transmission. Also featured was a spin-on cartridge-type oil filter in the charge pump suction line. A special free wheeling valve permitted maneuvering of the vehicle without starting the engine. The 912 carried a Kohler 12 horsepower engine and the 916 used a Kohler 16 horsepower model.

Entering production in early 1979 was the 914 lawn and garden tractor. The 914 featured a semi-automatic transmission with four forward ranges and reverse — all with single lever control. Like the hydrostatic drive models, the 914 was equipped with a controlled traction differential to limit slippage. The 14 horsepower engine was supplied by Kohler — its single cylinder design included a 3.5 x 3.25 inch bore and stroke with a rated engine speed of 3600 rpm. The tractor alone weighed 777 pounds, but adding a 42 inch mower raised the weight to 893 pounds.

During 1971 and 1972 Allis-Chalmers offered three different trail bikes. Shown at the left of this photograph is the Outrider Mini 4 Bike, the smallest member of the Outrider family. Built for trail riding, the Trail 5 Bike featured chrome wire wheels. To the right is the Mini 5 Bike, the top of the line model which included a headlamp as standard equipment. Although the entry of Allis-Chalmers into the trail bike business might seem peculiar, the early 1970s saw the entry of most farm equipment builders into diverse and oftentimes, highly lucrative lines that included everything from trail bikes to snowmobiles.

During 1966 Allis-Chalmers introduced their Kaddie Kart line of golf carts. Two of the models are illustrated here, with the Red Attache standing to the left, and the Lord Chamberlin on the right. The Lord Chamberlin featured a fold-down seat for resting between shots and also included a butler for golf balls, tees, and other items. There is no question that Allis-Chalmers was determined to make major inroads into the outdoor and recreational equipment fields at this time. Although the golf cart line did not apparently last very long, they at least saw limited production.

Outdoor & Recreational

For those desiring a "high wheel" rotary mower, Allis-Chalmers offered several different sizes, including this 21-inch model, during the 1978-1981 period. The 21-inch size was equipped with a Briggs & Stratton 4 horsepower engine. A heavy 14-gauge steel blade housing was used, and cutting height was variable from 1 to 3 inches. A belt-drive to the blade prevented serious damage in case an obstruction was encountered. Although the "high wheel" design was probably not the most popular overall, except in the southern states, it nevertheless was a component of the complete lawn mower line.

From 1968 comes a photograph illustrating the Allis-Chalmers Nineteen power lawn mower. After acquiring the Simplicity line in 1965, Allis-Chalmers made a determined effort to offer a lawn mower line that was not only complete in terms of sizes and styles, but also gave as much or more innovation than anyone in the industry. The original Simplicity line was well known and had an excellent reputation. In addition, Allis-Chalmers operated Simplicity as a wholly owned subsidiary, so the existing sales force and local dealer organization remained virtually unchanged.

A 1982 illustration shows the walk-behind lawn mower series. Included is the 19-inch push-type, a 21-inch push-type, and a 21-inch self-propelled model. These models were equipped with vertical shaft air cooled engines by Briggs & Stratton. While very simple, lawn mowers have been and remain a household necessity with a limited life span. A fairly constant demand remains, year after year, and as part of the overall strategy by Allis-Chalmers, these small models were part of the total offering.

Originally intended as a "shop mule," the Carry-All of 1966 was built in several varieties, including this 1966 version with a 500 pound carrying capacity, plus two passenger seats. A 1966 company publication noted that in addition to eight lift trucks, S. S. Kresge Company was also using two of the electric-powered Carry-All vehicles at its Fort Wayne warehouse. Lightweight, fast, and versatile, the Carry-All was quick to make an appearance in factories all over the United States.

A strong reinforced fiberglass body tilted back out of the way for easy access to the complete engine and power train. This particular model was intended for the golf course, with company advertising noting that it could carry enough gasoline for "two full days of 36-hole golf." Power was delivered by a Kohler 8 horsepower air-cooled engine. Electric starting was standard equipment, and the tiller bar of this model could be replaced with a conventional steering wheel if desired.

Outdoor & Recreational

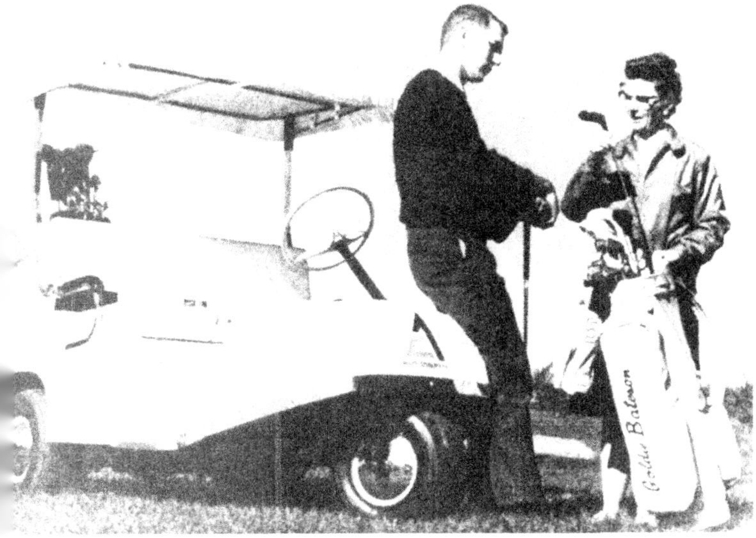

In what was probably an attempt to make a serious entry into the golf hobby, Allis-Chalmers offered its 1966 golf car with an original design. Built over the Carry-All chassis, it featured sleek styling, from the tailored canopy to the chrome hub caps. During this same period Allis-Chalmers also introduced a golf caddy in at least two different models. While the original Carry-All design used an 8 horsepower air-cooled engine, these compact little carriers could also be furnished with an electric battery drive.

In May, 1965, Allis-Chalmers pictured one of the futuristic towing tractors and passenger car units for delivery to the Milwaukee County Zoo. The prototype shown in this photograph had been demonstrated and tested at the zoo during the summer of 1964. The 1965 announcement noted that the zoo had ordered five of these rubber-tired land trains with a value of about $75,000 for delivery in June, 1966. Curiously, the same news item also indicates that "Allis-Chalmers Sports Products Inc. sees other applications, including shopping centers, airports, and industrial plants," for the new land trains. This is one of the very few references to a separate Sports Products Division of the company.

Between 1972 and 1974 Allis-Chalmers offered compost shredders and leaf mills as part of the overall outdoor equipment line. The early 1970s saw a resurgence of the "back to nature" movement, creating new markets for these machines as a method of hastening the process of returning organic matter to the soil. The leaf mills also presented a unique method of eliminating huge piles of leaves which were either burned or hauled to a landfill. Proper shredding and composting permitted their return to the earth as a fertilizer and soil conditioner.

As another part of the lawn and outdoor equipment line, Allis-Chalmers presented several different rotary tillers between 1971 and 1983. These hand-operated models were intended for small gardens as a replacement for the old-fashioned hoe. For garden plowing, and for large plantings, the mounted rotary tiller attachments as available for A-C garden tractors was the machine of choice. While perhaps not of a unique character compared to other rotary tillers then on the market, it is nevertheless significant that A-C offered these units as part of a total lawn and garden product package.

Paper Processing

The late 1950s and early 1960s saw the acquisition of several small firms by Allis-Chalmers. Ostensibly, the plan was to further diversify the A-C operations as one goal. Valley Iron Works of Appleton, Wis. was acquired in 1959 and operated by Allis-Chalmers as a wholly-owned subsidiary to manufacture papermaking machinery. In 1965 this plant became Appleton Works. A 1966 photograph illustrates a forming machine built at Appleton Works. Nearly 300 feet long, the *Valley* machine is 197 feet wide, and produces 500 tons per day. Installed for Georgia-Pacific in their Samoa, Cal. plant, this machine was designed to make bleached kraft pulp from a blend of redwood and Douglas fir.

Another 1966 product of Appleton Works was the "Paper Merchant," a paper processing machine designed to work with the "Valley" machine. The machine shown here was intended to boost the production of Nekoosa Paper Company's fine papers by about 33,000 tons. Due in part to increasing demands for fine papers, the Paper Merchant was designed to produce papers of 12 through 28 pounds in specific grades such as bond, duplicator, opaque, and numerous other categories. Despite some innovation in paper processing machinery, Allis-Chalmers does not appear to have made major inroads in the field of paper manufacturing machinery after the late 1960's.

Planters

Favorable responses to the mounted implements designed for the early WC tractors led to the development of improved mounted implements. This 1939 illustration shows a planter attachment designed specifically for the 60 Series cultivator as used on the Allis-Chalmers Model B tractors. Certain crop practices dictated the use of this apparatus, and Allis-Chalmers was at the ready with an implement designed for this specific task. Despite occasional criticisms leveled at an occasional A-C design, there is no question that Allis-Chalmers was highly responsive to building equipment not only for the ordinary, but also for the extraordinary requirements.

As part of a complete equipment line, Allis-Chalmers began marketing check row and drill planters about 1938. These were of more or less conventional design. The tractor planter shown here was in reality adapted from an ordinary horsedrawn planter of the period. A peculiar quirk of agricultural history is that the transition from horse machinery to tractor farming was greatly hastened by World War Two, making this unfortunate event a dividing line between two distinct periods in the development of American farm equipment.

A 1946 photograph illustrates an Allis-Chalmers Model C tractor, complete with bedding attachment and planter. Comments in advertising of the day, plus various notes in company literature and other sources support the inexorable opinion that while Allis-Chalmers was indeed interested in the welfare of the average midwestern farmer, company engineers seemed to take a special interest in the unusual and oftentimes unique needs of the southern farmer. In this admirable regard, Allis-Chalmers was perhaps more aggressive than their competitors.

Planters

Beginning in 1948, Allis-Chalmers offered a front-mounted planter. These were intended for mounting on the standard cultivator frame as can be seen in this photograph. These units were intended for planting in beds, on flatland, or in furrows, and could be adjusted anywhere from 36 to 44-inch row spacing. A number of variations were possible, but most were equipped for planting either corn or cotton, and almost all units could be equipped with a fertilizer attachment. Production ended in 1968.

The 400 Series rear mounted planters were built between 1948 and 1968. These planters were designed for use on a tool bar frame or the rear furrowing bar on the A-C cultivator. For earlier rigid shank cultivators another series of rear-mounted planter units was available. The 2 and 4-row rear-mounted bedder planters as shown here, carried a 1960 price of $567. This unit was available for WD-45, D-14, and D-17 tractors. A bedder was required for mounting. Production was centered at LaCrosse Works.

A photograph at the Allis-Chalmers Research Farm near Racine, Wis., illustrates a D-17 tractor with a 4-row wheel-track planter. This particular unit is equipped with an A-C preemerge spray band unit — one of the sprayer tanks is evident on the right hand side of the tractor. The unit-type planter was built in the 50, 60, and 100 Series between 1958 and 1960. Production was centered at the LaCrosse, Wis., Works of Allis-Chalmers.

Although rather popular in the United States, the Allis-Chalmers Unit-type planters also found an excellent market in Australia. This 1964 photograph shows Australian farmers with D-19 Allis-Chalmers tractors and unit planters at work on 4,000 acres of cotton. This particular scene was photographed on the Austcott cotton farm in New South Wales, just outside of Narrabri. Production was expected to be close to two bales of cotton per acre.

Photographed near LaCrosse, Wis., this mounted 6-row beet and bean planter is being pulled by a D-17 Series III tractor. The No. 55 planting units are set for 28-inch rows. A 1967 price book indicates that the 50 Series planter units were available in several configurations, including the No. 51, a drill-type unit with runner openers; the No. 52 drill unit with double-disc openers; No. 53 High Speed Drill with runner opener; No. 54 High Speed Drill with double disc opener; and the illustrated No. 55 Beet & Bean Planter with double disc openers.

Planters

Shown here in operation is a No. 413 Allis-Chalmers Cotton Planter. This four-row unit is being pulled by a D-17 tractor near Mattson, Miss. The 413 was a high speed hill planter with runner openers. Others of this series included the No. 411, a standard machine with runner openers and the No. 414 high speed hill planter with double disc openers. List price of the unit shown here was approximately $900. Production ran from 1960 to 1964.

Photographed near Milan, Tenn., this Series 500 planter is equipped with optional fertilizer, insecticide, and herbicide boxes. The Series 500 planters were built between 1962 and 1966, using several different configurations that depended on local conditions and the type of crop to be planted. Built as a drill planter, the Series 500 could be furnished either as a pull-type unit or with 3-point mountings. In addition to the standard configurations, a No. 505 Beet and Bean Planter was also available for these special crops.

While the 411, 413, and 414 unit-type planters were intended for cotton planting, the No. 400 and No. 401 unit-type planters were designed for corn. These units also differed from the cotton planters in that the latter were designed for rear-mounted use, using the Snap-Coupler system, while the planter shown here was of the conventional pull-type design. The No. 401 with runner openers listed at approximately $675, and the 500 Series fertilizer attachment shown here added another $188.

The 500 Series planters were followed in 1966 with the 600 Series planters, which remained in production until 1978. The 700 Series planters were offered in 1972 and 1973. Shown here is a 500 Series planter equipped with a 3-point hitch. It is hooked to an A-C D-15 Series II tractor, and is operating near Milan, Tenn. In late 1967, the No. 501 drill planter with runner openers sold at $1230, and adding the fertilizer attachment put another $200 on the list price.

A 1973 photograph shows the 12-row No-Til planter of Dick Rolfe of Sabina, Ohio, who was one of many participants in the Purdue University "automatic crop budget" computer program. The computer advised him to replace his two 8-row No-Til planters with this 12-row No-Til unit. It also advised replacement of an Allis-Chalmers 110 horsepower tractor with the 165 horsepower A-C 440 tractor. Mr. Rolfe is shown here planting in an alfalfa field.

Planters

The 70 Series No-Til planter units made their debut about 1971 and were built for use with the A-C 600 Series planter frame. This unit had the strength and ruggedness for No-Til work and offered the room for such special attachments as might be required by the No-Til system. The fluted No-Til coulter was designed for unsurpassed performance in any No-Til crop situation. A patented rubber torsion mount provided accurate and ample penetration, even in hard soils. The 70 Series No-Til units replaced the early 50 and 60 Series No-Til planting units.

Designed for better and more accurate high speed planting, the 78 Series air-type planter units were among the most recent A-C introductions. Double disk openers were featured to clear trash and assure accurate seed placement. The 78 Series planting units also featured seed boots with replaceable, encapsulated seed sensors located only 3 inches from the ground at the seed trench. Depending on the tool bar selection, any number of units from 4 to 12 could be selected.

A major difference between the 78 Series and 79 Series planting units was the latter's feature of a stainless steel Accu-Runner to work in lighter soils without lifting and throwing the soil. A replaceable cast wear strip provided longer runner life. Both units featured oversize 1¼ bushel seed hoppers. Other important features included roller chain drives for increased planting accuracy, quick change sprockets for changing population rates, and five different press wheels depending on soil conditions and planting practices.

Allis-Chalmers Model 330 planters featured an all-welded, extra-heavy frame for high speed planting. This unit could be purchased with a common drive using the 78 Series air planting units, or as a unit drive with plate type or air type planting units. Either a four- or six-row frame could be selected initially, and among the standard equipment was a seed monitor system. Hydraulic or mechanical row makers were optionally available.

Allis-Chalmers offered the Model 333 planter in a standard configuration, or as the Model 333 No-Til Special. The frame design of the 333 was such that planting units could be placed as close as 15 inch centers, or as wide as 40 inches. Two frame sizes of 144 and 176 inches permitted either four or six rows of Model 78 air units or Model 74 plate planter units. A host of additional attachments was available to tailor the planter to specific conditions and tillage practices.

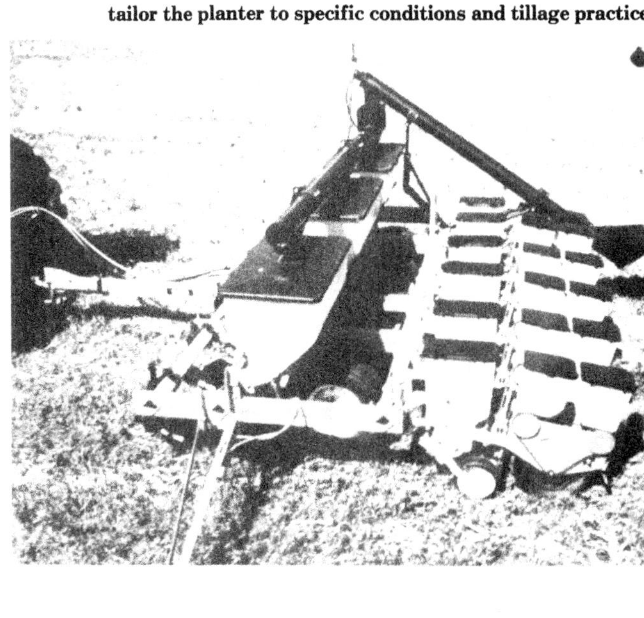

Plows

When Allis-Chalmers bought out LaCrosse Plow Co. in 1929 they instantly acquired an established and well-respected plow line. The No. 1 tractor plow was available only in a one-bottom configuration, but could be ordered in a width selection of 14, 16, or 18 inches. Additional options included a 26-inch landside as shown on this model, or with a rolling landside. An instruction manual for this plow includes detailed information on hitching the plow to the tractor for proper operation. Several linkages could be adjusted to prevent the plow from overcutting or undercutting the furrow.

Although many of the old line implement manufacturers started business as plow makers or had acquired an existing firm early in their career, Allis-Chalmers did not have a tillage line, primary or secondary, until the acquisition of LaCrosse Plow Company in 1929. Initial offerings were a virtual continuation of the LaCrosse line, and in fact, a 1930 advertising piece ascribes the "Allis-Chalmers-LaCrosse" title to the newly acquired line.

The LaCrosse acquisition brought a line of tillage implements into the fold but brought little in the way of a sales organization. Monarch Tractors, acquired about the same time, likewise added little to the pressing need for an established sales force. This problem was however, rectified with the 1931 acquisition of Advance-Rumely. Not only did the Rumely purchase bring an established marketing force into the picture, but this particular purchase represented all those intangible ideas of quality and reliability that farmers looked for in their equipment.

During the first few years of LaCrosse Works and the new plow line, few noticeable changes were made. A-C engineers, convinced that not moving forward actually meant moving backward, were not content to remain with old lines for long. With the introduction of new tractors in 1938 and 1939 came an extensive series of new plows, many of them of radically different mounted designs.

The 1950s and 1960s saw the introduction of many new plow models, reflecting in part the incredible changes taking place within the agricultural equipment industry. These changes included the widespread adaptation of tractor hydraulics and the 3-point hitch system to a wide variety of farm implements, including of course, the plow.

With the 1970s came a call from conservationists and others to refrain from moldboard plowing wherever possible as a means of helping to minimize soil erosion. Many farmers who had previously plowed all of their cropland changed to some form of minimum tillage, and in some case no-till farming was adopted. The implement that had once turned over the prairie and made farming of these regions a possibility was doomed to an unpopular status and overall sales of moldboard plows quickly reflected the change of thought. By 1980, Allis-Chalmers, like most other agricultural equipment builders, had largely left the plow business.

The No. 2 Allis-Chalmers tractor plow was identical in design to the single-bottom No. 1, but instead used two bottoms. Although originally equipped with a mechanical lift, Allis-Chalmers continued to build this plow until at least 1950, even offering it with an option hydraulic lift option. A 1947 price book indicates that while the No. 116, 16-inch plow sold for $99, the No. 2 with 16-inch bottoms listed at $152. Detailed instructions with these plows detailed the rule stating that the center of draft was one-fourth the width of one bottom to the left of the center of the total cut.

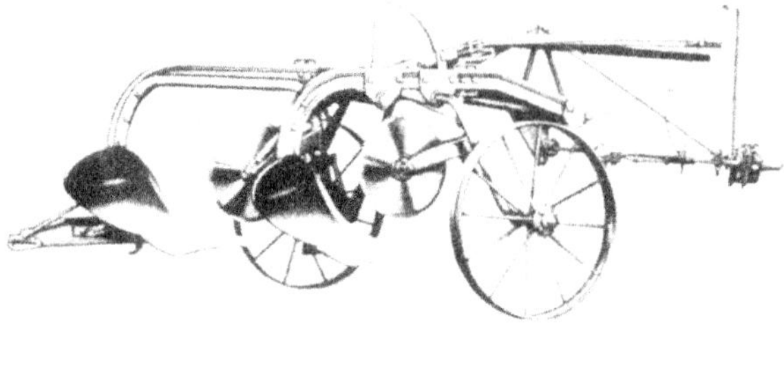

Built during the same general time period as the No. 1 and No. 2 tractor plows, the No. 4 series was available only in a two-bottom size using either 12 or 14 inch moldboards. A 15 inch rolling coulter was standard equipment on this entire series, although an 18 inch coulter was optionally available. In 1947, the No. 4, 214 plow sold for $131. It weighed 681 pounds. Many different moldboard designs were available to suit local conditions.

Plows

Allis-Chalmers pioneered the use of rubber tires on the farm tractor. A 1945 photograph illustrates how, in 1932, an experimental set of 48 x 12 inch airplane tires were installed on this Model U Allis-Chalmers tractor. Owned by Albert Schroeder of Waukesha County, Wis., the tires had been in use for some 14 years when this photograph was taken, and were still serviceable. Initially, almost every farmer in the country, and virtually every implement dealer and manufacturer said that rubber tires simply would not work. Very few demonstrations were required before the transformation to rubber tires began.

Allis-Chalmers pioneered the use of rubber tires on tractors, and was also very aggressive in promoting the use of tractors for plowing and a virtually every field task that had formerly been dominated by animal power. Allis-Chalmers advertising of the 1930s used many different illustrations of the portrayal shown here — an Allis-Chalmers WC tractor at work with a plow. This strategy undoubtedly sold a lot of WC tractors!

From the Allis-Chalmers Archives comes this unidentified photograph of the 1930s. Apparently this immense plow was intended as a brush breaker, with this example probably being a prototype due to the built up frame of structural steel components. Many different reclamation practices were required before some large areas could be farmed. Open ditches and tile drainage were necessary in some areas, while irrigation was imperative in others. Drainage of swamps and low ground was but one step of the process, with huge brush plows being necessary to turn over the almost impenetrable underbrush and begin the process of regular cropping.

Already in 1938 Allis-Chalmers announced the No. 5, one-bottom plow and the No. 52, 2-way plow for the Allis-Chalmers B and C tractors. Using the power lift system of these two tractors, the No. 5 plow shown here was fully mounted. It could be ordered with either a 14 or 16-inch moldboard. Total weight was 225 pounds, and the 1947 list price was $43. The drawbar and depth control equipment required with these plows added another $10 to the initial cost.

The "B-C" tractor-mounted plows could, in addition to the single bottom version, be purchased as a 2-way pickup plow, as shown in this photograph. The 2-way style permitted the plowing of an entire field with no dead furrows or back furrows. The 2-way was also used to great advantage on contoured and terraced fields. Allis-Chalmers offered these plows with a great many different moldboards and jointers to suit local conditions. Most were a minor cost item, compared to the 1947 list price of $96 for the plow itself.

Plows

Continuing the pick-up plow design of earlier years, Allis-Chalmers adapted it to the CA tractor. A major difference was the use of the tractor hydraulics for raising and lowering the plow. The fully mounted design eliminated gauge wheels and support wheels for a substantial reduction in price over pull-type implements. Fully mounting the plow gave some added traction on the rear wheels. For small acreages, especially those with irregular borders, a small mounted plow was an ideal answer.

An exact production period has not been determined for the "WD" moldboard plow, although it is listed in a 1957 price book at $105. Shown working near Waukesha, Wis., this model was hydraulically controlled and was equipped with either 14 or 16-inch bottoms at the request of the buyer. The list price did not include the plow bottoms, since widely varying conditions required any one of several different styles. This particular unit includes a wide-front WD-45 tractor, in itself a bit unusual, since the vast majority were furnished in a conventional tricycle design.

A 1957 Allis-Chalmers price book lists over twenty different plow bottoms, and within each of these major styles remained the choice of up to three different widths. Other variations included a choice of solid steel, chilled steel, or soft center; short landside or long landside; the slat bottom shown here was but one variation — others included stubble bottoms and other styles, all forged for a specific purpose and for use under certain conditions.

As farm implement builders began the transformation from heavy pull-type implement to lightweight mounted or semi-mounted implements, Allis-Chalmers initiated their own approach with their patented Snap-Coupler system. Several advantages were apparent — a major one being the ease of hooking and unhooking from any one of several different implements virtually unaided by a hired hand. The unique mounting method boosted traction considerably, so the A-C Traction Booster system also improved operating efficiency. This farmer is about to hook up a 3-bottom plow to the Snap-Coupler.

The 1950s saw the 60 Series fully mounted plows designed specifically for Allis-Chalmers tractors equipped with the Snap-Coupler system. This requirement precluded their use on any other tractor makes or models — a disadvantage later remedied by adoption of the standard 3-point hitch system. This 60 Series 4-bottom plow included the small transport wheel as standard equipment. Fully equipped, it weighed 978 pounds, and carried a 1957 list price of $478. This particular WD-45 tractor is in itself unique, since it is one of those specially equipped for propane fuel.

Plows

The 60 Series plows appear to have remained in production until the mid-1960s, although specific data on the plow line has been difficult to locate. Shown in this scene is a D-15 tractor with a fully mounted No. 63 plow — a three-bottom version of the 60 Series line. This particular variety was built specifically for the WD, WD-45, D-14, and D-15 tractors equipped with the Snap-Coupler system, and could not be used on tractors without this special hitch. The 1957 list price of this plow was $313.

Working in sod is this No. 73 mounted plow with 3-point hitch. Originally built with the Snap-Coupler as the only attaching means, it was eventually offered with either hitch arrangement and could then be used on a wide variety of tractors. For 1967 this model could be supplied with Snap-Coupler hitch for use with D-19, D-17, D-15, D-12, and D-10 tractors. Front end weights were recommended with the three smallest tractors. In addition, the No. 73 could be purchased with the choice of Category I or Category II 3-point hitch. The list price was $490.

Offered between 1964 and 1968, the 9000 Series plows were offered in semi-mounted or pull-type designs. A 7-bottom semi-mounted model is shown here behind an Allis-Chalmers D-21 tractor. Ordinarily the semi-mounted version was supplied with Category II or Category III 3-point hitch, although the Snap-Coupler system could be furnished on order. The 9000 Series plows were ordinarily furnished in 4, 5, or 6-bottom sizes, but for each of these models an extra bottom could be added. Thus, the 7-bottom unit shown here originated as a 6-bottom plow, plus the single-bottom attachment. As furnished, this plow sold at approximately $1,900.

The 70 and 80 Series plows were first offered in 1964 and remained available as late as 1983. While rigid beams were standard for the 70 Series plows, the 80 Series featured spring trip beams, a feature that prevented serious damage to the plow if an immovable object was encountered. Like the 70 Series plows, the 80 Series was built with either a 3-point hitch design or with the Allis-Chalmers Snap-Coupler hitch. In the late 1960s the four-furrow No. 84 mounted plow sold at $825.

Another Allis-Chalmers design was the Monoframe system as shown on this 3000 Series pull-type plow. The Monoframe used a single, structural steel tube to provide ample strength for holding the bottoms in alignment and absorb shock loads. This design also eliminated the proliferation of crossbraces seen in orthodox designs. For any given plow size, another bottom could be added as an attachment if so desired. Although shown here as a pull-type plow, the 3000 Series Monoframe plow was also available as a semi-mounted unit. The six-bottom base unit could be expanded to seven or even eight bottoms with adapter kits.

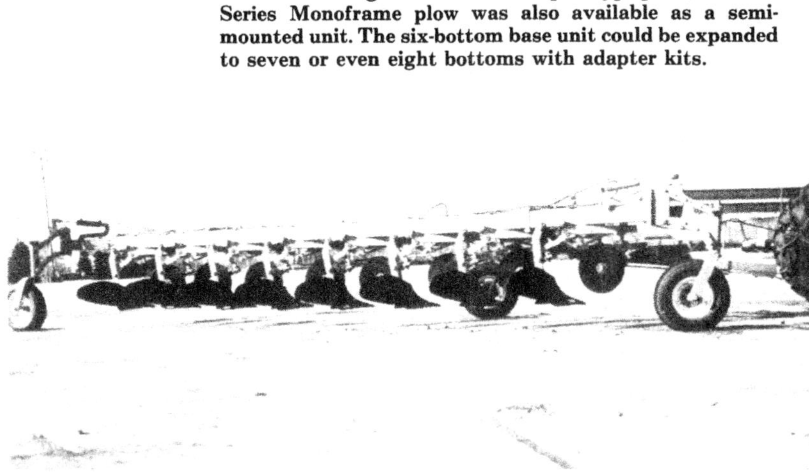

Plows

Two-way plows permit throwing all the furrow slices in the same direction. This keeps irrigated fields level and eliminates dead furrows on flatlands while reducing water runoff on hillsides. Allis-Chalmers offered a number of two-way plow models over the years, with the age of hydraulics bringing about great design improvements in what was a rather complicated design. By the mid-1960s, A-C two-way plows, also called spinner plows or rollover plows, were available in 3, 4, and 5-bottom sizes. Shown here is a D-17 Series IV tractor with a 3-bottom spinner plow working at the A-C Test Farm near Racine, Wis.

As noted previously, Allis-Chalmers engineers were quick to design special equipment for the unique requirements of certain farming practices. Another example is this special ditching plow built for cutting irrigation ditches in a field to be planted in cotton. Unfortunately, some of these manufacturing projects proved to be little more than a gesture of good will, leaving precious little on the plus-side of the books after engineering and production costs were paid. A WD-45 is at work in this scene near Cutter, Cal.

Certain farming practices were predicated on the use of the disk plow, particularly for the tillage of wheatland. To this end Allis-Chalmers offered a series of disk plows as an adjunct to a far more extensive line of moldboard plows. Production was apparently centered in LaCrosse Works as well as in the A-C plant at Oxnard, Cal. A 1961 price list indicates that a 4-disc fully mounted unit similar to the one shown here carried a list price of $712.

As part of a complete tillage implement line, Allis-Chalmers built a rather extensive line of field cultivators and chisel plows, the former usually being used for secondary tillage, and the latter ordinarily seeing duty as a primary tillage implement. Chisel plows have had various uses, one of which has been an ability to break up hardpan which otherwise would prevent drainage and root system development of a growing crop. The chisel plow also has found special favor in recent years as a primary tillage implement ideally suited for minimum tillage systems aimed at soil and water conservation.

Pneumatic Blowers

Long before the 1901 organization of Allis-Chalmers, Edward P. Allis Co. had a reputation for various types of blowers and compressors. In the early stages, reciprocating blowing engines were the rule, but by the 1930s these cumbersome machines were being replaced with blowers. During 1937 Allis-Chalmers built their first blast furnace turbo-blower. It was designed for use on a 500 ton blast furnace at the Hamilton Coke & Iron Division of American Rolling Mill Co. of Hamilton, Ohio. Another significant blower built in 1937 was this copper converter blower of 20,000 cfm at 15 lb. G. Built for the Phelps-Dodge Corp. plant at Douglas, Arizona, it was powered by a 1500 horsepower Allis-Chalmers motor.

In the Annual Report of Sales & Engineering Developments for 1937, Allis-Chalmers reported that The Whiting Corp. was using Allis-Chalmers blowers exclusively for their foundry cupolas. This unit was demonstrated at the 1937 convention of the American Foundrymen's Association, and was subsequently sold to Jamestown Malleable Iron Co. of Jamestown, New York, for use on a Whiting cupola in their foundry. During 1937 Allis-Chalmers made an interesting installation of a large single-stage blower in the plant of Nordberg Mfg. Co., also of Milwaukee. This particular blower was set up to supply scavenging air to two-stroke diesel engines on the Nordberg test floor.

A 1938 photograph illustrates a huge squirel cage fan in its final stages of construction. Its immense size may be gauged from the workman in the lower right hand corner of the photograph. As new technologies developed, Allis-Chalmers engineers stood at the ready to apply these new designs to new and existing situations. The substantial amount of business which developed from sales of centrifugal compressors, blowers, and fans led to the formation of a separate Blower and Compressor Department on January 1, 1935.

Although the intervention of World War Two suspended many important engineering developments for several years, others of even greater importance were taking place. A 1947 photograph illustrates a bank of rotary sliding-vane vacuum pumps installed in the maintenance terminal of United Air Lines. These pumps were required to obtain the low pressures encountered at high altitudes. One application of this simulator was to test carburetor performance in aircraft at altitudes up to 50,000 feet.

Another application of Allis-Chalmers axial blowers was for sewage aeration. The first axial air blower in the United States used for this duty was an Allis-Chalmers unit installed in 1957 for the Sanitary District of Chicago's West-Southwest Sewage Treatment Plant at Stickney, Ill. The blower shown here delivers 200,000 cubic feet in free air at 8½ psig. During the 1950s Allis-Chalmers pioneered new types of compressor designs with special emphasis on the chemical and petroleum fields — rotary gas units saw service for boosting pipeline pressures in natural gas mains.

Pneumatic Blowers

During the 1950s Allis-Chalmers pioneered the use of rotary vacuum pumps on large steam condensers above 17,000 square feet of surface. Two units are shown in this 1956 photograph, with either pump capable of sustaining the load, and the other machine serving as a spare. The addition of automatic controls permitted the unattended operation of the pumps, going so far as to alternate the duty cycle from one pump to another at regular intervals.

Post Hole Diggers

During the 1960s Allis-Chalmers offered several styles of post hole diggers, continuing to do so into the 1970s. From all appearances the various models were not actually manufactured by Allis-Chalmers but were furnished to A-C by others, probably to A-C specifications however. The No. 10 post hole digger shown here was but one of several different varieties. A great labor saver, the post hole digger was however a low-cost item that represented such a small percentage of total sales that its inclusion in the line was probably more in the nature of good customer relations than for monetary reward. An important point is that the post hole digger was yet another implement offered as part of a complete equipment line.

Potato Harvesters

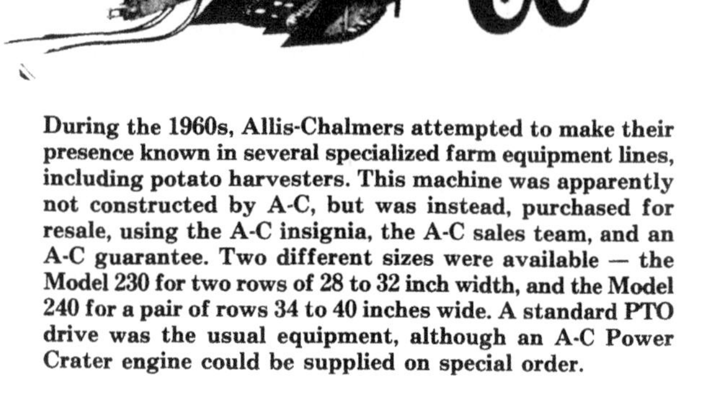

During the 1960s, Allis-Chalmers attempted to make their presence known in several specialized farm equipment lines, including potato harvesters. This machine was apparently not constructed by A-C, but was instead, purchased for resale, using the A-C insignia, the A-C sales team, and an A-C guarantee. Two different sizes were available — the Model 230 for two rows of 28 to 32 inch width, and the Model 240 for a pair of rows 34 to 40 inches wide. A standard PTO drive was the usual equipment, although an A-C Power Crater engine could be supplied on special order.

The major difference between the Model 230 and Model 240 potato harvesters was the row spacing for which each machine was designed. Both required on the order of 50 horsepower for efficient operation, with Allis-Chalmers recommending a tractor of at least 80 PTO horsepower. Although work platforms were provided for as many as eight people, these machines were designed to reduce manual labor to a minimum. The huge capacity of these machines permitted the harvesting of up to 3,500 bags per day.

Pumps

Pump manufacturing was an early claim to fame for Edward P. Allis Company. Through the ingenuity of Edwin Reynolds came pump designs previously unheard of, and which much of the engineering community scorned until Reynolds and Allis proved they were right. On the basis of the steam pumping engines of the 1870s came other designs that eventually used virtually every configuration from the standard reciprocating styles to huge centrifugal pumps designed to move immense quantities of water, but at a very low head.

The engineering experiments eventually led to the construction of the company's own pump testing laboratory as part of its larger Hydraulics Division which also included hydraulic turbines. Some of the most eminent hydraulic engineers of the early 1900s were at some or other employed by Allis-Chalmers.

Not content to build pumping machinery for city water systems and similar applications, Allis-Chalmers was early to design special pumps for large oil pipelines. These and countless other special designs were produced in addition to one of the industry's most extensive offerings of heavy duty reciprocating and rotary pumps. Constantly changing manufacturing technologies brought constant change to the pump industry as well. Allis-Chalmers did a remarkable job in staying abreast of new pump developments and continued to offer some innovations of its own. As late as 1977 the company introduced a new series of split-case pumps in a wide range of sizes for applications in building trades, water, waste water, and industrial applications.

A 1920s photograph illustrates a huge steam pumping engine designed, built and installed for the City of Harbor Springs, Mich. Designed for a capacity of 1 million gallons every 24 hours, it was built to work against a head of 196 feet (about 85 psi), and an operating steam pressure of 100 psi. Allis-Chalmers built a great many of these units, with a very few still in existence. Most pumping engines of this immense size operated at 50 rpm or less, and many operated non-stop for several years at a time — a record difficult to equal by today's "modern" machinery and technology. Because Allis-Chalmers built a great many pumping engines powered by steam, further information will be found under the *Steam Engines* heading.

During the 1920s Allis-Chalmers pioneered the development of high pressure rotary pump designs, with one early application being the movement of petroleum in long distance pipelines. Another application, as shown in this 1938 photograph, was for boiler feed pumps as used in high pressure power house boilers. Since these boilers often operated for months at a time without shutdowns, absolute reliability was imperative in all parts of the system, especially in the feed water section. With this though in mind, Allis-Chalmers developed a multi-stage rotary pump as shown here — in this case the pump is driven by a small steam turbine.

A 1937 photograph illustrates one of two 42-inch vertical propeller circulating pumps serving two condensers at United Power Mfg. at Davenport, Iowa. In addition to their application as circulating water pumps, Allis-Chalmers noted in their advertising that these pumps were especially well suited to irrigation systems and drainage service. These pumps, like other large components, were built to order from information supplied by the customer, and probably verified on-site by company engineers.

Pumps

The special requirements of wartime also applied to pumping equipment. The Industrial Pump Division of A-C stood ready with specially designed marine-type pumps having a capacity of a few gallons per minute to thousands of gallons every sixty seconds. Thousands of individual units were required in virtually every available size and application, both for shore duty and on-board uses. Major items like planes, tanks, and ships usually garner the limelight in histories of military production, yet the unglamorous support items were every bit as essential to the war effort.

Tailings and fine silica are being pumped by an Allis-Chalmers pump at a Florida rock crushing plant in this 1946 photograph. Since Allis-Chalmers was an industry pioneer in rock crushing and materials handling equipment, being able to supply this huge pump was yet another part of a complex organization that was able to equip an entire crushing plant with A-C equipment — everything from the electric switchgear and motors to the crushing and conveying equipment. This pump is handling 900 gallons per minute and is equipped with an 8-inch intake plus a 6-inch discharge.

Allis-Chalmers pumping machinery has been used all over the world. A 1962 photograph illustrates a huge sand barge working on the Volta River in Ghana, Africa. On this barge are several Allis-Chalmers multi-stage pumps used to create suction for sand pumping. Nearly ten years before, Allis-Chalmers supplied some of the largest circulating water pumps ever built for Wisconsin Electric Power Company's Oak Creek Station. They were rated at 110,000 gallons per minute against a head of 17½ feet.

In 1966 Allis-Chalmers announced a new line of rotary compressors as a packaged and moveable unit of in-plant air systems. A-C had long been involved with rotary pump design, so the development of a rotary air compressor was a logical progression of these experiments. The unit shown here has the motor, compressor, aftercooler, and air reservoir all mounted on a steel base. The packaged unit concept saved considerable time and expense on installation, and could be easily moved when requirements changed. This unit was part of the A-C Ro-Flo line.

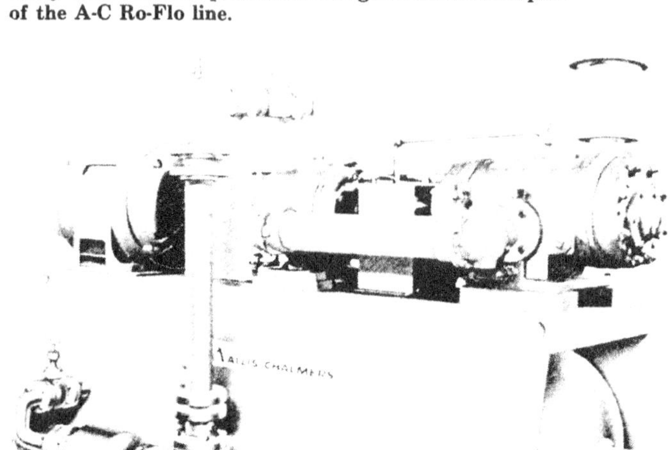

The 2000 Series pumps were announced in early 1977. Intended for the high volume heating, ventilating, air conditioning, and general industrial market, these frame mounted units were available in 29 sizes, ranging up to 2000 gallons per minute. In addition, these pumps could be supplied from the A-C Industrial Pump Division for service heads up to 360 feet, temperatures to 250 degrees, and pressures to 175 psi. The 2000 Series pumps were intended for clear to mildly corrosive liquids and were normally built in cast iron construction, but could be bronze fitted.

Pumps

New for 1977 were the 15000 Series pumps, a line of axial flow units designed for large volume pumping at low heads. Allis-Chalmers announced that these pumps were intended for fresh, brackish, or seawater applications, including such specific applications as storm water, flood control, waste water pumping, irrigation, and dry docks. Although each unit was built from standardized components, each pump was custom assembled to specific requirements for optimum efficiency.

In early 1977 Allis-Chalmers announced the completion of "North America's newest major industrial, large capacity pump testing facility... by the Allis-Chalmers Corporation Industrial Pump Division." Located at Cincinnati, Ohio this center was built to assure a complete and accurate test of pump performance according to industry standards. The new facility was designed to handle capacities up to 40,000 gallons per minute against a head of 920 feet.

Refrigerators

The January 10, 1969, issue of the *Wall Street Journal* carried a small article with the headline: *Allis-Chalmers Says It Plans New Ventures, Including Appliances.* In the fifth paragraph a source was quoted as stating that "Allis-Chalmers is currently holding 'delicate and continuing' negotiations with unidentified foreign companies to fulfill its ambitions."

When the above article appeared, Allis-Chalmers was attempting to ward off a takeover attempt by White Consolidated Industries. *Forbes Magazine* in a November 15, 1969, article cited the A-C move as a bold one, stating that "It's another controversial step in Dave Scott's worldwide strategy to save the ailing company by changing it from a middling-sized provincial U. S. outfit into a global giant." Scott had moved into the president's chair at A-C in 1968.

White Consolidated Industries had purchased Kelvinator from American Motors in 1968, so the A-C entry into the appliance business might have been intended to create an antitrust obstacle to White's takeover attempt of Allis-Chalmers. Another definite possibility is that Scott, who had spent a number of years at General Electric, saw some genuine opportunities for success in the major appliance business, and may even have had visions of a head-to-head confrontation with General Electric in the major appliance business, plus doing battle with Caterpillar in construction machinery, and the huge Deere, International Harvester, and J. I. Case interests in farm equipment. To have mounted successful offensives on even one of these fronts would in itself have been a triumph, but to do major battle on several fronts at the same time was not exactly the method prescribed by economic orthodoxy.

As time went by, so did the crisis surrounding the takeover attempt by White Consolidated Industries. Likewise, the A-C major appliances, heavily touted and widely publicized in 1969 and 1970 seem to have vanished from the scene shortly thereafter.

Priced at $119.95, the UR6 was intended for under-counter mounting as a standalone unit. Automatic pushbutton defrosting was a standard feature. In December, 1969, Allis-Chalmers announced the acquisition of 80% of the outstanding stock of Lantz International Corp. of Philadelphia, from Penn-Pacific Corporation. The purchase price was $900,000. A-C announced that Lantz would be operated as a subsidier of the A-C Appliance Division.

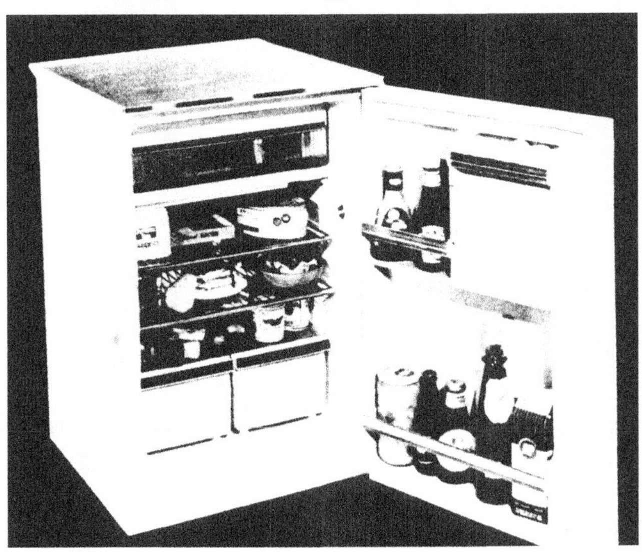

Refrigerators

Model BR3
Bar Type Refrigerator

Weighing but 55 pounds, the BR3 bar-type refrigerator was, as its name implies, intended for the home bar or other areas requiring a small and compact unit. This 2½ cubic foot unit included an extra large freezer compartment with six ice cube trays. Magnetic door seals were standard, as was the one-piece wrap around construction, and seamless one-piece polystyrene liner. It was available in white, coppertone, and avocado. The cart was an optional feature. List price for the BR3 was $99.95.

The attractive RF11 refrigerator-freezer was of the two-door design, offering true zero degree freezing plus moist cold refrigeration in the refrigerator compartment. Weighing but 143 pounds, it had a capacity of 11 cubic feet. Allis-Chalmers used its own small sales force to market these units in 1970, but their efforts apparently met with only limited success. Insiders within the appliance industry were apparently of the opinion that the Lantz takeover was motivated to help deflect an attempted takeover of Allis-Chalmers by White Consolidated. With A-C in the appliance business, and with White Consolidated heavily committed to appliances as well, questions of monopoly might well have been raised.

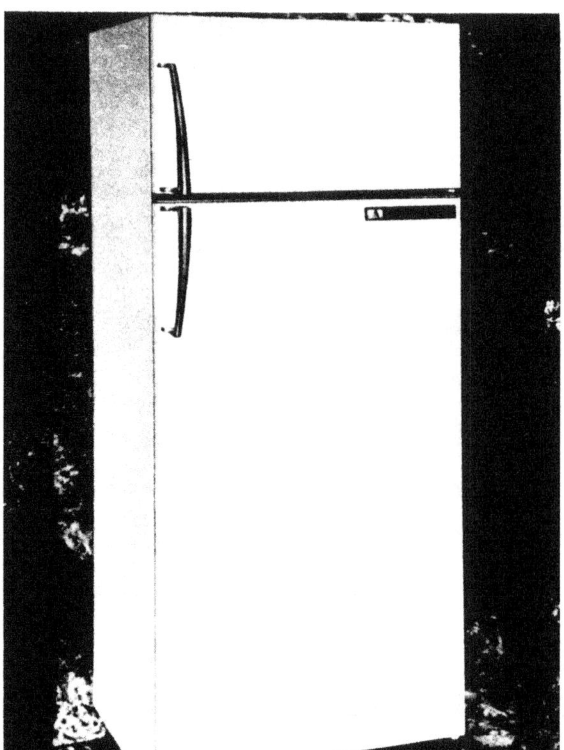

Model CR10
Conventional Refrigerator

Another of the original 1970 A-C refrigerator models was this CR10, a 10 cubic foot style listing at only $149.95 f.o.b. the New York dock. Additional shipping and handling charges boosted this figure slightly. The Allis-Chalmers refrigerators were actually built at Pordenone, Italy by Industria A. Zanussi S.p.A. Organized after World War Two, Zanussi grew so quickly that by 1969 they were producing 2.2 million refrigerators, 1.5 million washing machines, plus a number of other items.

Topping the 1970 A-C refrigerator line was the RF12, a 12 cubic foot model with many important features that included a moist cold refrigeration system which eliminated the need for defrosting. This unit carried a list price, f.o.b. New York of $189.95. Allis-Chalmers apparently sold the refrigerator business back to Lantz International in late 1972 for about $2 million, and had earlier designated the Lantz sales force to handle the sales and service from an A-C facility at King of Prussia, Penn.

Rotary Hoes

Like many other farm equipment manufacturers, Allis-Chalmers offered various rotary hoe designs, with the majority of them emanating from LaCrosse Works, and in later years from the LaPorte Works. This heavy duty, offset design was not intended for weed eradication in standing crops like the standard rotary hoe, but was instead, used primarily for stubble treading and for weed control in summer fallow ground. Comparitively little information is now available on these units, although they appear to have been available into the late 1960s.

For the row-crop farmer Allis-Chalmers offered a complete series of rotary hoes, including this 4-section mounted style of the 1960s. In this photograph, the operator is lowering one of the outside gangs from the transport position. Under certain conditions the rotary hoe has proved to be a valuable farm tool. In their effort to capture a larger market share of a complete farm equipment line, Allis-Chalmers offered this and many other tillage implements in the 1960s. A D-15 A-C tractor is carrying this rotary hoe unit.

Rotary Tillers

During the 1950s Allis-Chalmers once again attempted to market a rotary tiller. Just prior to World War One the company had made a similar attempt, using the design of a Swiss firm. Despite its advantages, the original venture was a failure. Mounted directly to the tractor, this rotary tiller design was intended primarily for vegetable gardeners, landscape work, and similar duties. For reasons unknown, the idea gained little support, so the project assumed a very low priority.

An action scene of the Allis-Chalmers power mulcher reveals its extreme pulverizing action. The Model 68 featured a cutting width of 68 inches, with the Models 84, 118, 132, and 148 each having a cutting width corresponding with their respective model number. The two smallest sizes were set up for Category 1 or 2 three-point hitch, but the three largest models were set up for either Category 2 or Category 3 hitches.

Sawmill Machinery

Without a doubt the A-C sawmill department owed much of its early impetus to George M. Hinkley. Beginning his work with Edward P. Allis in October, 1873, Hinkley has been characterized as being eminently successful, yet his sawmill department was the least profitable unit of the company in terms of sales and profits. Part of this situation was due to the intense competition in the sawmill business, whether it was for the sale of circular saws or band saws. Lumbering was a tough business — one that required the strength of Hercules as just one attribute. The injury and mortality rate in the industry was at times, no less than atrocious as lumbermen were hellbent on converting forests to lumber, and money, as quickly as possible.

Hinkley, during his career, received a total of 35 patents covering various sawmill improvements. As Peterson notes in his book, *An Industrial Heritage*, "Hinkley did not invent the band saw, but he is given credit for its perfection." During 1884 and 1885 when Hinkley was perfecting his design, there was no steel mill in the United States capable of rolling an endless steel band, and so the very first blades were made in France. In December, 1885, Hinkley and Allis announced an exhibit of the new band mill, going on to state that this mill would be taken on January 1 to the Jump River Lumber Co. at Prentice, Wis. for installation.

Improvements came, one after another in the following years. Although the initial market seems to have been in the Great Lakes Region, the Allis band mills quickly made their way to West Coast sites and other lumber producing regions around the world. About 1900 the company announced a new double-cutting telescopic band mill which permitted cutting lumber in both directions of the carriage. The idea had been tried many times before without success. With the telescopic mill design however, double-cutting was no problem whatever. The telescopic design permitted moving the wheels up or down to accommodate the log, always keeping the top wheel as close as possible to the log, and leaving the slack side of the band on the bottom side. The object being of course, to get a short, rigid blade and to keep the saw above the cut as short as possible.

In addition to the band mills, Allis built an extremely broad range of sawmill accessories — everything from lath mills to edgers, plus a variety of conveyors and rollers required in a working mill. Unfortunately however, the market eventually became saturated, and the quality of the Allis designs was such that the machinery was rarely replaced. By 1930 the sawmill department was but a ghost of its illustrious past, and in 1950 this division was abandoned completely.

The first Allis band mill of 1886, was offered in five different sizes. The smallest, called Model No. 3, used 8 foot wheels, could make a cut 44 inches deep, and weighed approximately 22,000 pounds. The largest, designated as the No. 0 mill, could make a cut 90 inches deep. This huge band mill weighed 38,000 pounds. Although the band mill existed prior to George Hinkley and Edward P. Allis Co., there is no doubt that it was this combination of talent and manufacturing ability which perfected it as a viable alternative to the circular mill.

As has often been stated, George M. Hinkley did not invent the band sawmill, but he must be given credit for developing the first successful machine. Prior to the band mill, circular sawmills were the rule of the day, and not too many decades earlier, man cut lumber by hand, with few improvements over all previous recorded time. Allis catalogs usually included several sawmill installations, including this huge layout of Pacific Lumber Co. in 1888. Chances are that this operation included an Allis band mill as original equipment.

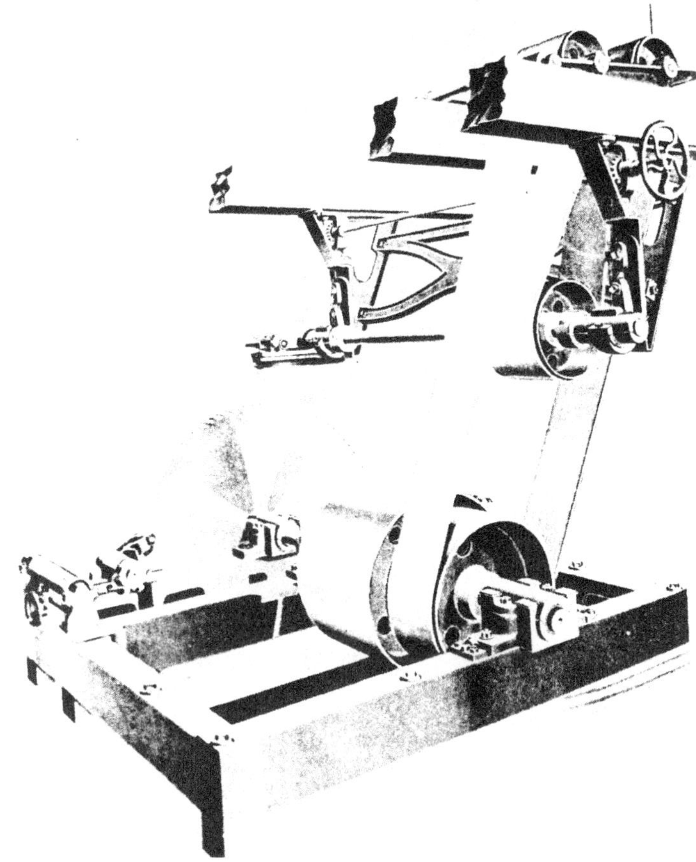

For several years prior to the introduction of the band mill, Edward P. Allis Co. offered a circular sawmill, and continued to build this mill at least into the early 1900s. For small sets, and under certain situations, the circular mill was an ideal choice. Recognizing this fact, Allis gave customers an alternative. The No. 1 circular mill shown here is equipped with a top saw arrangement to permit the cutting of logs somewhat larger than the capacity of the main blade. The Allis approach was for a permanent installation which utilized the building timbers to mount the top saw. Farm-type sawmills as offered by many of the steam traction engine builders were more or less self-contained, which rendered them semi-portable.

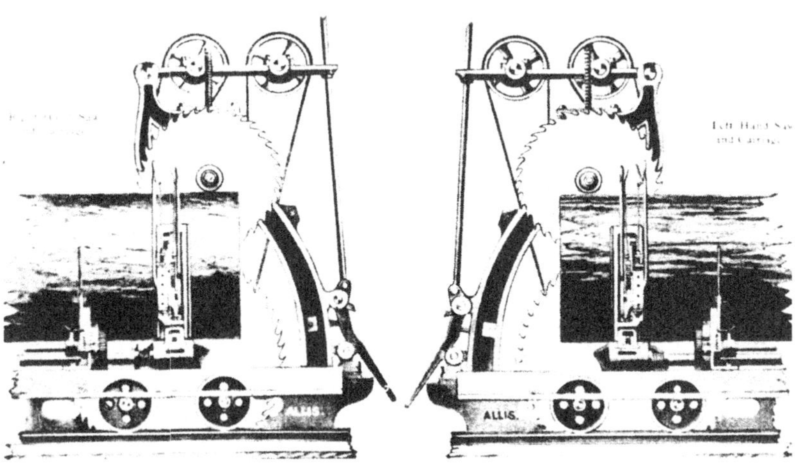

As with nearly every other sawmill builder, Allis-Chalmers offered their mills in the choice of a right-hand or left hand configuration. The choice was usually one of personal preference on the part of the sawyer, or the existing mill layout might have been more conducive to one hand or the other. This photograph from an early catalog illustrates right and left hand designs, together with the following instructions, "As you stand facing the saw, the saw running towards you, the hand the carriage is on indicates the kind of mill. All right hand mills have a right hand (mandrel) nut, and all left hand mills have a left hand nut to tighten saw to arbor."

George M. Hinkley brought all of his patents into the Edward P. Allis Co., and while in the company he added several additional patents to his repertoire of sawmill developments. Hinkley, his bombastic personality excepted, was extremely talented as an inventor. His many developments in sawmill machinery undoubtedly increased lumber production dramatically. But despite his inventive work, Hinkley is scarcely remembered for his achievements. The Hinkley saw guide, patented in June, 1892 is but one of his improvements for the sawmill. Built with exceptional strength, the Hinkley guide could be lifted away from the saw by a handwheel in front of the sawyer.

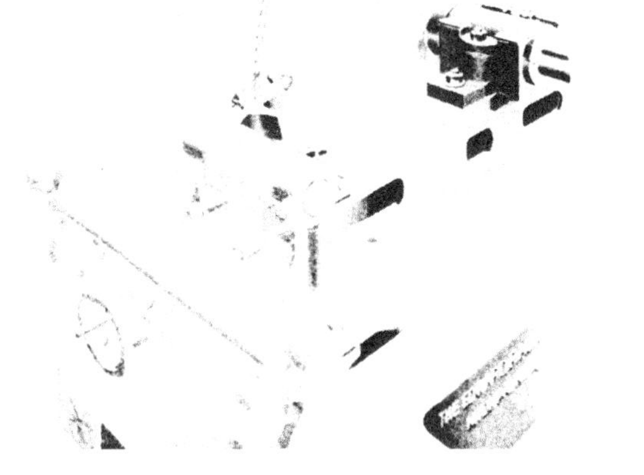

Sawmill Machinery

E. P. Allis, and later on, Allis-Chalmers also offered circular sawmills with a cast iron husk and a cast iron mounting for the top saw rig. This rendered the entire unit semi-portable, since the heavy cast frame was sufficiently rigid so as to withstand the strain of moving from job to job. Shown here with the blades removed, this semi-portable design utilized a semi-automatic belt tightener for the top saw. Presumably this design also permitted a rather accurate adjustment of the top blade to the lower blade. A large operating handle in the foreground loosened the drive belt to the top saw when it was not in use.

A poorly retouched photograph of 1905 illustrates a log of nearly 10 foot diameter being converted to lumber. Much of the original first cutting, particularly on the West Coast of the United States yielded logs of gigantic proportions, with some being so immense that even the largest band mills of Edward P. Allis Co. were inadequate for the task. In the early days of the lumber industry, logging machinery was woefully inadequate and horrendously unsafe. With the introduction of the telescopic mill came the perfection of an old idea, namely, that of putting teeth on both sides of the band and cutting in both directions. Prior to the introduction of the Allis telescopic band saw mill and its ability to handle heavy cuts without depending on the guides for maintaining alignment, double-cutting saws had not seen any success. As word spread of the Allis telescopic mill, large producers especially, were quick to install this design to raise their production.

Virtually any sawmill could be retrofitted with a top blade unit. Although this one was designed specifically for the Allis circular sawmills, there is no doubt that the same saw might have been adapted to almost any other make of mill, especially since this one depended on the building timbers for its mounting, rather than on specific mountings on the sawmill husk itself. Fortunately for Allis, the designs and improvements of George Hinkley were of such character that subsequent advancements were by comparison, minor indeed.

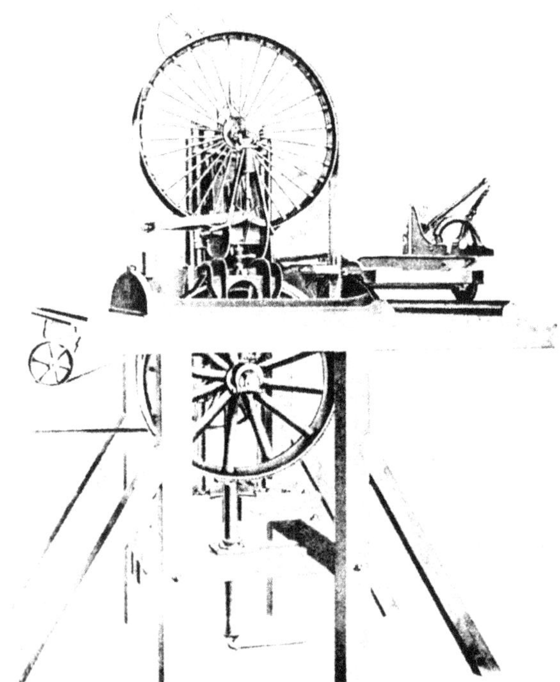

By 1898, or perhaps earlier, E. P. Allis Co. was offering a telescopic band saw mill. It differed immensely from the original design in that the band wheel unit could be raised or lowered to accommodate the diameter of each log. This permitted the sawyer to keep the top wheel as close as possible to the log, minimizing the bending and prying action on the saw band and permitting a cut as heavy as the capacity of the tooth gullets to carry away the saw dust. By bringing the pressure of the log against the saw at a point close to where the saw leaves the upper wheel, it was virtually impossible to crowd the saw backward on the wheels.

Sawmill Machinery

An 1890 catalog of Edward P. Allis Co. illustrates the band mill, noting that it was protected by a number of U. S. Patents. Some of the important features included a design in which the inside rim of the lower wheel was raised from the edge to the center on an angle of about 45 degrees. This raise in the rim prevented sawdust from lodging there. The wheel was also built with a patent hub which prevented casting strains on the rim. Allis noted that this design was patented, and any infringers would, if found, be vigorously prosecuted. The flat belt and pulley high on the saw was provided to raise and lower the saw guide to match the diameter of the log.

Allis-Chalmers horizontal band resaws were intended for the resawing of slabs and cants as desired. A necessary task, it might well have been accomplished by the head saw, and in many small mills, it was indeed. For larger mills however, the extra time required for resawing work slowed down production, making the resaw a necessity. Six foot wheels were used, along with a 10-inch wide saw band. An independent, variable speed, friction feedworks was standard equipment, and two heavy press rolls were used to hold the stock firmly on the feed rolls. Stock up to 12 inches thick could be fed into this machine.

This early 1900s photograph of Sierra Nevada Wood & Lumber Co. at Truckee, Cal. typifies an early western sawmill. The log pond is in the foreground — possibly the logs floated down in a wooden trough from a nearby mountain, or perhaps they were delivered by a logging railroad. The large ramp to the left of the illustration leads to the headsaw. Its second floor location was required so that normal service operations could be performed on the lower end of the saw, and so that the powered bottom wheel could be conveniently belted to a steam engine. Once the log reached the headsaw, the rest of the trip was academic — in just a few minutes the log was converted to boards and planks.

A 1908 Allis-Chalmers catalog illustrates the gigantic band mill with extended bed. Building the bed extension forward of the saw permitted the bed to take a bearing on the floor timbers forward of the saw and giving greater spread on the foundation. Five sizes were available, with the largest No. 1 model using 9-foot band wheels. This huge machine could saw a cant 72 inches wide. (The term "cant" ususally refers to lumber of varying thickness, and oftentimes as wide as the log itself. The headsaw might cut cants four or six inches thick, leaving the resaw machines to convert the cant into 2 x 4 or 2 x 6 inch lumber). Note the water spray device above the saw guide. Standard practice dictates the use of water on both sides of the band, both as a coolant and as a lubricant.

Sawmill

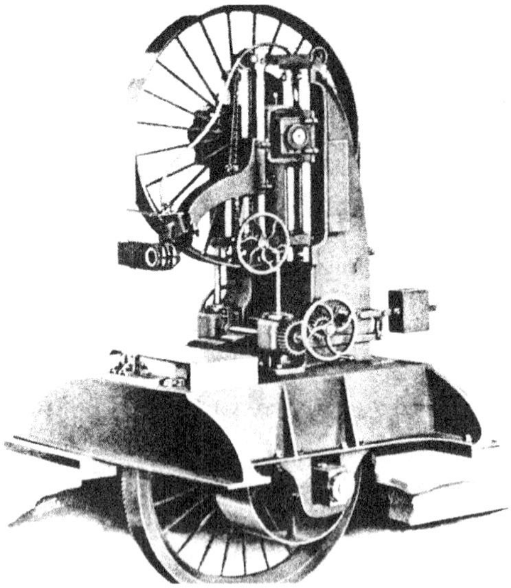

An Allis-Chalmers catalog of the early 1900s illustrates the 6-foot band mill. Designed for the small independent lumberman, this saw possessed all the durability of the larger mills, differing primarily in its physical size and cutting capacity. In the latter regard the 6-foot mill was no toy — it could cut through logs up to 40 inches in diameter. Band length could vary from 31½ to 34 feet, and the saw width could vary anywhere from 8 to 10 inches. The saw guide was raised and lowered by the large handwheel in the foreground. Just above the table is the blade tensioning device geared to the screws which were operated direct with mitre gears.

A 1937 photograph illustrates the Type "C" bandmill in the Pacific Northwest. A-C field representatives reported that in the early part of 1937 the lumber industry had reasonably good sales, but after October, sales of finished lumber dwindled due in part to a declining housing market. With this unfortunate event, sales of A-C sawmill machinery also dropped off severely. During 1937 however, A-C developed new bandmills with welded steel construction. Although the company hoped for brighter years on the horizon, the intervention of World War Two prevented a return of the profits necessary to keep the sawmill division afloat, and so it came to an end by 1950.

A 1930s photograph illustrates a large log coming through the band saw. The operator is cutting a large cant, probably six inches thick. The cant will drop onto powered rollers, traveling a distance to the resawing machines, and at the time will be converted into a stack of 2 x 6 boards. Additional operations will include the automatic cutoff saws or sizing machines that will cut the board to a specified length, plus the planing mill that will finish the board, and the huge kilns that will bring the moisture content within acceptable limits.

By the early 1900s Allis-Chalmers had developed an automatic lumber trimmer. Each individual saw blade is connected to a foot treadle. Depressing the selected treadle would bring up a specific saw, and it would remain in a locked position until another saw was selected. Power-driven feed chains carried the lumber past the saws, and on leaving the table, each board or plank was cut to a specific length. Hand levers could be used in lieu of foot treadles if so desired.

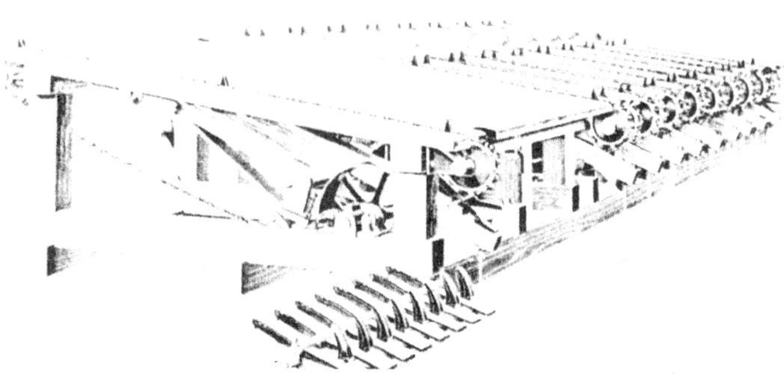

With the coming of gypsum board and similar building products, the use of wooden lath as a base for plaster has vanished. Prior to these developments however, wood lath was used in virtually every dwelling, and in a majority of public buildings. The thin slats were cut approximately ¼-inch thick and 1¼ to 1½ inches wide. In addition the thin wooden strips were used extensively in the manufacture of picket fence. With a tremendous requirement for wood lath, Allis-Chalmers offered this four-saw lath mill.

Sawmill Machinery

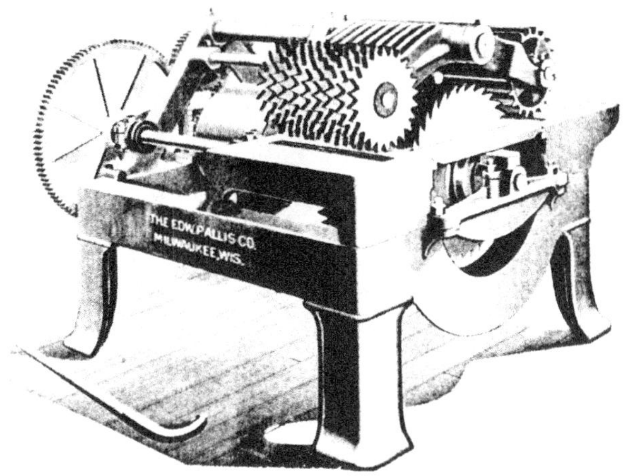

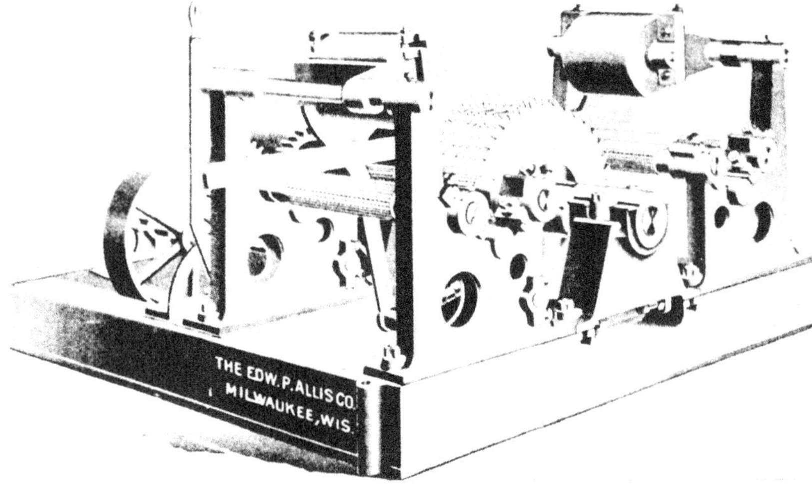

Recognizing the enormous market for lath and picket, Allis-Chalmers built this big ten-saw lath bolter for the large lumber operations. The feed end used a single corrugated press roll, while the individual powered press rolls in the foreground of this illustration were used to propel each individual picket out of the machine. Each saw was 22 inches in diameter, and the bridgetree evident in this illustration permitted reasonable easy access to the saws for sharpening or replacement.

The housing industry required enormous amounts of lumber, much of it falling within the specialty category. Designed specifically to cut wooden flooring, this circular gang flooring machine was offered by Allis-Chalmers at least by the early 1900s, and perhaps earlier. The saws were each 26 inches in diameter, and the machine was designed to handle a 6-inch thick cant from the headsaw. According the A-C advertising, this machine using 6 inch cants wide enough to make nine boards should produce 10,000 feet per hour. The machine illustrated here weighed 5,500 pounds.

Part of the required equipment for every sawmill is the edger. This machine takes material with bad edges or other faults and cuts it into useable lumber. The cutting width is adjustable by the operator, with the judgement call of setting the machine falling on his shoulders. Allis-Chalmers edgers were offered in several distinct styles, each tailored to local needs and conditions. Huge timber edgers were available to mills cutting primarily cant material, while smaller edgers could be secured by the less prominent installations.

An important part of the sawmill operation was the carriage. On it the log rested securely as the blade did its work. Depending on the type of mill, carriages assumed various forms and lengths. The Reliance 3-block carriage shown in this illustration typified the various models offered by Allis-Chalmers. Three headblocks were provided in this instance — projecting dogs on the headblock allowed the sawyer to firmly secure the log before starting the cut. The setworks shown toward the rear of this engraving permitted the sawyer to select the width of the cut with great accuracy.

Large lumber mills often used a power-feed setworks, together with an extremely heavy carriage. Although quite simple in external appearance, it was imperative that the carriage remain in alignment under the heavy weight of a large log, or while carrying no load at all. To this end, Allis-Chalmers built their mills with absolute accuracy and durability as paramount features. In retrospect, A-C may have overbuilt their machinery to the point that supplying even a few repair parts became a rarity.

Sawmill Machinery

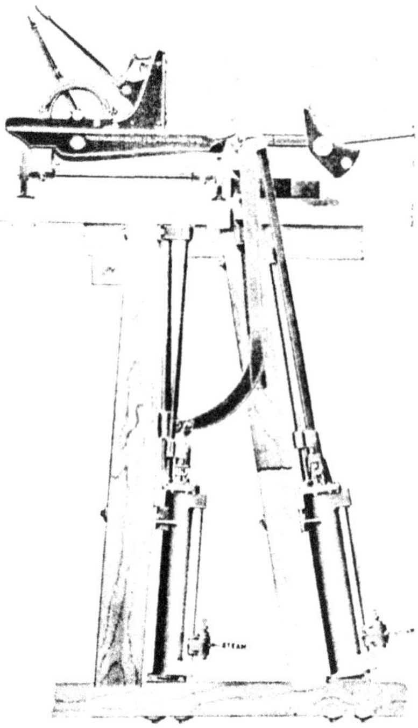

Log handling at the sawmill carriage is well illustrated by this advertising cut of the early 1900s. Allis-Chalmers, like other saw builders of the day, was equipped to provide complete sawmill equipment for virtually any type of operation. Shown here is the steam-powered "nigger" used to turn the log on the carriage after the first cuts had been made. Another steam cylinder operates a "gate" to keep the stockpile of logs in place, ready to be cut, out on the log deck. The term "steam nigger" was probably not connected with an ethnic slur, and is presented here "as it used to be," with no intent to perpetuate the use of the term. Old-time lumbermen were not always noted for tact and diplomacy in their diction, thus the probable origin of the term as applied to what we might now call a log turner.

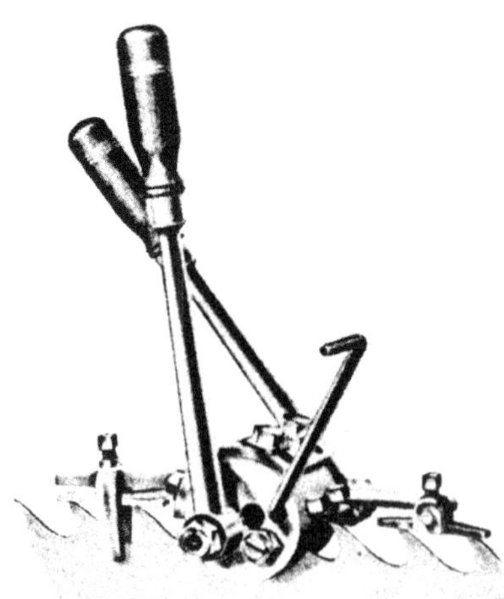

Whether for circular mills or band mills, some sort of swaging device was imperative to maintain the proper clearance between the log and the saw. To not do so would be a sure invitation to problems — these would include, but not be limited to, hot spots and similar ailments, all of which would effectively shut down the mill. Various sorts of hand-operated swags were built, and in fact, this particular one does not appear to be of Allis-Chalmers design, but was merely marketed by them as part of the overall service to its customers.

George Hinkley brought a number of patents into Edward P. Allis Co., and acquired several more during his tenure there. Shown here is an 1890s engraving of Hinkley's band saw swaging machine. The operator was to run the machine at something like 40 to 45 revolutions per minute, and after making various adjustments each individual tooth would be swaged with absolute uniformity. Swaging is the process of mechanically spreading the metal at the tip of each tooth so as to provide running clearance for the rest of the saw.

Except perhaps for having the position of sawyer, few jobs in the early lumber industry were as prestigious as that of becoming a saw filer. The huge band saws dulled rather quickly, and lumbermen soon found that it was poor economy to run a saw too long before resharpening its teeth. This early scene from a West Coast sawmill shows the saw filers at their work stations, with one huge band on the floor, and several more in the process of sharpening or repair. A master saw filer carried in his repertoire the ability to not only sharpen the teeth, but to properly tension the band, make such repairs as were possible, hammer our bumps and hot spots when they manifested themselves, and be an overall expert on band and circular saws.

Sawmill Machinery

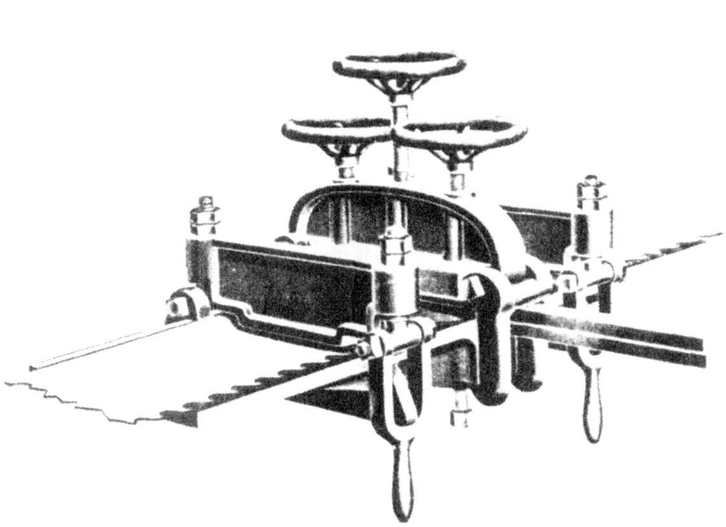

Having been heavily involved in band saw development, and in fact, in the development of the endless steel saw band itself, it is rather curious that Edward P. Allis Co. and its successor, Allis-Chalmers seems to not have had any interest in manufacturing saw bands. Allis was obviously involved in convincing American-based saw manufacturers to develop the necessary means for manufacturing saw bands, and in fact, appears to have found it necessary to import the first few, since no American builder was capable of this work in 1885. Allis did however, build many different devices for saw repair and maintenance, including this huge brazing clamp. It was intended to hold the two ends of a saw band together while workmen joined them together in a brazing process.

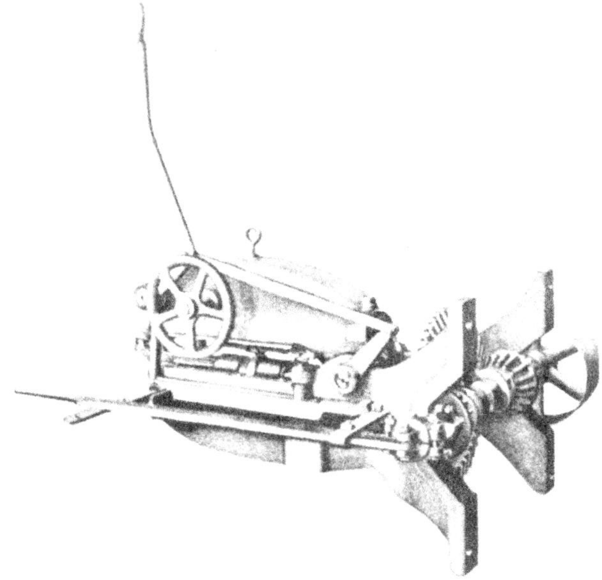

Stretching and tensioning saw bands is an art virtually impossible to learn from books — it must be learned in large part from someone already proficient at the trade. The stretching and tensioning process is designed in part to overcome the effect of centrifugal force while the saw is in operation. To some small extent at least, Allis-Chalmers offered this saw stretching roller machine in an 1898 catalog. As is evident from the construction of the machine, considerable pressure was required to have the desired effect on the saw band.

Double-tooth band saws, that is, those bands which could cut in both directions of carriage travel, required special machinery in the filing room. Shown here is one such machine designed to sharpen double-toothed saws of 14 inches or less in width. By machine sharpening these saws an element of accuracy was introduced that would have been virtually impossible by hand filing methods. Accurate tooth grinding meant more efficient sawing, and so, the saw filing room was indeed one of the most important areas of the entire plant.

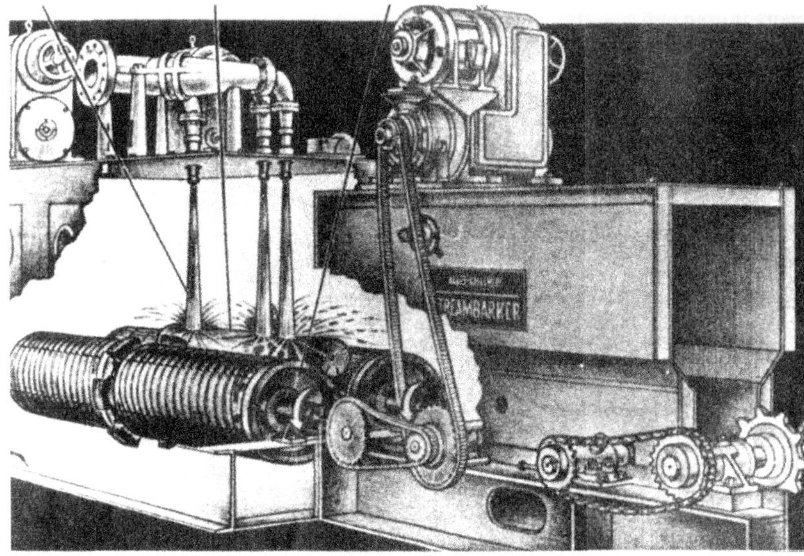

During the 1930s Allis-Chalmers developed the Stream-Barker. Intended primarily for U. S. pulp mills, the Stream-Barker was intended to clean pulpwood logs of bark, dirt, and other foreign materials with a water pressure of 650 pounds delivered through elliptically shaped nozzles. A cutaway drawing illustrates a log on the revolving rolls, while huge high pressure water jets from above continue their work of cleaning the log.

Intended to dramatically increase pulpwood quality while simultaneously reducing labor costs, the Stream-Barker ran into a lengthy delay due to World War Two. After hostilities ended the project was again revived, as illustrated in this 1952 photograph. Small logs coming up the buckhorn chain are completely debarked, and are perfectly clean, ready for the initial stages of pulpwood processing. Despite a significant 28% increase in quality, and a substantial 6 to 10% increase in production by use of the Stream-Barker, the system remained virtually without an ally, and finally was abandoned due to an almost complete lack of interest by the pulpwood industry.

Sawmill Machinery

This unidentified mill is illustrated in an early Allis-Chalmers sawmill catalog. The log pond is obvious on at least three sides of the mill. What appears to be a silo in the foreground is actually a huge incinerator designed to burn sawdust, bark and other refuse from the mill. A large conveyor is evident, leading from the mill building to the burner. Allis-Chalmers might well have supplied this as part of a complete equipment package. The well-screened dome at the top of the burner was intended to prevent flying embers from starting a fire. Today, virtually nothing is wasted at a mill site, including the sawdust and tree bark.

Service Elevator

Apparently with the 1927 acquisition of Nordyke & Marmon came the acquisition of the service elevator shown here. It consisted primarily of a wide rubber belt to which were attached handles a platforms at suitable distances. The continuously running belt permitted factory personnel to move easily from one floor to another, and when first built, the service elevator was suggested as a quick and easy method of moving various materials as well. Despite its convenience the service elevator was rather dangerous, especially if used indiscreetly. This photograph illustrates a workman "on board" and moving up to the next floor.

Snow Blower

In 1952 Allis-Chalmers pioneered what would later become an important machine in the snow belt — the tractor-powered snow blower. Using their well known and eminently successful forage blower as a base, and mounting it to the tractor, the machine was practically built except for the addition of augers to carry the snow into the fan. Allis-Chalmers actually used this machine for several years to clean snow in their parking areas. No answer has been found as to why it was never placed into production — had the idea come along twenty years later, the buying public might have been far more interested.

Sprayers

By the early 1960's Allis-Chalmers had developed a rather extensive line of tractor-mounted, pull-type, and self-propelled sprayers. At the small end of the line was a tractor-mounted design, with this D-12 Allis-Chalmers carrying a typical setup. The sidemounted tanks prevented the usual problems of too much weight on the rear of the tractor — the tanks and their mounting brackets could be easily mounted and detached as well. The sprayer pump was mounted directly to the PTO shaft. This operator is applying weed spray to undesirable vegetation in a fencerow.

For somewhat larger spraying jobs Allis-Chalmers offered this sidemounted sprayer, using a pair of 100 gallon tanks. Virtually all of the sprayer components were priced individually so that each unit could be tailored to specific needs. The tanks could be furnished in fibreglass, aluminum, or stainless steel. The spray boom could be furnished in several widths, mounting styles, and nozzle spacings. Also available was the choice of a gear-type spray pump at $155, a 6-roller pump for $85, or a centrifugal pump for $305.

A 1962 photograph shows a D-17 tractor pulling an Allis-Chalmers Red River Special 40-foot high pressure sprayer. Pull-type units were available with tanks ranging up to 400 gallons in capacity. In fact, the Model 400T pull-type unit with the large 400 gallon tank carried a 1968 list price of $555. By comparison, the Model 44 high pressure sprayer shown here with a 200 gallon fibreglass tank retailed at $810 for 1967.

The Allis-Chalmers Model 2300 self-propelled sprayer of 1967 listed at $3,425. Adding the hydraulic boom control raised the price another $240, power steering added $370, and a hydraulic boom leveler added another $260. The Model 2300 could be furnished with the choice of a gear-type pump, a piston pump, or a centrifugal pump having a capacity of 40 gallons per minute against a pressure of 30 psi. In fact, because of the choices involved in pump selection, its cost was not included in the base price for the machine. To permit better utilization of the machine Allis-Chalmers also offered a knifing-in attachment for liquid fertilizers at an additional cost of $625.

Steam Engines

The flour mills of William Dixon Gray and the sawmills of George Madison Hinkley made up only a fraction of the total sales revenue when compared with the steam engines of Edwin Reynolds. In 1890 for instance, the sale of Reynolds-Corliss engines yielded sales of $2.08 million against total sales for all lines of $3.3 million. As has already been outlined on page 11 of this volume, Edwin Reynolds had accumulated considerable experience with the Corliss engine before coming to Milwaukee.

After coming into the Allis organization in 1877, Reynolds promptly began work on the first "Reynolds-Corliss" design. The new engine embodied the radial-type cutoff mechanism, far superior to the earlier crab-claw design of the Corliss Co. This feature permitted substantially higher rotative speeds than before.

Reynolds immediately offered his new girder frame engine design which permitted substantially higher steam pressures than the bayonet frame characteristic of the original Corliss designs. Higher steam pressures and higher operation speeds both permitted a dramatic increase in horsepower output. Announcement of the Reynolds-Corliss engine first appears in October, 1877. The first unit was of 14 x 36 inch bore and stroke.

An eminent engineer, Reynolds was also the salesman extraordinaire, as is evidenced by the following narrative related by John Schuette of Manitowoc, Wis. "Mr. Reynolds told me he would build me a 150 horsepower Corliss engine for $3,000 and guarantee a savings 50% saving of fuel over the engine we were now running." Seeing my hesitation to the proposal Mr. Reynolds said, "What do you care? You have nothing to risk. If I don't save you 50 per cent you need not pay me a cent, and the engine will cost you nothing." "Yes," I said, "but such an intricate, complicated fine piece of machinery I fear my engineer will not be able to run." To which he [Reynolds] replied, "I will further guarantee that your engineer after he has run the engine a day will admit that he can run it easier than the old one, and the fireman will think firing is a picnic." That did it — Reynolds sold the engine, it did all he claimed, and more making Mr. Schuettte a big booster for the cause.

Reynolds made numerous improvements to compound steam engines, and Allis offered them in tandem or cross compound styles — the terminology meaning that the two cylinders were one ahead of the other or side-by-side. Compound engines were intended to take the exhaust of the high pressure cylinder and supply it to a second cylinder of somewhat larger diameter for further expansion. Although compound engines were the most popular, Reynolds designed and Allis built Corliss engines of triple and quadruple expansion.

Edwin Reynolds assumed his position as superintendent of the Reliance Works on July 1, 1877. In October of that year came the first announcement of "E. P. Allis & Co.'s Improved Corliss Engine" in the *United States Milling & Manufacturing Journal*. Unique to this engine was the girder frame design and a valve operating mechanism developed by Edwin Reynolds. The girder frame utilized a heavy H-section girder between the main bearings and the crosshead tunnel. This type of construction, although light in weight compared to standard designs, was nevertheless exceedingly strong.

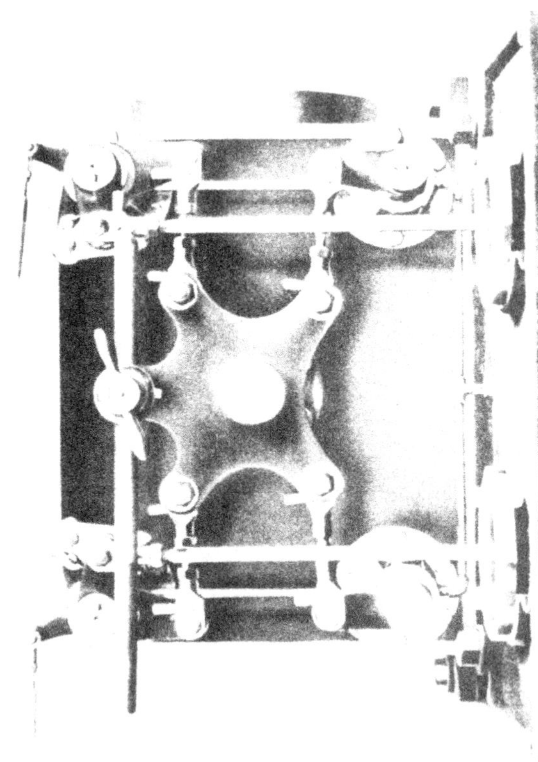

Although this illustration is taken from a 1905 catalog, the design of the valve operating mechanism was virtually identical to that designed and placed into practice by Edwin Reynolds in 1876. The Reynolds design was a substantial improvement over earlier valve operating mechanisms in that its action posed little or no work on the regulator. Due primarily to this fact, the Reynolds design was able to secure extremely close speed regulation. In addition, the Reynolds design permitted substantially higher operating speeds than had been possible before. The centrally located wrist plate moves back and forth in response to an eccentric on the crankshaft. By this means, movement is imparted to the two steam valves at the top of the cylinder, and the two exhaust valves at the bottom.

Steam Engines

In a radical departure from accepted practice, Reynolds built the first Allis blowing engine for Joliet (Illinois) Steel Co. in 1880. Priced at well over $30,000 this huge engine stood some 40 feet high. Its air cylinders were 76 inches in diameter, and were capable of a blast that would "blow a man as though he were chaff." The remarkable and unique design used steel valves at each end of the huge air cylinders instead of the well-accepted leather valves of the time. For this design Reynolds received protection under Patent No. 245,660.

During 1885 Reynolds completed a huge rolling mill engine for Joliet Steel Co. This engine was to drive "the first continuous plate mill in America." The engine was built with a wrought iron frame, weighed 119 tons, and had a 53-ton flywheel. The Allis blowing engines and the rolling mill engines were eminently successful. Even considering the rather limited need for such engines, by 1901 the company had built 184 blowing engines and 80 rolling mill engines.

Reynolds was not content to remain in one position for very long. During 1885 came the first Allis hoisting engines — a pair built for Osceola Consolidated Mining Co. of Opechee, Mich. Two years later, in 1887, Reynolds turned his ingenuity to the steam stamp used for reducing mineral-bearing ores to powder. The Reynolds design used a solid cast iron anvil block under the mortar bed instead of the spring timbers of the past. This novel design fully doubled production.

Reynolds built his first pumping engine for Milwaukee's North Point Waterworks Plant in 1880, and four years later in 1884 he built a huge centrifugal sewage pump for Milwaukee. Its capacity of 70 million gallons per day made it the largest such pump in the United States. It was of course, driven by a direct-connected Reynolds-Corliss tandem compound engine. In 1888 came the first Reynolds screw pump. Designed with a capacity of 500 million gallons per day, it was the largest pump in the world, and was used to move water from Lake Michigan to flush the Milwaukee River. Yet another triumph was achieved in 1889 with the completion of the gigantic 5 million gallon a day Cornish pumping engine for Chapin Mining Co. of Iron Mountain, Mich. Portions of this pump remain in place today as a permanent memorial.

Edwin Reynolds was a major proponent of compounding — the practice of utilizing as much of the expansive force of steam as possible. A two-stage expansion might take the form of a tandem compound or cross compound design. Occasionally a triple expansion was used, and Reynolds pioneered the first American-built quadruple expansion engine in 1886. This engine used cylinders of 24, 36, 52½ and 64½ inches, all with a common stroke of 72 inches. This engine was built for Warren Mfg. Co., a textile mill in Rhode Island.

Although it is well known that Allis-Chalmers could supply boilers for its engines, and could in fact, supply them with their own logo, what is not so well known is the fact that these boilers were built by Milwaukee Boiler Co., of whom Edwin Reynolds was president. With the Allis-Chalmers merger of 1901 came the well known Sederholm boilers of Fraser & Chalmers, giving the new firm unequalled boiler manufacturing facilities.

In 1884 Irving Reynolds, a nephew of Edwin, came to the Reliance Works. Although he had no formal education, Irving had an excellent background in mechanics, especially with marine steam engines.

Girder frame engines were generally used for initial steam pressures of 100 psi or less. This girder frame engine of 1880s vintage illustrates the valve operating mechanism. In the book, *An Industrial Heritage*, Walter F. Peterson notes that "the engineering of the 'Improved Corliss' engine was not very remarkable." In retrospect however, the Reynolds achievement was remarkable since it was one of the first successful attempts at conservation of energy (and money) in manufacturing operations. Even a cursory examination of early engineering texts involved with the design of Corliss engines leads inevitably to the conclusion that the Reynolds design was indeed a major engineering achievement.

A major disadvantage of the single-eccentric valve gear for Corliss engines was its limited range of steam cutoff. In order to achieve a maximum operating economy, the eccentric was usually set and the valves adjusted so as to give the greatest range of cutoff with the least possible angular advance of the eccentric. This scheme was not necessarily an ideal one for all situations, since if the engine was under a constantly heavy load, it was handicapped by its inability to bring the exhaust steam down to atmosperhic pressure by the beginning of the exhaust stroke. With separate eccentrics for the steam and exhaust valves, the exhaust eccentric could be given sufficient advance to provide an early release.

Steam Engines

The Reynolds-Corliss engine design was a raging success from the beginning. This success was reflected both in its impact on the manufacturing industries and within the company as well. The sawmill department and the flour mill department both took short shrift in comparison to the steam engine department at Reliance Works. By the early 1890s the Reynolds-Corliss engine was built in sizes up to a gigantic 48 x 72 model with a maximum output of 2,726 horsepower. Rated operating speed was 55 rpm. The huge flywheel was 30 feet in diameter and it alone weighed well over 50 tons.

Working almost non-stop for two years, Irving Reynolds designed a completely new triple expansion pumping engine, with the City of Milwaukee accepting the Allis proposal in mid-1886.

By the time of his death in 1909, Edwin Reynolds had garnered over 40 patents. In addition to those previously noted, other significant Reynolds patents include his heavy duty engine design of 1890; a vertical engine design in 1894; self-adjusting main bearings came in 1898, and were eventually adapted by almost every builder; the first horizontal-vertical engines came in 1899. One of the most significant Reynolds contributions was the first flywheel-type generator of March, 1892. No one, either in Europe or in the United States had achieved success in mounting the generator windings directly on the engine shaft before.

Reynolds' work was probably best epitomized with the "Manhattan" engines, bid in 1899 and built subsequently. Of the seventeen units, the largest had an output of about 12,000 horsepower, making these the largest steam engines ever built. With the successful installation of these engines however, the recripocating engine came upon hard times. Steam turbines were fast becoming the prime mover of the future, and despite their advantages, the Reynolds-Corliss design was sold less and less. The last steam engine was built in 1930, and except for the occasional sale of a few parts, the steam engine department was practically dissolved after that time.

Within a few months after announcing the Reynolds-Corliss design, Edward P. Allis Co. was offering these engines in nearly 30 different single cylinder configurations. These ranged from the smallest size with an 8-inch bore and 24-inch stroke, all the way to a huge engine with a 32-inch bore and a 60-inch stroke. Using an initial steam pressure of 100 psi and a ¼ cutoff, the 8 x 24 model delivered 33 horsepower, while the big 32 x 60 size under similar circumstances put out 816 horsepower. All of these engines could be adjusted for a ⅛ cutoff instead of ¼ cutoff. Doing so reduced the horsepower output slightly, but provided substantially improved economy due to the more efficient use of steam.

An 1891 proposal "from Edw. P. Allis Company, Reliance Works, Milwaukee," and written to the firm of Thompson, Blackstock & Co., Paxton, Illinois offers to furnish "one horizontal Reynolds' Improved Corliss Automatic Cut-off Steam Engine of best Material and Workmanship..." This engine was to be of 8-inch bore and 24-inch stroke. The 4-inch crankshaft was to be made of hammered wrought iron, and the flywheel was to be turned for a belt — it was 7 feet in diameter with an 11-inch face. The weight of the flywheel was to be 2,300 pounds. Included in the proposal were all necessary foundation bolts, engine trimmings, and similar items. The company was also to furnish one man to supervise setup of the engine, but if his work was delayed by the lack of preparatory work, then the buyer would be charged $4.00 per day additional. For all this, and with the freight prepaid, Allis agreed to deliver the above engine for the sum of $980!

A 1905 Allis-Chalmers advertisement notes that the company's Reliance engine "was designed for driving electrical and other fast running machinery." Apparently, 50% or more of the Reliance engines had been placed into generator service. In order to permit a somewhat higher crankshaft speed than was commonly used in the Corliss engines, the Reliance used a somewhat shorter stroke than was customary in the standard girder frame design. Note also that the frame design is entirely different from the girder engine, in that the Reliance used a huge casting which embodied the lower half of the main bearings and included the entire crosshead tunnel up to its attachment with the engine cylinder.

Another view of the Allis-Chalmers "Reliance" design illustrates massive shielding over the engine connecting rod. Barely visible is a sight glass oiler directly in front of the crank disk. A special bracket held this lubricator in position, and a brass tube conducted oil from this point to the crank pin bearing. A heavy pendulum below the lubricator maintained it in a vertical position. Reliance Corliss engines were commonly operated at speeds up to 150 rpm, but this speed approached the maximum, due to the operating time required for the valve mechanism.

Reliance Corliss engines were commonly furnished for steam pressures from 80 to 130 psi. A 26 x 36 inch engine for example, and using a ⅕ cutoff, would deliver 340 horsepower at 80 psi, but this figure rose to 585 horsepower by going to 130 psi at the throttle. Edwin Reynolds was a major proponent of the compound engine design whereby steam from one cylinder was passed through a receiver to a second cylinder of larger diameter for further expansion of the steam, and ultimately, the extraction of more power. Compounding also led to the use of higher initial steam pressures, often ranging up to 200 psi.

Edwin Reynolds was totally committed to the Corliss engine as the prime mover for his generation and those yet to come. Thus, it seems likely that Edw. P. Allis Company quickly disposed of any small slide valve steam engines once their new Reynolds-Corliss engine was ready for the market. Fraser & Chalmers however, used a different marketing approach, in that they continued offering self-contained slide valve engines to their customers, even though they also marketed the more efficient Corliss design. After the 1901 Allis-Chalmers merger the new company apparently continued to offer a few slide valve engine models.

Prior to the 1901 formation of Allis-Chalmers, the competing firm of Fraser & Chalmers had developed a rather extensive line of Corliss engines using many features of the expired George H. Corliss patents, plus innovations of their own. This vertical Corliss engine was designed and built for the De Beers Diamond Mines of South Africa in 1893. It was of the triple expansion design, and built to deliver 226 horsepower with steam of 120 psi and a speed of 75 rpm. Four of these engines were delivered to South Africa and the De Beers Mining Company.

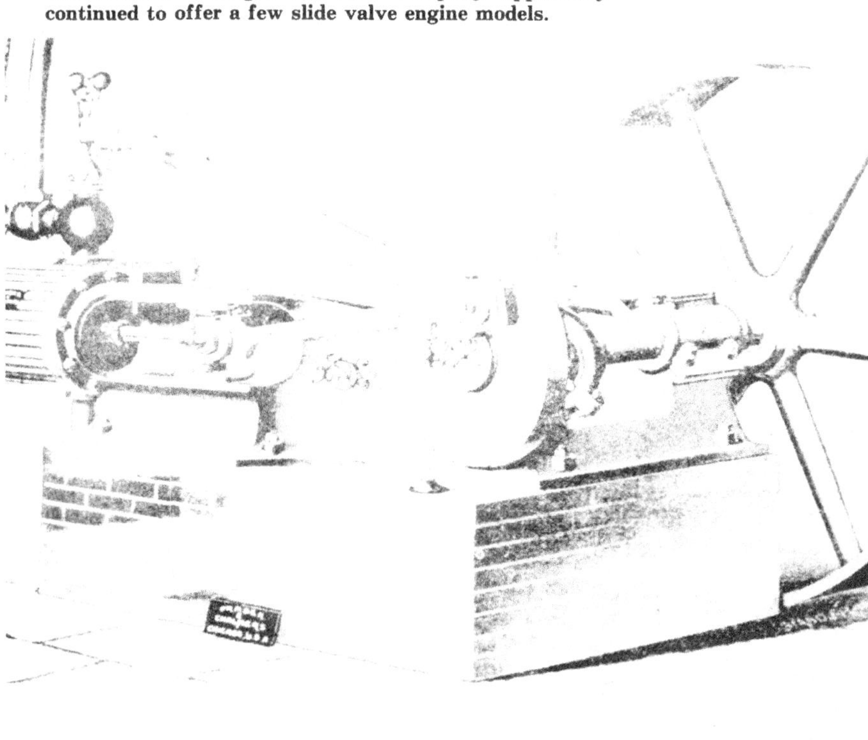

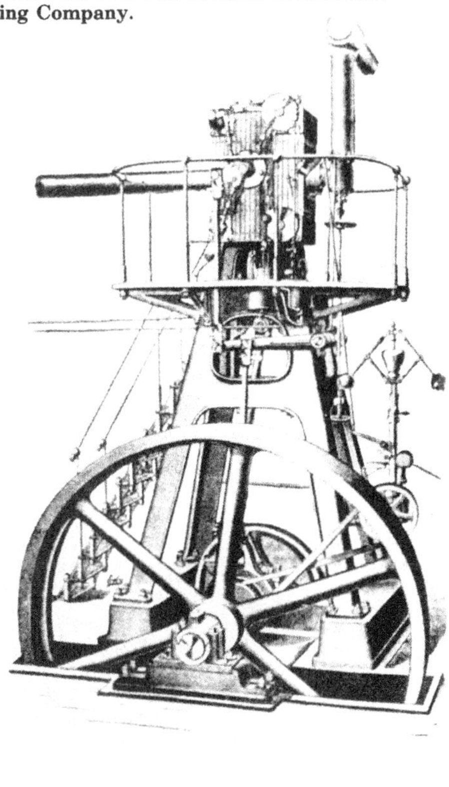

An 1890s model of the Standard Horizontal Corliss Engine of Fraser & Chalmers shows little in the way of distinctive features compared to competitive designs. The dashpots in the floor show control rods reaching up to the steam valves. When the governor mechanism tripped off the steam valves, the vacuum under the dashpots provided the quick closing so necessary for successful operation. Because of their low operating speed Corliss engines could, with reasonable care, operate for *years* at a time with virtually no down time.

Steam Engines

By about 1910 Allis-Chalmers had apparently discontinued the manufacture of small steam engines. This illustration, taken from a 1911 catalog, illustrates an Allis-Chalmers "NI" generator direct-coupled to a Type A engine from American Blower Company. Small engine-generator units of this type were often required in start-up operations within an electric plant, supplying the necessary direct-current power for control and pilot duty, among other purposes.

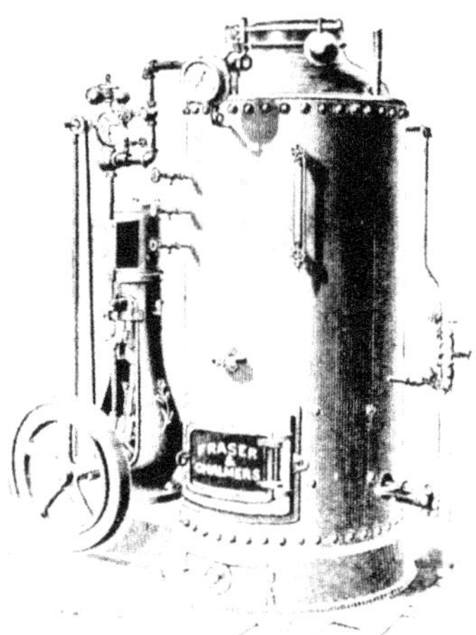

Fraser & Chalmers, one of the partners of the 1901 Allis-Chalmers merger, offered this small engine and boiler combination in the 1890s. Designed especially "for use in printing offices, wood working, and machine shops," this unit probably was outside of the realm of endeavor for either Edward P. Allis Company or Allis-Chalmers. Presumably, Fraser & Chalmers sold hundreds and perhaps thousands of these small engines to give the buyer a first lesson in the operation of engine-driven machinery.

Edward P. Allis Company did not build boilers, but the Milwaukee Boiler Works of the same city took care of this task. Interestingly, Edwin Reynolds was a major partner in this firm and did in fact, contribute several valuable improvements to boiler design. This vertical boiler illustrates a unique design of the feed water inlet and blowoff openings. The vertical tubular design found its greatest application where space was limited and where quick steaming was a requirement.

Horizontal return boilers were apparently a specialty of Milwaukee Boiler Works. In all likelihood, Allis-Chalmers salesmen carried the pricebook in hand for Milwaukee boilers when closing a deal for a new Reynolds-Corliss engine. Fourteen different models were available. The smallest No. 1 size was capable of handling a 20 horsepower slide valve engine or a 25 horsepower Corliss model. Weighing 6,890 pounds, complete with all fixtures and trim, it carried a base price of $182 for the boiler only, fixtures and trim coming at an additional cost. The huge No. 14 size could handle a 170 horsepower Corliss engine — it was priced at $720, and had a total weight of 31,000 pounds, complete with all fixtures and trim. It measured 18 feet long, with a 90-inch diameter.

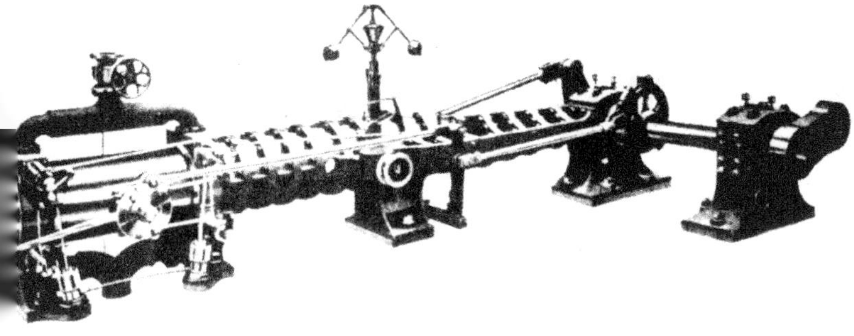

To accommodate the power requirements of remote manufacturing or processing plants, Allis-Chalmers offered a sectionalized engine in which no single piece weighed over 300 pounds. As early catalog data indicates, the methodology was built around mule-pack transportation of the entire engine! Major modifications were required to the steam cylinder to keep the weight within requirements. This included specially designed steam and exhaust passages. The crosshead tunnel and girder frame were likewise built in sections. Completing the engine was the flywheel not shown here, but also built in bolt-together sections.

So-called "working scenes" of the Reynolds-Corliss engine have been difficult to locate. This photograph, copied from an original in extremely poor condition, illustrates the 18 x 36 Reynolds-Corliss engine installed in the Amana Woolen Mills at Amana, Iowa. Thought to have been installed in the early 1900s, this engine operated a huge generator, which in turn supplied power to electric motors within the woolen mills.

A disastrous fire in 1923 destroyed virtually all of the Amana Woolen Mills complex in Iowa's Amana Colonies. In this rare and unique photograph, workmen are shown cleaning up the debris and assessing the damage to the Allis-Chalmers engine. Apparently the engine was not severely damaged, since it was rebuilt and again placed into service when the Amana Woolen Mills were rebuilt. This engine then remained in service until the early 1940s when it was replaced with a diesel engine. Assuming an initial steam pressure of 100 psi and a ¼ cutoff, this engine was capable of 270 horsepower. The huge flywheel was 12 feet in diameter with a face of 29 inches. It weighed 13,000 pounds. Under certain atmospheric conditions the huge leather drive belt was prone to create immense electric charges which could discharge sparks several inches long. (Photo courtesy Museum of Amana History).

In addition to an extensive line of horizontal engines, Allis-Chalmers was prepared to likewise furnish the Reynolds-Corliss design in a multitude of vertical models. Typical of the smaller models is this single cylinder style with a direct-mounted electric generator barely visible behind the flywheel. The vertical engine was an ideal solution when floor space was at a premium. Another major reason for specifying a vertical design was the judgement of the design engineer, since two schools of thought were prevalent, one giving great praise to horizontal engines, and the other stressing the superiority of design manifested in the vertical style.

Large size vertical engines were available in models up to a 48 x 72 inch bore and stroke. An engine of this size was capable of nearly 2,900 horsepower operating with a throttle pressure of 150 psi and a ¼ cutoff. The engine operating valve could be opened from the platform beneath the cylinder, or through an extension rod it could be operated from the floor. The spiral staircase was built to save space, and as previously noted, the vertical design was used chiefly where floor space was at a premium. Vertical engines were somewhat more expensive to build and install than comparable horizontal models.

Steam Engines

This engine in a horizontal design would be referred to as a tandem compound, but in the case of a vertical the title of steeple compound was ascribed. Steam was admitted first to the uppermost cylinder for the first expansion, and was then exhausted into the lower, or compound cylinder. Surprisingly, this engine, like almost any other, could be built up to specifically fit the customer's requirements. Sales engineers could select the various components from within their specification manual, proportioning virtually every part from tables and charts developed from tried and tested formulas.

Yet another approach to the vertical compound was this cross compound design. Although it utilized two separate crank pins, required two A-frames and two crossheads, this style was still far more popular with customers and with engineers than the steeple compound design. In this particular instance, steam from the smaller high-pressure cylinder was exhausted into a horizontally mounted receiver visible at the back of the engine. From this receiver the steam passed into the steam chest of the larger low-pressure cylinder, finally being exhausted at nearly atmospheric pressure.

Edwin Reynolds' earliest cross compound engines employed the famous girder frame design. This system immensely simplified engine design and manufacturing, since certain components were common to each engine within a specific horsepower range. In other words, the Reynolds design required fewer patterns, fewer castings, and fewer repair parts in inventory than would have been necessary had each engine required completely different parts than its comparable brothers. Tremendously strong, the girder frame design was however, limited to 100 psi throttle pressure, at least in the early years. By about 1910 steam pressures had risen dramatically, in some cases to as high as 200 psi.

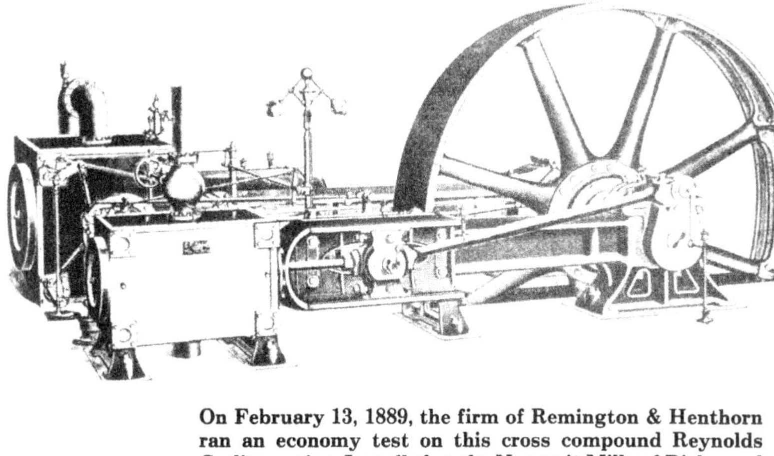

On February 13, 1889, the firm of Remington & Henthorn ran an economy test on this cross compound Reynolds Corliss engine. Installed at the Namquit Mills of Richmond Mfg. Co. of Providence, Rhode Island, this engine used a high pressure cylinder of 18 inches, a low pressure cylinder of 32 inches, and a common stroke of 42 inches. The 18-foot flywheel had a face of 31 inches for the driving belt. Operating at 71 rpm, this engine was capable of about 275 horsepower with a steam pressure of 100 psi. The company guaranteed the engine to develop one horsepower hour with water usage not to exceed fifteen pounds of water. At the conclusion of the test, a water usage of only 13.26 pounds per horsepower hour was recorded.

Already by the 1870s, mechanical engineers were assured of greater engine efficiency through the use of higher initial steam pressures, and with the practice of compounding. In order to extract the maximum possible work from each pound of steam delivered to the throttle, it was theoretically expedient to reduce the exhaust steam to atomospheric pressure. Various tests indicated fuel savings in the range of 25% through the use of a compound engine. Shown here is the "1890 Pattern," developed and first marketed that year by Edward P. Allis & Co., and later marketed by Allis-Chalmers.

Rated at approximately 1,200 horsepower, this huge tandem compound engine was installed about 1888 in the plant of Plymouth Cordage Co. of North Plymouth, Mass. Dimensions included a 30 & 60 x 72 inch bore and stroke. The detailed specifications for this engine include a displacement of 58.04 cubic feet per stroke for the high pressure cylinder and a similar figure of 233.08 for the low pressure cylinder. Piston rods of 5⅛ inch diameter were used in the Plymouth engine, along with a 10-inch diameter crosshead pin, and a 12½-inch diameter crank pin. A 19-inch crankshaft carried the 30-foot flywheel. Normal operating speed was 56 revolutions per minute.

Steam Engines

This overall view of the tandem compound girder frame engine illustrates the cylindrical receiver situated above the two engine cylinders. Steam from the boilers was piped to the flange on top of the receiver. A passage within the receiver carried the steam through it, and down to the steam chest of the high pressure cylinder. Live steam passing through a passage in the receiver tended to reheat the high pressure exhaust steam on its journey through the receiver to the low pressure cylinder. Note the pipe flanges at the bottom of the high pressure cylinder and on the side of the receiver. They accommodated a manifold between the two. Exhaust steam either passed into the atmosphere or into a condenser.

By 1915 Carnegie Steel Co. had purchased no less than 25 Reynolds-Corliss blowing engines for their Pittsburg, PA plants. These included huge engines having a 96-inch diameter steam cylinder and an identically sized air cylinder, both operating on a stroke of 60 inches. History has it that these huge engines could blow a man away as though he were a feather. The engine cylinder in this engraving is nestled between the flywheels, with the blowing cylinder at the top. A long crosshead with connecting pins on each end and external connecting rods provide communication with the flywheels. Assuming the workman on the stairway to have a height of approximately 6 feet gives a faint idea of this engine's huge size.

Specific details of this 1885 rolling mill engine are sketchy at best. Without a doubt, Edwin Reynolds was determined to build the type of engine required to fit the extremely severe duty required for a roller mill drive, using castings of immense size and strength wherever required. Not of the usual Reynolds-Corliss design, this particular engine appears to be fitted with a piston valve. Due to the engine's immense size the flywheel was cast in segments, as is evidenced by the huge connecting links shrunk into place between every spoke. Careful inspection of this engraving reveals a workman standing behind the engine.

Built before the days of tapered roller bearings, this huge geared reversing engine employees herringbone gears in the reduction drive. The herringbone design, while difficult to manufacture, was an effective means of cancelling the tremendous side thrust which could occur due to even the slightest misalignment. Ironically, Allis-Chalmers designed and built these huge engines at West Allis, while engineers at the company's Norwood Works in Ohio were developing huge electric motor drives which would eventually relegate the steam engine to the salvage yard.

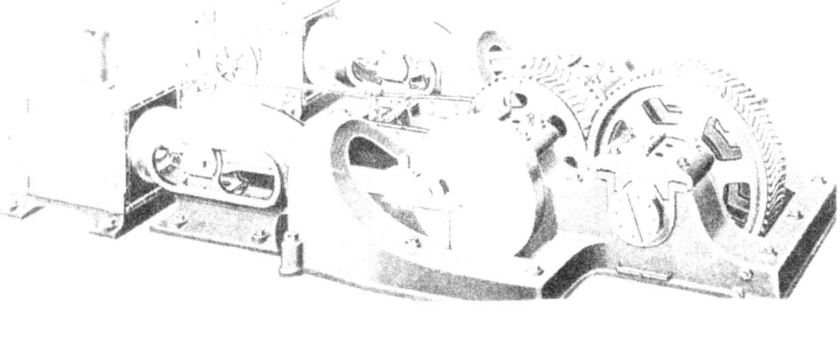

Steam Engines

Before the days of high-pressure blowers and similar devices, forced air for blast furnaces and steel works was provided by huge "blowing engines." An early Allis-Chalmers catalog illustrates this tandem design, with the air cylinder being connected directly to the steam piston rod. With this arrangement, the flywheel was generally not used for driving other equipment, but instead was used to minimize speed fluctuation. In additional to the single cylinder design shown here, the Allis-Chalmers horizontal blowing engines could also be secured in double simple or in cross compound designs.

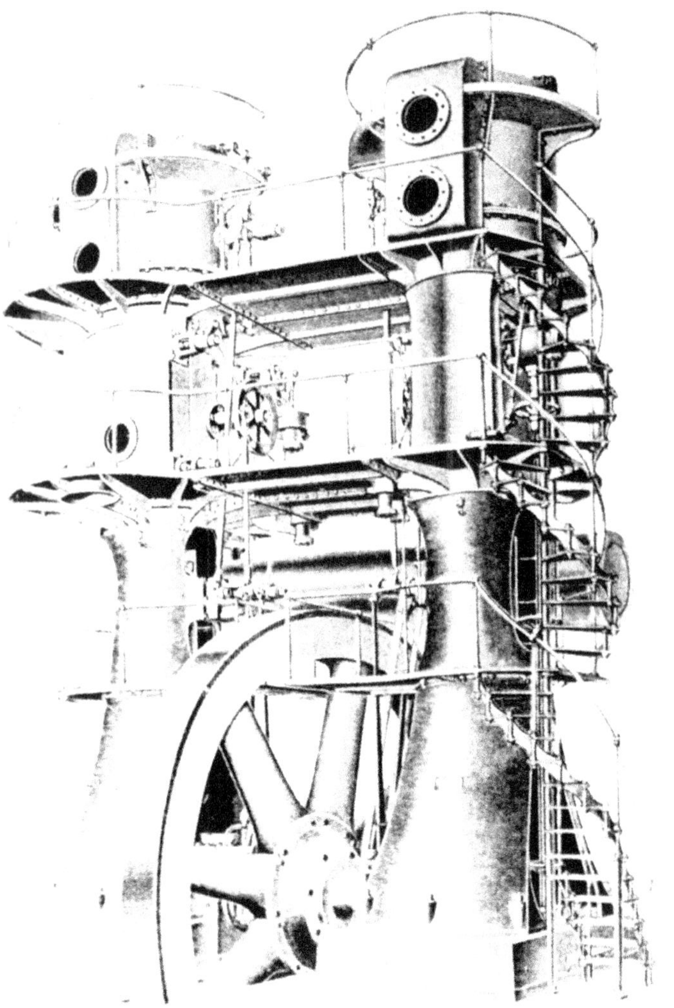

Apparently the single cylinder blowing engines were far more popular than the double cylinder design shown in this engraving. This particular engine assumed far greater height, since the steam cylinders were situated well above the crankshaft, instead of being nestled between the flywheels. Specific dimensions of this particular engine were unknown, but the engine is obviously of cross compound design, with the receiver being located to the rear of the engine. Ample stairways and operating decks were imperative since these engines often operated for weeks at a time without shutting down, even though manual lubrication of the moving parts was required on an hourly basis.

Viewed from a larger perspective, Edward P. Allis Co. and Edwin Reynolds did not operate without intense competition by any means. Fraser & Chalmers, a firm destined to become a part of Allis-Chalmers was offering this Riedler blowing engine in the 1890s. Having a capacity of 2,000 cubic feet per minute with a pressure of 10 psi, this compressor was driven in tandem from a plain Corliss engine. It was installed for the Mazapil Copper Mining Co.

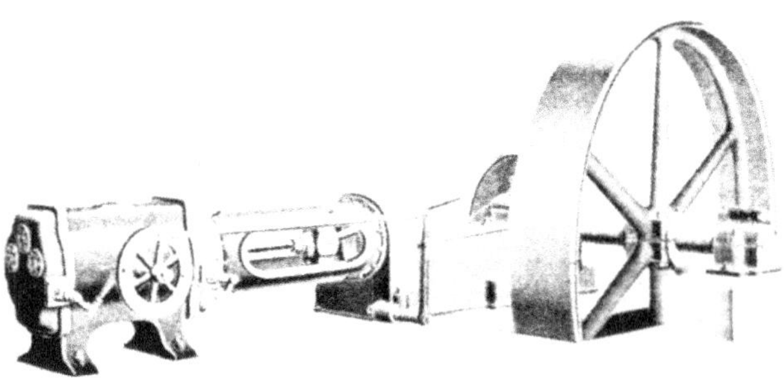

Somewhat resembling a standard Reynolds Corliss engine, this engraving actually illustrates a belted-type Allis-Chalmers air compressor. Typical of the Allis-Chalmers design is the use of the same rocker plate and linkage for operating the compressor valves as would be seen on a comparable steam engine. As has been pointed out previously, this scheme was an effective cost cutter, since it greatly reduced the number of foundry patterns, castings, jigs, and other items required for individual parts.

A 1905 Allis-Chalmers catalog illustrates this steam driven duplex air compressor. The engine is a standard Reynolds Corliss cross compound design, with the two air compressor cylinders mounted in tandem over each steam cylinder. Blowing engines usually operated at a pressure of about 10 psi. For air requirements higher than this, standard air compressor designs were used, and as usual, the volume of air was sacrificed for a higher pressure.

Steam Engines

By the early 1920s Allis-Chalmers had developed several improved engine designs, including the drop-piston valve design and the rocking valve engine shown here. The drop-piston arrangement retained virtually the same valve design as used in the Reynolds-Corliss, but instead of being driven by an eccentric and rod linkage, the valves were individually operated from a lay shaft driven off the crankshaft and extending alongside the engine frame. This arrangement, as with the rocking valve engine shown here, permitted substantially high crankshaft speeds than could be obtained with the standard Reynolds Corliss design.

Mining companies in particular had a need for steam-powered hoists. Edwin Reynolds designed and Edward P. Allis Company built an extensive line of hoists — not the common off-the-shelf variety, but specially designed by Allis' engineers to suit the specific application. This huge hoist was built for Quincy Mining Co. of Hancock, Mich. The engine cylinders were 48 inches in diameter with an 84-inch stroke. The huge drum is 26 feet in diameter, 12 feet wide, and can carry 7,500 feet of 1⅝ inch steel rope.

While not as large in terms of cylinder size as the Quincy Mining Company engine, this huge double-cone hoist was 36¾ feet in diameter at the center, and 13½ feet in diameter at each end. Overall, this double cone had a face of 25½ feet, and could carry 7,000 feet of 1¼ inch rope. The two engine cylinders were 32 inches in diameter with a stroke of 84 inches — 7 feet! It was furnished to Tamarack Mining Co. of Red Jacket, Mich. This mining site was also home to a pair of double-cylinder hoisting engines with a 42 x 84 inch bore and stroke. The hoist operator is scarcely visible on his operating platform several feet above the engine cylinders. A 1900 company price list indicates that a two-cylinder 22 x 60 inch hoisting engine, complete with steam-operated post brakes and reverse gear, plus steam operated friction clutches weighed a total of 220,000 pounds! Unfortunately, the space for the final price was blank.

This huge reel hoist was installed for Bi Metallic Mining Co. of Granite, Mont. Company records indicate that two of these units were placed in service for Bi Metallic, each one having an engine with a 22 inch bore and 60 inch stroke. The huge braking system is obvious on the hoisting drum, as is an additional brake working on the crank disc. Allis-Chalmers hoisting engines could be fitted with cut-offs that could be applied by the operator, and in addition could be furnished with a steam-operated reversing gear, and steam-powered brakes.

Prior to the 1901 merger that formed Allis-Chalmers, the Fraser & Chalmers operation at Chicago was capable of furnishing a most extensive line of hoisting engines. After the 1901 merger process, several years had to go by before old competitors could work together for a common cause. This 1890s example illustrates a double-cylinder Corliss hoisting engine supplied by Fraser & Chalmers to the Grand Central Station in Minneapolis, Minn. Although the floor plates are removed for the photograph, the operator's platform is evident in the center foreground.

Steam Engines

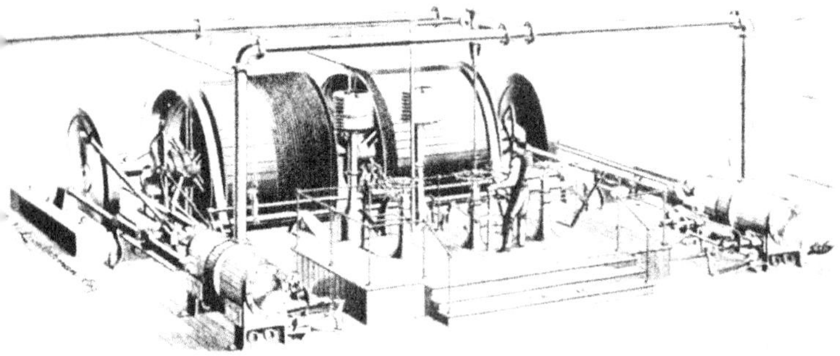

This double-drum hoisting engine was designed and built for Osceola Mining Co. of Houghton, Mich. This mining operation used nearly a dozen hoisting engines in their facility, many of them having 20 x 60 inch bore and stroke for the engine. A single unit was built using a bore of only 8 inches, but having a stroke of 84 inches! This particular unit is of early vintage, probably dating back to the late 1880s. An operator is shown on the platform between the engine cylinders. Note also the wooden lagging which covers the hoisting drums. An almost illegible paper discovered in the A-C Archives indicates that in 1900 the company sold a double-cylinder 16 x 30 engine, together with hand-operated brakes and reverse gear for a total price of $9,000. This complete unit weighed 85,000 pounds.

The 1920s continued to see the occasional sale of steam hoisting engines, but by this time, electric power and electric motors had largely overtaken this work. Electric drives and electric controllers were far more simple than the complicated mechanisms of the steam drive. In addition, electric controllers saw the gradual introduction of safety devices previously unheard of. This particular steam-driven, double-drum unit is equipped with heavy duty brakes on each end, with a single driving gear in the center.

In March, 1892, Edwin Reynolds contracted to build the first direct-connected generator in the United States. The two horizontal cross-compound engines were to be delivered to Narragansett Electric Light Co. of Providence, R.I. Prior to this time, generators were belt-driven, it being thought impossible to run the engine fast enough or to design the generator to run slow enough for direct-coupling to be possible. Without a doubt, at least some of the engineering community met this idea with impudent scorn. In almost no time however, even the "doubting Thomases" had to grudingly concede to the Reynolds design.

Edwin Reynolds followed his heavy duty "1890 frame" shown here with the 1894 vertical frame design. In 1895 Reynolds secured a patent on a high capacity water tube boiler, and in 1898 he designed and placed into service the self-adjusting and self-aligning bearings that soon became the hallmark of virtually every engine manufacturer, whether in violation of the Reynolds patent or not. This 1905 engraving illustrates a tandem compound direct-coupled engine with the dynamo rotor mounted directly on the crankshaft. Partially evident is the outside view of the self-aligning bearing design of the outer pillow block.

Sears, Roebuck & Company of Chicago was home for these huge Allis-Chalmers cross compound engines driving 750 kilowatt Allis-Chalmers direct-current generators. This photograph of March, 1912, typifies many early power house installations. Since these huge engines were a complete marvel to the average American, companies like Sears-Roebuck often offered tours for their customers. For this reason, tiled floors were often used, and workmen kept the engines spotlessly clean.

As has already been noted, vertical engines were the logical choice where space was quite limited or very expensive. Accordingly, these vertical engines and generators were installed in the Commerce Street Station of Milwaukee Electric Railway & Light Company about 1906. Eight of these vertical cross-compound engine-generator units were installed, having an aggregate capacity of 20,000 horsepower. Aside from their overwhelming size, these engines probably presented some formidable challenges to the company all through their construction, from the foundry, all the way to the machine shop.

A major proponent of double, triple, and even quadruple expansion, Edwin Reynolds was no doubt quite proud of this three-cylinder compound engine. It was unique, in that steam passed through the first high pressure cylinder, was exhausted into the receiver, then traveling to *two* cylinders of the same size as the high pressure cylinder. The engine probably would have worked just as well with a somewhat larger low pressure cylinder, although this approach might have had some slight advantages.

A most peculiar design was this double drum hoist set up in tandem, rather than the usual parallel design. Double connecting rods hooked one drum to another, terminating at the cylinder crossheads. Each hoist is likewise equipped with a heavy brake on each drum. Although the exact purpose of this novel and unique arrangement has not been ascertained, the specific requirements of its particular application probably were significant factors in the design.

Photographed at an undisclosed location, this power house scene typified an installation of the early 1900s. Evident here is the belt-driven exciter generator in the foreground. Open flat-belt drives were a frequent cause of annoyance, with Allis-Chalmers later developing the Tex-Rope multiple v-belt drive system. The notches in the flywheel are used with a special bar to manually roll the engine into its proper starting position. Special measures were necessary to insure adequate lubrication of the engine, since in electric power service it might run for months at a time without interruption.

A 1905 engraving yields an unobstructed view of the vertical cross compound Reynolds-Corliss engine, complete with direct-coupled generators. In this particular instance, two smaller generators are installed, rather than a single unit of larger capacity. Given the relative unpredictability of early 1900s electrical equipment, perhaps this was a built-in insurance policy that kept one generator alive, should something go awry with the other.

The first pumping engine to use the Reynolds Corliss design was placed into service at Milwaukee's North Point pumping station in 1881. Of compound design, the high pressure piston was vertically oriented, just above the floor. The low pressure piston was placed on top of the massive framework and directly central over the high pressure cylinder. The pistons were then connected with two separate piston rods, and these in turn to the rocking beam, used only to drive the flywheel. The pumping cylinder was located directly under the high pressure cylinder. This engine operated for some twenty years, delivering 12 million gallons of water per day. Dimensions included 34 & 66 inch cylinder bores with a common 60-inch stroke. Operating speed was 25½ rpm.

Few, if any of these huge tandem compound generating engines remain in existence. A great many remained in existence, but unused until World War Two, but the scrap iron drives of that conflict brought an end to the vast majority of these engines. Four of these huge tandem compound Allis-Chalmers engines are obvious in this photograph. Given reasonable care, it was virtually impossible to wear them out, but the advent of much smaller, and even more efficient steam turbines brought the Reynolds-Corliss engine to a sure and certain end. Some perspective of this engine's immense proportions can be compared against the operator barely visible in the left foreground.

In 1884 Edwin Reynolds devised what was at that time, the largest pumping engine in the world. Built for the city of Milwaukee, it pumped water from Lake Michigan through a 12-foot diameter tunnel, 2,500 feet long, into the Milwaukee River for flushing purposes during the summer months. Contract capacity of this unit was 32,000 cubic feet per minute, raised 3½ feet high, but an 1888 economy test of the engine noted that this unit was customarily operated at a capacity of 40,000 cfm, and was easily capable of 48,000 cfm, or over 500 million gallons in 24 hours. The engine used a 48 inch stroke, with 19 and 38 inch cylinder diameters. The huge pump impeller was 14 feet in diameter, used four blades with a pitch of 8 feet, and had a theoretical discharge of 909 cubic feet per revolution. Average engine speed was 52 rpm.

Edwin Reynolds undoubtedly beamed with pride when this huge 5,000 horsepower Allis-Chalmers engine took a Grand Prize Award at the St. Louis World's Fair of 1904. Direct-coupled to a Bullock 3,500 kva generator, it operated at 75 rpm. Bullock Electric Mfg. Co. eventually became the Norwood Works of Allis-Chalmers. This unusually detailed photograph of the St. Louis engine illustrates that it is of the vertical-horizontal design.

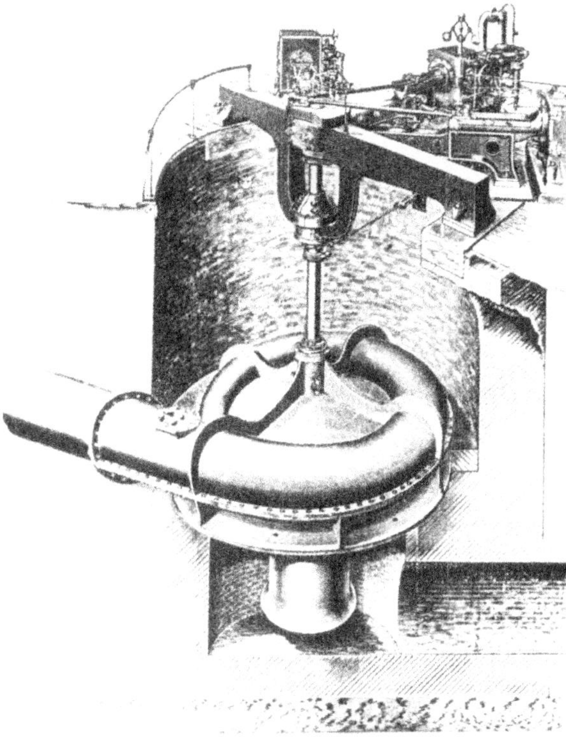

Steam Engines

The phenomenal success of the Milwaukee sewage pumping engine brought the idea to the attention of other cities as well. By 1893 Edward P. Allis Co. had at least six engines of the style shown in this illustration either installed or under construction for the city of Boston, Mass. The engines were of the triple expansion type, with the three cylinders set 60 degrees apart. This arrangement, plus the flywheel action of the pump imeller made a separate engine flywheel unnecessary. The connecting rods of all three cylinders drove on a single crank pin.

On July 5, 1893, tests were completed on three large triple expansion engines installed by Edward P. Allis Company in Chicago's Fourteenth Street Pumping Station. Each engine was required to develop a duty of 125 million foot pounds for each one thousand pounds of feed water evaporated, while pumping 15 million gallons of water per 24 hours against a head of 125 feet. These engines, like other triple-expansion models, were designed with the cranks 120 degrees apart. A common stroke of 60 inches was used, with the steam cylinders ranging from 27 to 46 to 70 inches in diameter. Each pump plunger was 32¾ inches in diameter. Engine speed was 16 rpm.

The famous Milwaukee pumping engine shown here was placed into service during August, 1891. The Milwaukee engine design was conceived by Irving Reynolds, a nephew of Edwin. From the beginning, its economy in the use of steam made headlines, coming in at a remarkably low figure of 12.17 pounds of water per horsepower hour. During 1893 Dr. Robert H. Thurston, one of the leading engineers of the time, ran an extensive series of test on this engine. The results of this work were presented in December, 1893, and form part of Volume XV of the *Transactions* of the American Society of Mechanical Engineers. This huge engine used cylinders of 28, 48, and 74 inch diameters with a common stroke of 60 inches. Operating speed was slightly over 20 rpm. The pump was rated at 18 million gallons per day, and probably had a maximum capacity greatly in excess of this figure.

A photograph of Chicago's Fourteenth Street Pumping Station shows the three Allis triple expansion pumping engines in place and ready for work. Looking rather small in this photograph, it must be borne in mind that the engines had a probable height of 25 feet, so the building itself was probably an immense structure. Irving Reynolds, nephew of Edwin, came to work for his uncle in 1884 at the age of 22. Irving was without formal training in engineering, but never felt handicapped since he felt he had nothing to "unlearn." Regarding some of his phenomenal engineering achievements, Irving Reynolds once said, "We didn't know that we wanted to do couldn't be done."

Steam Engines

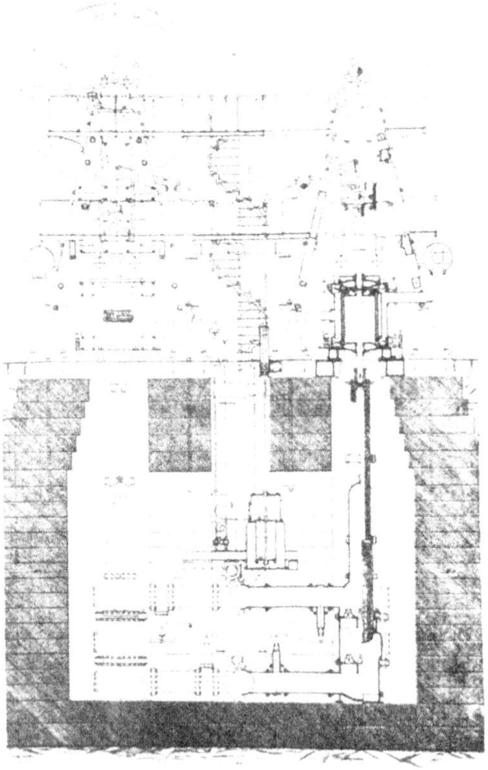

In addition to their usual direct-connected pumps, the company was also able to furnish geared pumping engines when necessary. This engraving typified the geared design, in this case, built for the city of Madison, Wis. A standard 14 x 36 inch Reynolds Corliss engine was used, and geared directly to a Duplex Power Pump. In this particular instance the pump was required to handle 1½ million gallons per day against a head of 225 feet. Another unit at Madison used a standard 18 x 42 Reynolds Corliss engine with a double acting Duplex Power Pump. Still another utilized the smallest of the Reynolds Corliss engines, the 8 x 24 inch size, together with a Duplex pump. The latter installation was made at County Hospital at Wauwatosa, Wis.

This cross-sectional view is the only pictorial evidence presently available for the huge Reynolds pumping engines at Allegheny, Penn. The two engines shown here were exactly alike, both in general arrangement and in detail. Each was of the vertical, three-cylinder compound type, with receivers between the high and low pressure cylinders. A common stroke of 48 inches was used, along with a bore of 31 inches for the high pressure cylinder and 43 inches for each of the two low pressure cylinders. Each of the 16-foot flywheels weighed 20,000 pounds and was carried on a 12-inch crankshaft. The engines were installed in 1883.

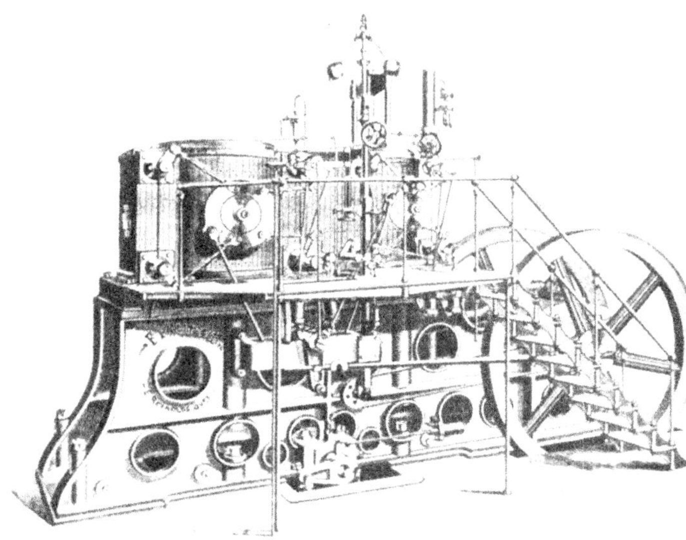

Where compactness of design was essential, Edward P. Allis Company offered this unusual engine. At least one engine using this design was installed for the city of St. Paul, Minn. The pumps were situated directly beneath the steam cylinders, and could be set at any desirable depth, according to local conditions. The steam piston rods extended through the bottom of the cylinder heads and connected directly to the pump plunger, with each plunger being connected back by means of links to a triangular rocking beam, which in turn drove the crank and flywheels. No work was carried through the beam except what was given to and taken from the flywheels.

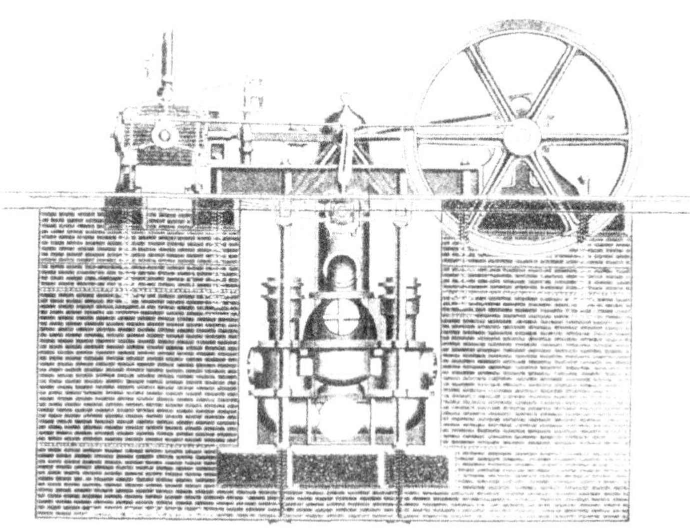

The so-called "Portage" design included a horizontal engine and vertical pumps. This arrangement was intended for locations where it was necessary to place the pumps some distance beneath the engine room floor. As can be seen from this cross-sectional view, a standard horizontal Reynolds Corliss engine was used — it operated a pair of vertical, single-acting pumps through a beam. Although this design required more floor space than the "St. Paul" design, it used standard components for most of the engine, thus making it far less expensive to purchase.

About 1884 this unusual pumping engine was installed at Decatur, Ill. The curious design, replete with a most unusual system of linkages, probably carried this as its greatest disadvantage due to lost motion from normal wear. During economy trials a boiler pressure of about 65 psi was employed, with engine speed variable from 15 to 40 rpm, depending on load requirements. Although the single cylinder design, the final efficiency of the engine was raised considerably through use of an exhaust condenser. A description of this engine is difficult without a cross-sectional view, but for the purposes of this volume it is sufficient to note that the crank and flywheels are mounted on the air chamber of the pump, and connected with a pin on the working lever. This pin describes a tangential arc from its center to the center of the crankshaft.

Steam Engines

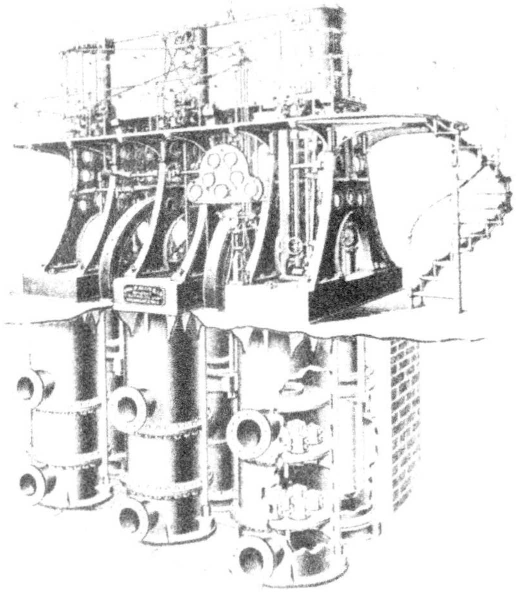

Yet another Reynolds Corliss triple expansion pumping engine was installed at St. Paul, Minn., in 1892. This engine, tested for economy in June, 1893, was placed at McCarron's Lake Pumping Station. A 36-inch stroke was used in this engine, with the steam-jacketed steam cylinders having bores of 21, 36, and 51 inches. The three single-acting pump plungers were 19¾ inches in diameter. A 24 hour test run indicated an average engine speed of 30 rpm, and a delivery of 6.268 million gallons of water against a head of 173 feet. This cross-sectional view shows the pumps situated directly beneath the engine. Nearly as much height was required below the operating floor for the pumps as was needed above for the engine.

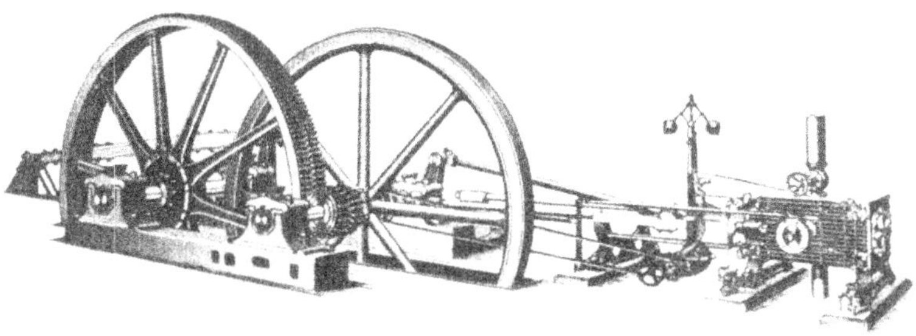

Geared mine pumps were yet another style built by Edward P. Allis Company. Since single-acting Cornish pumps were the usual choice for mine service, the shrouded step-tooth pinion and gear arrangement permitted a fair rate of speed on the engine with the pump plungers moving at a slower speed. The top of the pump "bob" is shown projecting through the floor — it could be built to suit local requirements. Rods connected the "bob" with the pumps, and could be made any size or length. Girder or wrought iron frame engines could be furnished in condensing, non-condensing, or compound designs.

In 1893 Edward P. Allis Co. installed a pumping engine at the Bay View Works of Illinois Steel Co. at Bay View, Wis. This announcement in itself was probably not of immense importance, except that Allis guaranteed Illinois Steel that they would pump the same quantity of water with one-fourth as much coal as was consumed by the existing equipment. In other words, Allis guaranteed to pump with 12,500 pounds of coal, as much water as was then being pumped with 50,000 pounds of coal. The engine supplied was this Reynolds Corliss cross compound model with non-jacketed cylinders. Performance tests after the installation saw coal consumption drop to approximately 12% of the previous usage.

Another distinctive Reynolds Corliss pumping engine was this handsome example built for the city of Superior, Wis. Steam jacketed cylinders were used in this compound engine. When tested in 1893, the average steam pressure was 80 psi, with an average engine speed of 20 rpm. A common stroke of 36 inches was used. The high pressure cylinder carried a bore of 24 inches, compared to a 42 inch bore on the low pressure or compound cylinder. Since this was a condensing engine, a vacuum of 13.4 pounds was carried on the low pressure exhaust. Clumsy, and perhaps crude by today's engineering standards, these engines were nonetheless incredibly reliable. With them, many cities and communities attained their first reliable source of water.

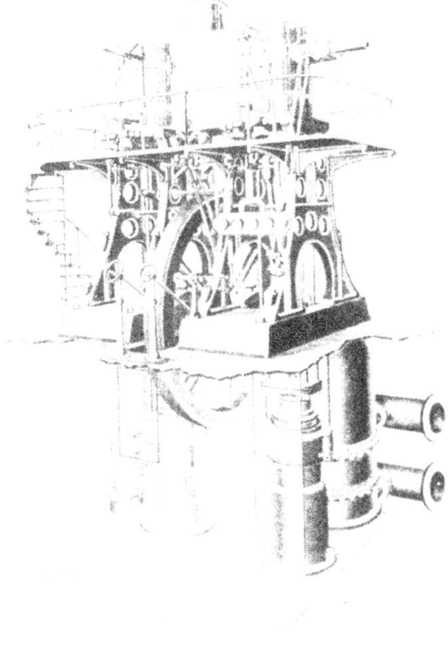

Steam Engines

By about 1910 the design of Allis-Chalmers horizontal cross compound pumping engines had stabilized to the point that few changes were manifested in ensuing years. This 1913 installation at Iowa City, Iowa, was probably responsible for a number of additional Iowa installations over the next ten years. One such engine, nearly identical in appearance to the one shown here, was installed at the Marshalltown (Iowa) Waterworks Plant in 1922. After operating there for nearly forty years, this engine was donated to the Midwest Old Settlers & Threshers Assn. at Mt. Pleasant, Iowa. Volunteers completely dismantled the engine and pump and reassembled the 130 ton behemoth within a museum building. This gigantic engine is probably one of the few Allis-Chalmers pumping engines in operation today.

An occasional contemporary report on the Chapin Pumping Engine usually illustrates the remains of this gargantuan machine at Iron Mountain, Mich. Seldom is the reader privileged to see a cross-sectional view of this engine. A man standing to the left of this engraving begins to give some idea of its size, but inspection reveals that the huge flywheel was 40 feet across! It alone weighed 164 tons. From ground level to the very top of the engine was 54 feet, and the total weight of the engine, exclusive of the pump work was over 600 tons. Designed to pump 3,000 gallons per minute from the mine depth of 1,500 feet, this engine used high and low pressure cylinders with respective diameters of 50 and 100 inches, together with a stroke of 120 inches. The speed could be varied from 4 to 10 revolutions per minute. Although the headframe and other parts were sacrificed to the scrap iron drives of World War Two, the engine remains today as a monument.

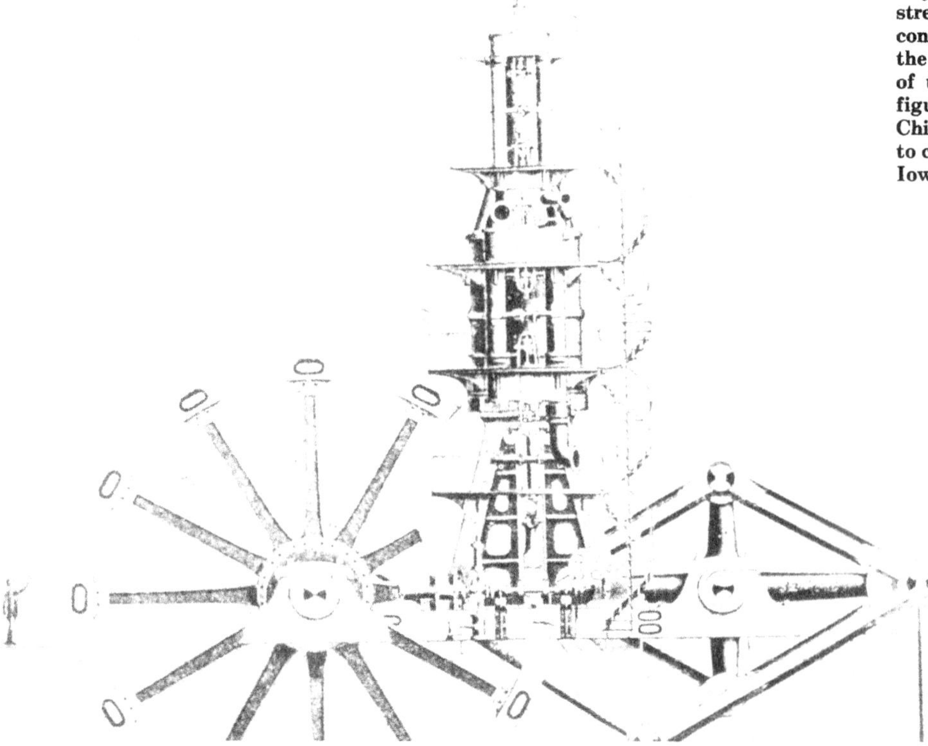

Photography at its best cannot fully describe the immense size of early pumping engines. By adding a "yardstick" to the illustration however, one can gain a better conception of these monstrous machines. In this case, a man is standing on the operating gallery at the left side of this illustration. This particular installation was made by Allis-Chalmers for the city of Louisville, Ky. By the 1920s the city of Louisville had purchased enough Allis-Chalmers pumping engines to have a daily capacity of 4 billion gallons. The triple-expansion engine is situated at the top, with the pumps all located on the ground floor. This 1920 model was equipped with the largest water end ever built to work against a high head. The main valve decks were 99 inches in diameter, and the distance from the pump floor to the operating deck is approximately 50 feet. This particular engine had a capacity of 20 million gallons in 24 hours.

Although there were other manufacturers supplying pumping engines to domestic and overseas users, there can be no doubt that certain of the Reynolds Corliss designs were among the best of their day. Aside from the famous Reynolds-Corliss valve gear which greatly simplified and streamlined the engine, the pump designs of Irving Reynolds contributed greatly to the company's overall success. By the close of 1893 the company had already built nearly 100 of these huge engines in many different sizes and configurations. Besides large cities like Milwaukee, Boston, or Chicago, the Reynolds-Corliss pumping engines also went to cities like Butte, Mont., Circleville, Ohio, and Ottumwa, Iowa to name but a few.

Although Edwin Reynolds was granted some forty patents during his career, Patent No. 811,520 issued on January 30, 1906 covered one of his most illustrious developments. Filed already in August, 1900, the Reynolds patent for a horizontal-vertical engine as shown here represented a new epoch in steam engine design. Even though compact, these engines could be built to tremendous sizes. The high pressure cylinder is horizontally oriented, with the huge vertically mounted receiver being plainly visible, and communicating with the vertical low pressure cylinder. Note the operator at the lower right hand corner of the engraving.

Another view of Reynolds' horizontal-vertical engine design shows a unit with the Reynolds-Corliss valve gear in plain view. Although the valve gear used the so-called single eccentric design, separate eccentrics were used for each cylinder. In this particular instance the receiver between the high and low pressure cylinders is absent, but was undoubtedly mounted over the high pressure cylinder. Owing to the extremely long stroke of these engines, rotative speeds were necessarily quite low so as to keep the piston speed within acceptable limits.

Designed by Edwin Reynolds while riding on a train from Chicago to New York, the "Manhattan Engines" were the largest steam engines ever built. Beginning in 1900, Allis eventually installed seventeen of these engines for the New York Rapid Transit Subway System, also known as Interborough Rapid Transit Co. Proportions of these engines overwhelmed the imagination. Of the horizontal-vertical type, each engine used two high pressure cylinders with a 44 inch diameter, with the steam then passing to a pair of low pressure cylinders with a diameter of 88 inches. A sixty inch stroke was common to all of the engines.

An unusual overhead view of a Manhattan engine shows the huge generator placed between the engine cylinders. Its rotating field was 32 feet in diameter and sufficiently heavy that no additional flywheel was required. Nine of these engines were installed at the 59th Street Station, making this the largest and most powerful steam plant of the day. When all nine engines were in operation they required on the order of 1,000 tons of coal per day, and over 1.8 million gallons of water to be converted to steam at a pressure of 150-175 psi.

Steam Engines

A rare view illustrates the Manhattan engines under construction at West Allis. Various components, large and small, have virtually taken over this entire bay of the factory. Having previously acquired a reputation for a huge foundry and machine shop capacity, these engines were probably no more formidable to build than had been some previous projects. These engines epitomized the large capacity steam engine and were likewise a major highlight of Edwin Reynolds' career. Not until December of 1924 did Allis-Chalmers receive an order for a steam engine approaching the size of the Manhattan engines. However in that year, American Steel & Wire Co. ordered a huge Reynolds-Corliss engine for their Donora, Penn. plant. This engine was of the twin tandem compound Corliss type having two 44 inch and two 84 inch cylinders with a 66 inch stroke, an operating steam pressure of 200 psi, and a speed of 72 rpm. Under these circumstances this engine was rated at 9,000 horsepower. The flywheel alone weighed 290,000 pounds; the crankshaft was 28 inches in diameter by 48 feet long, and total weight of the engine was over 1.6 million pounds.

Walking full circle 'round some of the Allis-Chalmers steam engine line makes it plainly evident that the inventive genius of Edwin Reynolds and the business acumen of Edward P. Allis worked to boost the company's fortunes tremendously. Some 1925 correspondence with the company from Carlyle Milling Co. at Carlyle, Ill. indicated that they were operating a Reynolds-Corliss engine built in 1878. Since that time it had run daily with an average mill output of 200 barrels of flour per 12 hours. During the period from 1878 to 1925 it had powered the milling of some 2½ million barrels of flour. In this same period the piston had traveled some 822,000 miles, but the cylinder had never been rebored. A conservative estimate stated that the engine had run approximately 620 million revolutions, and all this with a total cost of repairs of $18.

Steam Turbines

Although Edwin Reynolds had pioneered significant changes in reciprocating steam engines, he was not nearly so enthusiastic about the steam turbine. Although the Allis-Chalmers Board of Directors had initiated efforts to gain some steam turbine manufacturing rights in 1902, Reynolds, the company's Chief Engineer, seemed quite content building the super-efficient Reynolds-Corliss steam engines.

Edwin Reynolds was appointed to the post of Consulting Engineer in 1903, being immediately succeeded by his nephew, Irving Reynolds. The nephew's stint as Chief Engineer was measured in weeks rather than years — upon leaving Allis-Chalmers in April, 1903, he went to work for the famous steam engine builder, William Tod & Company.

Some writers have it that Edwin Reynolds accepted the rather nondescript title of "consulting engineer" due to poor health, but in his book, *An Industrial Heritage*, Walter F. Peterson notes that, "When the elder Reynolds did not move rapidly enough to suit the board, it appointed him, early in 1903, to the new post of consulting engineer. Irving Reynolds was apparently no better liked by the board, and he left later the same year in April."

Walter F. Peterson's book also gains considerable credibility from the exhaustive company history compiled in 1951 by Alberta Johnson Price. Her unpublished manuscript, *From Millstones to Atom Smashers* in the Allis-Chalmers Archives is the result of extensive interviews, plus an exhaustive study of company records, many of which no longer exist.

From the work of these two writers in particular, it seems evident that the newly formed company was quickly split into warring factions, and much of the contention rested on whether to get into the steam turbine business.

In February, 1904, the Executive Committee named Edward Dean Adams as its chairman. Adams was determined to bring in Benjamin H. Warren as president of the company, replacing Charles Allis, who represented the views of the Reynolds group. Thus, for the next few years, Allis-Chalmers suffered an almost never-ending internal battle between the "Mid-Western" and the "East Coast" factions.

Steam Turbines

The E. D. Adams plan included joining the Steam Turbine Advisory Syndicate of England, consisting of: Yarrow Shipbuilding Co., Tweedie (Vulcan) Shipbuilding Co., Willans & Robinson Engineers, and a former Parsons engineer, H. F. Fullagar. By obtaining a license from the Syndicate in 1904, Allis-Chalmers gained exclusive rights to manufacture steam turbines in the United States, Canada, and Mexico, with equal rights for South America. The following year American rights to certain Parsons patents were acquired.

The first Allis-Chalmers turbo-generator was installed during 1905 at Utica, N.Y. This turbine was built by Willans & Robinson of England, with the generator being built by Bullock. Sometime earlier, Sir Charles A. Parsons had worked at Willians & Robinson, but when he left the company, his steam turbine patents remained with them. Later however, Parsons recovered his earlier patents, and subsequently went on to acquire many others. A biography of Sir Charles could easily make an entire book on its own.

The first Allis-Chalmers steam turbine built at West Allis Works was a 5,500 kilowatt unit for the Williamsburg Plant of Brooklyn Rapid Transit Co. in 1905. Of the Parsons horizontal style, it operated at 750 rpm. The following year, 1906, saw the Reynolds-Corliss reciprocating engines reach their zenith, with A-C building more total horsepower that year than any other, before or after. After 1906 the steam turbine made steady inroads, with steam engine production ending entirely in 1930.

Although Allis-Chalmers embraced the Parsons design by 1905, actual experimental work at West Allis began as early as 1901 with a single-wheel radial turbine. J. F. Max Patitz of the Engineering Department then went to Escher-Wyss of Zurich, Switzerland, to study a new Zoelly impulse turbine of the Rateau type. After getting to Zurich it was discovered that the design was more akin to a Pelton style.

After returning to West Allis, Patitz proceeded to develop a six-stage impulse machine that proved to be quite efficient. This design was protected under U.S. Patent No. 971,555 of 1911. Following this success, Patitz began work on a 16-stage turbine, but meanwhile the company secured manufacturing rights from the Steam Turbine Advisory Syndicate, previously referred to. For the time at least, the Patitz design took second place behind the newly acquired manufacturing rights to the Parsons turbine. Eventually, many of the Patitz improvements found their way into practice.

Although the company was plagued with internal strife until about 1911, slow but steady progress was made in the development of Allis-Chalmers steam turbine-generators. To be sure, General Electric and Westinghouse took a large portion of the developing market, but Allis-Chalmers maintained a relatively secure position for several decades. By 1962 however, Allis-Chalmers only accounted for about 10% of the turbine-generator market, hardly enough to support the huge plant facilities at West Allis. With a trend to fewer and larger turbine-generators, the total United States market for 1960 was barely forty units per year, and 10% of this figure meant that only about forty units per year were leaving a plant large enough to produce several times as many turbines. Under these circumstances, Robert S. Stevenson, the president of Allis-Chalmers announced in December, 1962, that Allis-Chalmers was bowing out of the turbine-generator business except for servicing units already installed. Subsequently, in 1968, the company once again entered the field using steam turbines and related equipment built by Siemens AG of Germany. In 1969 Allis-Chalmers and Kraftwerk Union of Germany formed a joint venture. The latter owned Siemens AG and AEG Telefunken, both of which built turbine-generators and related equipment.

David C. Scott, the president of Allis-Chalmers engineered the joint venture with Kraftwerk Union, and this resulted in a jointly owned engineering and marketing company known as Allis-Chalmers Power Systems Inc. The role of this company was to market and service turbine-generator systems across the United States. Through this arrangement, Allis-Chalmers was actually able to re-enter the turbine manufacturing business by 1971, using the Siemens designs.

Plagued with a fickle market for turbine-generator units and considerable losses therefrom, Allis-Chalmers divested itself from the overhead and underground transformer business in 1975. Subsequently, Scott negotiated an agreement with Siemens AG for technical assistance, manufacturing and sales rights under a license from Siemens. Despite guarded optimism over this venture, it too met with an uncertain fate.

Allis-Chalmers commenced building steam turbines in 1905, with the first machine being installed at the Williamsburg Station of the Brooklyn Rapid Transit Co. While the major competitors, General Electric and Westinghouse, were having excellent sales to newly formed electric utility companies, Allis-Chalmers seems to have pursued industrial users very aggressively. Possibly this was because their previous sales base was oriented toward industry rather than toward public utilities. The turbine-generator unit shown here is of 1915 vintage and was installed in a plant of the Weyerhaeuser Lumber Co.

Steam Turbines

Allis-Chalmers used the Parsons turbine design initially, as did Westinghouse. General Electric Company used the Curtis design. In the Parsons steam turbine, expansion of the steam occurs in both the moving and the stationary blades throughout the machine. Therefore a pressure difference exists between each row of blades and the succeeding row. In the Curtis design, expansion occurs only in the stationary nozzles, so that the pressure in any given stage is constant at all points. Each design had certain advantages and disadvantages, but both designs were sold with considerable success. A 1915 photograph shows two Allis-Chalmers turbine-generator units installed for an unnamed steel company.

Turbo blowers became a major part of company operations during the early 1930s, and resulted in the formation of the Blower & Compressor Department in 1935. During 1937 the first blast furnance tubo-blower was built by Allis-Chalmers. Destined for a 500 ton blast furnance at the Hamilton Coke & Iron Division of American Rolling Mill Co. of Hamilton, Ohio, it featured a four-stage turbo-blower. Rated at 40,000 cfm at 30 psig, this unit operated at 3,600 rpm and was powered by an A-C condensing turbine of the reaction type, and rated at 4,600 horsepower. Units like this quickly replaced the gargantuan blowing engines illustrated in the *Steam Engines* section of this volume.

Allis-Chalmers sold 24 steam turbine generating units in 1937. Included was this 30,000 kw machine at Chicago's Fisk Station plant. This unit was of 25-cycle design. A-C built its first 25-cycle steam turbine unit in 1907 for the Illinois Steel Co. plant at South Chicago. Subsequently Allis-Chalmers built a considerable number of 25-cycle machines specifically for steel mill service. In 1925 the company reported that two similar units had been installed for the Calumet & Hecla Mining Co., one in 1913 and another in 1918. Both of these machines had operated with a repair bill averaging less than $150 per year over a seven-year period. One machine had run for six years without ever being opened for inspection.

This 1939 turbine-generator had assumed the general form that would characterize A-C machines for the next two decades. By 1937 all turbines above 10,000 kw in size were equipped with a motor-operated turning gear. This device was used to start the shafts rolling from a standstill, and was also used to keep the shaft turning when the machine was shut down. A high pressure oiling system was developed which could actually float the shafts to insure an oil film in the bearings during starting and stopping operations. This standard equipment was extremely important to prevent a breakdown of the oil film from the heat transmitted along the turbine shaft.

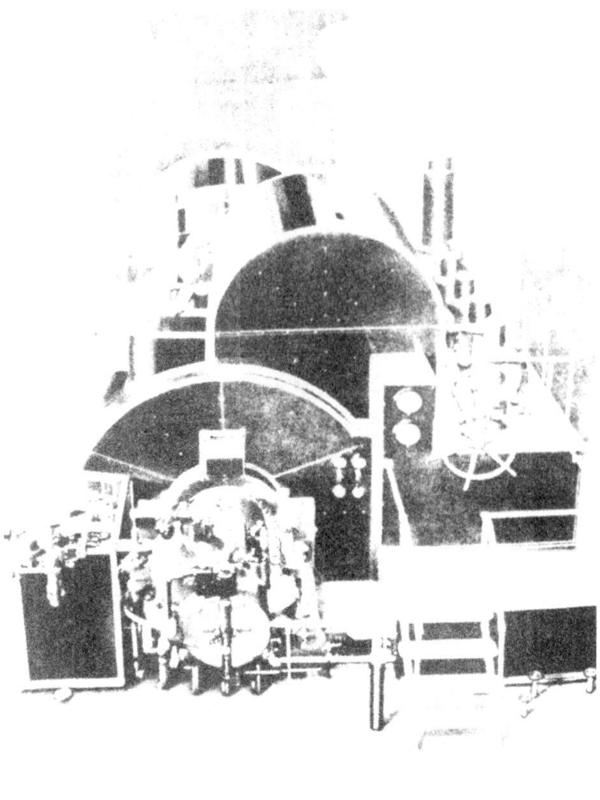

Steam Turbines

Throttle pressures for steam turbines ranged as low as 60 psi in 1900. A quarter century later, 350 psig and a 700 degree superheat seemed to be the general rule, with developmental work looking at initial pressures of 500 to 600 psi, steam temperatures of 700 degrees F., and reheating at a pressure of 120 psi after expansion from 550 pounds, back again to 700 degrees. The second expansion was completed to a vacuum slightly over 29 inches. Within ten years the industry would be accepting initial pressures of 800 psi and a throttle temperature of 850 degrees F. A few machines were even proposed with an initial pressure of 1,200 psi.

During the late 1930s Allis-Chalmers installed this 80,000 kilowatt steam turbine-generator in the Port Washington Station of Wisconsin Electric Power Co. The alternator was designed for an output of 22 kilovolts. Boiler pressure was 1290 psig, with a throttle inlet temperature of 850 degrees F. After the high pressure stage, the steam was returned to the reheat section of the boiler, where it once again gained a temperature of 850 degrees. After converting this energy to work within the low pressure section of the turbine, the steam was exhausted into the condenser at 29½ inches of vacuum. Since total numbers of these units were very small, design changes meant huge additional costs for new drawings, new or modified patterns, and additional machine shop work.

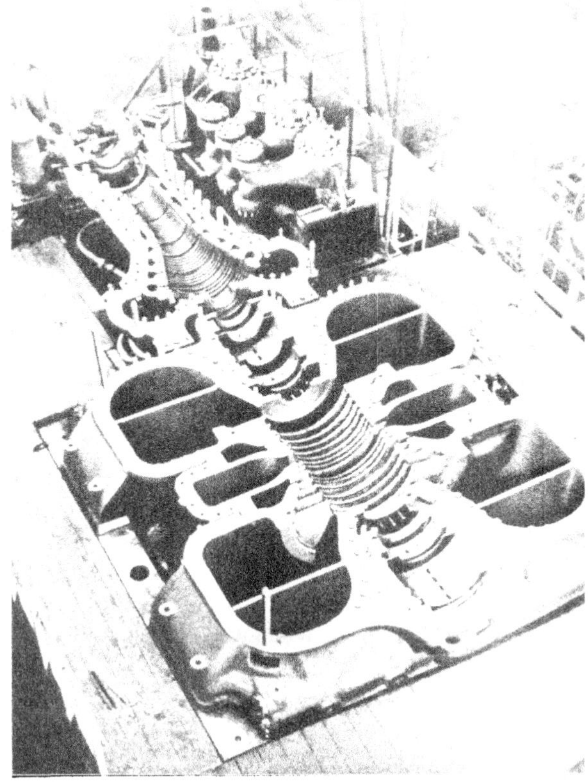

A 1938 photograph illustrates the bottom half of a turbine case during erection. The same scene would also appear in case of a major breakdown, or at the time of an overhaul program. The workman at the upper right is working on the oil pump section of the turbine. Immediately below the workman is the high pressure stage of the turbine. Massive bearings and seals intervene, and in the foreground is the low pressure stage of the unit. Steam initially enters the high pressure stage through throttle valves, and after expending energy through the first stage, the steam returns to a reheat area in the boiler, coming back again to the much larger low pressure stage. From there the steam exhausts to a massive condenser which it reverts back to water and is again pumped into the boiler for a repeat of the cycle.

A 1943 photograph illustrates an A-C auxiliary turbine-generator unit with a capacity of 300 kw. The company noted that large quantities of these machines were being built following a standardization of design. Presumably, these machines were intended for the war effort, with a major portion probably being used on ships. Throughout the years, Allis-Chalmers assumed a progressive stance in turbine design, always looking for new and better alloys that would better withstand the severe duty imposed by high-pressure and high-temperature steam. In this the company was eminently successful.

Steam Turbines

In addition to steam turbines for power house duty, turbo-blowers, and other applications, Allis-Chalmers also built ship propulsion turbines for merchant and military uses. During World War Two, some merchant ships were converted to fighting vessels, and oftentimes, these were furnished with A-C propulsion turbines. No data has been located to indicate the full extent of A-C involvement in this phase of war production, but there can be no doubt whatever, that with the company's patriotic stance, any required assistance was rendered without delay.

As might have been expected, virtually all new turbine-generator construction during World War Two was limited to those instances where the war industries were directly affected. In fact, the scarcity of new materials during this period left some rural areas without electric utility service until after hostilities ended in 1945. This huge 147,000 kilowatt turbine generator unit was illustrated in a 1944 Allis-Chalmers publication, with the company noting that it was the "largest turbo-generator since Pearl Harbor, [and is] serving midwest war industries."

A 1945 photograph from the A-C files illustrates an industrial application for steam turbine units ranging from 1000 to 10,000 kilowatts, with A-C indicating that a number of these units were being built for war industries. This particular turbine was installed in a large chemical plant. Rated at 6,000 kw, it operated at 3,600 rpm. Prior to, and during the World War Two period, Allis-Chalmers gave extensive promotion to the benefits derived from proper treatment of the boiler water. Through proper treatment, turbine blade corrosion could be drastically reduced, or practically eliminated. Allis-Chalmers promoted this feedwater system under the "AKON" trademark.

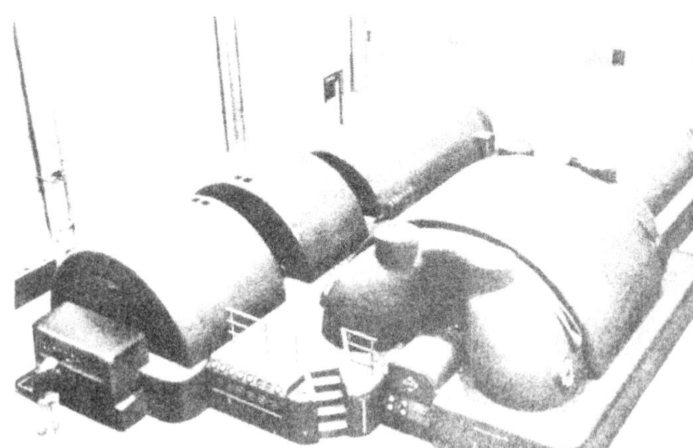

During 1953, the Wisconsin Electric Power Co. put this huge 120,000 kilowatt unit on line. Designed by Allis-Chalmers to have the world's lowest all-year vacuum, 0.5 in Hg abs., this close-coupled, cross-compound reheat steam turbine-generator utilized both a 3600 rpm element, and an 1800 rpm element. An 85,000 square foot Allis-Chalmers condenser was required to handle the exhaust steam. Despite some major installations such as this one, Allis-Chalmers was not acquiring enough orders to keep the huge West Allis Works operating at full capacity.

Steam Turbines

A 1956 photograph illustrates Unit No. 3 at the Oak Creek Station of Wisconsin Electric Power Co. Units 1 and 2, shown in the background, went into service during 1953 and 1954, respectively. Each of these three units was capable of delivering 120,000 kilowatts, for a total of 360,000 kw from the three machines. Despite the huge size of these machines, extremely close tolerances are required, with one example being the spacing between the stationary and the revolving blades of the turbine. To machine these huge components within critical tolerances required a huge shop, huge machinery, and talented personnel.

During 1957 Allis-Chalmers reported that construction was nearing completion on a huge 321 megawatt unit. This photograph illustrates the generator end of the 175 mw (megawatt), 3,600 rpm section. The remaining 146 mw came from the low-pressure section of this close-coupled, cross-compound unit. Building these large units required immense forgings. Already in 1929 the company reported that a 115 mw unit for the Waukegan Station of the Public Service Co. of Northern Illinois would require a forging 42 feet in length and weighing 269,000 pounds! At that point in time, Allis-Chalmers considered this to be the largest forging ever made.

The turbine test pit at West Allis Works was designed for actual test operation of newly built machines. After successful completion of the tests, the machine was then dismantled for shipment. Shown here is the bottom case and rotor for a 300 megawatt (300,000 kilowatts) unit being built for Detroit Edison Company. A close inspection of the photograph shows a massive series of bolts installed in the bottom half of the case in the foreground, while the case to the back does not yet have these installed. This myriad of huge bolts serves to hold the two halves of the turbine case together.

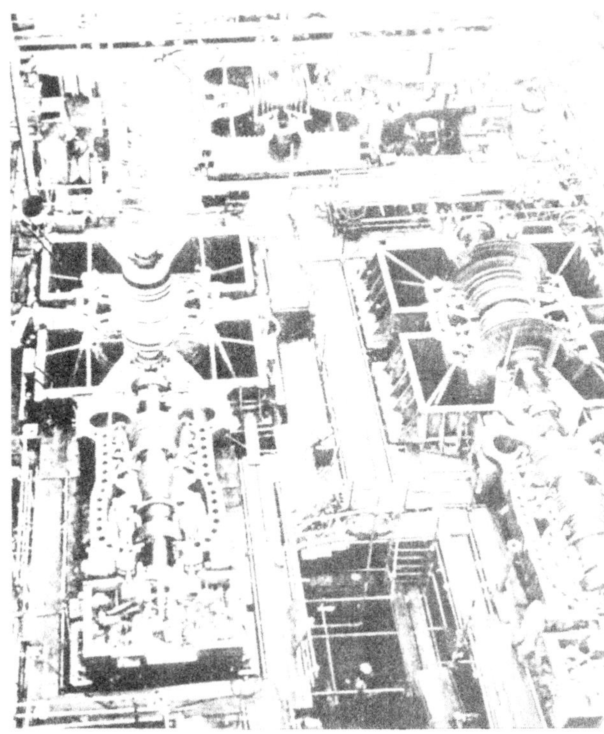

These two units of 100,000 and 150,000 kilowatt capacity are shown on the test floor at West Allis Works in 1958. At this point in time, Allis-Chalmers had units under construction for five different states. The two units in the foreground are reheat-tandem machines, with the unit on the right illustrating the smaller high-pressure blading in the foreground. After leaving this section the steam returns to the reheat section of the boiler and comes back to the second stage, obvious with its much larger blades. Operating the test floor alone represented an immense cost that was of course factored into the bidding process.

Steam Turbines

During 1963, Allis-Chalmers announced that Consolidated Edison Company had ordered the "world's first MILLION-KILOWATT unit... big enough to serve 3,000,000 people." This announcement does not seem to square with a December, 1962, announcement that Allis-Chalmers was leaving the steam turbine-generator business except for servicing units already built. The only logical assumption that current research can draw is that the 1-million kilowatt unit was in the works prior to a withdrawal from the turbine business. Despite any apparent conflict in dates, Allis-Chalmers bowed out of the turbine business for several years, returning again in 1968.

This huge installation at Commonwealth Edison's Fisk Station plant was built at West Allis. As previously indicated, A-C left the steam turbine business in 1962, only to return in 1968 as an installation and service agent for the German-built turbines of Siemens AG. In 1971 A-C once again began building turbines at West Allis, using the designs and engineering knowledge obtained from their joint venture with Kraftwerk Union of German. By 1974 the situation had changed to the point that actual manufacturing was once again transferred to Kraftwerk Union. And this time the indomitable giant could not recover — Allis-Chalmers was no longer building steam turbines at West Allis.

Stephenson-Adamson Co.

In 1974 Allis-Chalmers purchased Stephens-Adamson Co. from Borg-Warner Corp. This Canadian affiliate had long been identified with materials handling equipment. Of particular interest is the company's 1978 announcement of completion of a huge coal handling system at Thunder Bay Terminals in Ontario, Canada. The project included a 6000 ton per hour bucketwheel reclaimer as shown here. Valued at approximately $8.5 million, the reclaimer was part of a system that also included a $3.6 million belt conveyor and stockpiling system for coal at the Thunder Bay location.

Subsoilers

This photograph illustrates an Allis-Chalmers HD-6 tractor with a tool bar designed specifically for this tractor model. Although the tool bar could accommodate a wide range of implements, this particular unit is equipped with subsoiler shanks. Production of heavy tillage equipment was centered at the company's Oxnard (California) Works acquired from the Brenneis Co. Since subsoiler units are scattered throughout various sections of this volume, continued coverage of this rather specialized implement would be redundant.

Sugar Beet Harvesters

For a time in the mid-1960s, Allis-Chalmers offered several sugar beet harvesters under their own name and trademark, even though there are indications that A-C did not actually manufacture these machines. Typical of the line was the Model 150 beet harvester shown here. This tractor-drawn unit was equipped with a 4,000 pound holding tank, and harvested one row at a time. The machine shown here was priced at approximately $5,500, with this figure depending substantially on optional attachments. One, two, and three-row models were offered until about 1970.

Terra Tiger

In early January, 1968, Allis-Chalmers announced the Terra Tiger, a six-wheel-drive unit with a water-tight fiberglass body. Built for sports, recreational, commercial, and military markets, the Terra Tiger was intended to be a "go-anywhere" vehicle. Production was to be centered at the Lexington Works, a new facility announced in February, 1968. This plant was intended to be the center of production for A-C lawn and garden lines, in addition to Terra Tiger production. In this photograph, A-C executives are looking over the new Terra Tiger.

Allis-Chalmers advertising of 1968 hailed the new Terra Tiger as "A new concept in off-the-road vehicles!" Weighing but 400 pounds, the Terra Tiger utilized a leakproof fiberglass body. Specifications for this machine indicated the hull to be waterproof with 11 to 18 inches freeboard. Seating was provided for two passengers with 11 cubic feet of area available for storage or payload. With only a 44 inch wheelbase, the Terra Tiger had a turning radius of under four feet. Ground pressure was only 0.96 psi at zero penetration.

Allis-Chalmers intended the Terra Tiger for several markets, including commercial applications. One such instance is shown here with a construction foreman using a Terra Tiger on a large project. Since it was ostensibly an all-terrain-vehicle, the Terra Tiger simplified access to areas otherwise difficult to negotiate, except perhaps, on foot. The headlight was standard equipment that permitted 24-hour operation.

Allis-Chalmers claimed their new Terra Tiger with its 10 horsepower engine could carry a 600 pound load and negotiate a 45 degree slope. The water-tight fiberglass body made the Terra Tiger useful in swampy areas, and in fact this one is in use by a hunter for placing decoys. The Terra Tiger was also advertised as an ideal vehicle on beaches, and for other recreational uses. Presumably it was demonstrated for the military as well, but no information has been located to indicate whether American military forces used it to any extent.

Terra Tiger

The Terra Tiger was equipped with a 10 horsepower air cooled engine. It powered a torque converter with automatic variable speed control. No suspension was required due to the use of super soft, obstacle-absorbing tires. A single T-bar lever controlled steering, throttle, and brakes. Special high flotation Terra Tiger tires were used. These 11 x 20 tubeless tires carried a 2 lb. air pressure. Fuel capacity was 4½ gallons, and a positive 6-wheel drive was featured. Speeds up to 25 mph were possible on land, with a 4 mph maximum on water.

To facilitate moving the Terra Tiger from place to place, Allis-Chalmers offered a trailer designed especially for this machine. Weighing only 300 pounds, the trailer included lights, tie-down chains, and other requirements for simple loading and unloading. Unfortunately, the Terra Tiger venture did not meet with the hoped-for success that the company desired. Indications are that sales of the Terra Tiger ended in 1971, although actual manufacturing may have ended even sooner. Had the company developed this machine even five years later, it might have met with greater success. No production figures on this unit have been located.

Texrope Drivers

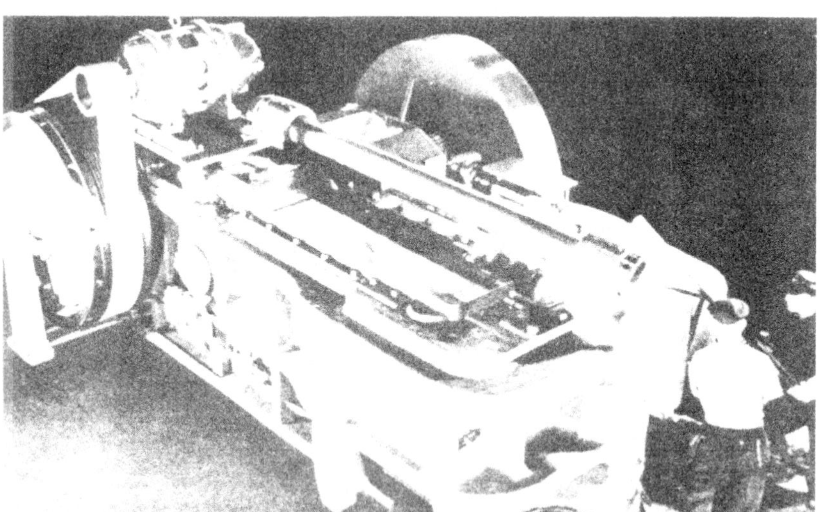

Originally intended to obviate problems with textile mill drives, the Texrope Drive was placed on the market in March, 1925. Developed by Walter Geist, who would later become the president of A-C, the Texrope Drive was given patent protection under No. 1,662,511 of March 13, 1928. This gave Allis-Chalmers exclusive rights to the multiple v-belt system for seventeen years. During that period, several manufacturers obtained a manufacturing license from Allis-Chalmers. With the Texrope Drive came a speedy end to the troublesome flat belts found in virtually every American factory. This huge 190,000 pound horizontal forging machine is being operated by a 60 horsepower Texrope Drive.

The Texrope Drive was an immediate success in the textile industry, and was showing great promise for other industries as well. Not until April, 1926, however, did Allis-Chalmers officially announce the potential for Texrope Drives in virtually every industry. So quickly did the Texrope concept take hold, that within five years after being announced this drive system had seen over 75,000 installations. During 1931 the Texteel line of pressed steel sheaves was introduced — these were followed in 1934 with the Duro-Brace Texteel sheave. Texrope Vari-Pitch Speed Changers were announced in 1938. The immense success of the Texrope Multiple Drive system played a major role in keeping the company afloat during the otherwise lean years of the 1930s.

Threshers

Production of Allis-Chalmers steel threshers was centered at LaPorte Works, and continued to follow the same basic lines as were used by Advance-Rumely Thresher Co. A-C took over the Rumely plant in 1931, continuing to market the 22 x 36 machine during 1932 and 1933. Only 108 of these machines were built during that time. Some 89 of the 28 x 46 steel threshers were produced during the same 1932-33 period, and as with the smaller 22 x 36 inch model, these were probably built from the parts stock on hand. Production records also show that Allis-Chalmers built 9 of the 28 x 50 threshers in 1933, followed by another 160 units in 1935. After that time, production ceased for this model.

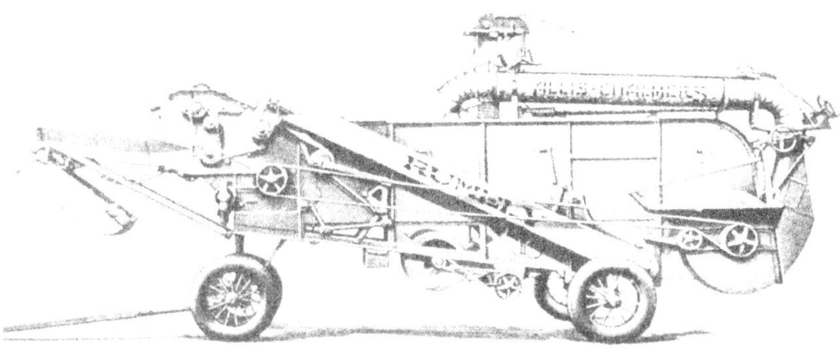

Designed for use with a 2-plow tractor, the Allis-Chalmers-Rumely 22 x 38 inch thresher appears to have been the only model built as a distinctively A-C machine. Spring-mounted on front and rear axles, the 22 x 38 was available with steel wheels, or as shown here, with rubber tires. A total of 32 high-grade ball and roller bearings were used, making the 22 x 38 an easy pulling machine. Production for 1932 totaled only two machines, peaking in 1936, the last year of production, with 499. During this period the 22 x 38 saw a total run of 709 units. Since Allis-Chalmers was heavily involved in promotion of their All-Crop combine at the time, continued production of threshers was probably done as an accommodation to their customers, and as a means of using up the parts inventory acquired from Rumely, rather than as an attempt to pursue a longstanding sales effort for these machines.

Tool Bar Equipment

The Model 600 Tool Carrier shown here behind an HD-11-B Allis-Chalmers crawler listed at $1,429 in 1967. It was one of several styles available for use with large crawler tractors, and could be equipped with the choice of a 10-foot or a 15-foot bar. This 6 x 12 inch steel bar added $289 or $370 respectively to the list price. Hydraulic control equipment added yet another $283. Subsoilers, chisel cultivators, and coil shank cultivators were all available at extra cost for use with the Model 600 Tool Carrier.

Tool Bar Equipment

The D-19 tractor with Snap-Coupler hitch was required for this tool bar frame and carrier assembly. Listing at $129 for the bare tool bar, the extra attachments raised the total price considerably. A 4½ x 8 inch tool bar added $82 to $245, depending on its length. Gauge wheels as shown with this unit, raised the price by another $260, and subsoiler shanks listed for $78 each in 1967. Pricing the components individually had the distinct advantage of allowing the purchaser to buy this unit tailor made to his needs instead of buying a package unit that included expensive attachments which would never be used.

A photograph taken near Eaton, Col., shows an Allis-Chalmers D-15 tractor equipped with a tool bar with three ditcher shanks attached. This outfit is ditching potato rows for irrigation. The tool bar system permitted the operator to set the shanks for specific row spacings, and also allowed the use of the tool bar with several different implements. Somewhat larger ditcher shanks than shown here were available for use with crawler tractors.

A photograph taken at the Allis-Chalmers Farm near Racine, Wis., illustrates a D-15 tractor equipped with a single-shank subsoiler. These units have been, and remain, quite popular in areas where the soil has a tendency to form a hardpan that resists free downward percolation of water. The use of a subsoiler tends to loosen and aerate the soil. By building units like the one shown here, even the small farmer could benefit from subsoiling on his own farm, without the need to hire a commercial operator for the job.

A-C toolbars could easily be equipped with various implements, and in this case a D-15 tractor is shown working with a springtooth harrow, rejuvenating an alfalfa field near Eaton, Colorado. Lister beams, and bedding discs were also available for use with A-C wheel-type tractors. These tools are indispensable in certain agricultural regions. A 1967 Allis-Chalmers price book indicates that at least half a dozen different ground working tools were available for use with this bar — included were alfalfa points, furrowers, and Alabama sweeps.

Tractor Tillers

On April 8, 1912, Delmar W. Call and Otto H. Falk were appointed as receivers of Allis-Chalmers. Company affairs continued to worsen after the financial panic of 1907, with various historians pointing to a shortage of corporate direction as a contributing factor. Mr. Call soon resigned, leaving General Falk in command. Falk recognized the potential for the farm tractor, and in less than a year a manufacturing license had been secured for this tractor-tiller of Swiss design.

Motoculture Ltd. of Basel, Switzerland, had developed this rotary tiller in nearby Austria. On December 31, 1913, the A-C Executive Committee, ostensibly at the request of General Falk, authorized a contract for payment of $10,000 cash, plus a 6% royalty on all rotary plows sold in the United States. Allis-Chalmers tested this machine in Wisconsin, Virginia, and Colorado during 1914 and early 1915. This rear view of the tractor-tiller illustrates the spring steel retarder fingers mounted directly behind the digger tines.

An extremely rare view of the earliest tractor-tiller design shows it working at an undisclosed location. Although the rotary tiller design was probably ideal for truck gardening, it seemed to gain virtually no acceptance on the ordinary grain and livestock farm of 1915. Absolutely no specifications, dimensions, or any other data pertaining to this machine has been located.

Although the tractor-tiller has generally been considered as a single-duty machine, capable only of tilling the soil, this rare photograph shows that the tractor-tiller was indeed equipped with a belt pulley, making it available for other duties as well. Quite possibly the engine was crossmounted on the structural steel frame to minimize the need for troublesome bevel gear drives to the tiller mechanism. Unfortunately, no data whatever has been located to indicate the type of engine used, its horsepower output, or any other specifics.

Tractor Tillers

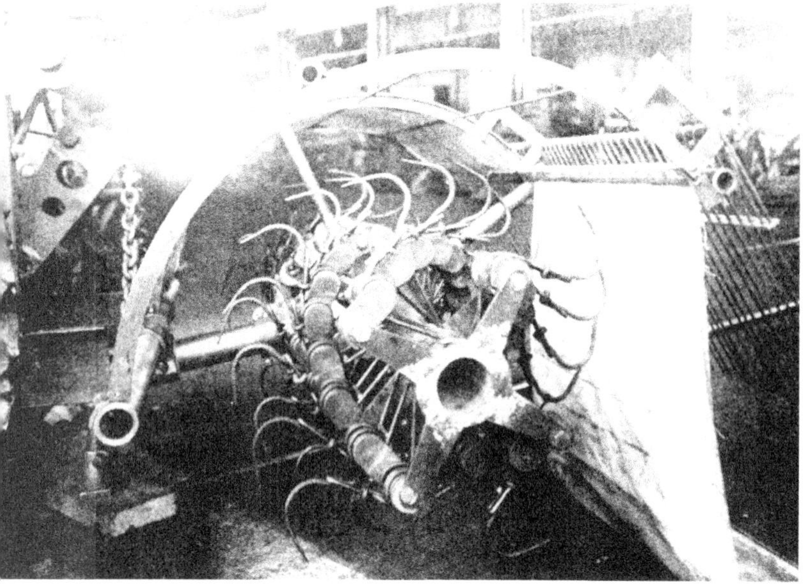

Yet another rare photograph illustrates the business end of the Allis-Chalmers tractor-tiller. A mechanical lifting device is partially shown, but once again, in insufficient detail so that any firm conclusions can be made regarding its construction. This entire series of extremely rare photographs was copied by the Author in the Tractor Engineering Department of Allis-Chalmers during a visit in February, 1985. Most of these photographs have probably never been published prior to this instance.

Despite the unimpressive showing of the Swiss-designed tiller-tractor, work nevertheless continued on improved versions. This activity seems to have had its major impetus with Johann F. M. (Max) Patitz. Better known for his contributions to hydraulic and steam turbines, Patitz nevertheless filed patent applications in May, 1918, that eventually resulted in Patent No. 1,476,917 in 1923 and No. 1,504,495 in 1924. Both of these patents unmistakeably cover the rotary tiller design shown on this "25-45" tractor. The original photograph gives the tractor this rating, despite the fact that the largest model for 1918 was rated as an 18-30 model. Chances are that the tractor-tiller of this photograph represents the only such unit ever built. The Patitz patent applications of 1918 give substance to the thought that Allis-Chalmers experimented with the tractor-tiller design several years longer than was previously supposed.

Tractor-Truck

In early 1915 Allis-Chalmers finished the first tractor-truck. Ostensibly designed for pulling plows and other drawbar loads in addition to its obvious purpose as a cargo carrier, the tractor-truck was intended to fill a vast potential market. Unfortunately however, the market never materialized. Perhaps it never existed. The farmers to whom this vehicle was targeted would not buy it. Probably the initial list price of $5,000 was too high, at a time when a T-model Ford sold for under $300. Another factor often overlooked is the simple fact that the Allis-Chalmers of 1915 had virtually no sales or distribution network aimed specifically at the agricultural markets.

In 1915 Allis-Chalmers assembled a fancy catalog entitled, "Some Interesting Pictures of the Allis-Chalmers Tractor-Truck." This photograph, probably taken at West Allis Works, illustrates the individual flexibility of the track members. The track in the foreground has mounted a large block, while the unit on the opposite side remains flat on the ground. Otto Falk envisioned a three-pronged approach with the rotary tiller, the tractor-truck, and a small tractor. Several other companies attempted to market a similar tractor-truck design, but like Allis-Chalmers, they met with a very limited success.

Tractor-Truck

In order to permit maximum utility of the tractor-truck, even on a rough or hilly terrain, the front axle, like the rear tracks, was designed with the utmost flexibility. This photograph illustrates the tractor-truck negotiating a steep grade. The cargo space is loaded with scrap iron to demonstrate this machine's capabilities even under severe conditions. C. Edwin Search, works manager at West Allis, was the first to propose the tractor-truck to General Falk, who immediately initiated plans for its manufacture. Work on the actual construction came under the control of Ray C. Newhouse. The latter individual was identified with an extensive number of improvements relating to rock crushers.

In order to prove that the tractor-truck was capable of negotiating very rough terrain, this photograph illustrates the machine with the front wheels just past the crest of a steep incline. The great flexibility of the track system permits the left hand track to grip the slope, even though the right hand track is not so steeply disposed. Much of the original design for the Allis-Chalmers tractor-truck appears to have been conceived by C. L. Tolles of Eau Claire, Wis. In fact, indications exist which show that A-C purchased a half-interest in the Tolles patents, along with the latter's "Centipede" tractor. No specific information has thus far been located on the Centipede as an individual entity.

Yet another view of the tractor-truck illustrates the machine ascending a steep embankment just after crossing a double set of railroad tracks. The earliest Tolles patent located thus far is No. 1,106,595 filed in November, 1912, and granted in August, 1914. Other Tolles patents involved with the tractor-truck include: 1,106,596 (1914); 1,135,559 & 1,135,560 (1915); and 1,146,537, also of 1915. After this time, Tolles appears to have turned to other inventive pursuits, including Patent No. 1,232,997 of 1916 for a turret tool post as used on a lathe.

Loaded with 9,200 pounds of scrap castings, the tractor-truck is shown here ascending a steep incline. Intended for just such duty, the tractor-truck failed to take hold. During 1915 it was demonstrated before representatives of the Allied military forces, but only Russia placed an order. They bought ten of the tractor-trucks at $5,500 each. After being assembled in Russia, they were never used. One additional unit, probably the original prototype, was used at West Allis Works for some years, but was finally scrapped.

Tractor-Truck

In order to demonstrate its pulling power, the tractor-truck is shown here moving a huge 30-ton, Model 30 Marion shovel. Granted, the move is being made on a hard and fairly level surface, but also to be remembered is the fact that steam shovels of 1915 were anything but lightweights. Curiously, Charles Edwin Search, the manager of West Allis Works, filed an application on November 29, 1915 that resulted in Patent No. 1,227,005 on May 22, 1917. By this time however, the tractor-truck project had become hopeless.

Another likely task for the tractor-truck was hauling trailers on road grading work. In his book, *An Industrial Heritage*, Walter F. Peterson notes that, "This prototype of the "half-track" of World War II was born nearly a generation too early." While this statement probably echoes a certain truth, the virtual absence of a marketing organization targeted to potential buyers was certainly a major factor. Allis-Chalmers, while well known in heavy industry, was not at all identified with agricultural equipment. However, the company's fame in ore and materials handling industries should have provided no less than a limited market for the tractor-truck. Obviously however, even this hope of establishing the tractor-truck as a viable profit maker proved to be disheartening.

Shown here in deep snow, the tractor-truck is hauling a load of huge gear castings at West Allis Works. Although the tractor-truck was built under the protection of several patents controlled wholly or in part by Allis-Chalmers, there are indications that manufacturing rights were secured on other patents owned by Monarch Tractor Co. of Watertown, Wis. To what extent these manufacturing rights extended is not presently known. Monarch would later move from Watertown to Springfield, Ill., and in 1928 Allis-Chalmers would make an outright purchase of the Monarch factory.

A close inspection of this photograph reveals that "Allis-Chalmers" has been painted on the sideboards of the cargo deck on the tractor-truck. General Falk, the president of Allis-Chalmers, along with some of his department managers were initially convinced that the tractor-truck could be marketed extensively. Despite their hopes, such was not the case. No data has been located indicating the make or model of engine used in the tractor-truck. Likewise, other data concerning its top speed, operating weight, and similar specifications cannot presently be located.

Tractors

Although Allis-Chalmers had by 1913 become a major builder of flour mills, sawmills, steam turbines, and countless other machines, the company was in fact, at a considerable disadvantage. Until General Otto Falk took over as President of Allis-Chalmers in 1913, Allis-Chalmers was in reality, a loosely organized firm with a multitude of manufacturing operations and little in the way of firm direction. Several historians, including Loretta Price and Walter F. Peterson are in agreement that prior to General Falk, A-C management spent much of its time arguing about almost everything, and this to the great detriment of company affairs.

General Falk was born into the famous Franz Falk brewing family and his interest in the military led him to Allen Military Academy of Chicago. By 1895 Falk has assumed a promotion to adjutant general for the Wisconsin National Guard. When he retired from the Wisconsin National Guard in 1911 it was with the rank of brigadier general. Then in 1912 Falk was appointed as one of the receivers for the bankrupt Allis-Chalmers Co.

Falk quickly saw a major problem and moved to correct it. Allis-Chalmers, while highly diversified, had that diversity almost entirely within the heavy manufacturing industries. Economic factors that affected one portion of the industry would almost inevitably affect the remainder. To Falk's mind, diversification into entirely different industries was required, and thus began the long odyssey of the Allis-Chalmers tractor. Failure followed failure initially as the company went through the rotary tiller design of the Swiss firm of Motoculture Ltd. Then came the equally unimpressive Tractor-Truck.

More failures were due — Allis-Chalmers then came along with the 10-18 tractor which never endeared itself to American farmers. The 6-12 which followed was little better. Absolutely determined to find a successful tractor design, A-C engineers finally developed a prototype of a more-or-less conventional farm tractor design — the 15-30.

The 15-30 was an almost immediate success, but sales were not as high as might have been hoped, the reason being that A-C had no dealer organization. Sawmills, steam turbines, and gigantic rock crushers were sold by company representatives directly to the end user. Once again, the company infrastructure was demonstrated to be hidebound. In order to sell tractors it would be necessary to build a sales organization, complete with branch houses and dealer outlets. The big break did not come until the 1931 acquisition of Advance-Rumely Thresher Co. Acquiring Rumely came at almost the perfect time, coinciding very closely with the Model U and Model UC tractors, and coming just before Allis-Chalmers pioneered the use of rubber tires on farm tractors.

A major advantage of the Rumely acquisition was that while it did of course give Allis-Chalmers a ready-made dealer and branch house organization, the house of Rumely had impeccable credentials with farmers generally. In those days of often intense dealer loyalty by the farmer, maintaining an identification with Rumely was a most positive factor for Allis-Chalmers.

Of all the capable managers in the Tractor Division, there are probably none who achieved more fame than Harry Merritt. His 1926 appointment as Manager of the Tractor Division marked the beginning of a new epoch in the A-C adventure into the tractor business. Merritt's enthusiasm for tractors was unbounded — he quickly gained the reputation of "disrupting more weekends than any other Allis-Chalmers man of the time." Although Merritt was ostensibly hired by General Falk to close out the Tractor Department after ten years of continual losses, Merritt's price reduction of the 20-35 from $1,950 to $1,295 proved that the market was there but Allis-Chalmers was pricing its goods too high.

Merritt and his engineers took a 20-35 tractor completely apart and spread it out on the floor. Each part was studied carefully, and wherever the part could be lightened or remodeled to weigh less and cost less, this was done. By the time this process was completed, Allis-Chalmers was in the market with a $1,295 tractor, and able to make money at that price.

Harry Merritt was absolutely convinced that in order for a tractor to get the farmer's attention it first had to be attractive. Allis-Chalmers had been using a dark green enamel with red wheels. In the book, *An Industrial Heritage*, Walter F. Peterson reports that in 1929 Merritt saw "acres of brilliant orange wild poppies" in California. These were visible for miles, so Merritt brought some back to Milwaukee and asked experts at Pittsburgh Plate Glass Co. to duplicate the color. This was the origin of "Persian Orange" as used on Allis-Chalmers tractors. Peterson's research also indicates that the United farm tractors were the first to use the orange paint. Presumably this also included the identical Model U tractor. Various inquiries by the Author have established the following as comparable colors to original Allis-Chalmers finishes:

Allis-Chalmers Green — Martin-Senour 90T-11511

Allis-Chalmers Red (for wheels — Martin-Senour 99N-4359

Allis-Chalmers Orange (Through 1960) — Martin-Senour 90T-3723

Allis-Chalmers Orange (1961 & Later) — American Parts 1204-6

While the above numbers are thought to be comparable color matches for the original, this information is presented solely for the information of the reader and no guarantees, express or implied are made in this regard, either by the Author or by the Publisher.

The 1928 purchase of Monarch Tractors instantly created a new manufacturing facility called Springfield Works. The perfection of crawler tractor designs put Allis-Chalmers solidly into this field with the Monarch acquisition. In 1929 A-C bought LaCrosse Implement

Co. and this too proved to be a valuable asset, since LaCrosse Works had the expertise to render valuable assistance in developing tractor-mounted implements. The former Advance-Rumely plant became LaPorte Works, and from this factory Allis-Chalmers would begin the production of the All-Crop Harvester in the 1930s. Of all visible catalysts which brought an end to the grain binder and thresher, the All-Crop Harvester was probably one of the most significant factors. Combines were on the market prior to this time, but the lightweight and low-cost All-Crop made combining a one-man operation.

The third member of General Falk's farm tractor vision was the 10-18 tractor introduced to the world in November, 1914, in anticipation of the 1915 crop season. It boasted being "the only tractor that has a one piece steel heat-treated frame — no rivets to work loose — will not sag under the heaviest strains." General Falk apparently was induced to take on the Bull tractor designed and built in Minnesota. Certain design similarities appear to exist between the 10-18 and models built by other companies. These include the Bull tractor, the Lion tractor, and the Happy Farmer, all having their first life in the Minneapolis-St. Paul area.

Whether a connection actually existed between the 10-18 and other look-alike models on the 1915 market is now unknown. Time would soon show that the design was unpopular with farmers — the offset front wheel simply did not lend itself to easy steering when operating under a heavy drawbar load. Allis-Chalmers equipped the 10-18 with its own two-cylinder opposed engine. Using 5¼ x 7 inch cylinders, the 4,800 pound 10-18 tractor offered only a single forward and a single reverse speed of 2⅓ mph. Also featured was a Detroit force-feed lubricator, Kingston double carburetor, and a high tension magneto.

Allis-Chalmers tractors were, from the very beginning, some of the most innovative in the industry. A-C engineers were early to use high strength steel and drop forgings in tractor design. This cut the weight and cut the cost. One tractor engineer of the World War One period complained that tractor builders were "selling too much iron for too much money." Although it appears that the very first 18-30 tractors may have been built without removeable cylinder sleeves, these soon were a standard feature, remaining so, even though this raised the initial cost somewhat.

Innumerable design achievements must be ascribed to A-C tractor engineers. Among A-C "firsts" are:

— The first square engine, 4-inch bore and stroke in the WC tractor.

— Rubber tires were first used and sold on Allis-Chalmers tractors in 1931.

— The Model B tractor which was the answer to Horace Greeley's Locomotive. In 1871 Greeley wrote, "What our farmers need is not a steam plow as a specialty, but a locomotive that can travel with facility, not only on common wagon roads, but across even freshly plowed field, without embarassment, and prove as docile to its manager's touch as an average span of horses. Such a locomotive should not cost more than five hundred dollars, nor weigh more than a ton....." The Model B weighed less than 2,000 pounds and sold for $495!

— A-C pioneered power adjustment of the rear wheels with its 1948 WD tractor.

— A-C was one of the first companies to use a dry air cleaner for tractor duty.

— The D-19 Diesel was the first agricultural tractor to be equipped with a turbocharger.

— Power steering attachments were introduced on the 1956 model of the WD-45 tractor.

— Drive-in cultivators were pioneered by A-C.

— Positive seal track idlers were announced for A-C crawlers in 1939.

These and many other "firsts" were designed and applied to practice by Allis-Chalmers engineers.

To delineate the factors leading to the 1985 announcement that Allis-Chalmers was selling its farm implement business would inevitably lead to subjective judgements not properly a part of what is ostensibly a historical narrative. Of major significance however, was the calamitous decline of the agricultural economy in the early 1980s, following an era of relative prosperity.

Already in 1974 Allis-Chalmers and Fiat of Italy merged forces in the crawler tractor and construction machinery business in Fiat-Allis. Since A-C was a minority shareholder in the new company, the Allis-Chalmers presence in this industry dimmed by 1975. Then on March 29, 1985, Allis-Chalmers sold its farm equipment business to Klockner-Humboldt-Deutz AG of Germany.

Tractors

Charles Edwin Search of Allis-Chalmers filed an application in April, 1917, that finally resulted in Patent No. 1,385,721 in July, 1921. This patent appears to have covered the unique drawbar arrangement of the 10-18, as evidenced in this photograph. Beyond this single patent, no other patent references to the 10-18 have been located. However, manufacturing rights may have been secured on already existing patents of other parties, or certain patents may have been purchased outright. Although Walter F. Peterson's title, *An Industrial Heritage* indicates that "early in 1921 the 12-20 tractor was introduced, apparently to replace the 10-18" the A-C Sales Bulletin of November, 1920, does not include the 10-18 as part of the then-current line.

Fortunately, General Falk was undeterred by a virtual failure of the first three tractor presentations — the tractor-tiller, the tractor-truck, and the 10-18 tractor. In retrospect, the 10-18 probably was fully as useful to the farmer as the competing Bull tractor, to name one of many. Despite its quality construction, sales of the 10-18 were a disappointment. In fact, the limited number of tractors built were apparently held in stock for several years, finally moving out a dramatic reduction in price. Precise production data on this tractor is not now known to exist. Few remaining examples of the 10-18 survive.

Comparatively little advertising data remains in existence for the 10-18 tractor. However, the A-C Advertising Department seized on the World War One cry of "keeping the boys on the farm" by noting that the 10-18 made plowing a "joy for the boy." With General Falk at the helm of Allis-Chalmers, tractor development continued. Falk was determined to get into the tractor business as a means of broadening the company's base. While the Allis-Chalmers of 1912 was diversified within the realm of heavy industry, almost all of its products were aimed in this direction. Therefore, a slowdown in virtually any phase of the manufacturing industry was detrimental to a positive cash flow. Diversifying into the farm tractor business was an approach which would later pay big dividends for the time and effort expended.

Although production of the 10-18 had begun just a few months earlier (in November, 1914) it appears that by April of the following year, work was already under way for an entirely different tractor called the 6-12. Altogether inconsistent with this date is a patent application filed March 8, 1920, for what ostensibly is the 6-12 tractor. Even more mysterious, this application was made by Johann M. F. (Max) Patitz, the eminent A-C engineer responsible for many developments in steam engineering. The Patitz application resulted in No. 1,419,113 being issued for a tractor on June 6, 1922.

Designed as an all-around farm helper, the 6-12 could be readily adapted to grain binder duty as is illustrated by means of an actual photograph. Eliminating the horses from the binder also eliminated frequent rest stops, thus increasing production substantially. Peterson notes in *An Industrial Heritage* that following the 1915 development of the 6-12, "Falk hoped to place it on the market in the near future, but for unexplained reasons it was not produced in any quantity until 1918." Quite possibly the delay was due to opposition from the Executive Committee. This theory seems to be tenable, since the majority were involved with heavy industry, were not attuned to an agricultural market, and likely resented any intrusions.

Tractors

An excellent side view of the 6-12 and a hayrake illustrates the simplicity with which ordinary horsedrawn implements could be adapted to the 6-12 system. All that was necessary was to shorten the tongue, and then drop the rear mounting (just ahead of the steering wheel) into the hole formerly used by the doubletree pin. A LeRoi Model 2-C engine was featured. Its four-cylinder design offered a 3⅛ x 4½ inch bore and stroke. This very popular engine was used in a number of small tractors and motor cultivators of the period.

A rare photograph from the Allis-Chalmers Archives illustrates the 6-12 coupled to a mower. The 6-12 could be adapted to virtually any farm implement with a minimum of necessary attachments. The same concept was used by the Moline Universal tractor, the Indiana tractor, and several others of more or less similar design. In fact, Moline Plow Co. made an agreement with the Indiana Silo & Tractor Co. as early as 1919 for the use of certain patents. An internal company memo by Moline Plow Co. of October, 1919, indicates that "we have already notified Allis-Chalmers Co. that their tractor infringes on our patents."

An original, and probably heretofore unpublished photograph from the A-C Archives illustrates a 6-12 tractor combined with a corn binder. This rather impressive combination was yet another possibility presented by the company to its dealers, and subsequently to the farmer. Allis-Chalmers made a valiant attempt to establish a retail sales organization during the 1920s, meeting with only a limited success. After all, some of the so-called "old line" companies had been dealing exclusively with farmers for half a century or more.

Although most of the precise details concerning the 6-12 tractor are now lost to time, it appears that Allis-Chalmers did indeed secure a manufacturing license from Moline Plow Co. and/or Indiana Silo & Tractor Co. in relation to certain features of the 6-12 tractor. Presumably the royalty payments were retroactive to the beginning of production. The 6-12 combined with a grain drill, as shown in this original A-C photograph was probably an excellent combination. Unfortunately, the 1918 agricultural equipment market was not yet ready for such an innovative design.

To prove the point that the 6-12 was also a ready servant for belt work, it is shown here at West Allis Works belted to a small burr mill. General Falk's determination to enter the burgeoning tractor market resulted in the organization of a separate tractor department within the company during 1917. Fred W. Kamm was the first manager. By December, 1918, the Tractor Department gained new status with space in the *Sales Bulletin*, an internal organ published for the benefit of A-C representatives and salesmen.

Although exact production figures cannot now be located, the research of Walter F. Peterson in his book, *An Industrial Heritage* indicates that 700 of the 6-12 tractors were produced. Initially the price was set at $850. Due to a market much poorer than anticipated, the last 200 of these tractors were sold at a tremendous reduction in price. In fact, by 1922 it appears that the company had dropped the price to a mere $250. Although this figure represented a substantial loss, continuing to hold the remaining inventory at a higher figure would have resulted in even greater losses to the company. This 6-12 is equipped with the recommended single 16-inch plow.

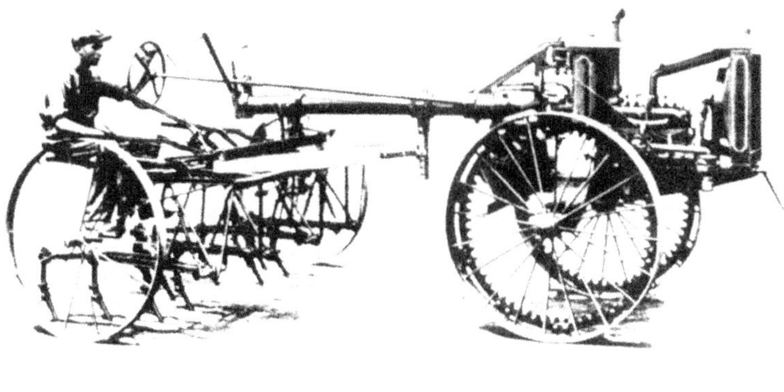

Yet another application for the 6-12 was the two-row cultivator. Many of the major farm equipment manufacturers attempted to market a motor cultivator of some type, and others spent considerable sums on similar designs, finally deciding the effort was not worth the cost. This particular design had the distinct advantage of not requiring an investment for new cultivating equipment — all that was required was to saw off the excess length of the wooden tongue, drill a couple of new holes, and presto! a motor cultivator was born.

Probably to prove the point that the 6-12 could be adapted to almost any farming task, this tractor is coupled to a potato digger. The 48-inch center-to-center dimension of the drive wheels fit exactly into 48-inch rows, and could work without a problem in 40 or 42 inch rows as well. Placing the traction wheels in front had certain advantages, and like most things also possessed certain disadvantages, one of them being a difficulty of backing up, especially if the drive wheels were lower than the truck wheels of the attached implement.

An extremely rare, and probably unpublished photograph from the Allis-Chalmers Archives illustrates a 6-12 tractor connected with a spring tooth harrow. In this particular instance the rear truck wheel remained in place on the tractor, with the harrow being situated between it and the drive wheels. Even though the 6-12 tractor saw disappointing sales, and the earlier 10-18 was something less than successful, General Falk persuaded the Executive Committee to authorize $800,000 for a new tractor foundry in October, 1919. This new state-of-the-art facility was completed a year later.

As early as 1917 General Falk and those involved in tractor production had begun work on a 6-12 Orchard tractor. Designed especially for orchards, vineyards, and truck gardens, this specialized unit differed from the General Purpose version mainly in the addition of extensive shielding to protect both the tractor and the crop from each other. Such a machine had the potential of eliminating much of the hand and horse cultivation required in orchards and vineyards. Despite an attempt to enter this specialized market, the 6-12 found this door, like many others, latched and bolted.

Tractors

Although current research has failed to locate any specific Allis-Chalmers records pertaining to 6-12 production and discontinuance thereof, a representative of Moline Plow Co. wrote in August, 1923, that "Allis-Chalmers has discontinued this type of tractor...." At what specific time Allis-Chalmers introduced the slightly modified and improved Model B, 6-12 tractor had likewise not been determined with certainty. The background of this photograph probably places this tractor in a yard at West Allis Works.

A seldom published photograph of the 6-12 Industrial tractor shows little difference in design from the conventional General Purpose model. Obvious however, are the much heavier drive wheels with flat steel spokes replacing the conventional round spokes of the 6-12 GP model. Also evident are the bolted rubber lugs designed especially for use on concrete and other hard surfaces. Since sales of the 6-12 were very low, it is quite possible that the demonstration model of this photograph was the only such style ever built.

Another potential market for the 6-12 tractor was seen with small road graders. The post-World War One period heard a clamor for better roads, and while many rural roads were to remain without gravel for years to come, at least an occasional pass with a grader would eliminate the ruts until the next thunderstorm. This 6-12 is coupled to a Galion Light Premier grader. Although small in size, this tractor-grader combination was probably a very practical machine. Unfortunately however, it, like the many other combinations, failed to bring A-C any measure of new sales.

While it has been generally assumed that this twin tractor setup was purely an experimental venture, Allis-Chalmers offered this combination as the 6-12 Duplex. Even those tractors already in the field could be converted with the specially designed Duplex Hitch. This permitted the farmer with two 6-12 tractors to couple them as shown here for much greater power than before. Since production of the standard 6-12 General Purpose tractors remained very low, it seems highly unlikely that any substantial number of the Duplex Hitch systems were sold. Johann F.M. Patitz filed a patent application on November 3, 1920 to protect this design. Finally, on September 15, 1925, the Patent Office issued No. 1,553,324 covering the Duplex design. By then, any thoughts of continued production had been long since abandoned.

Although at first glance the Indiana tractor shown here seems to be immensely different than the A-C 6-12 model, a closer look shows a similar chassis arrangement, and this may well have been the point of contention between Moline Plow Co. and Allis-Chalmers. From all indications, Moline Plow acquired a controlling interest in the Indiana tractor, and from all appearances, Allis-Chalmers later agreed to pay royalties to Moline Plow, with the latter firm then paying half the monies received to the Indiana people. This tractor is illustrated strictly for comparison purposes.

Tractors

Allis-Chalmers announced the new 15-30 tractor in December, 1918. Its sleek lines bore a resemblance to auto and truck designs, and in fact, this was a major goal of A-C engineers. However, by February, 1919, the company had given it a new 18-30 rating to better correspond with its delivered output. Although the general consensus exists that the 15-30 and 18-30 tractors were identical, save for the new model designation, a comparison shows this notion to be in error. One notable difference is the sheet metal radiator of this model, compared to the cast iron style of the 18-30. Also evident is an entirely different front wheel design.

Undoubtedly the majority of the 15-30 design was carried into the 18-30 tractor, especially the unit frame and chassis. Totally out of character however, is the engine design, with the magneto plainly visible in this photograph, while the opposite side of the 18-30 shows it to be on that side of the tractor. Despite repeated attempts to locate specific data on the 15-30, nothing can be located thus far to indicate precisely the type of engine that was used. Since the 15-30 was built as such for only a few months, it seems entirely likely that any photographs at all of this model represent the very first prototype tractors.

This photograph, among others in a file of 18-30 illustrations, could probably be either a 15-30, or one of the first 18-30 tractors. Despite the company's notice to the public that the 15-30 to 18-30 rating change was just that, evidence from the company's own files in the Allis-Chalmers Archives seems to indicate some major differences between the two tractors. As has been previously noted, the radiator style and front wheel design are two major differences, with others probably existing, but not nearly so apparent.

Another photograph, probably of early 1919, illustrates what could have been either the 15-30 or the 18-30 tractor. This particular photograph illustrates the manifold side of the engine quite well, and a comparison of this tractor with later 18-30 models indicates a major difference — in fact, it seems logical that two entirely different engines were used. At this point in time the front wheels with round spokes remain, as does the sheet metal radiator. The roll-down sidecurtains do not appear to have followed on the regular production models as standard equipment — the same holds true for the fabric roadster-type canopy over the operator.

A rear view of the Allis-Chalmers 15-30 tractor shows it to be quite similar to the later 18-30 model. Two forward speeds were provided, Allis-Chalmers priced these tractors initially at $2,100. Standard colors included a deep green body with yellow striping, plus bright red wheels. The heavy steel fenders were no luxury. Especially in plowed ground, the fenders were required to keep loose dirt away from the operator. As otherwise noted, specific data on this particular tractor has not been located.

The Allis-Chalmers engine used in the 18-30 featured a three-bearing crankshaft. The 1923 *Tractor Red Book*, a widely used industry reference, noted that the 18-30 tractor was rated at 830 rpm. Page 22 of that volume also lists the Allis-Chalmers 20-35 tractor as a separate entity. A careful study of the specifications reveals that the 20-35 was rated 100 rpm faster, at 930 rpm. This appears to be the only difference between the two tractors.

The original photograph in the Allis-Chalmers Archives denotes this tractor as the 18-30 model of 1921. By that point in time, Allis-Chalmers had standardized the construction, and an 18-30 had successfully completed testing at Nebraska's Tractor Test Laboratory. (See: *Nebraska Tractor Tests Since 1920, page 29*). It was equipped with the company's own four-cylinder engine. Rated at 830 rpm, it used a 4¾ x 6½ inch bore and stroke. Total weight of this tractor was 6,000 pounds. Note that the engine on this tractor is entirely different than its immediate predecessor, the 15-30.

A so-called worm's eye view of the 18-30 for 1921 shows an exceptionally strong and compact design. In fact, the basic chassis design of the 18-30 would remain virtually unchanged for several years to come, even though the tractor's horsepower rating would continue to see upward adjustments. With the unit frame design of the 18-30, Allis-Chalmers brought farmers a chassis design that was acceptable and eminently useful on their farms. Styles like the 10-18 and the 6-12 were rapidly growing in disfavor.

In February, 1919, the 18-30 listed at $2,100, a fairly stiff price for a 3-plow tractor. Under Test No. 55 at the Nebraska Test Laboratory, the 18-30 using kerosene fuel delivered 20.19 drawbar and 30.58 brake horsepower. Test No. 83 of September, 1921, saw virtually the same tractor, but using gasoline fuel. This test yielded 23.62 drawbar and 43.73 brake horsepower. Thus, the company announced in 1922 that, "the 18-30 A-C Model was superseded by the present 20-35." However, company production records indicate that no 18-30 or 20-35 tractors were built during 1922. In fact, by the end of 1921 only about 160 tractors of this series had been built. Unfortunately, the production records do not differentiate between the various style changes.

This photograph from the Allis-Chalmers Archives was taken in 1921 and illustrates an A-C employee with an engine block for the 18-30 tractor. The block itself has been machined, and the removable cylinder sleeves are set in place for a final fitting. General Falk had taken the farm tractor under his wing as a pet project. In his role as president of the company, Falk was determined that Allis-Chalmers make a noticeable entry into the farm tractor business. This goal was eventually achieved, even though real success was still a decade away.

Tractors

The carburetor and manifold side of the 18-30 engine clearly illustrates the required equipment for burning kerosene and other low-grade fuels. In this instance two individual manifolds damper adjustments are obvious — both of them affecting manifold temperature control. The dual carburetor system was another necessary requirement, in this instance the carburetor was supplied by Kingston. Although somewhat less expensive to buy, kerosene required extra equipment on the tractor, and many of these special devices were subject to premature failure since they were required to operate in the intense heat of the engine exhaust.

A head-on view of the 18-30 tractor for 1921 illustrates the huge cast iron radiator used on this tractor, compared to the sheet metal style featured on the earlier 15-30. A careful study of existing data certainly gives credibility to the theory that the upward rating from 15-30 to 18-30 was achieved with an increase in engine speed from 830 to 930 rpm. Unlike the supposition that the rating change came about purely as a substitution of terms, the mechanical end of things also required tightening the governor spring a couple of turns.

Development work on a smaller partner for the 18-30 probably began during 1920, with the company announcing a new 12-20 tractor in January, 1921. Designed primarily after the larger 18-30 tractor, the 12-20 began actual production in March, 1921. One of these tractors was submitted to the Nebraska Tractor Test Laboratory, with Test No. 82 being run in September of the same year. On the rated load tests the 12-20 delivered 15.13 drawbar and 27.28 belt horsepower.

Lacking the finished engine exhaust system, this 12-20 might well have been one of the first off the line, and indeed, might have been one of the hand-built prototypes. The 12-20 was introduced in early 1921 and tested at Nebraska in September of that year. Since its output was substantially above the initial rating, Allis-Chalmers re-rated the 12-20 upward to a 15-25 model. Unfortunately no records have thus far been located to indicate precisely when this change took place, but it seems plausible that the 15-25 rating was in full force by the beginning of the 1922 season.

A head-on view of the 12-20 shows it to be a full cousin to the larger 18-30 model. Uniformity in the line was advantageous in several ways. From the sales and promotion aspects, it was possible to show a logical progression from small to large models. From the manufacturing aspect it was possible to have a certain interchangeability of parts, even though the tractors were of different sizes. Note the words "Allis-Chalmers" cast into the top radiator tank.

The 12-20 A-C tractor offered a roomy platform for the operator, and included a fender and firewall arrangement that kept the majority of dirt and dust away from the operator area. By April, 1921, a list price of $1,495 was announced for the 12-20 tractor. Despite intense competition, (there were some 170 tractor builders listed in 1920) the 12-20 was moderately successful, due in part to its conventional and immensely practical design. With the 12-20 and 18-30 models, Allis-Chalmers had finally gained the rites of passage in the tractor business.

While the 18-30 used an engine built by Allis-Chalmers, the 12-20 was equipped with a four-cylinder Midwest engine. Built by Midwest Engine Co. of Indianapolis, this four-cylinder style used a 4⅛ x 5¼ inch bore and stroke. The cylinders were cast in pairs. Total engine weight was 1,000 pounds. A Dixie Model 46 magneto is evident. It like the water pump, was driven from the cam gear. Always extremely proud of their equipment, whether farm tractors or rock crushers, the company had "Allis-Chalmers" cast into the side of the valve cover.

As previously noted, the 15-25 and the 12-20 tractors were virtually identical, except of course for the upward horsepower rating. Seemingly this resulted from Test No. 82 at Nebraska, although no specific announcement in this regard has been located. Many competing tractors of the early 1920s still used the heavy, slow-speed designs of the past. Allis-Chalmers however, opted for the smaller high-speed designs used in automotive practice. Despite heavy resistance to this trend by much of the industry, this trend was a precursor of a new generation in tractor development.

Announced in February, 1924 the 15-25 Orchard Tractor was, to quote Allis-Chalmers, "Specially Built for a Special Job." Although the tractor itself was the same 15-25 as before, special fenders and wheel guards were provided to protect both the trees and the tractor. Unfortunately, the tractor operator was the highest object, forcing a certain measure of acrobatics when working under low branches.

Although the Midwest engine of the 15-25 was ordinarily furnished for use with gasoline fuel, it could be fitted with a special manifold for kerosene on special order, and of course, at extra cost. This rare photograph from the Allis-Chalmers Archives illustrates the general layout of the kerosene manifold. Even though the majority of tractors built in the arly 1920s could be equipped to burn kerosene fuel, the trend gradually moved toward gasoline fuel by the early 1930s.

One major addition on the 15-25 Orchard Tractor was individual wheel brakes to facilitate short turns encountered in orchard and vineyard work. The individual brake pedals are evident in this platform view of the 15-25 Orchard model. Announced concurrently with the 15-25 Orchard Tractor was a 15-25 Special Road Maintenance Tractor. It, like the orchard model, was built on the same chassis as the general purpose line, but featured additional accessories required for efficient road work. Production of the 15-25 tractor line ended subsequent to an announcement of January, 1927. A-C production records do in fact attest to this fact — only 23 of the 15-25 tractors were produced in that year.

The original photograph in the Allis-Chalmers Archives indicates this to be a 20-35 tractor. From all appearances, it could be classed either as an 18-30 or as a 20-35. Without a doubt it is early in the production run, as evidenced by the round spokes in the front wheels. The 18-30 and 20-35 were almost invariably equipped with front wheels having flat iron spokes. The exhaust manifold of this tractor terminates in an unusual rectangular cast iron passage which discharges just ahead of the belt pulley. Additional information regarding this particular photograph has not been located.

An advertising circular copyrighted in 1921 illustrates the 20-35. The air intake duct penetrates the hood and extends well above the tractor, its bottom end terminating in the huge cast iron air washer. A predecessor of the oil-bath air cleaners, the air washer used water as the filtering medium. Although quite effective, the air washer required frequent attention, and was of course, virtually useless in freezing temperatures. Since precise documentation concerning various modifications on the 20-35 probably no longer exists, a precise narration is now impossible.

As word spread of the Allis-Chalmers 20-35, new dealers entered the fold. Depending to a degree on the dealer's aggressiveness, Allis-Chalmers tractors sold fairly well in certain areas. Through its monthly *Sales Bulletin* furnished to all A-C sales personnel, the company devoted a full page in the August, 1922, issue to a ten tractor deal at Valley City, N.D. All of the 20-35 tractors of this order were sold prior to delivery. Here a 20-35 is pulling a rack and hayloader in the days of loose hay.

During 1922 Purdue University, a recognized agricultural school, developed a tractor testing program. It differed materially from that of the University of Nebraska's Tractor Test Laboratory at Lincoln in that Purdue used a traveling belt arrangement for the drawbar tests, while Lincoln used an oval track for this duty. Interestingly, the experimental program at Purdue University used Allis-Chalmers tractors exclusively, to the great pride of the company.

Tractors

Early in 1925 Allis-Chalmers closed a deal with Nichols & Shepard Co. of Battle Creek, Mich. for the latter to handle A-C tractors. This achievement represented somewhat of a coup for Allis-Chalmers. First of all, the long-established Nichols & Shepard organization had an undisputed and enviable following in the thresher business. This fact alone was beneficial in the promotion of A-C tractors. Additionally, N & S had a well established network of dealers and distributors, thus giving Allis-Chalmers the benefit of a sales organization with virtually no additional investment. About the same time A-C had a similar sales arrangement for its tractors through Banting Mfg. Co. of Toledo, Ohio. The latter built the "Greyhound" line of threshers, and apparently the A-C tractors sold by Banting carried the "Greyhound" trade name on the radiator. To reciprocate, A-C sold the Greyhound threshers in 1925, and perhaps even later.

In August, 1924, Allis-Chalmers announced the 20-35 Special tractor, noting that it was their most powerful model. This tractor used a compression of 74-78 psi, some 10-15 psi higher than the regular models. This single change resulted in an output of 45-plus belt horsepower. Another unique feature was the use of the "Dot" lubrication system, replacing all the grease cups with high pressure fittings. Regarding the color of this and other Allis-Chalmers tractors of the period, an A-C *Sales Bulletin* noted, "[These tractors have]...a striking color combination finished with a coat of varnish. The motor and transmission are painted green, striped with yellow and the wheels are a bright red." The winter cab shown here was an extra-cost option available for any 20-35 tractor.

Early in 1926 Allis-Chalmers announced certain changes in the 20-35 tractor. These included placement of the air washer "to a place immediately back of the motor," rather than at the side of the engine as shown here. This photo illustrates the rather unusual air piping required for the Kingston dual carburetor. Also for 1926, the eight rocker arms were lubricated from the top of the hood by means of tubes extending downward to the valve covers. Not visible, but very important was the change to four-stud main bearings from the previous two-bolt configuration.

Already in February, 1922, Allis-Chalmers announced a 20-35 Special Road Tractor. Listing at $2,085 this model used special heavy rear wheels with ½ inch thick steel tires. Since farmers often were able to contract their services to townships or counties for road work, the sales appeal naturally was extended to them, since a summer road job could help balance the books at year's end. Except of course for the special front and rear wheels, the sun canopy, and other slight modifications, this special model was identical to the standard 20-35 model.

From every indication, 1926 was the last year that Allis-Chalmers built the 20-35 with "long fenders," that is, with fenders which extended down to the level of the operator's platform. Beginning in 1927 the so-called "short fenders" went into use, as evidenced by this original A-C photograph. During 1926 several changes took place, as previously noted, including dressy hubcaps for all four wheels. This feature, once thought to be unique to the later 25-40 tractor was actually initiated in late 1926 or early 1927. Due probably to older inventory on hand, the change from early style wheel covers to the later style illustrated here was not always consistent with the actual year built.

Tractors

Substantial improvements were made to the 20-35 in 1927. These included the addition of a Purolator oil filter as standard equipment. Also during 1927, the price was lowered from $1,885 to $1,495. In 1928 the price was dropped another $200 to $1,295. Although the 20-35 had been on the market since 1922 it was not submitted to the Tractor Test Laboratory for certification of this rating until June, 1928. Burning gasoline fuel, it delivered substantially more horsepower than the rating. (See Nebraska Test No. 151). A substantial point of confusion is the fact that the 18-30, 20-35, and 25-40 tractors were all designated as Model "E" by the company, even though the horsepower ratings, and in fact, the tractors themselves, differed substantially from each other.

In 1930 production of the Model E, 20-35 ended, while production of the Model E, 25-40 tractor began. Effective with Serial No. 24186 of 1930 the 25-40 appeared as a larger version of an earlier tractor. The increased horsepower output resulted from a somewhat larger 5 x 6½ inch bore and stroke, compared to a 4¾ x 6½ inch bore and stroke for the 20-35. Tested at Nebraska as the Model EK under Test No. 193 of June, 1931 this tractor delivered 27.69 drawbar and 47 brake horsepower. Distillate fuel was used for this test.

Production of the Model E, 25-40 tractor ran from 1930 to 1936. During this period 1,425 units were produced. Although the majority were furnished on steel wheels, a limited number were equipped with pneumatic tires from the factory. The 1931 acquisition of Advance-Rumely Thresher Co. gave Allis-Chalmers a ready-made sales organization. Since Advance-Rumely was still building grain threshers, it seems plausible that this new organization might well have been conducive to additional sales of 25-40 tractors as power for the Advance-Rumely threshers. By the time production of the 25-40 ended in 1936 heavy tractors of its style were becoming outmoded — small row-crop tractors were now the vanguard.

Already in 1927 Allis-Chalmers announced that the 15-25 tractor would be discontinued. In 1929 this style was followed with an entirely new design known as the Model "U" tractor. The original design used a Continental S10, 4-cylinder engine. Rated at 1200 rpm, it used a 4¼ x 5 inch bore and stroke. The Continental engine was used prior to tractor No. U-7405 which closed out 1932. The original version offered a 3-speed transmission offering a choice of 2½, 3½, and 5 mph.

In 1930, and perhaps earlier, A-C engineers had begun experiments with pneumatic tires instead of lugged steel wheels. In April, 1932, a pair of Firestone 48 x 12 airplane tires were mounted on an A-C Model U tractor belonging to Albert Schroeder of Waukesha, Wis. This test was eminently successful and was followed by the first public demonstration of an air-tired tractor on Labor Day, 1932. This test, made near Dodge City, Kan., was again a roaring success, with Allis-Chalmers announcing on October 13, 1932, that it was offering air tires as standard equipment for the Model U tractor. This rare photograph from the A-C Archives shows a Model U equipped with Firestone tires; 11.25 x 24 rear, and 6.50 x 16 inch front tires. These were mounted on French & Hecht rims. Note also that this tractor is equipped with the Continental engine.

Although A-C made rubber tires standard equipment for the Model U in 1932, it was of course, still available on steel wheels if preferred. In this case, the additional "road gear" of 10.75 mph was locked out and could not be used. Although the use of pneumatic tires on a tractor was sensational news, farmers were reluctant to buy — partially because of the hard times of the Great Depression, and partially due to an inbred skepticism of something so revolutionary as rubber tires! Due to several factors this resistance was gradually overcome. In order to continue the sensationalism of rubber tires, A-C installed special high-speed gears in stock model tractors, and 1933 became known as the year of the tractor races. Barney Oldfield, the famous racing driver, drove tractor 999 at a world record speed of 64.28 mph on September 17, 1933.

Tractors

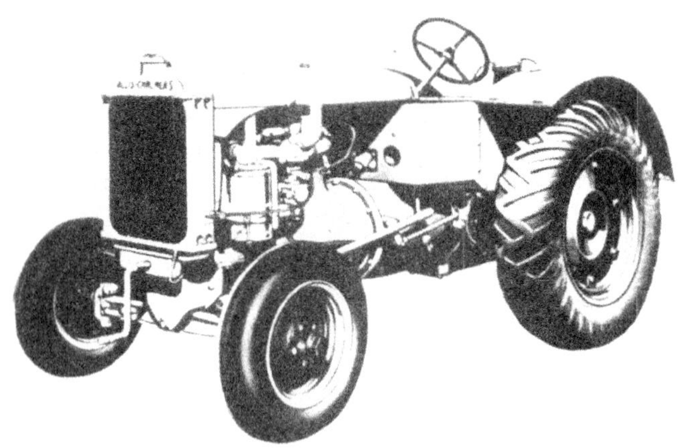

In 1933 Allis-Chalmers began using their own "UM" engine in the Model U tractor. Rated at 1200 rpm, it used a $4\frac{3}{8}$ x 5 inch bore and stroke. Tested previously with the Continental engine, this style with the UM engine was tested at Nebraska in July, 1935. Since it could be furnished either with steel wheels or with rubber tires, dual tests were performed. Significant improvements in fuel economy were noted with pneumatic tires. Although exact figures have been impossible to locate, it appears that the Model U on steel wheels listed at a price of $840 to $895 when originally announced. Later models on rubber tires were priced as high as $1,250.

Model U tractors of 1936 and later were equipped with an A-C "UM" engine with a slightly larger $4\frac{1}{2}$ inch bore. Various industry publications including the *Tractor Red Book* show that production of the Model U tractor ended in 1943 with serial number U-17470. Company records indicate however, that 17 additional tractors were built in 1944 with the "U" prefix. Beginning in 1945 an entirely different set of serial numbers appears, with limited quantities of the Model U tractor apparently being built until at least 1952. Presumably these tractors were built on special order, and quite possibly some were exported. Even more curious is the fact that at least some of these tractors were equipped with steel wheels.

The United Tractor & Equipment Co. was organized in 1929. Composed of some 32 independent implement manufacturers and distributors, United marketed the Allis-Chalmers Model U tractor under their own name, both in the general purpose and industrial models. Since the tractors differed primarily in the name cast into the radiator, illustrations of both the Allis-Chalmers Model U and the United tractors may appear among the industrial versions. Shown here is the A-C Ind"U"strial tractor equipped with steel wheels.

Although precise figures have not been located, total production of the "U" Industrial version appears to have been rather low. For industrial applications the solid rubber tires of this Model IU-7 tractor were probably the most common. The Model U Industrial tractor was equipped with individual rear wheel brakes by Bendix — a feature not found on the general purpose model. Although a separate listing of the IU tractors may have been maintained, existing serial number lists do not indicate this modification. In fact, a company parts book reports, "....U and IU tractors are marked U-1 and up in consecutive order as built..."

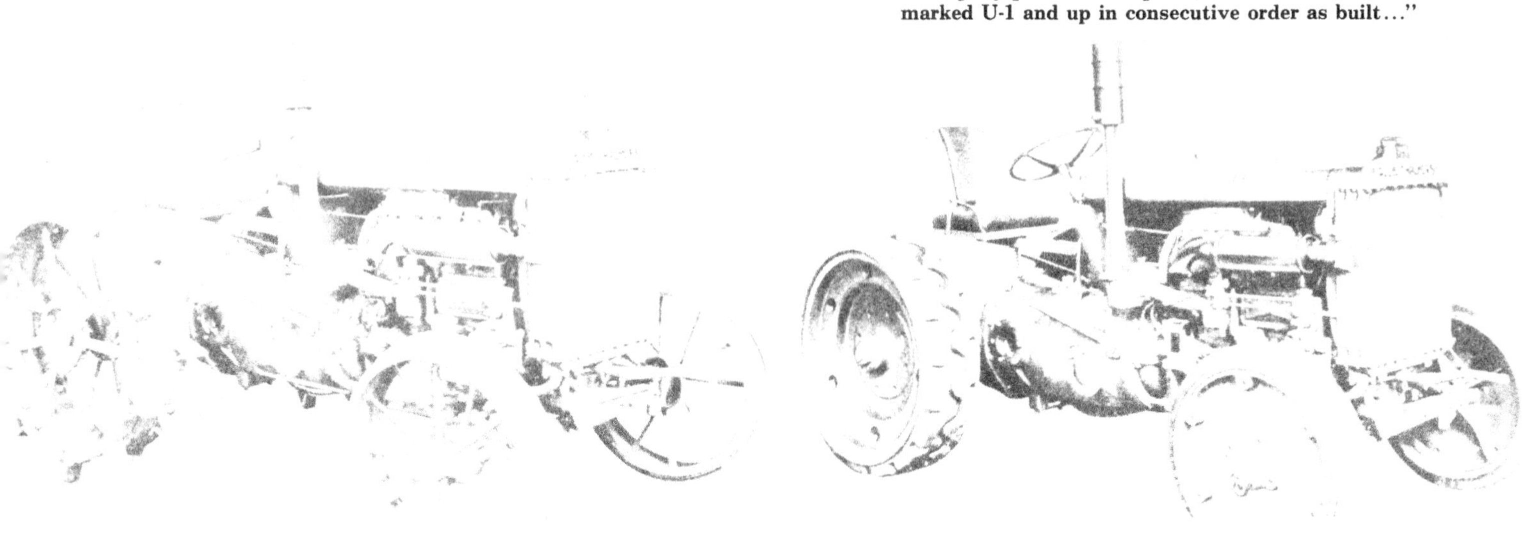

Tractors

Allis-Chalmers offered a tremendous variety of accessory items for the IU tractor. With a full understanding of changing weather patterns, the Highland Cab as shown in this illustration was available at extra cost. Built by Highland Body Mfg. Co. of Cincinnati, this cab went a long way toward protecting the operator from the elements. During the summer of course, no cooling was available, and more than likely, no heating was available in winter. Even though the Highland Cab of 1930 was little more than a windbreak, this was an immense improvement over the full force of a winter storm.

Allis-Chalmers IU tractors were built in almost any required configuration. This obviously staged photograph shows a IU tractor with dual rear wheels. The front and rear tires were all of the solid rubber type commonly used in industrial applications. Tractors like the IU were an immense boon to manufacturing and processing plants. Much of the work formerly done by manpower or horsepower could now be done in a fraction of the time and with far less effort than ever before.

Allis-Chalmers pioneered the use of pneumatic tires on tractors, and in particular, the farm tractor. Experimental work obviously took place however, on pneumatic tires for the industrial tractors. This IU is equipped with huge Goodyear Air-Wheels, of 26 x 11 front and 46 x 20 rear. Undoubtedly these aircraft-type tires smoothed the ride considerably over steel wheels or even the conventional hard rubber tires pressed over cast iron wheels. Beyond building the so-called "United" farm and industrial tractor line, the role of Allis-Chalmers in United Tractor & Equipment Co. is unclear. From all appearances however, A-C was a dominant force in the latter firm's activities.

The Allis-Chalmers Model U tractor was built in many different configurations. An early operator's manual indicates that in addition to the General Purpose style came the IU Industrial, UO Orchard Tractor, and as shown here, the UN Narrow Tread model. Designed specifically for use in manufacturing plants and other applications were narrow aisles and doorways were still the rule, the UN was probably not a highly recommended machine for ordinary duty. The narrow rear wheels made fenders imperative. This tractor also used a spring-mounted front axle.

A 1930 catalog of the Ind"U"strial line manifests a breathtaking array of applications for the IU tractor. Shown here is the W-K-M winch designed especially for work in oil fields. The entire unit was superimposed over an IU tractor, requiring only removal of the front axle and the addition of various mounting brackets. A roller chain drive operates the winch mechanism. At this point in time Allis-Chalmers was involved in certain phases of oil field machinery that included the pumps and engines for transfer stations. The winch shown here was built by W-K-M Co. of Houston, Texas.

Tractors

A 1930s approach to portable welding was this self-contained unit offered by Allis-Chalmers. Consisting of an IU tractor with a mounted welding generator, this approach was probably unique in its day. The welding generator itself was operated from the belt pulley on the opposite side of the tractor. Since this setup had no effect on the rest of the tractor, it could serve dual duty for drawbar work when required. Note that this particular tractor has "Allis-Chalmers" cast into the radiator, yet it carries a "United" decal on the rear wheel. The Hansen arc welder was supplied by Northwestern Mfg. Co. of Milwaukee.

Understanding the severe duties placed on industrial tractors, Allis-Chalmers could supply Detroit bumpers for the IU tractors. Built by Detroit Harvester Co. of that same city in Michigan, these units made the IU a good "pusher" whenever required, such as on docks, or in railroad yards and industrial plants. In addition, mounting this outfit on the front end probably saved on radiator repairs caused by driving into pipes, poles, and other items which happened to jump into the driver's path. This particular tractor appears to be of the UN narrow-tread variety.

Perhaps the application of the IU tractor to an extremely diverse line of equipment was calculated to bring in extra sales of IU tractor components as original equipment by other firms. A close look at this illustration reveals the "United" trademark on the front of this road roller, although there is no doubt that an IU engine and drive train is mounted within the structural steel chassis. Wehr Co. of Cudahy, Wis., equipped the Model Y-5 rollers with the IU, although the extent to which this was done is now unknown. While initially this was a 3-ton roller, it could be converted in steps all the way to a 7-ton unit by adding weights to the rear rolls. The frame was 10-inch channel steel, which should have been strong enough for the work intended.

Adding a special fifth-wheel adapter to the IU tractor turned an ordinary industrial model into a heavy duty hauler. This model is pulling a big trailer load of scrap iron. For this duty the IU of this illustration is equipped with pneumatic tires. Ironically, General Falk of A-C had made the tractor his pet project, but even his determination to succeed finally began yielding to pragmatism by 1925. To clear out the remaining tractors he hired Harry Merritt. When Merritt dropped the price of the 20-35 from $1,950 to $1,245, sales were dramatic. This extraordinary stroke of fortune placed A-C solidly in the tractor business and was largely responsible for the experimental work that led to the entire Model U series.

Walter F. Strehlow is identified with many of the early Allis-Chalmers tractor developments, as is Conrad E. Frudden, Willard H. Tanke, and several others. Although A-C engineers seem to have contributed materially toward the concept of a tractor-mounted sickle mower, this side-mounted model of 1929 is not an Allis-Chalmers mower, but according to the company's own catalog is a Detroit Mower, "adaptable for use with Allis-Chalmers Model "U" tractor." This device was built by Detroit Harvester Co. of Detroit, Mich. About 1914 this firm had marketed a most unusual tractor of its own. Detroit claimed this unit would replace three horse-drawn mowers, and would cut as much as 30 acres of heavy hay in a day.

Tractors

Special applications for the IU tractor were probably endless, and for every illustration still in existence, there are probably many others of which nothing is now known. Although this IU tractor is equipped with dual solid-rubber tires, a special spring mounting is evident. Unfortunately, particulars of this device are not now available. This rear view of the IU tractor illustrates the swinging drawbar that was standard equipment. Also evident is the flange for mounting the optional power take off shaft.

Mount this Schramm compressor behind the IU tractor, and presto! — a self-propelled tractor-compressor unit was born. The 1930s saw dramatic changes in manufacturing and construction, with power-operated tools, especially pneumatic tools, making major advances. Due in part to companies like Schramm, the portable air compressor became a reality and permitted the relatively easy use of air hammers and other pneumatic tools. This belt-driven unit was powered from the tractor belt pulley. The compressor was built by Schramm, Inc. of West Chester, Penn.

Aimed primarily at airports, municipalities, and similar applications was the Detroit Street Sweeper and Snow Brush offered with the Allis-Chalmers IU tractors. Company advertising noted that this fine machine weighed less than 700 pounds and could be easily attached or removed by one person. Allis-Chalmers made a strong bid for various aspects of road construction and maintenance machinery, beginning primarily with the IU tractor, and following with various other acquired product lines. Note that this tractor carries "Allis-Chalmers" on the radiator and "United" on the front wheel.

No long haul machine, the United locomotive was equipped with an IU engine and drive train. Four models were built, from the 4-ton Model BL-4 up to the big 8-ton Model BL-8. Designed specifically as a switcher and a car mover, this particular application, while seeming to be entirely foreign to the series, was in fact, tied in by virtue of the company's numerous industrial accounts. After all, Allis-Chalmers in 1930 was a major supplier of sawmill equipment, flour mill machinery, mining machinery, and extensive equipment lines for other industries. The locomotive was built by Brookville Locomotive Co. of Brookville, Penn.

A significant addition to the "U" Industrial line was the Hughes-Keenan Iron Mule shown here. The 2-yard capacity of the Iron Mule made it a fair-sized hauler, especially efficient for short haul dirt moving. The complete unit was mounted over an Allis-Chalmers Model "U" tractor and required elimination of the standard operator's platform and a relocated deck as shown in this illustration. The Iron Mule was built by Hughes-Keenan Mfg. Co. of Mansfield, Ohio.

Tractors

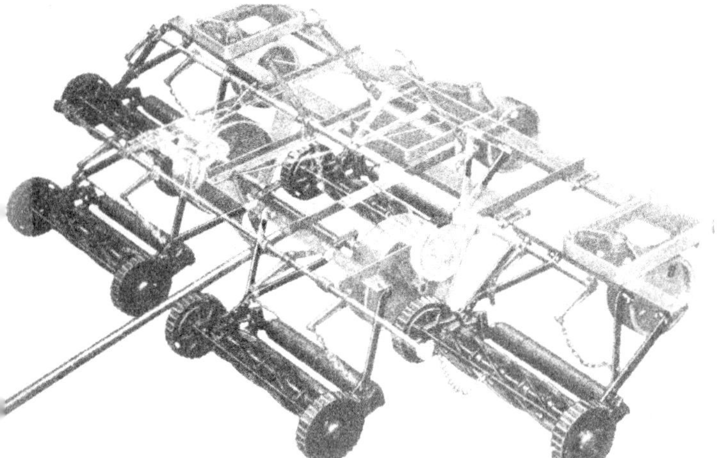

A study of the Allis-Chalmers Ind "U"strial tractor catalog gives the impression that there was virtually no endeavor untested for the IU tractors. Another interesting application was that of huge gang-type lawn mowers as used on golf courses, parks, and large estates — these of course being pulled by an IU tractor. The move into this field was probably a good one, as the competing International Harvester Co. was also spending a great deal of time and money developing an industrial equipment line during the 1930s. This particular mower was built by Ideal Power Lawn Mower Co. of Lansing, Mich. and was sold under the name "Bulldog."

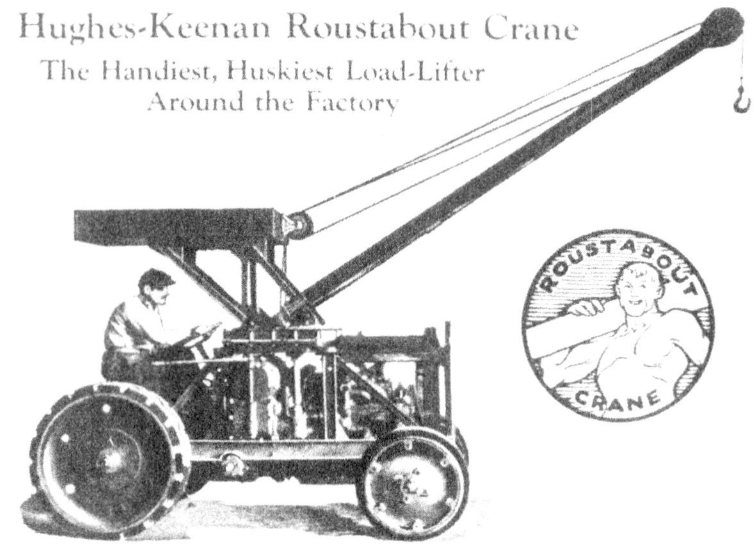

Considerable steel was required in the addition of the Hughes-Keenan Roustabout Crane to the IU tractor. Called "The Handiest, Huskiest Load-Lifter Around the Factory," the Roustabout was probably an extremely useful addition, especially for work outside of a factory area covered with overhead lifting equipment. Although this photograph illustrates the winch equipment rather poorly, it appears to have been chain-driven from the tractor belt pulley shaft. In addition to the Roustabout, IU tractors could also be equipped with side boom units and various other modifications.

While it is commonly supposed that the Allis-Chalmers Ind-"U"strial line was a highly publicized but seldom seen entity, this is not entirely true. There seems to be little doubt, given the overall production of Model U tractors, that sales of this highly specialized equipment were quite low. However, construction jobs of the 1930s were not noted for their extensive use of power equipment, so any use at all of powered equipment was in itself rather significant. This UI tractor and mounted winch are lifting concrete to an upper floor of this new building.

In what might be termed, "backfilling the hard way," A-C could furnish the Waterman Twin Backfiller as a package unit and mounted on any IU tractor. In operation, one cable payed out as the other was drawn in, so that lost time was minimized, and at least one worker was busy all the time. Designed specifically for filling ditches and trenches, the backfiller shown here was eventually made obsolete by other mechanical methods. The builder of this backfiller system was The Sidney Steel Scraper Co. of Sidney, Ohio.

Since Allis-Chalmers was a well established manufacturer of sawmill machinery, little wonder is it that their 1930 Model IU industrial line included skidders. The Model "F" double-drum skidder shown here could be attached with relative ease to the IU tractor. The Dorsey unit shown here was built by Dorsey Bros. of Elba, Ala. Single drum skidders were also available, and a slightly modified version was offered as a stump puller.

Tractors

A further step in the industrialization of the Model U tractor was the addition of a crawler attachment. The Model GU tractor shown here was furnished with equipment from Trackson Co. of Milwaukee, Wis. In a day when many crawler tractors were extremely expensive compared to wheel-type tractors, perhaps the crawler tractor conversion was advantageous. Once again, making a determination of the IU tractors so equipped has been impossible, since no available information relates to precise production figures of the various modified versions.

Differing slightly in design was the Trackson-Allis-Chalmers Model LU Crawler Tractor. A-C noted that this setup gave the lowest cost per drawbar horsepower, and also gave the Model LU the benefit of a four-speed transmission with speeds of 2.2 to 9.5 mph. Despite the fact that this particular Model LU is illustrated in an Allis-Chalmers catalog, the top radiator tank nevertheless has "United" cast in place. Even though specific data concerning United Tractor & Equipment is virtually impossible to locate, it appears that for all practical purposes, Allis-Chalmers was the dominant player within this organization.

Allis-Chalmers IU tractors equipped with Tracksons could also be furnished with a hydraulically operated bulldozer. For 1930, the use of high pressure hydraulics was unique and represents yet another innovative design. Throughout its long history, Allis-Chalmers used a business philosophy predicated on "getting there first." A careful study of this converted bulldozer reveals some interesting aspects far ahead of the times, including serrated grouser bars. The company's 1928 purchase of Monarch Tractors would also lead to a number of industry first's within a decade.

Trackson Co. built not only the track conversion for this IU tractor; they also furnished the high-lift shovel which is about to unload into a Hughes-Keenan Iron Mule — it too is built around the Model IU tractor. A wide range of loaders and buckets were available, including the Blair Hydraulic Digger, a machine designed for the IU wheel-type tractors. The bucket, of ½ yard capacity, automatically stopped at its highest and lowest position. Allis-Chalmers claimed that it could be mounted onto the tractor in less than four hours.

Available for either the Model GU crawler, or for the Model "U" Industrial tractor as illustrated here, the Baker Model 30 was advertised as "The Backfiller With Many Uses." Raising and lowering the backfiller blade was achieved by a large handwheel — in other words, it was manually operated! The blade itself could pivot on its mounting axis. Allis-Chalmers had a long working relationship with Baker Mfg. Co., finally purchasing this Springfield, Ill. firm in 1955.

Tractors

The 1928 purchase of Monarch Tractors, plus the continuing acquisition of manufacturing or sales rights to various lines of construction machinery certainly provided a strong impression that Allis-Chalmers was opting for a share of the road construction business. Despite these moves, A-C heavily promoted their IU engine and drive train as the power source for a number of motor graders. Included was this rather unusual United crawler-powered, center-control grader. The Model U engine and drive train are barely visible at the rear of the machine, but the operator's platform was located directly over the blade. An accessory scarifier attachment is situated directly behind the front wheels.

The United wheel-type grader also featured center control, with the operator standing high above the blade in order to maintain his perception of depth and distance. The wheel-type machine was virtually identical to the crawler model, except of course for the type of drive. All blade control for this grader was hand operated. For everyday use, the addition of power-operated blade control was a godsend to the operator. This center-control machine was built by Wehr Co. of Milwaukee.

The tumult for improved roads certainly bolstered the market for motor graders. In an earlier day, harrows and road drags were sufficient to occasionally close the ruts resulting from wet weather. Newer and faster cars however created the need for faster roads, and that meant more sales for the grader. Allis-Chalmers was able to approach this opportunity on two fronts — the sale of their own motor graders, or the sale of IU chassis units to Wehr Co. of Milwaukee; the Baker Mfg. Co. of Springfield, Ill., or the Shaw-Enochs Tractor Co. of Stillwater, Minn. The latter firm started out at Minneapolis in 1918 with a small tractor not unlike the A-C 6-12 model. Its limited success as an all-around tractor was mitigated by somewhat better satisfaction when equipped with a road grader blade.

An old established firm, Galion Iron Works & Mfg. Co. of Galion, Ohio, equipped their E-Z Lift Motor Patrol Grader with an IU engine and drive train for this 1930 model. Recognizing the crying need for a power-lift system on motor graders, Galion responded with this and other grader models. Eventually, Galion graders would become a major competitor for the developing Allis-Chalmers motor grader line. Curiously, this grader is marked, "Galion" as might be expected, but this is followed by "Allis-Chalmers-United," giving credence to the notion that A-C was from the beginning, the dominant force in the United organization.

Even though the A-C Monarch line of crawler tractors was eminently better prepared to move through heavy snow drifts, this Allis-Chalmers-United snow plow setup was available in a 1930 catalog. Maine Steel Products Co. of Portland, Maine, was listed as the builder of this Sargent snow plow. Although many of the Model IU Industrial machines were probably never sold to any extent either by Allis-Chalmers or by United, the immense variety of designs reflects an intense interest on the part of A-C to enhance its existing line with new and innovative designs.

Tractors

Pull-type graders were also available for the Model IU tractors. Although graders like this Ryan could be pulled by other tractors as well, those equipped for use with the Model IU had a special tongue arrangement for close coupling with the tractor. In addition, the tractor itself was modified with the seat occupying a sideways position so that the operator could attend to both the tractor and the grader. Mounting the grader controls forward made this setup truly a one-man operation. Ryan Dual Blade graders were built by the Ryan Corp. of Chicago.

First built in 1930, the Model UC tractor was nearly identical to the standard-tread Model U, except that a tricycle front end was substituted for the wide front of the latter model. Some historians have it that the UC came about because sales of the Model U were mediocre and the market for a row-crop model seemed more promising. While this may be partially true, there is abundant evidence within the U. S. Patents for the 1929-31 period to indicate that Walter F. Strehlow and others were indeed very active in perfecting a row-crop design and building a series of implements, each of which was tailor-made for use with the UC tractor.

Probably one of the most curious graders built for use with the Model IU tractor was the Baker One-Man Automatic Road Maintainer. As its lengthy title implies, this machine was intended to be completely automatic in its job of grading country roads. Power for operating the various mechanisms was secured from the grader wheels through malleable chain drives. Chances are that this outfit saw very few sales, especially since Allis-Chalmers and other companies were offering conventional machines that were far more practical.

Except for this photograph, no information has been secured regarding the adaptation of the Model IU tractor to the Sure-Trac crawler. This extremely rare photograph found in the Allis-Chalmers Archives bears no identification whatever, nor can any information regarding Sure-Trac be found either in A-C files or in the Author's research files. A close inspection of this photograph shows what appears to be a rubber track on this machine. Top and/or bottom track rollers are conspicuous by their total absence. Strong backlighting of the original makes this somewhat less than a perfect photograph, but no other images of the Sure-Trac have been located.

United Tractor & Equipment Co. was announced to the public in late 1928 or early 1929. Ostensibly, this combine of farm implement manufacturers and distributors would be in a position to market a very extensive line of farm implements and tractors. The Allis-Chalmers Model U tractor was selected to carry the "United" banner on its radiator. United tractors would be sold by United's member partners, while the identical tractor with "Allis-Chalmers" cast in place would be sold by A-C dealers. Full details have not been located concerning United, although the venture seems to have vanished within a couple of years after some glowing announcements at the beginning. Note that although the "United" name appears at the radiator crest, the name "Allis-Chalmers" is cast into the vertical side plates of the radiator shell.

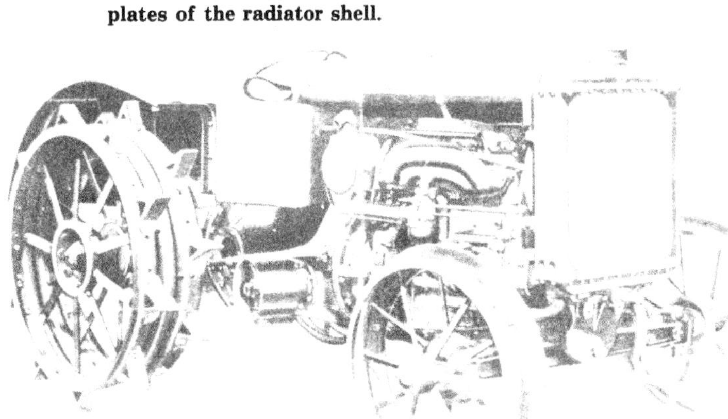

Tractors

Like the early Model U tractors, the UC was originally equipped with a Continental S10 engine. These tractors through Serial No. UC-1268 of 1933 were so equipped, followed then by the A-C built Model UM engine. Apparently the unit frame design for the Model U and Model UC tractors was the brainchild of Walter F. Strehlow, an A-C engineer. Patent No. 1,868,474 was issued to Strehlow and assigned to A-C in July, 1932. This patent was filed in November, 1930. On this same date Strehlow also filed patent applications for an "Agricultural Implement." These generally protected the drive-in cultivator design of Patents 1,906,430 and 1,906,431 issued in May, 1933.

Specific data regarding various changes requires a careful study of various A-C parts manuals, instruction books, and such other information as can be located. From all appearances, the UC line was streamlined in 1936, effective with tractor Serial No. 2701. Various engine modifications followed the same general pattern of the Model U standard-tread version, with hood changes and other obvious modifications being primarily of a cosmetic nature. The UC was tested at Nebraska under No. 238 of 1935.

The Model UC tractor could be equipped with a revolutionary new cultivator design that permitted mounting or detaching in five minutes time and required only a pair of pliers. Competing tractor cultivators contained plenty of extra iron and had to be installed a piece at a time. In fact, some competing designs would retain this mounting method for at least another decade. Allis-Chalmers in the meantime was actively searching for ways to make the concept of tractor-mounted implements more palatable to farmers.

Specific details remain a mystery regarding the so-called Model UC Cane Tractors. This special high-clearance design was apparently intended for work in sugar cane and other crops requiring a high-clearance tractor, but the precise number of units built for this rather specialized duty remains a mystery to current research efforts. This particular UC Cane Tractor on steel wheels was probably built during the 1930s. Official company serial number records indicate that production of the UC tractor ended in 1941. Other records indicate however, that at least a limited number were produced subsequently as Cane Tractors, so it is entirely possible that this particular unit was built during World War Two when a great many tractors were equipped with steel wheels, regardless of the purchaser's preference.

Another illustration of the Model UC Cane Tractor shows rubber tires used instead of steel wheels. This was probably the most popular configuration, although once again, no specific production figures have been located. The high arch of the front axle was probably designed to straddle growing crops. In addition, these special tractors appear to have been equipped with individual wheel brakes, a feature not available on the general purpose Model UC, but standard equipment for the Model IU industrial model. UC Cane Tractors were apparently built as late as 1952, and possibly may have been built subsequent to that year. These tractors were however, built only on special order for certain companies. They were then suitably modified for their intended purpose. An A-C Master Parts Index notes that parts for these tractors were available only through these special vendors, and not through the regular A-C parts network.

The 1928 purchase of Monarch Tractors resulted in the company's Springfield Works. From this plant Allis-Chalmers produced the majority of their construction machinery line. In January, 1935, Allis-Chalmers announced their new "Speedster," a rubber-tired outfit built not at all unlike this earlier dirt hauler. The Speedster and the later "Speed Ace" versions actually consisted of a Model U or IU rear end, together with an Allis-Chalmers Model K engine. It was fitted with a fifth wheel attachment for towing a large dirt trailer. Significantly different than the rubber-tired version is this steel-wheeled version with both the tractor and the trailer thus equipped. Also of special significance is the Allis-Chalmers elevating grader shown here loading an early Speedster.

The Speedster is shown here in a turning position, ready to return for another load. The special fifth-wheel system permitted very short turns, a frequent requirement in construction work. Whether the design included a built-in safety stop to protect the operator in case of too close a turn is unknown. Allis-Chalmers serial number records indicate that the Speed Ace began life with TW-1 in 1935 and ended with TW-87 in 1937. Occasional illustrations of the Speed Ace with rubber tires are found, but rarely is this unit seen with all-steel wheels. The term "Speedster" was dropped in favor of "Speed Ace" in January, 1936.

Already in 1930 Allis-Chalmers announced the Trail-It Dump Wagon for the Model IU tractors. In this particular case the IU is equipped with the Trackson crawler attachment. The dump wagon was built by Trail-It Co. of St. Paul, Minn. To what extent Allis-Chalmers capitalized on the Trail-It design for their so-called Speed Ace is unknown. By the late 1930s, clouds of war hung on the horizon, and postwar construction activities saw the almost universal use of motor scrapers which made the Speed Ace totally obsolete.

During 1935 Allis-Chalmers introduced their "Speedster," an attempt to mechanize the task of dirt moving. Already in 1931 the company had announced a "wagon crawler" which consisted of a steel box mounted on a crawler-type chassis. The Speedster introduced in 1935 became the "Speed Ace" for 1936 and 1937, but during a three year production run, only 87 units were built. The wagon was capable of 7½ yards capacity, and the tractor was a combination of the Model K engine with the "U" transmission. These early A-C machines did little except to prove their potential value, but the intervention of World War Two forced suspension of these activities until 1946.

Although it is often presumed that the Allis-Chalmers WC represented the company's first row-crop tractor, the UC tractor shown here carries this honor. With the UC came the company's revolutionary new drive-in cultivator, a brand new power lift system, and numerous other unique features. The marketing strategy was based in part on the company's "All-Crop" slogan, and in fact, the UC was originally introduced as the Allis-Chalmers All-Crop tractor. Despite its many features, sales of the All-Crop were not as high as expected. This led in part to the design of a smaller row-crop model. The new "W" tractor appeared in late 1933 and was formally announced to the public in 1934 as the WC model.

Tractors

The WC tractor was the first in the industry to use a so-called "square" engine — industry parlance translating to an engine having the same bore and stroke. In the case of the WC, the bore and stroke figure came in at 4 inches. For 1933, even the thought of using a square engine was anathema to most of the industry, since this practice was held to be sacred to the automotive scene, and even in this field, had only a limited following by 1933. The WC resulted largely from the inventive efforts of Walter F. Strehlow.

An A-C publicity photo shows an early WC tractor superimposed on a big Advance-Rumely steam engine. A-C bought out Rumely in 1931, providing the company with the manufacturing capacity of their LaPorte Works, and more importantly, giving Allis-Chalmers the established dealer organization they needed to sell and service their tractors. A major feature of the WC design was the extensive use of high tensile strength steel wherever possible. Allis-Chalmers, through its long experience with steam turbines and other machinery, had been instrumental in developing the use of high tensile steels, but initially at least, farmers were reluctant to accept the notion that a tractor weighing less than 3,200 pounds could pull a 2-bottom plow!

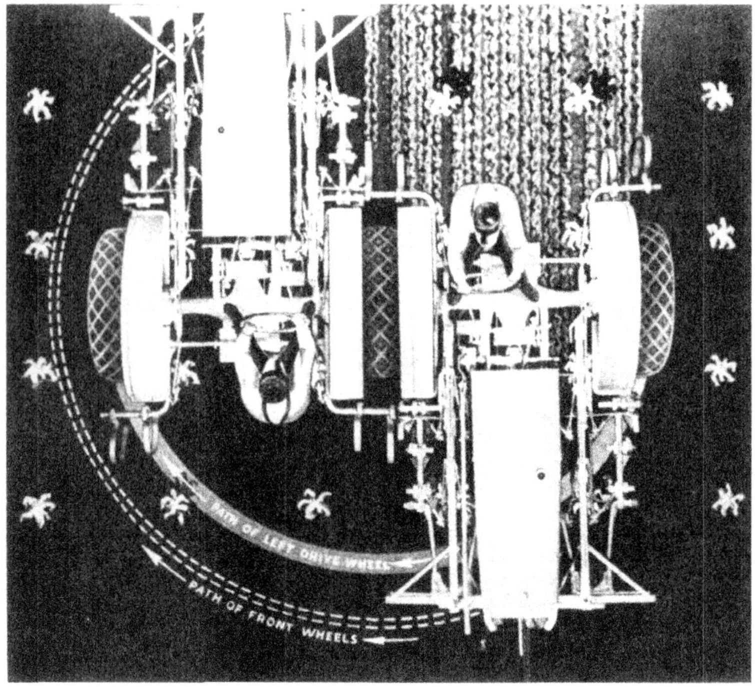

This unique overhead view was intended to illustrate the ease with which the WC and its drive-in cultivator could negotiate turns at the end of a corn field. By turning properly, not a single stalk of corn would be damaged. The WC was the first tractor offered to the public with pneumatic tires as ***standard equipment***. Early advertising of 1934 speaks of the WC as having, "Full 2-Plow Power — only $675 Steel Wheels; Only $825 Air Tires, f.o.b. Milwaukee." Despite its low price, acceptance came slowly, at least for the first couple of years.

A 1937 Allis-Chalmers catalog illustrates the state-of-the-art for WC tractors in that year. The company lost no time in submitting the WC to the Nebraska Tractor Test Laboratory at Lincoln. In May, 1934, testing began, and is recorded under Test No. 223. This history-making tractor was the first to be tested at Nebraska using pneumatic tires. Since no official tests had ever been run with this wheel type, dual tests made using first the pneumatic tires and following with steel wheels. The results were dramatic — an economy of 5.62 horsepower hours per gallon of fuel was realized on steel wheels, but this figure jumped to an amazing economy of 8.18 on rubber tires. This test alone was sufficient to convert a good many farmers to the merits of the WC tractor.

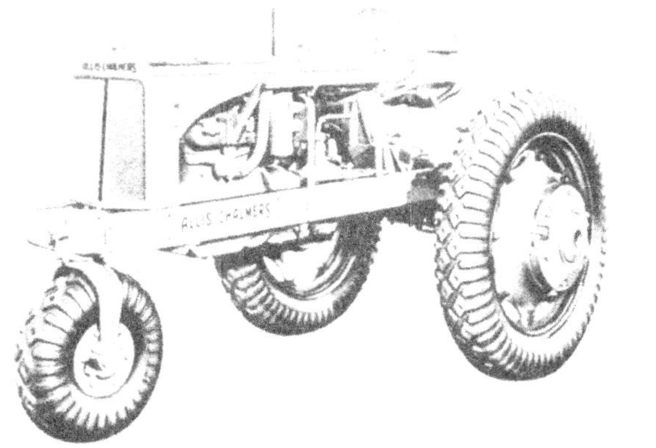

Although rubber tires were standard equipment, the 1937 WC shown here could also be furnished on steel wheels if so desired. The majority of pre-1930 tractors were sold for use with kerosene fuel, but the fact was that of these, most were poorly suited for the task. A-C was instrumental in promoting the use of gasoline fuel in their revolutionary new and lightweight engines during the early 1930s. Within the industry however, a new craze developed for the use of distillate — a low grade fuel that was offered at a lower price. In order to maintain its sales of WC tractors when this was a requirement, Allis-Chalmers tested a WC at Nebraska in 1938 using distillate fuel. This data is recorded in Test No. 303.

Recognizing the special needs of certain crop growers, Allis-Chalmers went all out to accommodate these special needs. In fact, Allis-Chalmers quickly became a recognized leader in adapting the row-crop tractors and implements to special requirements. A 1937 catalog illustrates an extensive series of front and rear wheel equipment that included reversible rubber-tired wheels as shown here to provide the proper width for several standard row spacings. In all, rubber tire equipment was available for a total of seven different tread widths.

U. S. Patent Office records indicate that the Allis-Chalmers "Hydromatic" tire trademark was first used on March 20, 1937. In another industry first, A-C announced in late 1937 or early 1938 that all rubber-tired WC tractors would come equipped with hydromatic tires on the drive wheels. To quote an A-C catalog, "Instead of being filled entirely with air, as in the past, these tires are three-fourths filled with a special freeze-resistant liquid ballast, that has a preserving effect upon the inner tube rubber." In addition to providing the extra weight necessary for efficient traction, the hydromatic tires also lowered the center of gravity for improved safety.

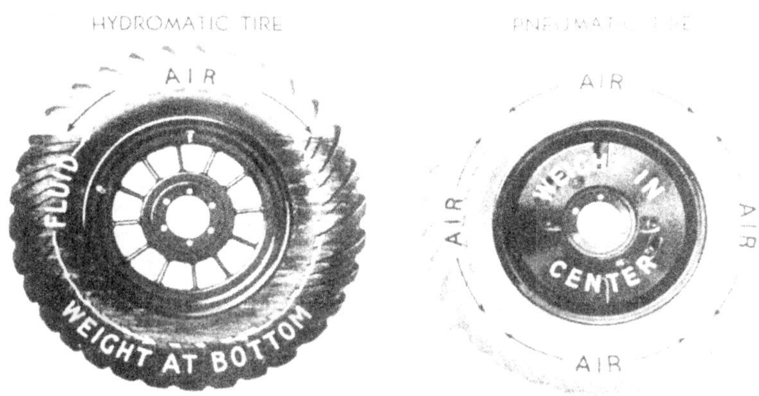

Although precise data has not been located, from all appearances the WC tractor underwent a facelift in 1938, effective with No. 74330. This late-year change saw a new hood, gas tank, and grille, along with the inclusion of electric starter and lights as standard equipment. The steel-wheeled version shown here without starter and lights was priced at $970. Adding the convenience of a complete electrical system added only another $25 to the list price. Only a few streamlined WC tractors were produced in 1938, and chances are that some of the unstyled models might have been assembled in early 1939 to use up the parts inventory.

For those who preferred the ease of steering afforded by rubber tires, but still adhered to the idea that steel wheels were necessary for good traction, Allis-Chalmers offered this rubber-steel combination of the WC tractor as late as 1940. Some farmers had purchased their very first tractor in the 1930s, so the change from horses to a steel-wheeled tractor, and from there to a bright orange tractor with electric starter and rubber tires was probably too much to expect within a decade. This catalog illustration graphically portrays the accommodation that A-C was willing to make for its dealers and ultimately for its farmer customers.

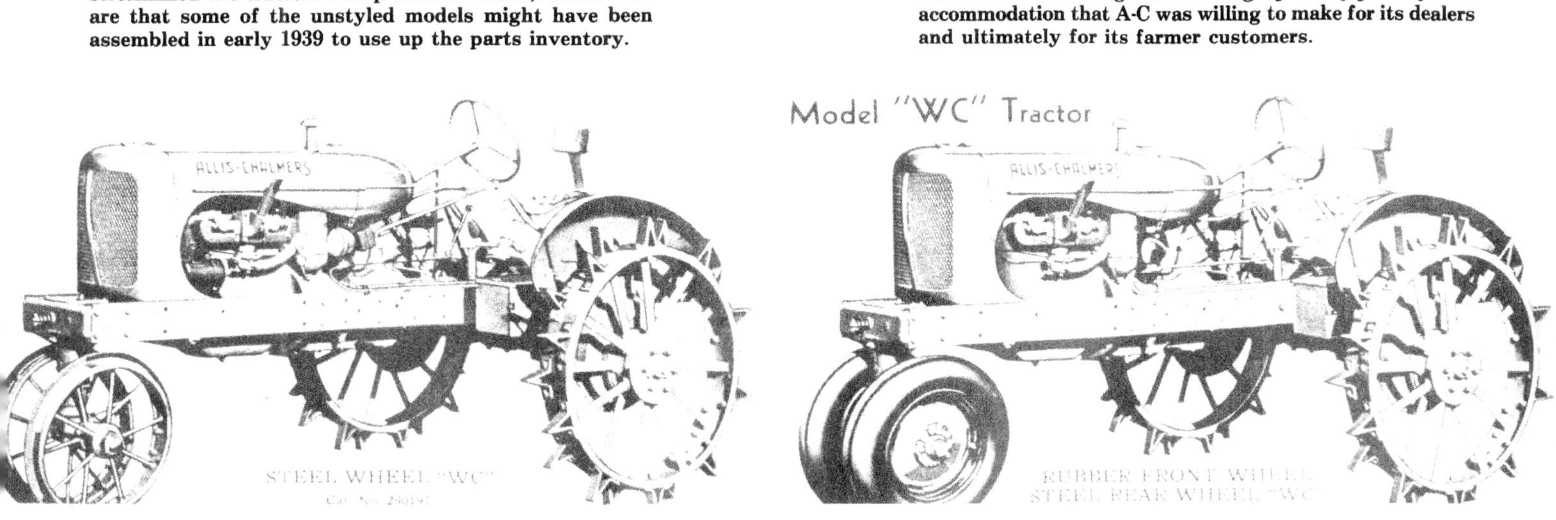

Tractors

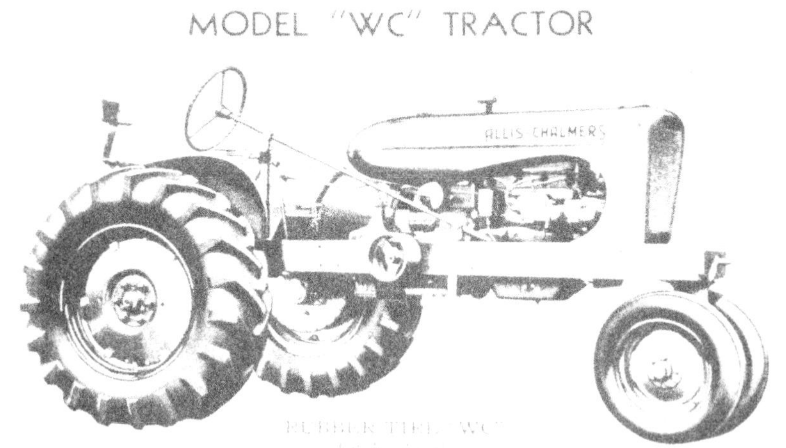

The Allis-Chalmers WC tractor of 1947 was virtually identical to the pre-war design. In fact, limited quantities of WC tractors were built throughout World War Two, despite the relative scarcity of materials during this period. In order to provide these tractors, some WC's of the World War Two period were built with a pressed steel rear end, compared to the usual casting found before, during, and after global hostilities. For 1947 this WC tractor with electric starter, lights, and full rubber was priced at $1,290. It was listed as having a factory weight of 3,310 pounds. These tractors used 11-28 rear tires, compared to 11.25-24 tires used with spoke wheels on the early WC tractors.

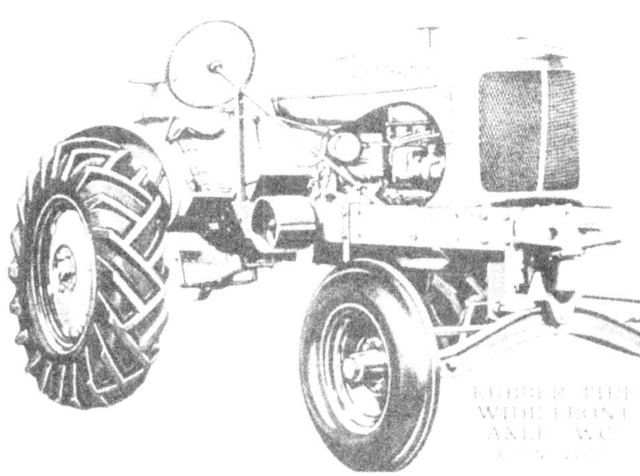

Production of the Allis-Chalmers WC tractor ended in 1948. Through this tractor Allis-Chalmers had become a prominent force in agricultural machinery development. During the 1930s the sale of WC tractors and other farm machinery accounted for a major part of company profits, and the total impact on the American market gave A-C for a time at least, third ranking in terms of output, of American farm machinery manufacturers. Despite a long and illustrious career, the WC had outlived its usefulness, and would be replaced by the newly designed WD series. Curiously, even though A-C was extremely innovative in applying new technology to their tractors, the WC continued to use hand-operated brakes to the very end!

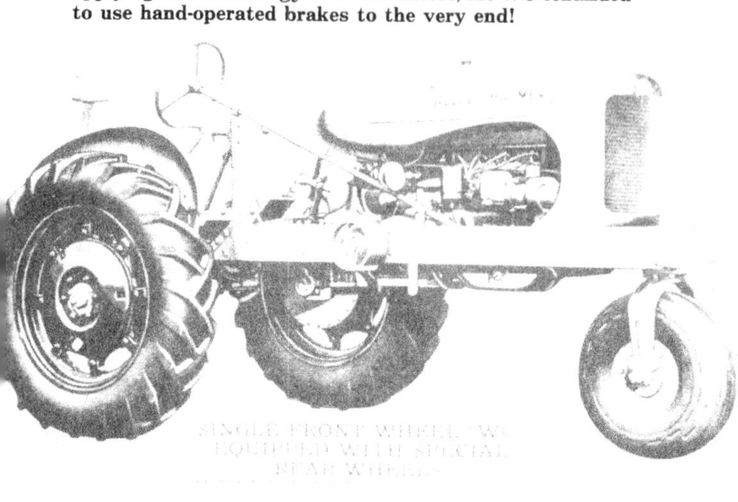

The contribution of Allis-Chalmers to the war effort of 1941-45 is well documented. During this period the company received considerable recognition from the United States Government for its patriotic efforts. At one point, President Franklin D. Roosevelt visited the West Allis Works. This photograph illustrates a portion of an advertising campaign launched by Allis-Chalmers about 1943 or 1944 in behalf of its agricultural customers. The company called this their "Farm Commando Issue" and went so far as to make brightly colored decals with the Farm Commando slogan. These were to be used on various A-C machines.

The Wide Front Axle "WC" was a distinctly separate version for specialized farming operations. The arched front axle was intended to straddle growing crops, and although the front tread was locked in at 60 inches, the rear tread remained adjustable from 65 to 76 inches. The extent of production for this wide-tread version of the WC is presently unknown since those records researched for this project do not differentiate between various styles. The tractor shown here carried a 1947 price of $1,380. An A-C Dealer Price Book also indicates that the wide front end was available for field installation at a cost of $198.

By the time production of WC tractors ended in 1948 the metamorphosis was completed — this lightweight 2-plow tractor had moved from a neophyte to one of America's popular farm tractor models. Even the diehards were forced to concede that the big cast iron behemoths of the 1930s were no match in terms of useable horsepower output. The sleek hood lines of this late model WC tractor are little different than the WD model which followed. Numerous patents were issued during the 1930s and 1940s to protect various features of the WC design. Many of these were initiated by Walter F. Shrehlow, an Allis-Chalmers engineer.

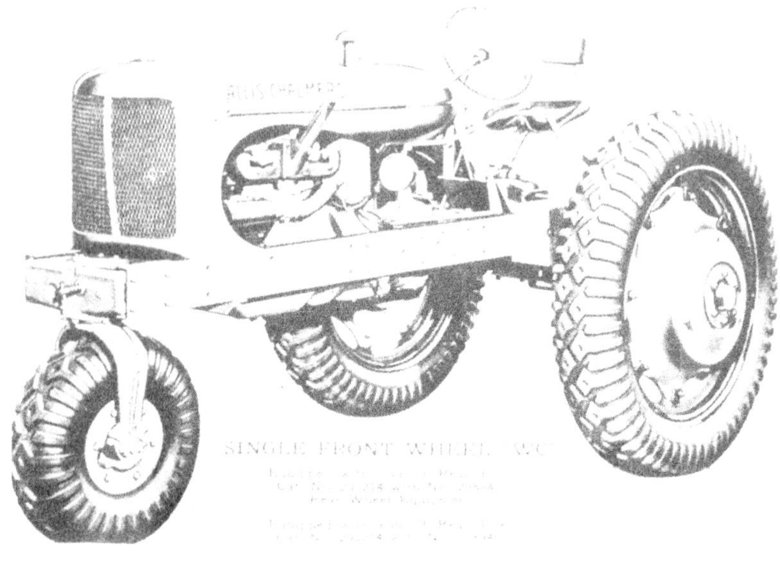

Tractors

Commonly thought of in terms of its agricultural uses, the WC also saw some application to specialized duties. This undated photograph from the Allis-Chalmers Archives illustrates a WC moving an airplane out of a hanger. WC tractors were also built in limited numbers as a special Cane Tractor, and the same WC chassis was slightly modified for the WC Road Maintainers. These are illustrated in the *Motor Grader* section of this volume. Beyond these special applications were many other requirements too numerous to mention.

Like other A-C tractor models of the late 1930s and 1940s, the WF was available with full steel instead of the standard pneumatic tires. During World War Two, the choice was not always up to the dealer or the farmer. Since rubber was in short supply, steel wheels were often furnished. Many of these tractors were then equipped with standard rubber tires when they became available for civilian use after 1945. The WC tractor of Test No. 304 at Nebraska delivered a rated output of 26.8 belt horsepower, with a comparable figure expected of the WF as well.

Sales of the standard-tread WF tractor were but a fraction of those enjoyed by the immensely popular WC row-crop model. Due in part to the lack of materials caused by World War Two, no WF tractors were built in 1943. Production resumed in 1944, and ended entirely in 1951. This WF tractor is shown working in an orchard. For 1948 the WF on steel wheels listed at $971; the same tractor with rubber tires, plus electric starter and lights sold for $1,210. Those built through No. 1903 of 1941 were not furnished with starter and lights as standard equipment.

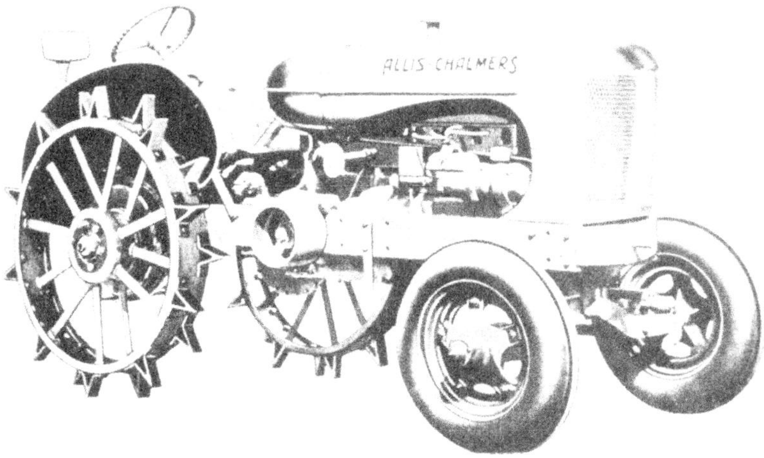

Production of the Allis-Chalmers WF tractor began in 1940. The WF was a standard-tread version of the row-crop WC tractor — nothing more, nothing less. Both tractors used the same A-C Type "W" engine with a 4 inch bore and stroke. Like the WC, the WF was rated at 1,300 rpm. The WF tractor was never tested at Nebraska. A twofold possibility exists in this regard. First of all, since both tractors were virtually identical save for the tread design, it may not have been deemed necessary. Secondly, all tractor testing at Lincoln ceased in November, 1941, and did not again resume until 1946, due to World War Two.

The standard-tread Model A tractor sold very poorly compared to the popular WC tractor, or in fact, to most other models of the A-C line. Rated at 1000 rpm, it used a 4¾ x 6½ inch, four-cylinder engine. Four forward speeds of 2½, 3½, 4½, and 9½ mph were provided. Total weight was 7,120 pounds. Production of this rubber-tired model bagan in 1936, finally ending in 1941. The last published price was $1,495. This tractor was not tested at Nebraska.

Tractors

Introduced in 1937, the Model B tractor was tested at Nebraska under No. 302 of May, 1938. Apparently, those tractors under No. 43220 of mid-1940 were not furnished with starter and lights as standard equipment. Weighing only about 2,000 pounds, the Model B was initially priced at $495. A 1940 price book shows this tractor weighing in at 1,860 pounds. Without electrical equipment it sold for $518. The same tractor with 9-24 rear tires, muffler, radiator shutter, starter, and lights listed at $570. An adjustable front axle substituted at the factory for the standard axle added another $19.75. The rear-mounted belt pulley and PTO shaft came as a $35 extra.

An action view of the Model B tractor illustrates what was probably a homebrew combination of snow blade on the front and buzz saw on the back of this tractor. Adding rear wheel chains boosted traction somewhat, and aided its adaptation to snow removal. The rated belt output of 14 horsepower was more than adequate for sawing wood, and the little four-cylinder engine was very responsive to load changes. This particular Model B is equipped with a complete electrical system. Without a doubt, the lightweight design of the Model B had a dramatic effect on future tractor designs, both by Allis-Chalmers and by their competitors.

Allis-Chalmers offered not just the Model B tractor — they also offered a virtual potpourri of implements tailor-made to fit this tractor. The overall strategy included making the Model B a complete power farming system for the small farmer, and in fact, this is exactly what happened in many instances. With slightly over 10 drawbar horsepower available, the Model B was perfect as a cultivator tractor. Its high-arch front axle, along with excellent rear axle clearance made the Model B well suited to the majority of special crops.

A catalog view of the Model B illustrates its clean lines and its no-nonsense design. This tractor was intentionally built without any frills. Allis-Chalmers had pioneered the use of forgings and high-strength steel components, not only in their tractors, but likewise in A-C steam turbines and other machinery as well. The company's remarkable achievements in the use of high-strength components for tractor manufacturing served the company's interests initially through greatly increased sales. In the greater perspective however, the entire tractor industry benefitted since competing manufacturers were forced to follow suit.

While too small for many farms, the Model B was a dream come true for truck farmers and nurserymen. Much of the cultivation previously done by hand or with draft animals could now be achieved in a fraction of the time with a Model B tractor and a matching cultivator. The standard front axle could be set for 43 or 50-inch tread spacings, but an additional cost the adjustable front axle provided ten different spacings from 38 to 60 inches. Rear wheel treads could likewise be changed by adjusting rims or clamps and reversing the rear wheels.

This 1943 photograph from the Allis-Chalmers Archives illustrates a rather rare combination — the Model B tractor on full steel! Because of the company's apparent policy that gave customers the maximum number of options, it is of course, quite possible that steel wheels may have been available in lieu of rubber tires. The severe shortage of rubber during World War Two hindered the use of rubber tires on farm tractors to a considerable extent, so steel wheels were necessary. A-C noted that farmers were turning timber into a cash crop, particularly walnut for gun-stocks.

The extent to which various special models of the Model B tractor was built is not known with any certainty. Odds are however, that relatively few of the Asparagus Special and Potato Special tractors shown here were built. The tractor itself was unmodified, but the special front axle and rear wheel equipment consituted the significant changes in these two tractors. Production of the Model B tractor ceased in 1957 after twenty years of production. Detailed engine specifications for this and most other A-C tractors are not included, since this information is already contained in the Author's *Nebraska Tractor Tests Since 1920*.

THE MODEL "B" ASPARAGUS SPECIAL

THE MODEL "B" POTATO SPECIAL

Among the special equipment built for the Model B was this two-row bean cutter and windrower attachment. It could be fitted to the mainframe of the standard cultivator, or could be furnished with the necessary mounting supports. It was designed to cut and lift two rows of beans at a time and move them into a single windrow, ready for the final harvest. A vine turner attachment was also available if needed. An $18 hydraulic pump assembly was another option for the Model B tractor. This eliminated hand lifting of various implements.

With this photograph Allis-Chalmers announced the Model IB tractor in the November-December, 1939, issue of *Timber Topics*, a journal printed by A-C for their lumbermen clients. In speaking of the "IB" Allis-Chalmers noted that it "drives like a car as it has similar controls with a foot controlled throttle, clutch, and brake. In addition there are individual hand brakes on each rear wheel." The low, compact construction of the IB industrial tractor made it ideal for work in small, cramped areas. It used the same four-cylinder, 3⅜ x 3½ inch engine as the general purpose Model B tractor.

Tractors

Another view of the IB industrial tractor illustrates special bumper equipment. It was designed to protect the tractor as well as providing a substantial means of pushing empty rail cars and similar duties. A-C serial number listings indicate that the IB tractors built prior to 1946 used regular "B" serial numbers and no special prefix. Beginning in 1946 however, and continuing through the end of production, probably in 1957, the "IB" prefix was used. Unfortunately, this research project did not uncover any pricing information on the IB tractors.

A February, 1943 issue of *Timber Topics* illustrates a home-made automatic hitch attached to an Allis-Chalmers Model B tractor. First use of this attachment was made by Merit Oak Flooring Co. of Birmingham, Ala. The funnel shaped guide, fabricated from sheet iron is bolted to the tractor frame. The wagon tongue visible in the lower right hand corner of this illustration is poised for entry into the automatic hitch. To unhitch the load, the tractor operator simply pulls a pin and drives away. Our research has not discovered a connection between this setup and a strikingly similar system adapted to later tractors within the A-C Snap-Coupler system.

While the Model B used a wide front axle, the Model C introduced in 1940 featured a narrow-front as standard equipment. This immensely popular tractor used a four-cylinder engine having a 3⅜ x 3½ inch bore and stroke for a displacement of 125 cubic inches. For 1940, the Model C shown here with electric lights retailed at $595. When equipped with an adjustable wide front axle the price jumped only slightly to $610. The engine could be secured with either high or low compression ratio. Standard compression engines could burn distillate tractor fuel, or low octane gasoline. High compression engines could use gasoline only. The hog houses, here being pulled into place, were not an A-C accessory.

Although electric starter and lights came as standard equipment, the rear power take off and clutch-type belt pulley came as extra cost options — in this instance, an additional $38.50. Likewise, the hydraulic pump assembly added another $9.50 and a hydraulic ram assembly for raising and lowering implements was priced at $5.50. Prices rose substantially from 1940 to 1948. This Model C, priced at $595 in 1940 carried a 1948 list price of $980! In addition to other attachments, those applications requiring a standard ASAE hookup for PTO-operated machines needed an additional drawbar assembly priced at $3.65.

As with many other models of the wartime A-C line, the Model C was produced in limited numbers at least, with steel wheels. Current research has not revealed whether steel wheels were available on special order outside of this time period, but if they were in fact available, the odds are that very, very few tractors were thus equipped. Row-crop tractors of the 1930s with their steel feet had bumped and jolted virtually every farmer to pieces, so rare was the time that rubber tires were passed up in favor of steel wheels.

As with the Model B tractor, a multitude of implements were tailor-made for the Model C as well. Allis-Chalmers was often ahead of its time in announcing new machines — this 1947 mounted lister-planter was probably several years premature. The Model C was tested at Nebraska in October, 1940. Probably in response to the then-current craze of distillate and tractor fuel as the ultimate answer compared to gasoline, A-C submitted the Model C in Test No. 363 equipped to burn distillate. Test No. 364 saw another Model C, but this time using gasoline fuel.

A 1949 Allis-Chalmers catalog for the Model C tractor makes more than a passing comment in regard to the company's Quick-Hitch cultivators and companion equipment for the Model C tractor. Once it was initially set up and adjusted, much of this equipment could be attached or detached in ten minutes time — a far cry from the back breaking task of attaching mounting implements through sheer muscle power and boundless determination. As might have been expected, A-C used high strength, light weight steel components wherever possible in the construction of mounted or towed implements. Construction of the Model C ended in 1950 when it was replaced with the Model CA tractor.

Many different implements were available for use with the Model C tractors. The 80 Series cultivators for the Model C were offered in several different models and styles to suit specific crops and tillage practices. Generally, these were priced in the $90-$100 range. One-way and two-way pick-up plows were available, as was a semi-mounted disc plow. Bedder Planters of the 300 Series could be purchased in no less than five different distinct models. Seven streamlined styles of the 800 Series planters were also available. To accommodate the specialized needs of truck farmers the No. 860 vegetable planter was built in 4-row and 6-row units.

Ostensibly intended to be a competitor to Harvester's concurrent F-14 tractor, the Model RC tractor was to all intents and purposes, a WC chassis and drive train powered by a 125 cubic inch engine as used in the Model B and Model C tractors. Introduced in 1939, the ill-fated RC tractor was out of production in 1941, and did not resume after World War Two. Advertising literature on this tractor is relatively scarce, but one photocopy in the hands of the Author notes that the RC offered "2-Plow Power for the Thrifty Buyer." Since the RC tractor was essentially using a WC chassis, the specialized series of implements developed for the WC would likewise fit on this tractor.

Tractors

Allis-Chalmers probably expected reasonably good sales from their new RC tractor, so it was tested at Nebraska under No. 316 of April, 1939. The reduction in price however, between the RC and the somewhat larger WC engines was not of enough significance to attract the attention of more than a relative handful of buyers, especially when the vast majority knew that they were in fact, buying an underpowered WC to begin with. Due to very disappointing sales, plus the onset of World War Two, production of the RC tractor ended in 1941.

Allis-Chalmers pioneered the option of power-adjust rear wheels in 1948. Within ten years, all but one major American tractor builder made this feature either standard or optional equipment. The CA tractor shown here was so equipped, and in fact embodied a number of new features found on the contemporary WD tractor. Although the CA was basically the same tractor as its immediate predecessor, the Model C, the 125 cubic inch engine was speeded up to 1650 rpm. On Nebraska Test No. 453 this gave a rated output of 22.97 belt horsepower, substantially higher than was achieved with the 1500 rpm on the Model C. Production of the Model CA tractor continued into 1957 when it was finally superseded by the new "D" series tractors.

The Allis-Chalmers Model G tractor announced in 1948 heralded a new era in farm power. Designed especially for small farms, vegetable gardeners, and nurserymen, the Model G was a small compact tractor which A-C called the "central unit of a new system of motorized farm tools." This so-called new system did in fact include an impressive array of implements that included plows, planters, and cultivators. Produced at the company's Gadsden (Alabama) Works, the Model G featured a rear-mounted engine. This provided the necessary traction for the rear wheels and at the same time gave the operator unequalled vision while planting and cultivating delicate crops.

The massive array of G-implements included a one-bottom plow — this one is working in heavy sod. A Continental N-62 engine was featured in this tractor. Its four-cylinder design and a 2⅜ x 3½ inch bore and stroke yielded a displacement of 62 cubic inches. On Nebraska Test No. 398 of July, 1948 this tractor delivered 9.6 drawbar and 10.9 belt horsepower. Extra cost options included a belt pulley for $19, master tool carrier at $25, and a hydraulic pump system for $99. In 1955, the last year of production, this tractor listed at $970.

In addition to its use by truck farmers, Allis-Chalmers also attempted to promote the Model G for the soybean farmer. A special catalog devoted to this purpose illustrated a special two-row drill planter for soybeans. One, two, and three-row tool bar cultivators were also available in a number of configurations to suit specific row widths and cultivating practices. Because of their limited production, their extreme usefulness in certain applications, and due to their unique construction, the Allis-Chalmers Model G tractors have become a highly sought after machine by collectors.

During the early 1950s Allis-Chalmers engineers attempted to adapt the Model G to other duties beyond its usual planting and cultivating chores. This rare photograph from the Allis-Chalmers Archives illustrates a Model G tractor equipped with a front-mounted windrower. Other schemes included a front-mounted hay rake. This outfit worked fine so long as the wind was at the operator's back. With a headwind, operating the "self-propelled" rake was virtually impossible, since the operator was caught in a choking cloud of dust from which there was no escape. Due to these, and other factors, Model G tractors saw little further development.

This 1943 photograph is one of the few to illustrate the "D" experimental tractor of 1943. Chances are that this style was developed in anticipation of the end of World War Two. When hostilities finally ceased in 1945, production of the WC continued, and nothing at all was heard of the "D" tractor. One major advantage of the "D" tractor was the ability to offer fourteen different wheel widths ranging from 52 to 82 inches. Presumably this tractor used the standard 4 x 4 engine found in the WC.

Shown here is the Allis-Chalmers Model F tractor of 1947. It is coupled to an All-Crop 60 combine and harvesting rye near Elm Grove, Wis. Only a few of these tractors were built, and of these, little information is available. The hood and grille design would be evident in later A-C tractor models, but for the time being at least, Allis-Chalmers would continue with the WC tractor, and follow it in 1948 with an improved version called the WD. Since so few Model F tractors were built, the usual sources of serial number listings are no help in determining their total production.

Introduced in 1948, the Allis-Chalmers WD tractor embodied several new design features and included the best aspects of the earlier, and immensely popular, WC tractor. The adjustable front axle shown on this particular tractor could be purchased as an option, but more important perhaps was the addition of power-adjustable rear wheels. The rather simple device was extremely quick and easy — so much so, that by 1959, all but one American tractor builder was using power-adjusted rear wheels.

Photographs of A-C tractor assembly lines have been extremely difficult to secure, although this particular illustration from a magazine clipping illustrates part of the line. Allis-Chalmers used a system of sub-assemblies on the tractor line. This plan of assembling small sections of the tractor, assembling these to other components, and continuing the process on a gradual basis finally led to a completed tractor. Overhead conveyors carried the various sub-assemblies to the main line. Only about 150 feet long, the main line terminated in cleaning areas and finally in the spray painting line.

Tractors

Allis-Chalmers WD tractors were first built in 1948 with production ending in 1953. Like its immediate ancestor, the WC, these tractors were immensely popular. Recognizing the need for improvements, A-C offered the power-adjustable rear wheel system as standard equipment, along with introducing a Traction Booster system on the WD series. A wet clutch between the PTO shaft and the transmission provided continuous PTO power totally independent of ground travel. The main clutch did however, disengage both drives simultaneously.

Introduced in 1953 the WD-45 tractor followed the same general lines as the earlier WD tractor, but could deliver substantially more power than its predecessor. This was due to the introduction of a new four-cylinder "Power Crater" engine with a 4 x 4½ inch bore and stroke. Rated at 1400 rpm, it had a displacement of 226 cubic inches. Test No. 499, begun at Nebraska in July, 1953 demonstrated 45.27 maximum belt and 40.01 maximum drawbar horsepower, significantly higher than was achieved with WD. Shown here is a WD-45 and an A-C Roto-Baler working near Carlinville, Ill.

Allis-Chalmers WD tractors were available with three front-end styles — the single front wheel shown here, dual front wheels of the usual tricycle style, or an adjustable front axle version. As was true with the earlier WC tractor, a wide variety of implements were available for the WD, including many that were tailor-made for this tractor. Nebraska Test No. 399 of July, 1948, was run using so-called "tractor fuel." Popular for only a short time, tractor fuel fell into disfavor. Test No. 440 of 1950 was run using gasoline fuel and delivered far better horsepower outputs on both the drawbar and in the belt.

Among other tasks, the WD was extremely popular as a cultivator. Its short wheel base and overall compactness made it easy to handle and easy to turn at the ends of the field. A-C had pioneered the use of hydraulics, and the addition of a hydraulic lift system enhanced this tractor even further. During 1953, the last year of production, Allis-Chalmers incorporated their newly developed Snap-Coupler system into the WD and CA tractors. With the Snap-Coupler came an entire series of implements developed especially for use with this system.

In 1953, the last year of production, the WD listed at $1,830. Weighing slightly over 4,000 pounds, it was regularly equipped with a gasoline engine, dual front wheels, power-adjustable rear wheels, 4-speed transmission, starter, lights, PTO shaft, and hydraulic control system. Extra cost items included a belt pulley for $28; and adjustable wide front axle in lieu of narrow front for $130 additional; and a remote hydraulic ram for $45. The 4-cylinder, 201 cubic inch engine used a 4 inch bore and stroke, with a rated engine speed of 1,400 rpm.

Tractors

Working near Fayetteville, N.C., this WD-45 tractor is pulling 2 30 Series Bush Hog disk. During its 1953-1957 production period the WD-45 saw the addition of new features, including the 1956 enhancement of factory-installed power steering. This feature quickly proved to be one of the biggest labor savers in tractor operation. Saving labor also helped minimize fatigue, and this led to greater operator safety. In order to accommodate customers in all areas, the WD-45 could be furnished with a regular gasoline manifold or with a two-fuel manifold suitable for use with either tractor fuel or gasoline. Other features included the Traction Booster and Snap Coupler systems.

This photograph from an original A-C negative shows a WD-45 LPG tractor with a two-row mounted corn picker. This specially equipped butane model was tested at Nebraska under No. 512 of October, 1953. At this particular point in time, lp-gas was a widely acclaimed substitute for gasoline and other liquid fuels. Certain changes were required, including raising the compression ratio from 6.5:1 up to 7.2:1. Special equipment was also required to store and handle butane fuel, both on the tractor and in farm storage. For 1957 the WD-45 gasoline model listed at $2,575. LP-gas equipment added another $333 to this price.

The first WD-45 Diesel was built in 1955, and this style remained in production until 1957. An Allis-Chalmers six-cylinder diesel engine was used. Its 3$\frac{7}{16}$ x 4$\frac{1}{8}$ inch bore and stroke gave a displacement of 230 cubic inches. Rated at 1625 rpm, it used a 15.5:1 compression ratio. Like other WD-45 models, it was available with three different front-end conversions for maximum useability, regardless of the row width. When the WD-45D tractor hit the market in 1956 it marked the first instance where Allis-Chalmers had furnished wheel tractors with a diesel engine. It carried a list price of about $3,400. Testing of the WD-45 Diesel is recorded in Nebraska Test No. 563 of October, 1955.

Very few photographs of the Model H-1 experimental model have been located, but this particular view shows an H-1 suitably equipped with orchard-type fenders working in a California orchard. Very little specific information has been located on this model, but one potential application was in orchard and vineyard work. During 1945 two of the H-1 tractors were shipped to the company's Oxnard Works in California. They were equipped with cultivating equipment from LaCrosse Works, and apparently were modified at Oxnard Works to the specific requirements found in that area. For any one of many reasons, the H-1 tractor project was abandoned. Unfortunately, Allis-Chalmers was a full decade ahead of the market — eventually a strikingly similar design would create an entirely new industry manufacturing skid-steer loaders and tractors.

This photograph of 1945 illustrates the Model "H-1" experimental tractor. It used a Model C engine, suitably modified for the task. Rockford disc-type steering clutches were used — these were operated with hand levers. Clark Equipment Co. furnished the transmission — it featured four speeds forward or revese. Roller chains and spur gears were used in the final drives. Contracting band brakes were used on each steering clutch. This experimental tractor was also equipped with a hydraulic pump.

Introduced in 1957, the D-14 heralded the beginning of an entirely new A-C tractor line. A comparison of the D-14 with the earlier WD-45 illustrates some similarities, and of course, included all the new features of the latter model. Included was factory-installed power steering, first used on the WD-45 and quickly adapted by nearly every tractor builder. The D-14 was built for either gasoline or LP-gas, and a very few experimental models may have been equipped with diesel engines. Nebraska Test No. 623 covers the gasoline model, and Test No. 645 details the specifications and fuel economy on the LP-gas version.

As previously noted, the D-14 LP-gas model used the same engine as the gasoline version, except for having a higher compression ratio, and of course, this tractor was equipped with the necessary special equipment to handle liquefied petroleum gas. A D-14 tricycle model using gasoline fuel listed at $2,875, but the addition of LP-gas equipment as shown here raised the price by another $310. Power steering was a $153 option, and the belt pulley shown on this tractor added another $58. The hydraulic system, including the Snap Coupler and Traction Booster came as standard equipment, but if these items were not wanted with the tractor a deduction of $225 was made from the list price.

A 1957 photograph illustrates the D-14 and D-17 tractors side by side. These two models were the beginning of an entirely new tractor line for Allis-Chalmers. The D-14 used the company's own G-149 engine, a 4-cylinder style having a $3\frac{1}{2} \times 3\frac{7}{8}$ inch bore and stroke. The same engine had an excellent track record and was widely used for industrial applications. Produced at the company's Harvey (Illinois) Works, this engine was virtually an off-the-shelf model. A 7.5:1 compression ratio was used for gasoline, but a ratio of 8.5:1 was used for LP-gas.

In addition to the standard tricycle style, Allis-Chalmers offered the D-14 in several additional configurations. A D-14 high-clearance tractor was priced at $3,275. The D-14 Orchard tractor shown here listed at $3,525. It was equipped with the company's Roll-Shift front axle, a wide-front that could be moved in or out using the power steering cylinders for the task. The A-C Power Director system provided a high and low range in each gear — two over-center, triple plate clutches ran in oil for the longest possible life. Hand-operated, the Power Director also had a neutral position for added safety.

All the necessary parts, including special fenders and other accessories for orchard work could be supplied for field installation. The overhead view of the D-14 Orchard model illustrates the special screening supplied for the protection of the operator. Actual production of the D-14 appears to have ended in 1960. However, D-14 tractors are listed in the June, 1962, Dealer's Price Book. The logical assumption is that these were units built prior to the end of 1960, although no specific closing date has been located. A limited number of D-14 tractors were sold with special rice equipment which included 7.50-16 inch front and 18.4-26 inch rear tires.

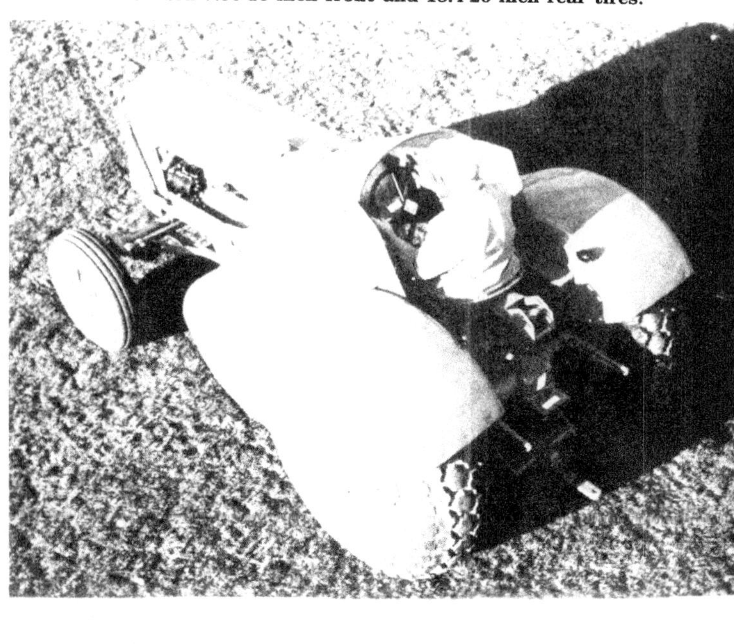

Tractors

The immensely popular D-17 tractor was introduced in 1957. Production of this tractor included the original version shown here, plus three other revisions named Series II, Series III, and Series IV. The Series II began in 1959, followed by the Series III in 1964, and Series IV in 1965. These variations are included elsewhere in this section. The 4-cylinder engine with its 4 x 4½ inch bore and stroke gave a displacement of 226 cubic inches for the gasoline version. Priced at approximately $3,550, the standard gasoline model could also be furnished for propane gas at an additional cost. The D-17 line also include a 6-cylinder diesel version. The entire D-17 series was of the 60-plus horsepower class.

Although the base price for the D-17 was based on the tricycle version shown here, a great many of these tractors were equipped at the factory with the Roll-Shift front axle — a power adjusted axle of extreme simplicity. Also available was a D-17 High Clearance model — the gasoline version was priced at $4,115, but the D-17 High Clearance Diesel listed at $5,010. Yet another style was the D-17 Orchard, available with a gasoline engine at $4,260 or in a diesel style at $5,150.

The D-17 was tested at Nebraska in 1957, with the results appearing under No. 635 for the gasoline version, No. 636 for the diesel tractor, and No. 644 for the propane version. Reference to these tests will show the specifications and economy ratings of the three different models. This 1957 photograph illustrates the relative size of the D-17 tractor against the somewhat smaller D-14 model in the background. In addition to the standard farm tractor design, A-C furnished the D-17 Utility tractor. This series was intended for use with many different machines, including backhoes, brooms, half-tracks, rotary tillers, loaders, and mowers.

The D-17 Series II tractor differed little from its original progenitor except for slightly different decals and other changes of a minor nature. In retrospect, the changes were so slight that industry reference manuals seldom distinguish between Series I and Series II tractors. The Power Director clutch provided on-the-go shifting from high to low range when the going got tough. Optional rear wheel weights were of sectional design. Each of the four segments weighed 70 pounds and required only a single bolt. This D-17 is pulling an A-C No. 64 plow near Plainview, Neb.

A composite photograph illustrates the D-17 tractors with special Wheatland fenders. To the left is a D-17 Wheatland gasoline style; in the center is an identical tractor except that it uses propane fuel; and on the right is a D-17 Diesel Wheatland model. The gasoline version of the D-17 used a 4-cylinder, 4 x 4½ inch engine rated at 1650 rpm. A 262 cubic inch, 6-cylinder engine was used in the diesel version. Rated at 1650 rpm, it was designed with a 3$\frac{9}{16}$ x 4$\frac{3}{8}$ inch bore and stroke.

Shown here is a D-17 Wheatland tractor equipped with special propane fuel accessories. It is chisel plowing near Yuma, Colo. Information is rather scarce on the Wheatland version, although its most evident difference from standard models seems to be in the extra-wide fenders, designed especially to protect the operator from flying particles of dirt. This particular tractor also appears to be equipped with rear tires somewhat wider than ordinary.

Allis-Chalmers built the D-14 and D-17 tractors primarily for farm use, but also offered them as Utility tractors, suitable for farm or industrial applications. Various company advertising indicates that an extensive number of attachments were available for these tractors. Included was the backhoe attachment of the Henry Corp., along with a host of other items, all of which were approved for use with the D-14 and D-17 tractors. This heavy duty push-blade is one such attachment. Note also the pie-shaped wheel weights used on the D-17. Each segment weighed 70 pounds and could be attached with a single bolt.

The D-17 Series III tractor appears to have been built in the 1962-1964 period. Differences between this and earlier models of the D-17 were primarily cosmetic in nature. All D-17 tractors retained the same engine specifications throughout production, and all models of this tractor qualified under the initial Nebraska Test results. A notable difference is that the hood decals now used a white background which ran the full length of the hood. Also, the radiator grille was now finished in white instead of the customary black finish of earlier production.

This D-17 Series III tractor is working at the Allis-Chalmers Farm near Racine, Wis. Behind is an A-C No. 84, 4-bottom plow. Rear-mounted implements often necessitate the use of front-end weights, and in this case they have been added to the front wheels. Specific data regarding the change from one tractor series to another does not appear to be well documented, but the manufacturing period for each series is presented here using data from several sources, including dealer trade-in guides.

A 1964 photograph illustrates a D-17 tractor at work in Australia. This farmer is pulling a mulcher-packer to produce the finely finished seedbed required for the planting of cotton. Standard serial number listings do not differentiate between tractors produced for domestic use and those shipped overseas, therefore current research offers no clues regarding the extent of foreign shipments. Note the single front wheel, high-clearance design, yet another option in addition to the standard tricycle front or the Roll-Shift front axle. Also note the waterbag slung from the tractor radiator.

Tractors

A photograph from the mid-1960s illustrates the tractor yard at West Allis Works. At this point in time Allis-Chalmers, although not the leading producer of American farm tractors, was nevertheless, firmly in the market, offering a wide range of models. The rather full range of Allis-Chalmers "D" series tractors was complemented by an extensive line of farm implements, many of which were designed specifically for use with these tractors and the Snap-Coupler system.

A photograph of November, 1965, shows an Allis-Chalmers D-17 Series IV tractor working with a box scraper and scarifier in a feed yard near Calexico, Cal. Built between 1964 and 1967 the Series IV tractor ceased production in June of the latter year. By this point in time the gasoline version was priced at $5,393 and the diesel model listed at $6,175. The A-C Snap-Coupler hitch could be supplied, or in its place the Series IV could be equipped with a standard 3-point hitch.

This D-10 tractor is shown mowing the lawns at Milwaukee's Currie Park Golf Course. The D-10 and D-12 tractors were built concurrently, and were quite similar in design. An excerpt from the company's own Instruction Manual for these tractors states, "The Model D-10 and D-12 tractors are similar in most respects with the exception of the tread. The Model D-10 has the narrower tread and is designed for one row cultivation. The D-12 has the wider wheel tread and is designed for two row cultivation. Both models are available in high clearance design."

D-10 high clearance tractors carried a 1960 list price of approximately $1,995. Basically the changes required were in the use of different rear wheels, rims, and tires, plus the addition of special front axle equipment. The high clearance design is essential for the cultivation of certain crops but finds almost no use in other crop growing areas. For 1960 the standard D-10 tractor listed at $1,810. Various other options were available for the D-10 tractors. This D-10 high-clearance model is pulling a 2-bottom A-C plow.

Another view of the D-10 high-clearance model illustrates the Snap-Coupler system and a mounted 2-bottom plow. Later models of the D-10 were available with a 3-point hitch as an extra cost option. Original equipment included a 4-cylinder A-C engine with a 3⅜ x 3⅞ inch engine having a 138.7 CID. Effective with Serial No. 3501 of 1961 a 3½ inch bore was substituted — this increased the displacement to 149 cubic inches. This change resulted in an output of 33.46 PTO horsepower, compared to 28.51 horsepower in the earlier model. Details are given in Nebraska Tractor Tests, No. 724 and No. 812.

Tractors

The D-10 Series III tractor was introduced in 1964 and remained in production until 1967. Its external appearance was enhanced over the earlier style through the use of new hood decals, a white radiator grille, and other highlights which served to make the D-10 a very attractive little tractor. For 1967 the Series III was priced at $3,292. Accessories included liquid ballast for front or rear tires, front and rear wheel weights, dual rear wheels, PTO-driven belt pulley, and numerous other items. Although the base price included a fully equipped tractor, the separate pricing of many different accessories allowed the purchaser to have a tractor built virtually "to specifications."

As has been pointed out for the D-10 tractor, it and the D-12 shown here were the same except for the tread width. The D-12 was designed to work in 2 rows, while the D-10 was intended for one-row work. Test No. 723 at Nebraska, shown here underway, indicates 28.56 PTO horsepower, virtually identical to the output shown for the D-10. When the earlier 3⅜ x 3⅞ engine was given an increased bore of 3½ inches, effective with Serial No. 3501, the D-12 was again tested at Nebraska, with Test No. 813 showing an output of 33.32 PTO horsepower.

This photograph taken near Natchez, Miss., shows a D-12 and an 80-S sidemounted mower working in Johnson grass. The low-down design and open lines of the D-12 permitted excellent visibility for the operator. Although the D-12 was built in standard and high-clearance models, only a gasoline engine was available — the D-10 and D-12 tractors could not be equipped with a diesel engine. Little difference is evident in the original D-12 tractor and the later Series II models. The original D-12 carried a price of $2,766, while the Series II listed at $2,849.

As has been previously noted, the D-10 and D-12 tractors were virtually identical save for the wheel tread design. The D-10 was primarily for single-row activities, while the D-12 was designed for two-row-duty. This composite photo shows the D-10 in the foreground, side-by-side with the D-12 model. Note that one major difference in the two tractors is the use of different rear wheels. Plain rear wheels could be furnished in lieu of power-adjustable rear wheels. Using plain wheels permitted certain specific width changes, but the power adjustable wheels were much easier to handle and provided greater versatility.

Typical applications for the D-12 included this rear-mounted 2-row cultivator. As previously noted, the compact design of the D-12 made it ideal for use with a 2-row cultivator. A-C built the 120 Series cultivators especially for the D-12 tractor — the similar 100 Series cultivators were matched to the D-10. Eventually however, the Snap-Coupler gave way to a 3-point hitch system as preferred equipment. This D-12 is equipped with Power-Adjustable rear wheels and is working in a field of young cotton.

The D-12 Series III tractor was produced from 1964 to 1967. Standard equipment included a 12-volt electrical system, adjustable front axle, hydraulic system with Traction Booster, PTO shaft, headlights, fenders, and 3-point hitch. Priced at $3,390, this tractor could be enhanced with a deluxe seat for an additional $27. Note that while the earlier models mounted the headlights on each side of the radiator, this Series III tractor mounts them on each fender. Specific production figures for the D-12 have not been located.

Introduced in 1960, the A-C D-15 tractor was tested at Nebraska under No. 795 of May, 1961. Standard equipment for the gasoline style used in this test included a four-cylinder, 149 cubic inch engine having a $3\frac{1}{2}$ x $3\frac{7}{8}$ inch bore and stroke. Rated at 2000 rpm, it delivered a maximum PTO output of 40.00 horsepower. This tractor, equipped with 12.4-26 rear tires, and the Roll Shift front axle carried a 1962 list price of $3,315. This photograph illustrates the D-15 tractor being used in Nebraska Test No. 795.

The D-15 Diesel tractor was tested at Nebraska in May, 1961, and the results are compiled in Test No. 796. Actually announced in 1960, this tractor used the company's own diesel engine. Of the 4-cylinder variety, it was rated at 2000 rpm and carried a $3\frac{9}{16}$ x $4\frac{3}{8}$ inch bore and stroke for a displacement of 175 cubic inches. Initially the D-15 Diesel sold at approximately $3,950, although this price moved upward substantially in ensuing years. All options available for the gasoline model could of course be supplied for the diesel tractor. These options included a Category I 3-point hitch — an extra cost item listing at approximately $125.

Production figures for the D-15 LPG tractor have not been located, but this photograph of the tractor used in Nebraska Test No. 797 illustrates such a model. In earlier times, Allis-Chalmers mounted the propane tank on the front end of the tractor, but for this model it is situated in the same general position as formerly occupied by the gasoline tank. The tractor was virtually identical to the gasoline model except for an altered compression ratio and the installation of appropriate equipment to handle this type of fuel.

A photograph from the Allis-Chalmers Archives illustrates the D-15 Utility tractor. Virtually identical to the comparable general purpose model, it was equipped with appropriate tires and other accessories required for commercial and industrial applications. In this particular instance, the D-15 Utility is equipped with an A-C drop hammer and demolition head. The D-15 tractors could be equipped with a shuttle clutch to provide quick forward and reverse. This $52 option likely required a special $67 oil cooler for the Power Director clutch.

The D-15 tractors prior to Serial No. 9001 of 1962 were built with a 4-cylinder engine having a $3\frac{1}{2}$ x $3\frac{7}{8}$ inch bore and stroke. After the above serial number, an increased bore of $3\frac{5}{8}$ inches was used on the gasoline models. The D-15 Diesel apparently went through the entire production run with the same engine. Using a larger engine required new tests at Nebraska, and these are recorded under No. 837 for the gasoline version and No. 838 for the propane model. Both tests indicate substantially higher horsepower than with the original models. Although some company documents indicate that the D-15 Series II tractors did not appear until 1965, various other references point to the increased horsepower output resulting from the 1962 engine modifications, as the likely beginning of the Series II tractors.

A 1968 price list shows this tractor with a base price of $4,148 with a gasoline engine or $4,834 when equipped with a diesel. The standard Roll-Shift front axle shown with this tractor came as regular equipment, but in lieu of this style the D-15 could be furnished with a standard tricycle front end or a single front wheel. A metal seat was included in the base price — addition of a deluxe seat added $26. The operation meter which indicated elapsed time and engine speed and ground speed was another $24 option. A standard clutch was supplied but a heavy duty clutch could be furnished for an additional $8. The heavy duty clutch was required with the diesel-powered tractor.

About 1965 Allis-Chalmers announced a new grove tractor built over the D-15 chassis. This unit was designed to inflict minimal damage to growing trees and likewise kept the operator safe from low branches. The unique feature of this tractor was the ability to raise the entire unit much like raising the hood on a car. This permitted easy servicing and eliminated the need for access doors — the latter inevitably being prone to loss or damage beyond repair. With this system, servicing the tractor was extremely simple, especially when compared to previous methods. Production of the D-15 tractors ended in 1968.

Announced in 1961, the D-19 tractor entered the pre-production phase as the D-18 tractor. Various information indicates that D-18 was still the title as late as July, 1961. For reasons unknown, by August of that year, the final title of D-19 was ascribed to this model. Early in 1962 the D-19 was sent to the Nebraska Tractor Test Laboratory in Lincoln. Test No. 810 contains the data for the gasoline model. Test No. 811 covers the D-19 Diesel, and Test No. 814 gives the specifications and performance on the D-19 LPG version. This photograph of 1961 illustrates the D-18 tractor, together with a four-row cultivator.

Produced between 1961 and 1964, the D-19 gasoline model delivered 71.54 maximum PTO horsepower on Nebraska Test No. 810. On drawbar work, 63.05 maximum horsepower was recorded, making this 3-bottom plow easy work for the D-19. A 262 cubic inch engine was used — its six cylinders carried a $3\frac{9}{16}$ x $4\frac{3}{8}$ inch bore and stroke. Rated speed was 2000 rpm. Allis-Chalmers billed these as the Power-Crater "6" engines. This term was derived from the placement of the combustion chamber in the piston head. In the strongest possible terms, A-C stressed that "you get uniform output from each cylinder."

Tractors

Special LP-gas equipment was available for the D-19 tractors. A heat pad in the intake manifold evened out air temperatures throughout each intake port. This permitted a higher density charge of air and fuel to be drawn into each cylinder. The carburetor was offset in relation to the manifold porting so as to provide a completely balanced intake. These were just a few of the features that permitted the D-19 to achieve impressive fuel economy ratings.

The turbocharged D-19 Diesel used the same bore and stroke as the gasoline-powered version, but of course used a substantially higher compression ratio. Allis-Chalmers pioneered the use of turbochargers on farm tractors, and despite some opposition to this idea from other manufacturers, turbocharging eventually has proved to be an economical method of harnessing more diesel engine power. From the company's own advertising brochures A-C engineers cited several advantages that included broadening of the lugging range. Turbocharging added 20% or more to the horsepower output, compared to naturally aspirated engines.

Many options were available for the D-19 tractors. Although the Roll-Shift front axle was usually supplied, this tricycle front could also be furnished, as could several other front and rear wheel options. High-clearance equipment was available from the factory, as was a Category II 3-point hitch. Power steering was standard equipment. Larger tractors required improved braking systems, and while some manufacturers were leaning toward hydraulic brakes, A-C developed a self-energizing brake system which required very little force at the pedal to derive exceptional braking power.

This photograph from the early 1960's illustrates D-19 tractors rolling off the assembly line at West Allis Works. D-19 tractors were produced from 1961-1964. The A-C Snap-Coupler hitch was furnished as standard equipment. Its chief disadvantage probably rested with the fact that only A-C tools could be used. The optional 3-point hitch system permitted the use of 3-point equipment already on the farm, requiring only some new adjustments prior to operation. Prices of the D-19 ranged from $5,305 for the gasoline model up to $6,080 for the D-19 Diesel. The same tractors in a high-clearance version were approximately $1,000 higher than comparable standard equipment models.

In 1963 Allis-Chalmers offered the ED-40 tractor to the Canadian market. This model was equipped with a 138 cubic inch diesel engine. Rated at 2250 rpm, it used a $3\frac{5}{16}$ x 4 inch bore and stroke. Heater Plug starting equipment was standard, as was the Category I, 3-point lift. Eight overlapping forward speeds plus two reverse speeds allowed the operator to match speed and power to any task. Production of the ED-40 was apparently limited to the year of 1963. No serial number data has been located for this model, although it is known that the tractor serial number is located under the seat on the left hand side of the transmission.

On June 6, 1963, Allis-Chalmers announced the D-21 tractor. Powered by a 119 horsepower diesel engine, the D-21 was the largest tractor of the A-C line. Power was derived from the new A-C Model 3400 diesel engine. Rated at 2200 rpm, this pilot model was naturally aspirated. The 426 CID was derived from six cylinders having a 4¼ x 5 inch bore and stroke. When Allis-Chalmers announced this tractor in 1963, it was the only tractor in excess of 100 horsepower to be designed for both drawbar and varied row crop work.

Without liquid tire ballast the D-21 weighed approximately 9,700 pounds, but when furnished with ballast, this figure jumped to nearly 12,000 pounds. Standard equipment included 18.4-34 rear tires, although 24.5-32 tires could be furnished as an option. The Model 3400 diesel engine used in the D-21 featured a 6-cylinder, direct-injection, open-chamber-combustion engine with a 16 to 1 compression ratio. Direct-injection provided easy starting and excellent fuel economy. Seven main bearings were used, along with aluminum pistons and removable wet cylinder sleeves. Also featured was a forged steel crankshaft with induction hardened bearing journals.

In addition to the regular farm tractor version, Allis-Chalmers also supplied the D-21 as an Industrial tractor. Slightly different wheel equipment, plus the replacement of the usual Persian Orange with a bright yellow enamel characterized the major changes between the two models. Considering the fact that considerable highway and commercial construction was in progress during the 1960s, sales of the D-21 Industrial tractor were probably quite good. The D-21 naturally aspirated tractor was tested in Nebraska under No. 855 of October, 1963.

At one point in production, experimental D-21 tractors with a hydraulic front-wheel-assist were built, but no information can be located to substantiate whether any of these tractors were sold as regular production. Although front-wheel-assist tractors are commonplace in the 1980s, this option in the early 1960s represents some innovative design on the part of Allis-Chalmers. Presumably this system was powered with hydraulic motors within the wheel housings. Except for a very few photographs of this design, no substantive data can be located.

During the 1965-1969 period Allis-Chalmers built the D-21 Series II tractor. This style was powered by an A-C Model 3500 diesel engine. Virtually identical to the Model 3400 used in the earlier D-21 tractors, the Model 3500 had the same bore and stroke, the same rated speed of 2200 rpm, and the same displacement of 426 cubic inches. The difference however, was in the addition of a turbocharger to this model. Adding the turbo gave an output of 127.75 PTO horsepower and 116.41 horses at the drawbar. The D-21, like other A-C models of the period, used a dry air cleaner. Allis-Chalmers was one of the first tractor builders to feature dry air cleaners instead of the earlier oil bath styles.

Introduced in 1969, the 160 tractor was equipped with a Perkins Model AD 3.152 diesel engine. This three-cylinder design had a displacement of 152 cubic inches and used a 3.6 x 5.0 inch bore and stroke. At 2250 rpm, this gave an output of 40.36 PTO horsepower. Standard features included power shift rear wheels, differential lock, and double disc brakes. Also included was a tilt-up hood for easy servicing. Several sizes of front and rear tires were available, depending on specific requirements. The front-end weights on this tractor came as an extra cost option.

The 160 tractor was tested at Nebraska under No. 1028 of 1969. Six different models were available within the 160 Series. These included the 160 Basic for those wanting a 40 horsepower tractor at a minimum investment. The 160 Intermediate included additional operation features, while the 160 Standard was a fully equipped tractor in the 40-plus horsepower class. Both the 160 Intermediate and the 160 Standard were also available in a Narrow version. Completing the series was the 160 LCG model — the term translating to Low Center of Gravity. This tractor was built for work on side slopes and for use in orchards. It contained all the features of the 160 Standard. List price for the 160 Standard was approximately $6,200. Production of this tractor ran from 1969 to 1974.

Announced in 1967, the 170 tractors remained in production until 1974. The earlier versions used a hood decal which spelled the model designator — about 1971 this was changed to simply read "170." The diesel version shown here was tested at Nebraska under No. 965. The 235 cubic inch engine was supplied by Perkins — it carried a 3⅞ x 5 inch bore and stroke. Since Allis-Chalmers had their own diesel engine manufacturing facility within the Harvey (Illinois) Works, the move toward using engines provided by other manufacturers was perhaps viewed by some within and without the company as the purest form of heresy!

The 170 tractors could be furnished with an AC-built gasoline engine. This tractor, listed in the Nebraska Tests under No. 966, was rated at 1800 rpm and carried a four-cylinder engine having a 4 x 4½ inch bore and stroke for a 226 CID. Research, both forward and backward from this point in time, will see considerable use of this 226 CID engine in Allis-Chalmers tractors, along with its use in countless industrial applications. The 170 was also available as a cane tractor, and could be supplied with a shuttle clutch at additional cost. Some of these models were also built as the 170 Landhandler Series.

Introduced in 1970, the 175 tractor was available with the company's own G-226 gasoline engine or a Perkins diesel engine. Initially at least, a 236 cubic inch engine was used. Its four-cylinder design included a 3.875 x 5.00 inch bore and stroke. An Operator's Manual for the 175 indicates however, that effective with Serial No. 2151 and up, a Perkins 248 cubic inch engine was used. It was nearly identical except that it used a 3.975 inch bore. The diesel version was tested at Nebraska under Test No. 1043 of May, 1970. The 236 cubic inch engine was used in the test tractor.

Tractors

A-C offered the 175 tractor in gasoline and diesel models. The gasoline style used the famous A-C G-226 engine, and in Nebraska Test No. 1156 of 1974 the 175 Crop Hustler model delivered 60.88 PTO Horsepower. Special attention was given to advertising the merits of the Perkins diesel engine. Its features included an open chamber design for easy starting and even combustion. Five main bearings were used, and two counter-rotating weights were used to balance the engine for smooth running.

This One-Eighty Diesel tractor is equipped with a six-row planter having a 3-point hitch. This tractor model incorporated many important features, including a big three-section hydraulic pump with an output of nearly 20 gallons per minute. An additional feature was the external mounting of the hydraulic pump on all but the D-15 Series II tractor. This method of mounting saved considerable time and expense if service was required on the pump. The One-Eighty Diesel used an A-C Model 2800 engine with a maximum output of 64.01 PTO horsepower per Test No. 964 of October, 1967. Rated at 2000 rpm, this six-cylinder, 301 cubic inch design used a 3.875 x 4.250 inch bore and stroke.

The One-Seventy and One-Eighty diesel tractors were also available in an industrial version. This style differed little from the standard farm models, except that a bright yellow industrial enamel was used instead of the famous Persian orange. Since the industrial models were often used on heavy drawbar work, certain wheel options, ballast, and other features were available on these specialized models. An option of One-Eighty series was the choice of a 540 rpm PTO shaft, or a dual 540/1000 rpm system.

The One-Eighty Series tractors were produced from 1967 to 1974. Available in gasoline or diesel versions, the One-Eighty was in the four-plow category with an output of approximately 57 drawbar horsepower. The gasoline version used an Allis-Chalmers G-2500 engine — six cylinders with a 3.75 x 4 inch bore and stroke. Rated at 2000 rpm, it had a displacement of 265 cubic inches. A hydraulic power clutch permitted shifting between high and low ranges on-the-go. A neutral position stopped tractor travel, but kept the PTO running at full rpm.

Built only in a diesel version, the 185 Diesel tractor was introduced in 1970 and remained in the line until 1981. Standard equipment included and Allis-Chalmers Model 2800 diesel engine. The six-cylinder, 301 cubic inch design was built around a 3.875 x 4.250 inch bore and stroke. On Nebraska Test No. 1044 the 185 delivered 74.87 maximum PTO horsepower. The hydraulic Power Director clutch permitted easy starting of heavy loads and permitted on-the-go shifting between high and low ranges. A total of eight forward speeds were available, ranging up to 17 mph.

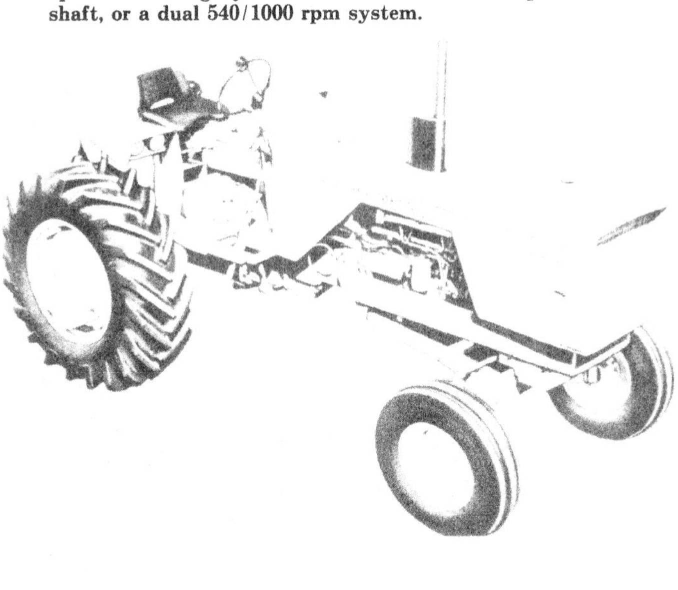

Tractors

The 185 tractors were available with a Category II 3-point hitch as factory equipment. Sway blocks were also a standard feature. When installed in the lower position with their thick ends downward, the implement was kept from having any side sway in either the raised or lowered position. With the sway blocks installed in the upper position with the thick ends upward, the implement was permitted full sway in the working position but no side sway in the raised position. Special adapters were available to adapt this system to Category I implements.

The One-Ninety tractor introduced in 1964 initiated an entirely new line of Allis-Chalmers tractors. At this point in time, farmers still maintained definite preferences for either gasoline or diesel engines, and at this point at least, fuel costs were low enough so that operating economy was not as important a factor as it would be in just a few more years. The gasoline version used a Model G2500, six-cylinder engine. Rated at 2200 rpm, it used a 3¾ x 4 inch bore and stroke for a displacement of 265 cubic inches. In Nebraska Test No. 928 this tractor yielded an output of 75.37 PTO horsepower.

Several different versions of the One-Ninety tractor were available, including the single-front wheel style shown here. From the 1964 through 1969 model years the One-Ninety gasoline model listed at approximately $7,175. Power steering was standard equipment, along with live hydraulics from a big three-section gear-type pump. An optional hydraulic-actuated clutch permitted independent PTO shaft operation. The One-Ninety tractors were also available in High-Clearance edition for special crop applications, and in fact, this single-front wheel combination appears to feature this special design.

The One-Ninety gasoline model was also available as the One-Ninety XT tractor. Virtually identical in outward appearance, the XT gasoline model featured an Allis-Chalmers Model G2800 Power-Crater engine, somewhat larger than the G2500 of the standard model. The six-cylinder, 301 CID design included a 3⅞ x 4¼ inch bore and stroke. Featurewise, it was virtually identical to the G2500 engine, having seven main bearings, replaceable cylinder sleeves, and a heavy duty diesel engine block. The One-Ninety XT gasoline version was apparently built as late as 1970. This particular model could also be purchased with the necessary equipment to burn propane fuel instead of gasoline.

A rear view of the One-Ninety Diesel illustrates the heavy duty 3-point hitch system featured on this series. Allis-Chalmers furnished this tractor with the standard 3-point system or their own patented Snap-Coupler hitch. (Snap-Coupler and Traction-Booster are trademarks registered to Allis-Chalmers). Also available was the optional Traction-Booster Drawbar. This device permitted the weight transfer system to be used for pull-type implements. The lower links sensed traction demand and the weight was automatically transferred to the drive wheels.

Tractors

Allis-Chalmers offered the One-Ninety with their own Model 2800 naturally aspirated engine — the latter term referring to a diesel engine which draws in its combustion air at atmospheric pressure. Test No. 886 at Nebraska indicates an output of 77.20 PTO horsepower for this engine. The One-Ninety diesel tractors were available in a high-clearance version and could also be supplied as Cane tractors for that special application. Weighing about 7,100 pounds, the 190 Diesel listed at approximately $8,850.

From 1967 to 1975 Allis-Chalmers offered the Model 918 Backhoe Loader. This machine was built over an industrialized version of the 190 Diesel tractor. With a dry weight of over 18,000 pounds, the Model 918 was one of the largest tractor-mounted backhoe units of the time. Accentuating its features was the ½-yard backhoe bucket capacity, along with a maximum reach of 20½ feet, plus a maximum digging depth of 18 feet. Standard equipment included a torque converter drive and shuttle clutch system. The second Model 918 to be built, Serial No. 1002 to be exact, is owned by the Author.

Highlighting the One-Ninety Series was the XT Diesel, a turbocharged design using the Allis-Chalmers 2900 diesel engine. Test No. 887 at Nebraska illustrates a PTO output of 93.64 horsepower for the One-Ninety XT Diesel, substantially higher than for other tractors of this series. Listing at about $9,500, this tractor remained in production until 1973. Slight modifications transpired about 1971 with the Series III tractors. Changes over the earlier style appear to have been largely superficial.

Introduced in 1972, the 200 Series tractors were not built with a gasoline engine, but only with the A-C Model 2900 turbocharged diesel. The 6-cylinder design with a 3⅞ x 4¼ inch bore and stroke gave a 301 CID. For reasons not immediately apparent, no official test report from Nebraska's Tractor Test Laboratory appears for this tractor, although from internal company communications, it appears that one of these tractors was sent to Lincoln for testing. The optional cab could also be equipped with a heater and air conditioner if so desired.

Features of the 200 Tractor included a seven-blade fan for big cooling capacity. Also included was a crankshaft-driven oil pump for the capacity required of heavy loads. During 1972 an internal company communication indicates that the Model 200 had delivered in excess of 100 horsepower during testing. This required an evaluation of risk factors associated with an increased horsepower rating. By adding the Mark II cooling improvement features, already available on the A-C 7010 tractor, doubts could be minimized in this regard. However, production of the Model 200 apparently ended in 1975, and no additional details of this tractor have been located.

Tractors

An A-C news release of 1971 noted that this Two-Ten Landhandler tractor was pulling a Series 2600, 6-bottom Allis-Chalmers plow. Power was provided by an Allis-Chalmers Model 3500 diesel engine. Its six-cylinders of a 4.25 x 5.00 inch bore and stroke yielded a displacement of 426 cubic inches. Adding a turbocharger gave an output of 122.40 PTO horsepower as noted in Nebraska Test No. 1065 of 1971. Disc brakes were standard equipment. With 212 square inches of face contact on each brake, each unit was capable of stopping and holding its load quickly and securely.

Built from 1970 to 1973 the Two-Ten Landhandler tractors featured a big 14-inch cera-metallic clutch to stand up to the pressures and temperatures created with heavy power duties. A four-pinion differential was used. Its size was designed to fully handle any load delivered by the 120-plus horsepower engine. Eight forward speeds from 1.93 to 18.17 mph were provided. Weighing 14,400 pounds ready for the field, the Two-Ten tractors could be equipped with several different rear wheel and tire configurations, depending on specific requirements.

Announced in late 1969, the Two-Twenty Landhandler tractor featured a turbocharged engine with an estimated 135 PTO horsepower, sufficient to handle any heavy drawbar job. The constant mesh, helical gear, double-range transmission gave eight forward and two reverse speeds. Power steering and adjustable front and rear wheels came as standard equipment. Also included was a four-position steering wheel and a 12-position cushioned seat. Safety features included a suspended clutch pedal that had to be depressed before the engine could be started.

Built 1966-1969, the I-400 tractors were ostensibly intended for various commercial and industrial applications. When equipped with a direction reverser transmission, gear shifting was minimized. These tractors could be furnished with an extensive number of industrial machines at the factory, and of course could be adapted to countless field applications. The A-C G-149 engine was featured — its 3½ x 3⅞ inch bore and stroke gave a displacement exactly matching the model designator. Factory equipment options included a front-end loader, backhoe, post hole digger, box scraper, and many other items.

Production of the Two-Twenty Landhandler Tractor ended in 1973. Optionally available was this front-wheel-drive system. It featured a planetary drive axle. The front wheel drive could be engaged with a lever on the operators platform. Test No. 1017 at Nebraska saw an output of 135.95 PTO horsepower, while Test No. 1065 of 1971 set the PTO output at 122.40 horsepower. Standard equipment included an Allis-Chalmers Model 3500 turbocharged diesel engine. Its 6-cylinder design, coupled with a 4.25 x 5.00 inch bore and stroke gave a displacement of 426 cubic inches.

Presumably, a considerable number of Allis-Chalmers I-400 tractors were equipped with the side-mounted sickle-bar mowers as shown here. Available in 5, 6, and 7-foot cutterbars, these mowers were used extensively for highway mowing and similar applications. Also available for use with this tractor was the A-C hydraulic hammer — it operated directly from the tractor's hydraulic system with suitable hoses. The tread width of the I-400 tractors was non-adjustable, but a three-point hitch was optionally available.

Intended primarily for industrial duties, the A-C Buckmaster was produced during the 1968-70 period. Production of this machine began with Serial No. 501 and ended in 1970 with No. 623, thus indicating that production was very limited to say the least. Ostensibly intended for pulpwood and similar industries, the Buckmaster with its mounted saw and hydraulically actuated holding mechanism probably saw little use by consumers, but may have had some application in commercial firewood enterprises.

Like the I-400 Industrial tractors, the I-600 series was built between 1966 and 1969. The I-600 gasoline version is shown here — it used an Allis-Chalmers Power-Crater engine of four cylinder design and having a $3\frac{5}{8}$ x $3\frac{7}{8}$ inch bore and stroke for a displacement of 160 cubic inches. The hydraulic system had a capacity of 15.5 gpm. A crankshaft-driven gear-type hydraulic pump was featured — it was a combination dual and single-stage unit. Within the I-600 Series came an entire fleet of tractors including the I-600 Forklift and several other unique machines.

I-600 Industrial tractors could be purchased with an Allis-Chalmers 4-cylinder, 175 cubic inch diesel engine. Rated at 1600 rpm, it used a $3\frac{9}{16}$ x $4\frac{3}{8}$ inch bore and stroke. Three distinct transmission types were available. The standard constant mesh used four forward speeds, plus reverse. The Shuttle-Clutch gave four forward and four reverse speeds. No foot clutching or pedal selection was required. The Power Director offered eight forward and two reverse speeds.

In the early 1960s Allis-Chalmers built a few T-16 Sugar Babe tractors. Of four-wheel-drive design, they featured a single-stage torque converter driven from the engine through a direct-mounted laminated flex-plate drive. Engine was increased on a 3.0 to 1 ratio in the converter and then transmitted to the full-power-shift transmission. Forward and reverse direction in any speed range was transmitted from the transmission through the universal drive shaft, to both the front and rear differentials and axles.

Tractors

Already in the 1950s Allis-Chalmers had the manufacturing and technological capabilities to produce massive four-wheel-drive tractors. This photograph of May, 1953, illustrates a huge new tractor-bulldozer unit for the U. S. Corps of Engineers. Built at the company's Cedar Rapids (Iowa) Works, this huge tractor weighed over 51,000 pounds. A 300 horsepower Cummins diesel engine was used in the "Bull Moose." With this huge amount of power and a torque converter transmission this gigantic tractor was able to achieve a top speed of 25 mph while pulling a 65,000 pound towed load!

Ostensibly intended for commercial and industrial applications, the T-16 tractor was also built in a farm version, with this T-16 pulling a 28-foot chisel plow. Standard equipment for this tractor included an Allis-Chalmers Model 7000 T16D diesel engine. Of turbocharged design, its four-cylinders of $4\frac{7}{16}$ x $5\frac{9}{16}$ inch bore and stroke yielded a displacement of 344 cubic inches. Options included a 4-speed transmission with a speed range up to 20.8 mph or a 6-speed transmission with increments to 16.5 mph. Production of the T-16 was rather limited, at least for farm duties.

The 440 tractor of 1974 featured a Cummins V-8 diesel engine with a displacement of 555 cubic inches. Rated at 2800 rpm, it used a 4,625 inch bore and 4.125 inch stroke. Ten forward speeds were provided, ranging up to 20 mph. Special features included a swinging power divider, transmission and divider oil coolers, outboard planetary axles, and disc caliper brakes. A 3-point hitch was optionally available. The roomy operator's cab was standard equipment, with air conditioning optionally available.

Built in the 1982-84 period, the 4W-220 tractor featured an Allis-Chalmers Model 670-HI 6-cylinder diesel engine. Turbocharged and intercooled, this engine was designed with a 4.250 x 5,000 inch bore and stroke for a displacement of 426 cubic inches. Rated speed was 2400 rpm. An A-C Power Director transmission provided 20 forward and 4 reverse speeds. An open combustion chamber engine design prolonged burning to maintain peak cylinder pressures and produce more horsepower. Replaceable wet sleeves were standard equipment. Forty-degree articulation with twenty-six-degree oscillation helped cut sharp corners and maintain excellent wheel grip under all conditions.

With the 4W-220 and 4W-305 tractors, the operator's cab moved from an enclosed dust protector to a full-fledged control center with many comfort features. These included large rubber cab mountings, external air intake filters, and heavy insulation and soundproofing. Even the color-matched optional luxury seat was fully adjustable for back angle and lumbar support. This huge four-wheel-drive model embodied all the features found in the 4W-220 model but featured a huge 6-cylinder turbocharged engine. Built by A-C and designated as the 6120T engine, it was rated at 2300 rpm and used a 5.250 x 5.625 inch bore and stroke for a displacement of 731 cubic inches.

Tractors

Introduced in 1982, the 5015 offered a 61 cubic inch diesel engine, Category I 3-point hitch, and a 10-speed transmission. Shown here pulling a rotary mower, the 5015 was ideal for large lawn mowing jobs, along with a host of other duties within its capabilities. The streamlined design included headlights as standard equipment, along with advanced features such as a differential lock, overrunning clutch, and hydraulic position control for the 3-point lift system. This tractor listed at approximately $6,500.

First built in 1978, the 5020 diesel tractor featured a 2-cylinder engine built by Toyosha of Japan. Rated at 2500 rpm, it carried a 3.62 x 3.74 inch bore and stroke (92 x 95 mm). This gave a displacement of 77.1 cubic inches or 1263 cubic centimeters. With a capability of nearly 22 PTO horsepower, this tractor was designed for a multitude of farm and commercial applications. A Category I, 3-point hitch system was standard equipment, enabling the 5020 to handle a wide variety of implements. Also available was the 5020 4WD version which incorporated a front-wheel-assist system.

Like the 5020 tractor, the slightly larger 5030 model featured a Toyosha 2-cylinder diesel engine. In this case however, a 3.82 inch bore and 3.94 inch stroke (97 x 100 mm) were used to give a displacement of 90.1 cubic inches and an output of 26.4 PTO horsepower. The impressive list of attachments for the 5000 Series tractors included a large rotary tiller with 3-point hitch mounting, several styles of mowers, 1 and 2-bottom moldboard plows, 2-row cultivators, and a 60-inch, 2-stage snow blower. The 5030 tractors listed at approximately $9,200.

Manufactured between 1975 and 1981 the 5040 tractor featured a 6-speed transmission, a Category II 3-point hitch, differential lock, and numerous other features. It was also available with a 9-speed transmission. Both styles used a 3-cylinder direct injection diesel. Rated at 2400 rpm, it carried a 3.74 x 4.33 inch bore and stroke for a 142.8 CID. The 5045, produced between April and October of 1981 featured a 44 horsepower, 3-cylinder diesel engine with a 3.93 x 4.33 inch bore and stroke for a 158 CID. The 5045 was equipped with an 8-speed transmission.

The 5050 diesel tractors were available with the choice of an 8-speed or a 12-speed transmission when built as a standard two-wheel-drive model. Adding the optional front-wheel-assist unit as shown on this tractor gave the single option of the 8-speed gear box. Built between 1976 and 1983, the 5050 featured a Fiat 3-cylinder, direct-injection diesel engine. Rated at 2700 rpm, it carried a 3.93 x 4.33 inch bore and stroke for a displacement of 167.8 cubic inches (2750 cubic centimeters). Standard equipment included a Category II 3-point hitch, power steering, differential lock, and many additional features.

Tractors

Nebraska Test No. 1195 manifests 106.44 PTO horsepower for the Model 7000 tractor. Built between June, 1975 and June, 1979 the 7000 featured an AC-built 301 cubic inch, 6-cylinder, 2900 Mark I diesel, the company's latest in turbocharged tractor engines. Several innovative features were standard equipment in this tractor, and several of them were devoted purely to uniform dissipation of heat generated within the engine. A new cylinder head casting boosted water flow capacity, and copper injector nozzle jackets increased heat transfer from the nozzles. A larger radiator and a 22-inch fan were used, and the intake and exhaust manifolds were located on opposite sides of the engine block to help keep the incoming combustion air cool. This 7000 tractor is being tested at the company's own research facility.

Production of the 7010 tractor ran from 1979 to 1981. Three transmission options were available. The 16-speed Power Director transmission was standard, and the 20-speed Power Director was optional. Both provided effortless control of all forward speeds through a shift lever, range lever, and two foot button for on-the-go shifting. A third option was the 12-speed Power Shift transmission which provided on-the-go shifting through six forward speeds in each of two ranges without clutching or stopping. Test No. 1345 at Nebraska used the 7010 with the 20-speed transmission, and Test No. 1346 used another 7010 equipped with the 12-speed Power Shift. These tractors were equipped with an Allis-Chalmers 649T turbocharged diesel engine with a 301 CID.

Allis-Chalmers equipped the 7020 with the 649-I turbocharged and intercooled diesel engine. This well known 6-cylinder in-line design used a 3.875 x 4.25 inch bore and stroke for a 301 CID. Built in the 1977-1981 period, the 7020 was tested at Nebraska under Numbers 1259 and 1260 of 1977. The 12-speed Power Shift transmission was used in the tractor of Test No. 1259, while the following test, No. 1260, was made with the Power-Director transmission. Except for these changes, the tractors were otherwise, virtually identical.

The 7020, like the 7010, was available in a special High-Clearance version as shown here. This photograph from an A-C catalog illustrates the actual difference in height between the high clearance tractor and a regular model as shown by the dotted line. Although the high clearance model was substantially more expensive than the standard unit, certain crops and cropping practices made the use of these special tractors mandatory. A most extensive and impressive line of accessories came as standard equipment on these tractors, with still more accessory items available as extra-cost options.

First built in 1973, the 7030 was one of the first A-C tractors built on the company's new Flexible Manufacturing System, a computer directed operation capable of performing complex machining operations on a variety of castings in the tractor power train. The 7030 featured an Allis-Chalmers Model 3500 Mark II turbocharged diesel engine. Its 426 cubic inch displacement resulted from a 4.25 x 5.00 inch bore and stroke in its six cylinders. Standard equipment included an A-C Power director 20-speed transmission.

Built between 1974 and 1977, the 7040 offered 14 forward speeds with the Power Director XX transmission. A hydraulically actuated differential lock came as standard equipment, as did the power-actuated, oil-cooled, multi-disc brakes. A dual 540/1000 rpm PTO came as standard equipment on the 7040 tractor, and A-C offered variable volume, variable pressure hydraulics exclusively on the 7040 and 7060 models. Power was derived from an A-C Model 3500 Mark II turbocharged diesel engine. Specific details on performance of the 7040 can be found in Nebraska Tests 1166 and 1196.

With over 125 drawbar horsepower, the 7045 featured 20 forward speeds in the Power Director model or 12 speeds in the 7045 Power-Shift series. Detailed performance statistics may be found in Nebraska Tests 1261 and 1308. Planetary final drives and a 1000 rpm independent PTO were standard features for this tractor. Produced in the 1977-81 period, it carried an approximate list price of $43,300. The 7045 diesel tractor was also built in a high clearance version with a suggested list price of $51,343.

Built in 1973 and 1974, the 7050 tractor was a virtual deadringer to the 7030 except for the engine. While the 7030 carried an A-C 3500 Mark II engine, the 7050 used an A-C Model 3700 engine. As with the 7030 model, the 6-cylinder design included a 4.25 x 5.00 inch bore and stroke for a 426 CID. Rated speed for both engines was 2300 rpm. However, while the 7030 was a turbocharged diesel, the 7050 was turbocharged and intercooled. Thus the 7050 was able to deliver approximately 155 PTO horsepower, compared to only 130 horsepower for the 7030.

Paralleling the 7040 and 7080 models, the 7060 used a Model 3700 Mark II engine. Turbocharged and intercooled, it was capable of about 161 PTO horsepower, compared to only 136 horsepower for the 7040 tractor. The latter was equipped with a 3500 Mark II engine, turbocharged, but not intercooled. Both operated at 2300 rpm. Production of the 7060 ran from 1974 to 1981. Standard features included a pressurized cab with heater. Air conditioning was optional, as was an AM-FM stereo radio or an 8-track stereo tape player.

Built during the 1975-1981 period, the 7080 was the third and largest tractor in a series which also included the 7040 and 7060 models. While the chassis was virtually the same in all three tractors, the indicated horsepower differed substantially. The 7040 delivered about 136 horsepower from its 3500 Mark II turbocharged engine. This figure rose to 162 for the 7060 and its 3700 Mark II turbocharged and intercooled style. Both of these tractors operated at 2300 rpm, but the big 7080 shown here had a rated speed of 2550 rpm. The higher engine speed, coupled with a turbocharged and intercooled design with a counterbalanced crankshaft saw an output of over 181 PTO horsepower from the 3750 Mark II engine used in this tractor. Performance data is recorded in Nebraska Tractor No. 1168.

The Allis-Chalmers 7580 four-wheel-drive tractor featured the same Model 3750 Mark II engine as found in the two-wheel-drive 7080 tractor. Of turbocharged and intercooled design, this engine featured a counterbalanced crankshaft and operated at 2550 rpm. In Nebraska Test No. 1229 this tractor developed 186.35 PTO horsepower. Reference to this test gives additional specifics on the 7580. Standard equipment included the 20-speed Power Director transmission plus a fully independent, 1000 rpm, PTO shaft.

At one point Allis-Chalmers offered a free front-wheel-assist with any new 8000 Series tractor. Throwing the FWD system in the bargain represented an immense savings and provided the purchaser with a mechanical FWD system that could be engaged at the flip of a switch for more traction, less compaction, and a short turning radius. The 8010 model was announced in December, 1981. It featured the Allis-Chalmers 649T turbocharged diesel engine. Test No. 1446 at Nebraska indicated a maximum output of 107.38 PTO horsepower.

The 8000 Series tractors included the 8010 and 8050 models in addition to the 8030 tractor shown here. Standard features included a 20-speed Power Director transmission on the regular models, although the special high-clearance models used the 12-speed Power-Shift transmission. Standard equipment for the 8030 included an Allis-Chalmers six-cylinder turbocharged diesel engine, and like others of the 8000 Series, it featured oil cooling of the pistons. An A-C 670T engine was used — its 4.25 x 5.00 inch bore and stroke gave a 426 CID.

Shown here is the 8050 tractor in a high-clearance version. By adding another 13 inches of operating clearance it was possible to work in cane and other tall crops without risk of snagging on the frame or twining around the axles. Beyond the special equipment reguired for the high clearance design, this tractor was virtually the same as the standard model. Due to the extreme height of the front axle, special bracing and stabilizing members were required.

Topping the 8000 Series tractors was the big 8070 model — 170 PTO horsepower ready for action! The standard version was of conventional two-wheel-drive, but an 8070 FWD style was also available. Allis-Chalmers once again used the basic 426 cubic inch, 6-cylinder diesel engine, and for the 8070 it was designated as the Model 670-I. It featured a turbocharged and intercooled design, piston cooling, and a counterbalanced crankshaft. The maximum PTO horsepower readings were observed with a rated engine speed of 2400 rpm, although the engine had an operating range of 800 to 2590 rpm.

Tractors

Built between 1977 and 1981, the big 8550 four-wheel-drive tractor featured an Allis-Chalmers 6120T diesel engine. Of 6-cylinder design, it used a 5.25 x 5.625 inch bore and stroke for a 731 CID. Rated at 2500 rpm, this huge engine was capable of 300 flywheel horsepower. Dual turbochargers were featured, along with a Power Director transmission having 20 forward speeds. The operator's cab, complete with heater and air conditioning came as standard equipment. Also included was a 1000 rpm PTO shaft as standard equipment.

Built during 1974 and 1975, the 6040 tractors offered features like an adjustable front axle, power-adjustable rear wheels, and a differential lock. Although other tractors of the 6000 Series appear to have available with an optional cab, the 6040 apparently was not built with this option. Weighing only 4,287 pounds, the 6040 was nevertheless equipped with a Category II 3-point hitch — it included the A-C hydraulic draft and position control as standard equipment. Suggested list price was approximately $6,900.

The 6060, 6070, and 6080 tractors all used the Allis-Chalmers Model 433 diesel engine. The 6060 shown here and the 6070 models both used the 433T turbocharged style, with the major difference between these two models being an 8-speed transmission in the 6060 and a 12-speed gear box in the 6070. A 433I engine in the 6080 delivered 83.66 PTO horsepower, compared to 63.83 horses in the 6060 model. In the case of the 6080, a turbocharger and intercooler were used to gain the extra output. Specific details regarding operating economy may be found in Nebraska Tests 1397 and 1398.

The last purely Allis-Chalmers tractor was built December 6, 1985. This Model 6070 tractor bearing Serial Number 1971 was thought to have been the last one, although its mate on the assembly line, No. 1972 was actually the last one built. Reports surfaced shortly after the tractor line was shut down that an 8070 tractor finished the A-C tractor production, but these reports were apparently erroneous. However, the last 8070 built was No. 3354, a four-wheel-drive unit with a Power Shift transmission. As this section is completed in February, 1988, it appears that the 6070, No. 1971 was the last A-C tractor to be sold at retail, with No. 1972 remaining in storage, along with an 8070 tractor bearing No. 3354. A hurriedly written placard on the tractor fender reads, "Dec. 6, 1985 — Last Tractor from Assy Line — 10:26 A.M."

Allis-Chalmers purchased Monarch in 1928. The latter had introduced their Model F, 10-Ton crawler in 1926, and by 1928, had sold 250 copies. Allis-Chalmers continued a similar design called the Model F, "75" crawler into 1931. These huge crawler tractors added a whole new dimension to tractive power and greatly reduced the time required for difficult work such as this road grading job. This photograph from the Allis-Chalmers Archives shows a "75" pulling a Mammoth grader while grading some 20 miles of new highway in Des Moines County, Iowa.

Tractors

This Allis-Chalmers file photo designates the Monarch tractor shown here as the "50" Blue Ox. The meaning of the latter designation in connection with the "50" tractor remains obscure to the present research, but the suggestion that this particular style was painted blue certainly gains credibility from the model designator. The heavy steel radiator grille suggests that this tractor might have been intended especially for the logging industry, but once again, details of the Blue Ox are almost totally lacking. Possibly too, it referred to the Blue Ox Babe who was a companion to the legendary Paul Bunyan, the fictious hero of the northern logging industry.

An A-C file photo illustrates a Monarch "50" tractor pulling three Allis-Chalmers 10-foot grain drills plus a 30-foot harrow on hilly ground near Ephrata, Wash. Monarch introduced this model as the "H" 6-Ton tractor in 1927. When tested at Nebraska in that year (Test No. 147), this model was equipped with a Stearns engine. Ironically, Allis-Chalmers bought out Stearns in 1930, although details of this purchase have remained unclear. Initially at least, A-C continued using the Stearns Model DU engine. Its 4-cylinder enbloc design included a 5⅛ x 6½ inch bore and stroke.

No details have surfaced regarding the unique features of the A-C Monarch "75" Logger tractor, although it is obvious that at least a few of these special tractors were built. A conspicuous addition is the heavy steel plate protecting the radiator. The "75" was equipped with a huge 4-cylinder LeRoi engine having a 6½ x 7 inch bore and stroke. Other specifications of this engine include a 3¾ inch diameter crankshaft, and valves 2¾ inches in diameter!

The 1932 edition of the *Cooperative Tractor Catalog* lists the Allis-Chalmers "50" crawler as using an Allis-Chalmers 4-cylinder engine, cast enbloc, and using removable cylinder sleeves. This engine was of valve-in-head design and carried a 5¼ x 6½ inch bore and stroke. Auto-Lite electric starting and lighting was standard equipment. Weighing 14,000 pounds, the 1932 edition of the A-C Monarch "50" tractor listed at $3,540. At this point in time A-C crawlers were basically a Monarch design, but within a few years an entirely new series of A-C crawlers would emerge.

An undated photograph from the Allis-Chalmers Archives illustrates an A-C "35" crawler pulling a pick-up type baler built by J. I. Case Co. A unique feature is that this baler is being operated from the tractor PTO shaft, and not with the auxiliary engine commonly found on pull-type machines of the 1930s. The "35" was the first crawler tractor to be distinctively a product of Allis-Chalmers — the "50" and "75" tractors were already in production when A-C bought out Monarch in 1928.

As is evident throughout the history of Allis-Chalmers products, innovative engineering designs often surfaced far ahead of their time. In the case of Allis-Chalmers at least, ideas occasionally outdistanced technology or the willingness of the market to accept radically new and different ideas. This Monarch "50" is equipped with special front-mounted winch equipment, suitably powered from the rear of the machine, and under the complete control of the operator's platform. This special winch equipment was supplied for American Telephone & Telegraph Co. Almost immediately after acquiring Monarch, Allis-Chalmers engineers began making new and important developments in this field. The *Patent Office Gazette* bears frequent tribute to new and important A-C designs, and in the case of crawler tractor development the name of Emil F. Norelius frequently appears.

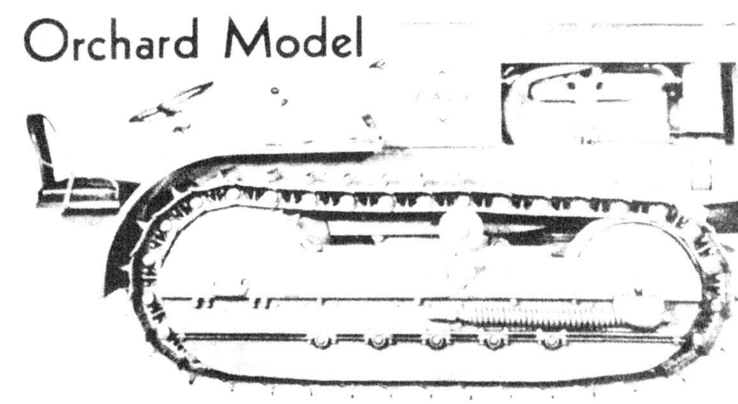

Although the majority of "35" crawlers probably spent their working life on roadbuilding projects and other commercial applications, a certain number were purchased for farm use, and a comparatively small number were equipped as Orchard Model crawlers. As is evident from this photograph, numberous modifications were made, especially the location of the controls and the operators seat. These were moved very low and well to the rear of the tractor. The engine and chassis remained virtually unchanged.

This A-C Monarch "35" tractor is pulling a Galion grader on a southern road improvement project. The "35" was tested at Nebraska under test No. 171 of 1929. A-C used their own 4-cylinder, 4¾ x 6½ inch engine in the "35." Since this model had no belt pulley it was impossible to run the brake horsepower tests, but using gasoline fuel it developed its rated 30 drawbar horsepower while pulling a 4,126 pound load at 2.73 mph. Model "K" 35 crawler tractors remained in production until 1943.

Tractors

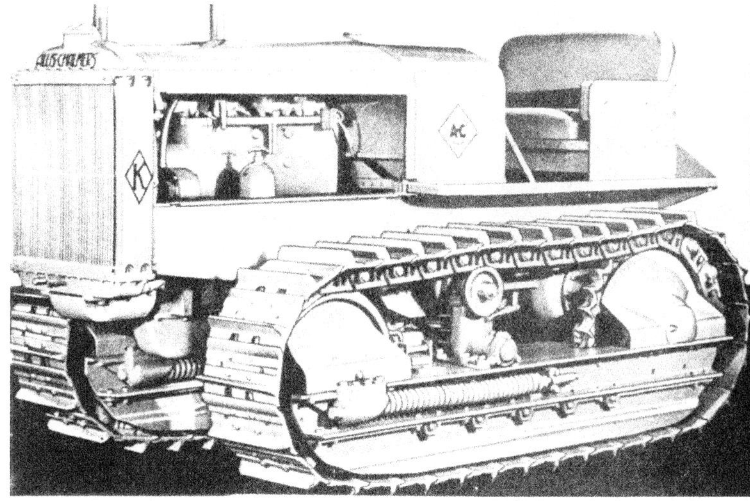

Already in September, 1933, Allis-Chalmers submitted a "Special 'K'" tractor to Nebraska's Tractor Test Laboratory. By this time the "K" had acquired a distinctive A-C appearance — one that would be evident in future years. Curiously enough, the Model "M" crawler tested immediately after (Test No. 216) is of the same general design as the "K" shown here — the wraparound hood of the "35" had been replaced with a far more attractive, and partially streamlined design. The test tractor used an A-C 4-cylinder engine with a 5 x 6½ inch bore and stroke.

Although current research data does not differentiate between various styles of the "K" series tractors, Nebraska Test No. 336 on the "WK" model was not run until November, 1939. The WK featured a wide track gauge for the extra stability required in very hilly terrain. This "WK" is building a logging road in Arkansas, and is pulling a No. 110 Allis-Chalmers Leaning Frame Grader. The grader is equipped with pneumatic tires and full mechanical controls. The "W" designator on this tractor series merely indicates that the tractor is built in a wide-gauge pattern.

The "K-O" tractor was built only in 1935, with a total of 18 copies being produced. The special oil engine in this tractor was designed and built by Allis-Chalmers. It used a combination of mechanical fuel injection as in a diesel, but due to the slightly lower compression ratio, used magneto ignition. This design, also referred to as the Hesselman system, was used by Waukesha, and perhaps a few other American builders. Current research has not however, made a definite determination whether this series of oil engines was in fact built under license from Hesselman.

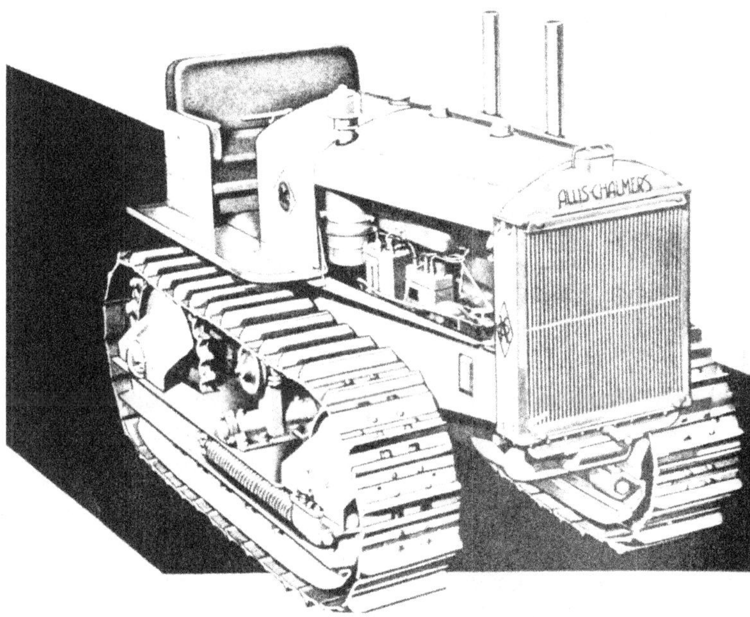

Built from 1931 to 1942 the Model "L" crawler was equipped with a gigantic 844 cubic inch gasoline engine. Its six cylinders were of 5¼ x 6½ inch bore and stroke. Between 1931 and 1935 the Model "L" was sold on the basis of Nebraska Test No. 200, but later production used the results of Test No. 338 conducted in 1939. The LP-SRP, industry jargon for Last Published Suggested Retail Price, was $5,300. Weighing 22,470 pounds, the "L" offered five forward speeds ranging from 1.48 to 6.47 mph.

"K" tractors with special oil engines were produced only in 1935, and serial number listings only indicate a total of 18 copies. From this total came a handful of "W-K-O" tractors — a wide-gauge version with the special oil engine installed. This A-C file photo illustrates a W-K-O tractor at work in a California orchard. A caption accompanying this illustration notes that "balanced horsepower weight ratio sold the "WKO" in a competitive demonstration with a much heavier tractor of slightly less power."

Only 20 copies of the "L-O" crawler were built in 1935. Of this production, one tractor was sent to Nebraska's Tractor Test Laboratory in Lincoln for testing. Curiously though, it was not tested until September, 1937. The special 6-cylinder, A-C engine used fuel injection with spark ignition. A maximum pull of over 20,000 pounds was recorded, as was a drawbar output of some 76 horsepower. This "L-O" is shown logging in the New Zealand jungle, making its own road over jungle swamps and up steep banks. Quite evident in this photo is the unusual manifolding which provided for a separate exhaust pipe to each pair of cylinders.

Apparently Allis-Chalmers intended to build a Monarch 65 tractor in the early 1930s. Although this unit apparently did not go into production, the engine at least made it into the prototype stage. This photograph, discovered in the A-C Archives is dates July 16, 1930. The original caption reads, "1931 6-Cyl. Gasoline Motor Built by Allis-Chalmers Mfg. Co. for Monarch 65 Tractor." The rather unusual exhaust manifold system is obvious, and was carried over into the special AC-built oil engines seen in certain of the company's crawler tractors during the 1930s.

A substantial number of Model "M" crawlers were used by the American military during World War Two. This photograph by the U.S. Army Signal Corps shows a Model M tractor equipped with a Buckeye dozer crossing a temporary bridge. Tractor-dozer units like this were used extensively during World War Two for building roads through places never before thought possible. Many others were used for carving airstrips out of the jungle or clearing the way for transport vehicles.

A-C Model "M" crawler tractors were produced between 1932 and 1942. Nebraska Test No. 216 of 1933 covered this model through the 1936 production. A new version of the Model "M" built 1937 and later, was not tested at Nebraska. Original equipment included a 4-cylinder A-C engine having a 4⅜ x 5 inch bore and stroke. Those built 1937 and later used an engine having an increased bore of 4½ inches. The standard-tread Model M was built on a 40-inch gauge, compared to a 50-inch gauge for the wide-track "WM."

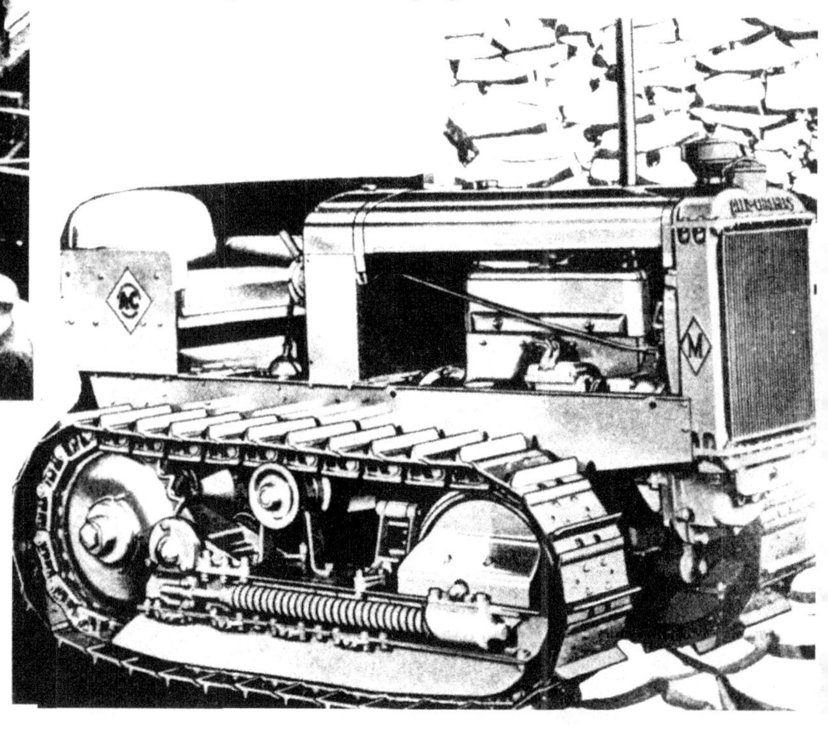

Tractors

In 1942, Hough Loader Co. announced this hydraulic loader built over the "WM" crawler tractor. Cumbersome and awkward by contemporary standards, this 11,500 pound machine nevertheless represented an important move forward in materials handling with speed and capacity. The "WM" crawler with its 50-inch track gauge afforded the necessary stability for raising heavy loads, and Hough built this loader with the choice of a 42-inch bucket having a ½ yard capacity, or a 66-inch, 1-yard bucket. These units were sold through Allis-Chalmers dealers.

Nebraska Test No. 286 provides detailed output and economy data on the Model "S-O" tractor announced in 1937. A huge 4-cylinder oil engine having a 5¾ x 6½ inch bore and stroke was featured. This engine was designed by A-C, and used mechanical fuel injection with electric spark ignition. Over 66 belt horsepower was indicated for this tractor. For all appearances, the "S-O" remained on the market longer than its counterparts, but all-in-all, this engine design did not achieve major cognizance.

In 1938 Allis-Chalmers announced the new Model "S" gasoline-powered crawler. Aside from being just another of a series, this tractor boasted an economy of 8.06 horsepower hours per gallon of gasoline at its rated drawbar load. By comparison, the Model "M" tractor of Test No. 216 in 1933 indicated only 7.5 horsepower hours per gallon of fuel. A-C built the big 4-cylinder engine — its 5¾ x 6½ inch bore and stroke gave a displacement of 675 cubic inches. This tractor listed at approximately $4,000.

Initially, this photograph was passed over as being just one of many similar illustrations. A close inspection indicates however, that this tractor has the letter "V" cast into the side of the radiator. Specially made blowups of this photograph leave no doubt whatever that it was a Model "V" tractor, although this raises the obvious question as to when it was built, and why no further information can be located on this unit. The original photograph in the Allis-Chalmers Archives contained no information whatever, but hopefully, continuing research work will provide some necessary clues to an A-C crawler tractor which up to now remains a complete mystery.

The General Motors 6-71 2-cycle diesel engine used in the HD-41 delivered 122 maximum drawbar horsepower in Nebraska Test No. 362 of October, 1940. A few HD-14 tractors were built in 1939, and production continued into 1947. Since many diesel-powered crawlers were used in high altitude, mountainous terrain, the power losses due to high altitude were addressed by the Model 71 G-M engine. A-C claimed that at 14,000 feet the Model 71 lost only about 13% of its normal power, compared to as much as 50% by 4-cycle diesels. This HD-14 and a Carco Arch are logging in Linn County, Oregon, for Swanson Logging Co. The photo was taken in 1941.

Tractors

An HD-7 is followed by an HD-10, and finally by the big HD-14 tractor in this rare photograph. Taken at the Nebraska Tractor Test Laboratory in Lincoln, these tractors are coming off the rail cars for testing. The HD-14, billed in 1940 as the "world's fastest and most powerful production crawler tractor," weighed 14 tons, and was the first of the A-C crawler line to use the G-M Model 71 diesel engines that would characterize the entire A-C crawler line for years to come. Note the fancy new grilles sported by this line of tractors.

Attachments were plentiful for the HD-14 tractor. Included was this huge Carco Model K Hoist on the rear of the tractor. Carco products were built by Pacific Car & Foundry Co. A-C offered a number of items built by other companies, including Gar Wood cable and hydraulic scrapers, Baker bulldozers, and sheepsfoot tamping rollers. Options were plentiful, and included a 3-man winter cab, completely enclosed and equipped with sliding doors. This tractor carried an approximate retail price of $9,250.

First built in 1947, the HD-19 tractor remained in production until 1950. When tested at Nebraska in 1948, the HD-19 broke all records for maximum drawbar pull at that facility, giving a performance that saw 37,536 pounds on the dynamometer! This immense tractor weighed 40,395 pounds and featured a torque converter drive with infinitely variable speed combinations up to 7 mph. This photograph taken during actual testing of the HD-19 tractor illustrates but a tiny portion of the special drawbar equipment required to run the test procedures.

Introduced in 1940, the HD-7 remained in production until 1950, when it was discontinued. Early production also saw an HD-7 Military model, but no data has been located to indicate how it differed from the standard production style. In 1943 the HD-7W wide-gauge track style was introduced — it too ceased production in 1950. A G-M 3-71 diesel engine was featured. It delivered approximately 65 horsepower at its rated 1,500 rpm. Beginning with the HD-14 tractor Allis-Chalmers incorporated their new Positive-Seal track wheel design. It used neoprene seals to keep lubricants in while keeping water and other contaminants out. This feature alone was a major improvement in crawler tractor design.

Improved Velve-Touch bi-metallic steering clutches were featured in the HD-7 tractor. Long a conundrum in crawler tractor design, these new steering clutches proved to be extremely satisfactory, especially when compared to earlier styles. For 1947 the HD-7 listed at $6,100. The HD-7W with its 63-inch tread (compared to 52 inches for the standard model), was available for an additional $80 over the base price. Extra equipment options were virtually endless, and even included the substitution of any paint color other than the standard A-C Orange for an additional $44.

Built from 1951 to 1955, the HD-9 was an intermediate model that continued using the same Series 71 G-M 2-cycle diesel engines as did other tractors of the "HD" series. These crawler tractors from Springfield Works, were sent primarily to commercial and industrial tasks, with a relatively small proportion being used for farm work. This 1953 photograph shows an HD-9 skidding logs near Oak Grove, La. Since Allis-Chalmers had much earlier established definite ties with the lumber industry, as evidenced by their sawmill machinery, it seems entirely logical that this sales base was beneficial in the sale of A-C crawler tractors.

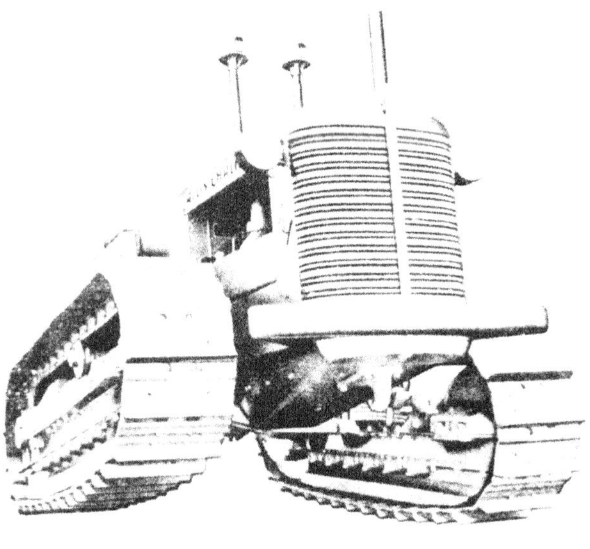

Billed as a "companion to the HD-14," the HD-10 was announced in 1940 and remained part of the A-C line into 1950. The standard model used a 62-inch tread, while the HD-10W was constructed over a 74-inch gauge. Both carried a G-M 4-71 diesel engine rated at 86.63 drawbar and 101.62 belt horsepower. This illustration from an HD-10 catalog illustrates what A-C termed as "controlled alignment which permits free track oscillation, yet holds the tracks rigidly in line." A newly designed spring tension and shock absorber track release was provided. This unit was sealed in oil to prevent rust and contaminants from seizing the precision parts of the mechanism.

A 1953 photograph illustrates the HD-9 tractor and a Baker 9-X bulldozer. The unique feature here is that two hydraulic cylinders replaced the equalizer spring normally used. The cylinders controlled movement of the vertical frame and permitted the body and blade to move anywhere from 13 inches below ground to 37 inches above ground. The entire unit was self-contained and had a maximum width of 8 feet. This made it ideal for ready transportation to forest fires and other requirements where frequent over-the-road transportation was necessary. This special bulldozer was also available on the HD-15 crawler.

This incredible photograph in the A-C Archives shows an HD-15 tractor and a giant tooth ripper actually tearing up a concrete roadway. When equipped with the hydraulic torque converter drive, this huge 14-ton tractor could develop a pull of 60,000 pounds based on zero track slippage. In actual practice however, maximum drawbar pull was generally based on a formula that included 90% of total tractor and equipment weight to derive the maximum pull.

Allis-Chalmers offered the HD-15 tractor with the choice of a standard transmission or a hydraulic torque converter drive. The sliding gear transmission offered six forward speeds ranging from 1.4 to 5.8 mph, while the torque converter used three gear ranges providing an infinitely variable speed range from 0 to 7.0 mph. A G-M 6-71 diesel engine was featured. Its $4\frac{1}{4}$ x 5 inch bore and stroke gave a displacement of 426 cubic inches. Nebraska Test No. 264 indicated a maximum pull of 29,400 pounds and an output of 105 drawbar horsepower using the sliding gear transmission. This test was conducted in July, 1951.

Built only in the 1951-54 period, the HD-20 tractor was available either as a standard tractor or as the endloader unit illustrated here. The 4-yard bucket offered immense capacity, and the optional torque converter drive permitted quick cycle times for maximum production. Huge crawler-loader units like the HD-20 found a ready market as the construction industries moved toward larger and more powerful machinery to replace much of the slow and underpowered equipment of earlier times.

Introduced in 1955, the HD-6 tractor was initially tested at Nebraska under Test No. 580 of July, 1956. The HD-6 featured an AC-built, 4-cylinder, 4-cycle diesel engine. Its $4\frac{7}{16}$ x $5\frac{9}{16}$ inch bore and stroke gave a displacement of 344 cubic inches. Production of the HD-6 and its derivatives continued into the late 1970s, apparently remaining viable for sometime after the Fiat-Allis merger of 1974. Initially at least, A-C offered this tractor as the HD-6A with a 44 inch tread and the HD-6B using a 60-inch tread width.

A 1960 price book indicates a suggested retail for the HD-6B of approximately $9,840. Numerous track and shoe combinations were possible, as were several different track and sprocket guards. Heavy duty lubricating and fuel oil filters added another $140 to the list price. Since these tractors were ostensibly intended for farm work as well as industrial duties, a rear-mounted belt pulley was available, as was a rear PTO shaft. Electric lights were standard equipment, as was the 24-volt electrical system.

Tractors

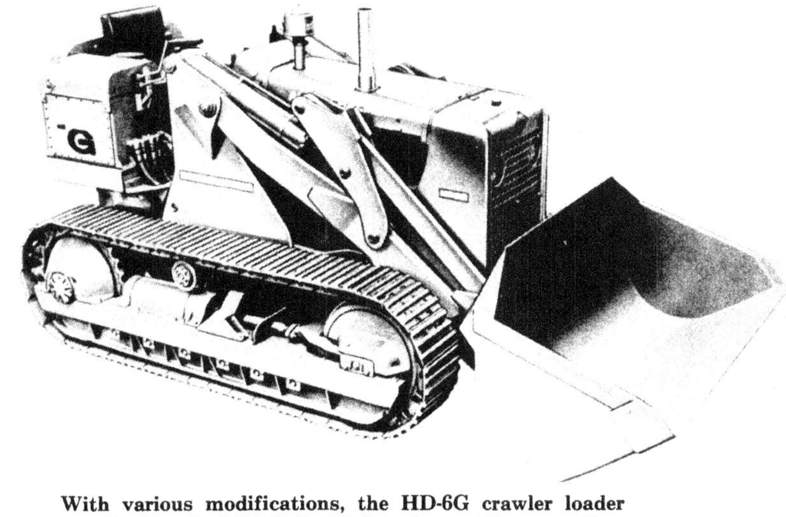

With various modifications, the HD-6G crawler loader remained part of the industrial equipment line sometime after the 1974 Fiat-Allis merger. This 1970 model featured an Allis-Chalmers Model 6000 engine. Of 4-cycle, naturally aspirated design, the 6000 offered a $4\frac{7}{16}$ x $5\frac{9}{16}$ inch bore and stroke with a 344 CID. Features included sealed track, hunting tooth sprockets, and steel-backed aluminum bearings. In addition to the standard bucket design, several special application buckets could be purchased, along with log and lumber forks, and a multi-purpose bucket.

First built in 1955, the HD-11 Series remained in production into the Fiat-Allis era after 1974. Standard equipment for this 122 horsepower tractor included a 6-cylinder, 516 cubic inch engine having a $4\frac{7}{16}$ x $5\frac{9}{16}$ inch bore and stroke. A turbocharger was standard equipment, and an oil-type clutch was featured. For 1968 the HD-11E tractor shown here listed at slightly over $25,000, just over $1 a pound. Hydraulic steering controls were featured, as were hydraulic track adjusters, and a pressure lubricated transmission.

Numerous configurations were possible for the basic HD-11 tractor. The commonly used bulldozer equipment was actually an extra-cost option, since to include it in the base price for those not needing this equipment would be absurd. In addition, hydraulic loaders were available, and these could be purchased with several bucket types, depending on the application. Log forks were also available, as were several rear-mounted winches. Likewise, a variety of track equipment was available instead of the standard grouser bar track pads shown here.

With 160 net engine horsepower, the mighty HD-16 tractor boasted a drawbar pull of some 38,000 pounds. Its power was derived by an A-C D-844 6-cylinder engine having a $5\frac{1}{4}$ x $6\frac{1}{2}$ inch bore and stroke. Rated at 1600 rpm, it featured six forward speeds from 1.4 to 5.8 mph. On large ranches a substantial number of HD-16 tractors were placed into service — they were able to make short work of tilling large acreages. With a 1968 list price of $32,500 the HD-16 was by no means inexpensive, but considering its work capabilities, this 34,500 pound tractor was a valuable asset for large farming operations.

Shortly after the formation of Allis-Chalmers International in September, 1957, came the acquisition of several foreign companies. Included was Vendor, S.p.A. in 1959. Shortly thereafter, Allis-Chalmers Italiana S.p.A. announced the A-C 250PL pipelayer tractor. The sideboom offered an 18-foot reach, and the huge counterweight weighed over two tons. Apparently, Allis-Chalmers fitted a number of these sideboom attachments to the Italian tractor at the company's Deerfield Works. This photograph illustrates the sideboom fitted to an HD-16 tractor.

Tractors

The gigantic HD-21 tractor sent to Nebraska for testing weighed in at 44,725 pounds. Test No. 550 of July, 1955, indicates that the 844 cubic inch, turbocharged engine delivered over 132 drawbar horsepower. Built for the very largest jobs, the HD-21 was equipped with a three-stage torque converter system that permitted infinitely variable speeds throughout the total range. An extensive line of attachments was available for the HD-21, but in 1968 the company added a huge sideboom attachment to make the conversion into a pipeline tractor.

Introduced in 1960, the H-3 crawler was powered with an Allis-Chalmers G-149 engine. Rated at 1650 rpm, it used a $3\frac{1}{2}$ x $3\frac{7}{8}$ inch bore and stroke. Standard features of the H-3 and HD-3 tractors included solid steel drive sprockets of "hunting tooth" design for maximum wear. An oil-type multiple disc shuttle clutch was used together with a 4-speed transmission. Special attention was given to accessibility of steering clutches and brake bands so that adjustments could be maintained with ease. The 1967 base price for the H-3 tractor was $5,885. Adding this hydraulic loader raised the price appreciably.

In 1962 Allis-Chalmers announced the "Spook," the world's first radio controlled crawler tractor shovel. All tractor operations were performed from radio signals sent by the operator from a hand-held console. Initially intended for steel mill slag removal applications, Allis-Chalmers cited its potential in military and defense operations as well. Unfortunately however, little more was heard of this unique device after the 1962 announcement. As can be seen in this photograph, the operator is standing some distance from the tractor, even though it is dumping a load of material. A second series of the HD-21 introduced in 1955 featured an A-C 21000 engine with an output of 204 flywheel horsepower. At the time, this was the world's largest and most powerful tractor. It weighed 44,000 pounds!

Production of the H-3 and HD-3 crawler tractors ran from 1960 to 1968. The HD-3 diesel model was equipped with an A-C 4-cylinder engine having a $3\frac{9}{16}$ x $4\frac{3}{8}$ inch bore and stroke for a displacement of 175 cubic inches. The HD-3 base unit weighed approximately 7,300 pounds and carried a 1967 list price of $6,530. When equipped with a 60-inch wide bucket having a $\frac{3}{4}$ yard capacity, the list price jumped to $8,950. This 1961 photograph taken at Port Arthur, Tex., illustrates the HD-3 tractor equipped with suitable sideboom equipment to convert it into a full-fledged pipeline tractor.

Built in the 1966-69 period, the HD-4 tractor was well suited to farm uses as well as industrial applications. The HD-4 diesel model featured an A-C Model D2200 diesel engine. Of naturally aspirated design, its four cylinders of 3.88 x 4.25 inch bore and stroke gave a displacement of 200 cubic inches. The HD-4, together with the hydraulic loader shown here carried an approximate list price of $12,150. Also available during the same period was an H-4 crawler, identical in all respects except that it used a G-2200 gasoline engine.

Tractors

Present research efforts have not uncovered any information on the T100 crawler shown here. From all appearances, it was built in late 1950s, since it appears to use the same black radiator grille as found on the early "D" Series farm tractors. Quite possibly it antedated the H-3 or H-4 crawlers; a definite possibility since much of the design appears to be very similar if not completely identical.

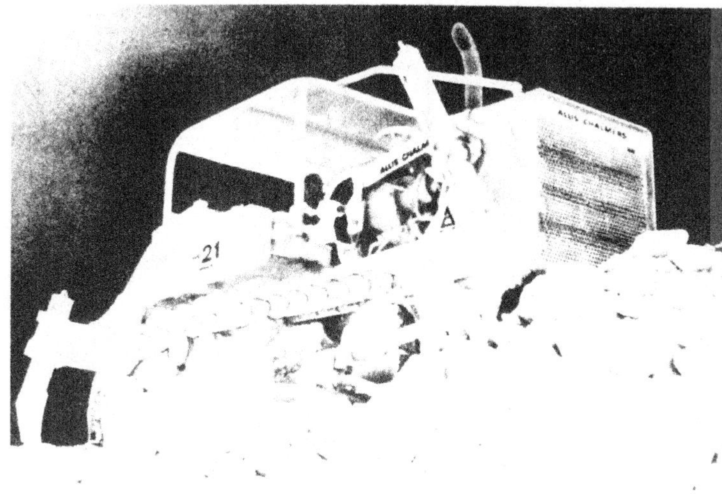

A 1971 advertising from Allis-Chalmers details the HD-21 Series B tractor — a gigantic machine with an output of 268 flywheel horsepower. A-C equipped this model with their own Model 21000 Mark II engine. Of 4-cycle, turbocharged, 6-cylinder design, it used a 5.25 x 6.5 inch bore and stroke for a displacement of 844 cubic inches (13.8 litres). Weighing 53,400 pounds, it featured a Power Shift transmission with three forward and three reverse speeds. Available with a wide variety of dozer blades, rippers, and other equipment, this second generation HD-21 incorporated several key design refinements that made it even more capable and reliable than its immediate predessor, the HD-21P tractor.

About 1970 Allis-Chalmers introduced the HD-41, the world's largest crawler tractor. This gigantic machine had a bare weight of 101,500 pounds with 524 flywheel horsepower. A variety of blades were available for this tractor, including a 30-foot land clearing blade. This huge device was capable of pushing down oak trees anywhere from 4 to 20 inches in diameter, leaving the debris in a huge windrow at the side of the machine. In average conditions, it was capable of knocking down and stacking two or more acres of trees per hour. Fiat-Allis tractors dominated the scene, effective with the 1974 merger between A-C and Fiat. Tractors built after that time can be found in various trade directories under the Fiat-Allis tradename.

Water Softeners

That Allis-Chalmers was diversified to a fault is evidenced by this water softener system announced in 1958. Completely automatic in design, this zeolite softener system was intended for commercial, industrial, and institutional applications, and was not built for the home water softener market. Important new features of this system included individual hydraulic valves which would open or close as required for system regeneration. The reasons for the company's obvious attempt to enter the commercial water softener market are presently unclear, but current research has not shown that this system was marketed to any great extent.

Windrowers

With the advent of the combine, or in the case of A-C, the All-Crop Harvester, came a parallel need for a windrower. Some crops required a windrower so that the grain and stalks could be partially dried prior to combining. In other situations the need arose from local practices or personal preferences, not necessarily in rank order either. This rather unstyled windrower design was built by Allis-Chalmers in 1941. Even a cursory examination shows that many of the components have been borrowed from the All-Crop harvester already in production at LaPorte Works.

First marketed in the late 1940s the A-C Speed Windrower was probably a continuation of the experimental model built in 1941. The tragic intervention of World War Two shelved a great many manufacturing projects for several years, and apparently the Speed Windrower was one such example. A-C claimed this to be the first modern windrower for use with small harvesters, and in fact, it was designed to work hand-in-hand with their existing All-Crop "60" Harvester. It had a capacity of 25 to 30 acres per day in good conditions.

Wood Preserving Machinery

Shortly after Allis-Chalmers was organized in 1901 the company was involved in the required equipment to preserve wood through a pressure creosoting process. The precise state of affairs in this section of company activity is unclear, at least in the period up to about 1915. By that time however, a rather extensive line of equipment was available, and in fact, it appears that Allis-Chalmers was prepared to erect a complete creosoting plant, tailoring it to the customer's needs. This 1916 illustration shows a special railroad tie car, loaded and ready to be wheeled into the impregnating cylinder.

A 1916 photograph illustrates an impregnating cylinder loaded on four flat cars enroute to its final destination. This huge cylinder was 7 feet in diameter and 132 feet long. A-C could supply cylinders up to 9 feet in diameter and 150 feet long. These were fitted with steel doors at both ends to permit entry and egress of the tie cars. When completely loaded the cylinder was sealed and creosote was forced into the cylinder, sometimes at pressures as high as 250 psi. The combination of heat and pressure forced the creosote into virtually every square inch of the rail ties or other material to be preserved. A-C further enhanced their position in this rather specialized field through the 1924 purchase of this line from Power & Mining Machinery Corp.

Epilogue

Thursday, March 28, 1985, was a cold cloudy day — one of those days when only a little persuasion was necessary for rain to begin, and by nightfall turn to sleet or snow. The chill in the air was ominous that day — not just the weather, but an inexplicable feeling of doom seemed to issue from the very hallways of the Allis-Chalmers Headquarters.

Ironically, Charles Wendel, the Author and George Dammann, the Editor of this history were in West Allis that day, and in fact, in the A-C offices. Even more coincidental, we were in the final stages of research and photography for a pictorial and historical narrative of Allis-Chalmers, with much of the work being embodied within this volume.

In mid-afternoon of that fateful day, March 28, 1985, the stock ticker in the A-C offices reported in effect, that the Allis-Chalmers farm equipment business had been sold to Klockner-Humboldt-Deutz AG of Germany for $100 million. For this writer, watching this momentous announcement come over the stock ticker was one of the most chilling and most unforgettable experiences of a lifetime. Before our very eyes, another familiar face in the farm equipment industry was disappearing.

The following morning, Friday, March 29, 1985, the *Milwaukee Sentinel* reported that the sale agreement included "a farm tractor manufacturing plant in West Allis; a farm combine manufacturing plant in Independence, Mo.; a components manufacturing plant in Topeka, Kan., and the credit corporation in West Allis."

By the end of 1985, Allis-Chalmers farm equipment was largely history. The new folks in town took the name of Deutz-Allis and continued with some machines previously familiar as having the A-C trademark. In addition, Deutz-Allis began introducing new machinery of their own designs.

The temptation for the Author to develop a supplementary section on Deutz-Allis and its other root, the famous German firm of KHD, has been great indeed. However, the illustrious history of Deutz is a book in itself, and in fact goes to the very beginnings of commercially produced internal combustion engines. To prove this point, the Author possesses an 1871 catalog of the Otto and Langen atmospheric engine as sold by Deutz of Cologne, Germany. Realizing that using this as a first step — the volume just completed could easily become half of a two-volume duet — we successfully resisted the temptation to digress further — at least for a while!

Serial Numbers and Production Data.

Since Allis-Chalmers purchased Advance-Rumely Thresher Company in 1931, known serial numbers and production data for this firm is listed herein.

Inclusion of serial numbers and production data is strictly for the convenience of the reader, and neither the Author nor the Publisher are able to substantiate the accuracy of the figures presented. This duty falls inevitably on the shoulders of the manufacturer, and in cases where precise information is required, the manufacturer(s) or their successors should be consulted.

In some situations various listings of serial numbers and production data disagree. This may result from simple typographical errors. Another cause may be the fact that most manufacturers did at times, reserve blocks of serial numbers for special equipment. Subsequently, perhaps only a few of these reserved numbers were used. The message here is this — taking the end-of-year serial number and subtracting the beginning serial number to derive the total production for the year may or may not represent an actual total of that year's production.

For certain models produced very early, including the 6-12 and 10-18 tractors, no serial number listings have been located, nor are any known to exist. Fortunately however, total production figures for these two models have been located. While this listing begins with the 1919 production year, it represents the grand total of all production information known thus far to exist.

Tractor serial numbers and engine serial numbers will probably not be the same in most cases, especially since many of these same engines were used for a variety of applications, agricultural and industrial. Where possible, this list includes information on the location of these numbers. As previously noted, in cases where specific information is required, it is suggested that the original manufacturer be contacted.

* * * * *

Tractor-Tiller 1914-16

Tractor-Truck 1915-16

10-18 1914-1923

(A company listing of 10-18 production shows 71 units for 1919, 31 in 1920, 45 in 1921, 38 in 1922, and 4 in 1923).

6-12 1918-1925

6-12 production breakdowns for 1919 and following years indicate:

Year	Model	Type	Units
1919	6-12	General Purpose	303
1920	6-12	General Purpose	447
	6-12	Cane Tractor	20
9121	6-12	General Purpose	66
	6-12	Cane Tractor	6
1922	6-12	General Purpose	243
	6-12	Cane Tractor	6
1923	6-12	General Purpose	198
	6-12	Cane Tractor	13
1924	6-12	General Purpose	99
1925	6-12	General Purpose	62
1926	6-12	General Purpose	8

(No other serial number information or production data has been located for this tractor, nor has any history been found relating to the 6-12 Cane Tractor).

* * * * *

Tractor Production: West Allis Works

MODEL L, 15-25

Year	From		To
1921	20001	to	20334
1922	20335	to	20497
1923	20498	to	20905
1924	20906	to	20995
1925	20996	to	21370
1926	21371	to	21681
1927	21682	to	21705
	None Since		

* * * * *

MODEL E, 18-30 & 20-35

Year	From		To
1918	5000	to	5005
1919	5006	to	5160
1920	5161	to	6014
1921	6015	to	6160
1922	None Built		
1923	6161	to	6396
1924	6397	to	6754
1925	6755	to	7368
1926	7369	to	8069

Serial number 8070 and ff., engine serial number is on left side of cylinder block. Tractor serial number is on transmission case ahead of gear shifter quadrant.

Year	From		To
1927	8070	to	9869
1928	9870	to	10000
1928	12001	to	16761
1929	16762	to	20250
1929	22001	to	23251
1930	23252	to	24185
	None Since		

* * * * *

MODEL E, 25-40

Engine serial number on LEFT side of cylinder block. Tractor serial number is on transmission case ahead of gear shifter quadrant.

Year	From		To
1930	24186	to	24842
1931	24843	to	24971
1932	24972	to	25023
1933	25024	to	25061
1934	25062	to	25308
1935	25309	to	25581
1936	25582	to	25611

(There are indications that 90 special 30-60 "E" tractors were built beginning with No. 25612 and ending with 25701).

None Since

* * * * *

MODEL U, 19-30

Tractor serial number is located on rear of differential housing for U, UC, and IU models. Engine serial number center of left side of block for S10 Continental engines; right side of block below oil filler on Model UM A-C engines. (U-1 — U-7404 equipped with Continental engines. U-7405 and ff. use A-C engine).

Year	From		To
1929	U-1	to	U-1974
1930	U-1975	to	U-6553
1931	U-6554	to	U-7261
1932	U-7262	to	U-7404
	None Since with Continental Engines		
1932	U-7405	to	U-7418
1933	U-7419	to	U-7684
1934	U-7685	to	U-8062
1935	U-8063	to	U-9470
1936	U-9471	to	----------
1937	----------	to	U14854
1938	U14855	to	U15546
1939	U15587	to	U16077

Year	From		To
1940	U16078	to	U16721
1941	U16722	to	U17136
1942	U17137	to	U17469
1943	U17470	to	U17801
1944	U17802	to	U17819
1945	18022	to	18324
1946	18325	to	20773
1947	20774	to	21021
1948	21022	to	22127
1949	22128	to	23028
1950	23029	to	
1951	22548	to	
	22561	to	22823
	20513	to	20585
1952	20586	to	20748
	22824	to	23056

Apparent discrepancies appear to exist in the 1951 and 1952 serial numbers. There are definite indications that production of this model may have continued as late as 1953, but these units were built to special order and not available through the regular A-C dealer channels.

MODEL UC, 19-30

(UC-1 — UC-1268 built with Continental engines. UC-1269 and following have A-C engines).

Year	From		To
1930	UC-1	to	UC-38
1931	UC-39	to	UC-1099
1932	UC-1100	to	UC-1231
1933	UC-1232	to	UC-1268
1933	UC-1269	to	UC-1293
1934	UC-1294	to	UC-1551
1935	UC-1552	to	UC-2000
1936	UC-2001	to	UC-2770
1937	UC-2771	to	UC-3756
1938	UC-3757	to	UC-4546
1939	UC-4547	to	UC-4769
1940	UC-4770	to	UC-4971
1941	UC-4972	to	UC-5037
1951	5806	to	5938

(UC tractors were built at irregular intervals as late as 1951, and possibly later. These were built on special order, and not available through normal dealer channels).

* * * * *

WC TRACTOR

For WC and WF tractor, serial number on rear of rear axle housing, just above operator platform. Engine serial number on LEFT side of engine, near the carburetor.

Year	From		To
1933	WC-1	to	28
1934	29	to	3126
1935	3127	to	13869
1936	13870	to	31783
1937	31784	to	60789
1938	60790	to	75215
1939	75216	to	91533
1940	91534	to	103516
1941	103517	to	114533
1942	114534	to	123170
1943	123171	to	127641
1944	127642	to	134623
1945	134624	to	148090
1946	148091	to	152844
1947	152845	to	170173
1948	170174	to	178202
	None Since		

(WC-1 to WC-74329 built without starter and lights as standard equipment, but became standard with number 74330 and following).

* * * * *

WF TRACTOR

Year	From		To
1937	WF-1	to	288
1938	289	to	1335
1939	1336	to	1891
1940	1892	to	2299
1941	2300	to	2703
1942	2704	to	3003
1943	None Built		
1944	3004	to	3194
1945	3195	to	3509
1946	3510	to	3747
1947	3748	to	4110
1948	4111	to	5499
1949	5500	to	7317
1950	7318	to	8315
1951	8316	to	
	None Since		

WF tractors WF-1 to WF-1903 were built without starter & lights as standard equipment, but became standard on WF-1904 and above.

* * * * *

WD TRACTOR

Tractor serial number on top of left differential brake housing.

Year	From		To
1948	7	to	9249
1949	9250	to	35444
1950	35445	to	72327
1951	72328	to	105181
1952	105182	to	126981
1953	127007	to	131242
1954	131243	to	160384
	None Since		

* * * * *

WD-45 TRACTOR

Year	From		To
1953	146607	to	160385
1954	160386	to	190992
1955	190993	to	217991
1956	217992	to	230294
1957	230295	to	236958

* * * * *

MODEL A TRACTOR

Tractor serial number on transmission case ahead of gear shift quadrant.

Year	From		To
1936	25701	to	25725
1937	25726	to	26304
1938	26305	to	26613
1939	26614	to	26781
1940	26782	to	26895
1941	26896	to	26914
1942	26915	to	26925
	None Since		

* * * * *

MODEL B TRACTOR

Tractor serial number on top of transmission above shift pattern.

Year	From		To
1937	1	to	96
1938	97	to	11799
1939	11800	to	33501
1940	33502	to	49720
1941	49721	to	56781
1942	56782	to	61400
1943	64501	to	65501
1944	65502	to	70209
1945	70210	to	72264
1946	72265	to	73369
1947	73370	to	74079
1947	75080	to	80555
1948	80556	to	85833
1948	87834	to	92294
1949	92295	to	102392
1950	102393	to	103578
1950	106579	to	114526

Serial Numbers and Production Data

1951	114527	to	118673
1952	118674	to	122309
1953	112310	to	124201
1954	124202	to	124710
1955	124711	to	126496
1956	126497	to	127185
1957	127186	to	----------
	None Since		

MODEL IB TRACTOR

(IB tractors, 260 in all, were built prior to 1946 and have a "B" serial number without the "IB" prefix).

1946	1001	to	1002
1947	1003	to	1009
1948	1010	to	1281
1949	1282	to	1555
1950	1556	to	1878
1951	1879	to	2218
1952	2219	to	2567
1953	2570	to	2847
1954	3068	to	-------
1955	-------	to	3372

* * * * *

MODEL C TRACTOR

Tractor serial number on top of transmission near shift pattern.

1940	1	to	111
1941	112	to	12388
1942	12389	to	18781
1943	18782	to	23907
1944	23908	to	30694
1945	30695	to	36377
1946	36378	to	39167
1947	39168	to	51514
1948	51515	to	68280
1949	68281	to	80517
1950	80518	to	84030

* * * * *

MODEL RC TRACTOR

1939	4	to	4391
1940	4392	to	5416
1941	5417	to	5504
	None Since		

* * * * *

MODEL CA TRACTOR

Tractor serial numbers on top of transmission next to shift pattern.

1950	14	to	321
1951	322	to	10538
1952	10539	to	22180
1953	22181	to	31423
1954	31424	to	32906
1955	32907	to	37202
1956	37203	to	38617
1957	38618	to	-------

* * * * *

MODEL G TRACTOR

Tractor serial number is located next to starter on engine. Engine serial number is located on starter side of engine, near starter.

1948	6	to	10960
1949	10961	to	23179
1950	23180	to	24005
1951	24006	to	25268
1952	25269	to	26496
1953	26497	to	28035
1954	28036	to	29035
1955	29036	to	-------

D-10

Year	Beginning Serial No.	Year	Beginning Serial No.
1959	1,001	1964	7,675
1960	1,950	1965	8,204
1961	2,801	1966	9,486
1962	4,511	1967	9,795
1963	6,801		

* * * * *

D-12

1959	1,001	1964	6,012
1960	1,950	1965	9,192
1961	2,801	1966	9,508
1962	3,638	1967	9,830
1963	5,501		

* * * * *

D-14

1957	1,001	1959	14,900
1958	9,400	1960	21,800

* * * * *

D-15

1960	1,001	1964	16,928
1961	1,900	1965	19,681
1962	6,470	1966	21,375
1963	13,001	1967	23,734

* * * * *

D-17

1958	4,300	1963	65,001
1959	16,500	1964	70,611
1960	28,200	1965	77,090
1961	33,100	1966	80,533
1962	38,070	1967	86,061

* * * * *

D-19

1961	1,001	1963	12,001
1962	1,250		

* * * * *

D-21

1963	1,001	1967	2,863
1964	1,417	1968	3,777
1965	2,079	1969	4,498
1966	2,408		

* * * * *

170

1968	2,721	1971	6,988
1969	5,374	1972	7,797
1970	6,369	1973	8,821

* * * * *

175

1970	1,001	1976	3,754
1971	1,477	1977	4,809
1972	1,624	1978	5,670
1973	1,740	1979	6,321
1974	2,153	1980	6,999
1975	3,255		

* * * * *

Serial Numbers and Production Data

180

1968	2,682	1971	10,561
1969	6,094	1972	11,729
1970	9,235	1973	12,447

* * * * *

185

1970	1,001	1976	8,292
1971	1,952	1977	10,003
1972	2,935	1978	11,631
1973	3,763	1979	13,160
1974	4,961	1980	14,672
1975	6,542	1981	15,648

* * * * *

190

1964	1,001	1969	23,234
1965	2,485	1970	25,901
1966	8,219	1971	29,136
1967	13,273	1972	31,118
1968	23,234		

* * * * *

190XT

1964	1,001	1968	19,262
1965	2,485	1969	23,234
1966	8,210	1970	25,901
1967	13,273	1971	29,136

* * * * *

200

1972	1,001	1974	6,294
1973	3,344	1975	9,250

* * * * *

210

1970	1,001	1972	2,082
1971	1,107	1973	2,469

* * * * *

220

1970	1,938	1972	2,626
1971	2,451	1973	2,860

* * * * *

4W-220

1982	1,001	1984	1,145
1983	1,081		

* * * * *

4W-305

1982	1,001	1984	1,176
1983	1,112		

* * * * *

5015

1982	1,001	1984	3,277
1983	1,727		

* * * * *

5020

1978	2,220	1982	7,034
1979	3,091	1983	8,388
1980	4,115	1984	8,734
1981	5,790		

* * * * *

5030

1979	2,005	1982	3,520
1980	2,255	1983	4,066
1981	2,976	1984	4,214

* * * * *

5040

1976	408,445	1979	473,000
1977	462,148	1980	474,000
1978	410,364		

* * * * *

5050

1977	573,461	1981	596,014
1978	579,632	1982	597,730
1979	584,000	1983	599,191
1980	591,000		

* * * * *

6060

1980	1,001	1983	3,894
1981	1,297	1984	4,572
1982	2,463		

* * * * *

6080

1980	1,001	1983	4,567
1981	1,152	1984	5,780
1982	3,002		

* * * * *

6140

1982	1,001	1983	1,851

* * * * *

7000

1975	1,001	1978	6,373
1976	1,645	1979	8,963
1977	5,037		

* * * * *

7010

1980	1,925	1981	2,806

* * * * *

7020

1978	1,317	1980	3,842
1979	2,732	1981	4,710

* * * * *

7030

1974 2,576

* * * * *

Serial Numbers and Production Data

7040

Year	Number	Year	Number
1975	1,303	1977	6,839
1976	4,089		

* * * * *

7045

Year	Number	Year	Number
1978	1,234	1980	3,399
1979	2,152	1981	4,225

* * * * *

7050

Year	Number
1974	1,688

* * * * *

7060

Year	Number	Year	Number
1975	1,299	1979	6,789
1976	2,741	1980	7,693
1977	4,580	1981	8,442
1978	6,001		

* * * * *

7080

Year	Number	Year	Number
1975	1,007	1979	3,268
1976	1,572	1980	3,648
1977	2,051	1981	3,954
1978	3,001		

* * * * *

7580

Year	Number	Year	Number
1977	1,287	1980	2,486
1978	1,605	1981	2,717
1979	2,486		

* * * * *

8010

Year	Number	Year	Number
1982	1,001	1984	2,266
1983	1,712		

* * * * *

8030

Year	Number	Year	Number
1982	1,001	1984	2,701
1983	2,093		

* * * * *

8050

Year	Number	Year	Number
1982	1,001	1984	2,596
1983	1,924		

* * * * *

8070

Year	Number	Year	Number
1982	1,001	1984	2,090
1983	1,430		

* * * * *

8550

Year	Number	Year	Number
1978	1,083	1980	1,553
1979	1,342	1981	1,723

* * * * *

SERIAL NUMBER LOCATIONS

160, 6040 — On plate beneath hood, ahead of fuel tank, on right side.
170, 175, 180, 185, 190, 190-XT, 210, 220 — At front of torque housing flange on either side.
440 — On plate on left side of instrument console.
5015, 5020, 5030, 5040, 5045, 5050 — On plate on the rear of the console under the steering wheel.
6060, 6080 — Right side of engine adapter housing flange.
6140 — Left side, left side of fuel tank.
7000 — Right side of differential housing.
7010, 7020, 7030, 7040, 7050, 7060, 7080, 8010, 8030, 8050, 8070 — Above PTO guard.
4W-220, 4W-305, 7580, 8550 — Rear left side of the front frame.

ALLIS-CHALMERS CRAWLER TRACTOR PRODUCTION

(This serial number listing includes certain Monarch tractors in production when Allis-Chalmers acquired Monarch in 1928). Almost all A-C crawler production was centered at Springfield, (Illinois) Works.

"F" 10-TON CRAWLER

Year	From		To
1926	10001	to	10081
1927	10082	to	10240
1928	10241	to	10250

* * * * *

"F" 75 CRAWLER

Year	From		To
1928	70001	to	70207
1929	70208	to	70654
1930	70655	to	70984
1931	70985	to	71066

* * * * *

"G" 5-TON CRAWLER

Year	From		To
1926	7001	to	7090
1927	7091	to	7117

* * * * *

"H" 6-TON CRAWLER

Year	From		To
1927	60001	to	60206
1928	60207	to	60297

* * * * *

"H" 50 CRAWLER

Year	From		To
1928	60298	to	60610
1929	60611	to	61558
1930	61559	to	62146
1931	62147	to	62297

* * * * *

"K" 35 CRAWLER

Year	From		To
1929	1	to	48
1930	49	to	1372
1931	1373	to	2333
1932	2334	to	2653
1933	2654	to	3045
1934	3046	to	3593
1935	3594	to	4792
1935	KO-6001	to	KO-6018
1936	4793	to	6335
1937	6336	to	7707
1938	7708	to	8017
1939	8018	to	8522
1940	8523	to	8956
1941	8957	to	9268
1942	9269	to	9393
1943	9294	to	9468
			None Since

Serial Numbers and Production Data

"L" AND "L-O" CRAWLER

1931	1	to	34
1932	35	to	498
1933	499	to	660
1934	661	to	887
1935	LO-4001	to	LO-4021
1935	888	to	1438
1936	1439	to	2136
1937	2137	to	2705
1938	2706	to	2943
1939	2944	to	3232
1940	3233	to	3251
1941	3252	to	3272
1942	3273	to	3357

None Since

* * * * *

"M" CRAWLER

1932	1	to	41
1933	42	to	401
1934	402	to	841
1935	842	to	1941
1936	1942	to	3841
1937	3842	to	7066
1938	7067	to	8126
1939	8127	to	9539
1940	9540	to	11379
1941	11380	to	12946
1942	12947	to	14524

None Since

* * * * *

"TW" SPEED ACE

1935	1	to	10
1936	11	to	37
1937	38	to	87

None Since

* * * * *

"S" AND "S-O" CRAWLER

1937	3	to	412
1938	413	to	584
1939	585	to	1084
1940	1085	to	1127
1941	1128	to	1211
1942	1212	to	1227

None Since

* * * * *

"LD" CRAWLER

1939	8	to	17

* * * * *

HD-3 CRAWLER

1942	3	to	32

(Numbers 20 and 30 not built. This tractor is in no way connected with the HD-3 tractor introduced in 1960).

* * * * *

H-3 AND HD-3 CRAWLER

1960	1001	to	1249
1961	1250	to	3198
1962	3199	to	6000
1963	6001	to	--------
1966	8856	to	9482
1967	9483	to	9699
1968	9700	to	--------

* * * * *

HD-4 CRAWLER

1966	1244	to	--------
1967	2690	to	--------
1968	3001	to	--------
1969	3906	to	--------

* * * * *

HD-5 CRAWLER

1946	1	to	6
1947	7	to	1357
1948	1358	to	4315
1949	4316	to	7498
1950	7499	to	11070
1951	11071	to	14289
1952	14290	to	17557
1953	17558	to	21836
1954	21837	to	25563
1955	25564	to	29255

None Since

* * * * *

HD-6 CRAWLER

1955	101	to	1146
1956	1147	to	6465
1957	6466	to	7947
1958	7948	to	10053
1959	10054	to	12505
1960	12506	to	13776
1961	13777	to	14897
1962	14898	to	16041
1963	16042	to	17104
1964	17105	to	18188
1965	18189	to	--------

* * * * *

HD-7 CRAWLER

(Includes HD-7, HD-7W, and HD-7 Military)

1940	3	to	502
1941	503	to	1136
1942	1137	to	2980
1943	2981	to	5752
1944	5753	to	8948
1945	8949	to	12090
1946	12091	to	13077
1947	13078	to	15121
1948	15122	to	16751
1949	16752	to	18085
1950	18086	to	18505

None Since

* * * * *

HD-7G CRAWLER

1961	22222	to	22360
1962	22361	to	22032
1963	22033	to	23503
1964	23504	to	24037
1965	24038	to	--------

* * * * *

HD-9 CRAWLER

1951	1	to	737
1952	738	to	1882
1953	1883	to	3590
1954	3591	to	5208
1955	5209	to	5850

None Since

* * * * *

Serial Numbers and Production Data

HD-10 CRAWLER

(Includes HD-10 and HD-10 Military).

Year	From		To
1940	2	to	641
1941	642	to	1445
1942	1446	to	2855
1943	2856	to	3651
1944	3652	to	4896
1945	4897	to	5531
1946	5532	to	6462
1947	6463	to	7601
1948	7602	to	8675
1949	8676	to	9630
1950	9631	to	10198

* * * * *

HD-10W CRAWLER

Year	From		To
1942	1446	to	2081
1943	2082	to	2338
1944	2339	to	4094
1945	4095	to	5963
1946	5964	to	6462
1947	6463	to	7601
1948	7602	to	8675
1949	8676	to	9630
1950	9631	to	———

None Since

* * * * *

HD-11 CRAWLER

Year	From		To
1955	101	to	1057
1956	1058	to	3254
1957	3255	to	4114
1958	4115	to	4767
1959	4768	to	5801
1960	5802	to	6447
1961	6448	to	6994
1962	6995	to	7869
1963	7870	to	8605
1964	8606	to	10532
1965	10533	to	———

* * * * *

HD-14 AND HD-14C CRAWLER

Year	From		To
1939	18	to	25
1940	26	to	548
1941	549	to	1165
1942	1166	to	2112
1943	2113	to	3137
1944	3138	to	4258
1945	4259	to	5454
1946 Std	5455	to	5487
1946 14-C	5488	to	5523
1946 Std	5524	to	5573
1946 14-C	5574	to	5575
1946 Std	5576	only	
1946 14-C	5577	to	5578
1946 Std	5579	only	
1946 14-C	5580	only	
1946 Std	5581	to	5583
1946 14-C	5584	only	
1946 Std	5585	to	5678
1946 14-C	5679	to	5722
1946 Std	5723	to	5747
1946 14-C	5748	to	5814
1947 14-C	5815	to	5822
1947 Std	5823	to	5847
1947 14-C	5848	to	5858
1947 Std	5859	only	
1947 14-C	5860	to	5880
1947 Std	5881	only	
1947 14-C	5882	to	5924
1947 Std	5925	only	
1947 14-C	5926	to	5930
1947 Std	5931	only	
1947 14-C	5932	to	5945
1947 Std	5946	to	5966
1947 14-C	5967	to	6022
1947 Std	6023	to	6047
1947 14-C	6048	to	6114
1947 Std	6115	to	6139
1947 14-C	6140	to	6203
1947 Std	6204	to	6222
1947 14-C	6223	to	6276
1947 Std	6277	to	6301
1947 14-C	6302	to	6339
1947 Std	6340	to	6356
1947 14-C	6357	to	6422

None Since

HD-15 CRAWLER

Year	From		To
1951	1	to	810
1952	811	to	1857
1953	1858	to	2855
1954	2856	to	3684
1955	3685	to	3909

None Since

* * * * *

HD-16 CRAWLER

Year	From		To
1955	101	to	1008
1956	1009	to	2462
1957	2463	to	2756
1958	2757	to	4143
1959	4144	to	4725
1960	4726	to	5097
1961	5098	to	5447
1962	5448	to	5731
1963	5732	to	6269
1964	6270	to	6964
1965	6965	to	———

* * * * *

HD-19 CRAWLER

Year	From		To
1947	4	to	120
1948	121	to	1195
1949	1196	to	2001
1950	2002	to	2606
1950	2617	to	2639
1950	2641	to	2653
1950	2607	to	2616
1950	2640	only	

None Since

* * * * *

HD-20 CRAWLER

Year	From		To
1951	3001	to	3827
1952	3828	to	4922
1953	4923	to	5736
1954	5737	to	6100

None Since

* * * * *

HD-21 CRAWLER

Year	From		To
1954	7001	to	7012
1955	7013	to	7915
1956	7916	to	9090
1957	9091	to	9301
1958	9302	to	11028
1959	11029	to	11681
1960	11682	to	11938
1961	11939	to	12259
1962	12260	to	12456
1963	12457	to	13249
1964	13250	to	13736
1965	13737	to	———

Serial Numbers and Production Data

650 SERIES CRAWLER

1970	6,501	1973	7,800
1971	7,183	1974	7,991
1972	7,438	1975	8,310

* * * * *

TRACTOR LOADERS, BACKHOE LOADERS, ARTICULATED LOADERS, DIRT WAGONS, GRADERS AND OTHER EQUIPMENT.

TL-6 TRACTOR LOADER

1951	1	to	5
1954	6	to	41
1955	42	to	181
1956	182	to	285
1957	286	to	344
1958	345	to	398
1959	399	to	458
1960	459	to	——

* * * * *

TL-10 TRACTOR LOADER

1950	1	to	3
1951	4	to	159
1952	160	to	384
1953	385	to	677
1954	678	to	951
1955	952	to	1288
1956	1289	to	1521
1957	1522	to	1749
1958	1750	to	1930
1959	1931	to	2065
1960	2066	to	2530
1961	2531	to	2563
1962	2564	to	2591
1963	2592	to	2641
1964	2642	to	——

* * * * *

TL-11 TRACTOR LOADER

1955	1	only	
1956	2	to	22
1957	23	to	60
1958	61	to	95
1959	96	to	122
1960	123	to	——

* * * * *

TL-12 TRACTOR LOADER

1954	1	to	166
1955	167	to	462
1956	463	to	821
1957	822	to	1179
1958	1180	to	1554
1959	1555	to	1813
1960	1814	to	2142
1961	2143	to	2253
1962	2254	to	4455
1963	4456	to	4688
1964	4689	to	4768
1965	4769	to	——

* * * * *

TL-14 TRACTOR LOADER

1959	101	to	359
1960	360	to	829
1961	830	to	1536
1962	1537	to	2003
1963	2004	to	3167
1964	3168	to	3369
1965	3370	to	——

* * * * *

TL-16 TRACTOR LOADER

1958	101	to	175
1959	176	to	410
1960	411	to	610
1961	611	to	688
1962	——	to	1461
1963	1854	to	——

TL-20 TRACTOR LOADER

1957	1	to	250
1958	251	to	460
1959	461	to	740
1960	741	to	990
1961	991	to	——
1962	——	to	1598
1963	1891	to	——

* * * * *

TL-30 TRACTOR LOADER

1961	101	to	178
1962	179	to	——
1963	——	to	691

* * * * *

TL-40 TRACTOR LOADER

1964	1244	to	——

* * * * *

TS-160 TRACTOR SCRAPER

1957	500	to	649
1958	650	to	726
1959	727	to	799
1960	800	to	849
1961	850	to	——

* * * * *

TS-200 TRACTOR SCRAPER

1950	1550	to	1585
1951	1586	to	1630
1952	1631	to	1725
1953	1726	to	1799
1954	1800	to	2095
1955	2096	to	2369
1956	2370	to	2700
1957	2701	to	2800
1958	2968	to	——

* * * * *

TS-260 TRACTOR SCRAPER

1956	2471	to	2730
1957	2731	to	2870
1958	2871	to	4074
1959	4075	to	4279
1960	4280	to	4374
1961	4375	to	4514
1962	4515	to	4897
1963	4928	to	5447
1964	5448	to	6012
1965	6013	to	——

* * * * *

TS-300 TRACTOR SCRAPER

1947	1	to	90
1948	91	to	191
1949	192	to	242
1950	243	to	545
1951	546	to	750
1952	751	to	865
1953	866	to	980
1954	981	to	1070

Serial Numbers and Production Data

TS-360 TRACTOR SCRAPER

Year	From		To
1955	1201	to	1390
1956	1391	to	1525
1957	1526	to	1640
1958	1641	to	1720
1959	1721	to	2020
1960	2021	to	2115
1962	2318	to	2377

* * * * *

TS-460 TRACTOR SCRAPER

Year	From		To
1962	3621	to	3633
1963	3634	to	3864
1964	3865	to	4068
1965	4140	to	-------

* * * * *

TS-562 TRACTOR SCRAPER

Year	From		To
1962	1733	to	1742
1963	1768	to	1831
1964	1832	to	2005
1965	2308	to	-------

* * * * *

TRACTOR DRAWN SCRAPERS

Model	Years
C-20	1946-53
C-40	1946-53
C-44	1946-48 1955-57
C-84	1947-48
C-106	1947-56
C-108	1946-57
C-114	1946-49
C-314	1949-56
C-315	1948-49 1955-57

* * * * *

AD-3 & AD-4 MOTOR GRADERS

Year	From		To
1946	381	to	780
1947	781	to	1510
1948	1511	to	2790
1949	2791	to	3420
1950	3421	to	3950
1951	3951	to	4310

* * * * *

AD-30 & AD-40 MOTOR GRADERS

Year	From		To
1952	1	to	500
1953	501	to	920
1954	921	to	1280
1955	1281	to	-------

* * * * *

BD-2 & BD-3 MOTOR GRADERS

Year	From		To
1948	1	to	550
1949	551	to	880
1950	881	to	1040
1951	1041	to	1290
1952	1291	to	1470
1953	1471	to	1549
1954	1550	to	1575
1955	1576	to	-------

* * * * *

MODEL D GRADER (Gasoline)

Year	From		To
1949	1	to	480
1950	481	to	1060
1951	1061	to	1900
1952	1901	to	3100
1953	3101	to	3605
1954	3606	to	4310
1955	4311	to	4770
1956	4771	to	5039
1957	5040	to	5300
1958	5301	to	5540
1959	5541	to	5739
1960	5740	to	5858
1961	5859	to	5985
1962	5986	to	6114
1963	6115	to	6243
1964	6244	to	6359
1965	6360	to	-------

* * * * *

MODEL DD GRADER (Diesel)

Year	From		To
1954	1000	to	1030
1955	1031	to	1220
1956	1221	to	1450
1957	1451	to	1789
1958	1790	to	1915
1959	1916	to	2025
1960	2026	to	2075
1961	2076	to	2328
1962	2329	to	2520
1963	2521	to	2938
1964	2939	to	3054
1965	3055	to	-------

* * * * *

MODEL W MOTOR GRADER

Year	From		To
1946	1026	to	-------
1948	2288	to	2813

* * * * *

MODEL AD-45 GRADER

Year	From		To
1955	1	to	210
1956	211	to	762
1957	763	to	1135
1958	1136	to	1435
1959	1436	to	1680
1960	1681	to	1837
1961	1838	to	1981
1962	1982	to	2127

* * * * *

MODEL AD-145 GRADER

Year	From		To
1959	1	to	350
1960	351	to	370
1961	371	to	382
1962	383	to	458
1963	459	to	497

* * * * *

M-60 GRADER

Year	From		To
1963	1	to	118

* * * * *

M-70 GRADER

Year	From		To
1963	1	to	140
1964	141	to	213
1965	214	to	------

* * * * *

Serial Numbers and Production Data

M-100 GRADER

Year	From		To
1962	1	to	138
1963	139	to	488
1964	489	to	786
1965	787	to	-----

Note: Complete serial number listings for some graders and other equipment have not been available after the 1965 production year.

* * * * *

10-K GRADER

Year	From		To
1932	501	to	525
1933	526	to	550
1934	551	to	640

Ended Oct. 22, 1934

ELEVATING GRADER

Year	From		To
1932	B-1	to	B-14
1933		None	
1934	B-15 and B-116	to	B-125

Ended June 21, 1934

MODEL "H" MOTOR PATROL

Year	From		To
1932	2001	to	2049
1933	2050	to	-------

Ended March 2, 1933

* * * * *

NO. 8 GRADER

Year	From		To
1935	8001	to	8060
1936		none	
1937	8061	to	8208
1938	8209	to	8210

None Since

* * * * *

NO. 9 TERRACER

Year	From		To
1935	9001	to	9030

* * * * *

NO. 10 GRADER

Year	From		To
1935	10-701	to	10-726
1936	10-743	to	10-800
1937	10-802	to	10-901
1938	10-902	to	11-001

None Since

* * * * *

NO. 14 POWER-CONTROLLED GRADER

Year	From		To
1933	501	to	525
1934	601	to	624
1935	625	to	635

NO. 14 GRADER

Year	From		To
1936	14-658	to	14-671
1937	14-672	to	14-675

* * * * *

NO. 12 POWER-CONTROLLED GRADER

Year	From		To
1936	518	to	525

* * * * *

NO. 110 GRADER

Year	From		To
1938	1-A-1	to	1-A-80
1939	1-A-81	to	1-A-200

NO. 112 GRADER

Year	From		To
1938	1-B-1	to	1-B-60
1939	1-B-61	to	1-B-116
1940	1-B-117	to	1-B-132

* * * * *

"WC" MAINTAINER

Year	From		To
1938	1-C-1	to	1-C-150
1939	1-C-151	to	1-C-400
1940	1-C-401	to	1-C-407
1941	1-C-408	to	1-C-411

* * * * *

"W" PATROL

Year	From		To
1940	1-E-1	to	1-E-466
1941	1-E-467	to	1-E-634
1942	1-E-635	to	1-E-678
1945	1-E-681	to	1-E-819
1946	1-E-820	to	1-E-1437
1947	1-E-1438	to	1-E-2278
1948	1-E-2279	to	1-E-3515
1949	1-E-3516	to	1-E-3746
1950	1-E-3747	to	1-E-3753

All machines through 1942 built at LaPorte Works; thereafter at Springfield Works.

* * * * *

WAGON CRAWLERS

Model	Years
Model X	1933-35
Model A	1931 only
Model W	1932-33

* * * * *

NO. 42 SINGLE (Axle) PATROL

Year	From		To
1933	2501	to	2642
1934	2643	to	2775
1935	2776	to	2941
1936	3031	to	3160
1937	3276	to	3325
1938	3455	to	3475
1938	3497	to	3516
1939	3517	to	3567
1940	3568	to	3575

* * * * *

NO. 42 TANDEM (Axle) PATROL

Year	From		To
1936	3161T	to	3199T
1937	3200T	to	3275T
1937	3327T	to	3361T
1938	3362T	to	3454T
1938	3476T	to	3488T
1939	3489T	to	2496T
1939	3576T	to	3625T
1940	3626T	to	3675T

* * * * *

NO. 54 SINGLE (Axle) PATROL

Year	From		To
1934	6501	to	6502
1935	6503	to	6597
1936	6628	to	6763
1937	6821	to	6824
1937	6878	to	6927
1937	6978	to	7002

A number of the above graders built in 1936 and following were built as tandem axle machines, and a few were built as tandems with the K-O diesel engine.

* * * * *

Serial Numbers and Production Data

NO. 54 TANDEM (Axle) PATROL

Year			
1936	6639	to	6777
1937	6825T	to	6827T
1937	6828T	to	6877T
1937	7003T	to	7020T
1937	7106T	to	7108T
1938	7021T	to	7052T
1938	7128T	to	7186T
1939	7187T	to	7227T
1940	7261T	to	7277T

Note: Numbers 7278T to 7315T built in 1938 with special snow blade and broom.

* * * * *

NO. 54 K-O SINGLE (Axle) PATROL
(Used A-C diesel engine)

Year			
1937 only	6817-0		
	7103-0	to	7105-0
	7109-0	to	7127-0

* * * * *

NO. 54 K-O TANDEM (Axle) PATROL

Year			
1937	6778-OT	to	6816-OT
1937	6818-OT		
1937	6928-OT	to	6977-OT
1938	7053-OT	to	7102-OT
1939	7228-OT	to	7269-OT

* * * * *

I-400 INDUSTRIAL TRACTOR

Year	Beginning Serial No.	Year	Beginning Serial No.
1966	1,001	1968	1,495
1967	1,281	1969	1,750

* * * * *

I-600 INDUSTRIAL TRACTOR

1966	1,001	1968	2,709
1967	1,747	1969	3,056

* * * * *

I-600FL (Front-end Loader)

1968	1,115	1969	1,715

* * * * *

615 BACKHOE LOADER TRACTOR

1969	501	1971	1,083
1970	660	1972	1,280

* * * * *

615 LOADER TRACTOR

1969	1,501	1972	2,503
1970	1,730	1973	2,745
1971	2,252		

* * * * *

715 BACKHOE LOADER

1973	1,001	1978	2,122
1974	1,223	1979	2,366
1975	1,494	1980	2,661
1976	1,759	1981	2,890
1977	1,938	1982	3,237

* * * * *

816 BACKHOE LOADER

1970	1,001	1973	2,632
1971	2,440	1974	2,781
1972	2,576	1975	2,852

* * * * *

918 BACKHOE LOADER

1967	1,001	1972	5,279
1968	1,005	1973	5,317
1969	1,027	1974	5,349
1970	5,107	1975	5,365
1971	5,234		

* * * * *

540 ARTICULATED LOADER

1975	1,001	1978	1,419
1976	1,072	1979	1,480
1977	1,289		

* * * * *

640 ARTICULATED LOADER

1980	1,501	1982	1,620
1981	1,593		

* * * * *

840 ARTICULATED LOADER

1970	1,001	1974	2,887
1971	1,536	1975	3,301
1972	1,960	1976	3,551
1973	2,465		

* * * * *

840-C ARTICULATED LOADER

1981	4,001	1982	4,012

* * * * *

940 ARTICULATED LOADER

1973	1,001	1975	1,478
1974	1,237	1976	1,661

* * * * *

500 SERIES FRONT LOADER

1968	501	1972	1,274
1969	687	1973	1,541
1970	857	1974	1,903
1971	1,124		

* * * * *

600 SERIES FRONT LOADER

1967	551	1971	1,900
1968	552	1972	2,005
1969	1,216	1973	2,257
1970	1,630		

* * * * *

630 SERIES FRONT LOADER

1971	3,055	1973	3,230
1972	3,120		

* * * * *

700 SERIES LOADER

1973	1,001	1975	1,521
1974	1,111		

Serial Numbers and Production Data

705 SERIES FRONT LOADER

Year		Year	
1979	2,101	1981	2,867
1980	2,520	1982	3,173

* * * * *

706 SERIES FRONT LOADER

Year		Year	
1975	1,001	1979	1,888
1976	1,032	1980	2,520
1977	1,252	1981	2,867
1978	1,534	1982	3,173

* * * * *

708 SERIES FRONT LOADER

Year		Year	
1980	2,661	1982	3,173
1981	2,867		

710 LOADER

Year	
1982	3,237

* * * * *

714 BACKHOE LOADER

Year	
1982	3,237

* * * * *

766 SERIES FRONT LOADER

Year	
1982	3,201

BUCKMASTER TRACTOR-SAW

Year		Year	
1968	501	1970	623
1969	513		

Advance-Rumely Thresher Co.

This section includes considerable production data on Advance-Rumely and its immediate predecessor, M. Rumely Company. Since Rumely acquired Advance, Gaar-Scott, and several other major companies, portions of production data concerning these companies are included. Much of this data is derived from Allis-Chalmers production records, while other sections are compiled from listings compiled by the original builders. Every effort has been made to check the accuracy of the figures herein presented, but neither the Author nor the Publisher can guarantee the authenticity of the data presented herewith.

Advance Thresher Company, Battle Creek, Michigan
Yearly Listing of Steam Engine Production.

Year	Beginning Serial No.	Year	Beginning Serial No.
1885	101	1902	6,237
1886	130	1903	6,951
1887	231	1904	7,732
1888	401	1905	8,463
1889	652	1906	9,192
1890	969	1907	9,926
1891	1,410	1908	10,578
1892	1,724	1909	11,134
1893	2,151	1910	11,690
1894	2,625	1911	1,2359
1895	2,915	1912	13,005
1896	3,304	1913	13,466
1897	3,707	1914	14,190
1898	4,016	1915	14,438
1899	4,423	1916	14,453
1900	4,957	1917	14,638
1901	5,512		last engine built

* * * * *

GAAR-SCOTT COMPANY, RICHMOND, INDIANA
STEAM TRACTION ENGINE PRODUCTION

Year		Year	
1885	4,158	1900	9,603
1886	4,331	1901	10,109
1887	4,666	1902	10,775
1888	4,817	1903	11,475
1889	5,118	1904	11,967
1890	5,519	1905	12,548
1891	5,861	1906	13,071
1892	6,362	1907	13,597
1893	no data	1908	14,052
1894	7,233	1909	14,363
1895	7,519	1910	14,738
1896	7,842	1911	15,228
1897	8,216	1912	15,608
1898	8,568	1913	16,182
1899	9,061	1914	16,544
			None Since

M. RUMELY COMPANY
STEAM TRACTION ENGINE PRODUCTION

Year	From		To
1895	2600	to	2725
1896	2726	to	2878
1897	2879	to	3049
1898	3050	to	3218
1899	3219	to	3297
1900	3438	to	3640
1901	3641	to	3855
1902	3856	to	4136
1903	4137	to	4374
1904	4375	to	4435
	4437		
	4439	to	4444
	4451		
	4458	to	4462
	4466	to	4468
	4473	to	4477
	4482	to	4483
	4506	to	4529
	4532	to	4539
1905	4436		
	4445	to	4450
	4452		
	4455	to	4457
	4469	to	4472
	4478		
	4480	to	4481
	4484	to	4505
	4530	to	4531
	4540	to	4628
1906	4438		
	4453	to	4454
	4463	to	4465
	4629	to	4764
1907	4479		
	4765	to	4952
1908	4953	to	5184
	5186	to	5195
1909	5185		
	5196	to	5587
1910	5588	to	5920
1911	5921	to	6287
1912	6288	to	6588
1913	6589	to	6714
1914	6715	to	7036
1915	7037	to	7038
1916	7039	to	——

Advance-Rumely Thresher Co.

M. RUMELY AND ADVANCE-RUMELY WOOD SEPARATORS

(Wood Separators not listed by sizes).

Year	From		To
1904	5397	to	5399
1905	5400	to	5537
	5387	to	5396
1906	5538	to	5815
1907	5816	to	6157
1908	6158	to	6544
1909	6545	to	7064
1910	7065	to	7676
1911	7677	to	8564
1912	8565	to	10202
1913	10203	to	11359
1914	11360	to	13439
1915	13440	to	13816
1916	13817	to	15107
1917	15108	to	16161
1918	16162	to	17730
1919	17731	to	20001
1920	20002	to	22661
1921	22662	to	24065
1922	24066	to	24935
1923	24936	to	26050
1924	26051	to	26120
	None Since		

* * * * *

ADVANCE-RUMELY STEEL SEPARATORS
Also includes
ALLIS-CHALMERS SEPARATORS 1931 AND LATER.

(Steel separators not listed by sizes, 1916-1928).

Year	From		To
1916	24958	to	25557
1917	25558	to	25658
1918	25659	to	25784
1919	25785	to	25896
1920	25897	to	26116
1921	26117	to	26287
1922	26288	to	26537
1923	26538	to	27352
1924	27353	to	28427
1925	28428	to	30427
1926	30428	to	30622
1927	32623	to	33406

22 x 36 STEEL SEPARATOR

Year	From		To
1928	33407	to	33706
	34157	to	34356
	34597	to	34746
1929	34892	to	35131
	35332	to	35431
1930	36420	to	36494
1931	None Built		
1932	36495	to	36574
1933	36778	to	36807

22 x 38 STEEL SEPARATOR

Year	From		To
1932	36675	to	36676
1933	36808	to	36817
1934	36873	to	37022
1935	37173	to	37323
1936	37674	to	38173
	None Since		

28 x 44 STEEL SEPARATOR

Year	From		To
1928	33707	to	33806
	None Since		

28 x 46 STEEL SEPARATOR

Year	From		To
1928	33807	to	34106
	34357	to	34596
	34747	to	34981
1929	35132	to	35331
	35432	to	35731
	35807	to	35946
1930	35947	to	36306
1931	None Built		
1932	36575	to	36674
	36677	to	36734
1933	36735	to	36752
	36754	to	36777
	36823	to	36872
1934	None Built		
1935	37023	to	37162
	37324	to	37373
1936	37374	to	37673

28 x 50 STEEL SEPARATOR

Year	From		To
1933	36753		
	36818	to	36822
1934	None Built		
1935	37163	to	37323
	None Since		

32 x 52 STEEL SEPARATOR

Year	From		To
1928	34107	to	34156
	None Since		

36 x 60 STEEL SEPARATOR

Year	From		To
1929	35732	to	35806
1930	36307	to	36419
	None Since		

ADVANCE-RUMELY CORN SHELLER

Year	From		To
1925	1	to	43
1926	44	to	320
	None Since		

ADVANCE-RUMELY CORN HUSKER-SHREDDER

Year	From		To
1901	101	to	150
1902	151	to	280
1903	281	to	422
1904	423	to	522
1905	523	to	626
1906	627	to	722
1907	723	to	861
1908	862	to	966
1909	967	to	1027
1910	1028	to	1101
1911	1102	to	1593
1912	1594	to	1779
1913	1780	to	2168
1914	2169	to	2180
1915	2181	to	2270
1916	2271	to	2309
1917	2310	to	2510
1918	2511	to	2579
1919	2580	to	2603
1920	None Built		
1921	None Built		
1922	2604	to	2703
1923	2704	to	2903
1924	2904	to	3203
1925	3204	to	3353
1926	3353	to	3570
1927	3571	to	3603
1928	3604	to	3653
	None Since		

ADVANCE-RUMELY NO. 1 PRAIRIE COMBINE

Year	From		To
1925	109	to	110
1926	121	to	320
1927	323	to	422
	None Since		

* * * * *

ADVANCE-RUMELY NO. 1 SIDEHILL COMBINE

Year	From		To
1925	101	to	108
	111	to	120
1926	121	to	320
1927	723	to	847
1928	1815	to	1914
1929	4866	to	4939
	None Since		

Advance-Rumely Thresher Co.

ADVANCE-RUMELY NO. 2 PRAIRIE COMBINE

Year	From		To
1926	321	to	322
1927	423	to	722
	850	to	1214
1928	1215	to	1414
	1615	to	1814
	2615	to	3014
1929	3116	to	3415
	4216	to	4865
	5031	to	5180
	6031	to	6155
	6231	to	6280
1930	6156	to	6230
	6281	to	6545

None Since

ADVANCE-RUMELY NO. 3 PRAIRIE COMBINE

Year	From		To
1927	848	to	849
1928	1415	to	1614
	1915	to	2614
	3015	to	3114
1929	3416	to	4215
	5331	to	5650
1930	5651	to	6030
	6606	to	6655
	6766	to	6815

ADVANCE-RUMELY NO. 3 HILLSIDE COMBINE

Year	From		To
1929	5181	to	5330
1930	6656	to	6765
1931	6546	to	6605
	6818	to	6833
1936	12398	to	12547

None Since

ADVANCE-RUMELY NO. 4 PRAIRIE COMBINE

Year	From		To
1928	3115		
1929	4931	to	5030

None Since

A-C "40" ALL-CROP HARVESTER

Year	From		To
1938	101	to	200
1939	201	to	6200
1940	6201	to	15200

None Since

* * * * *

A-C "60" ALL-CROP HARVESTER

Year	From		To
1935	6848	to	7397
1936	7398	to	15050
1937	15051	to	23800
1938	23829	to	26695
1938	26696	to	42015
1939	42016	to	49831
1940	49837	to	58586
1941	60105	to	71391
1942	71392	to	86336
1943	86337	to	91162
1944	95389	to	104449
1945	104450	to	120473
1946	120474	to	128763
1947	128764	to	149588
1948	149589	to	168563
1949	168564	to	183923

None Since

* * * *

A-C "60" SIDE HILL HARVESTER

Year	From		To
1940	58587	to	58836

* * * * *

ADVANCE-RUMELY OILPULL TRACTORS

"B" 25-45 OILPULL

Year	From		To
1910	1	to	100
1911	2101	to	2269
1912	2270	to	2936

"E" 30-60 OILPULL

Year	From		To
1910	101	to	236
1911	237	to	746
1912	747	to	1678
1913	1679	to	1787
1914	None Built		
1915	1819	to	2018
1916	2019	to	2100
1917	2997	to	8724
1918	8725	to	8902
1919	11500	to	11596
1920	2252	to	2351
1921	2352	to	2402
1922	2404	to	2453
1923	2454	to	2503

None Since

"F" 18-35 OILPULL

Year	From		To
1911	5001	to	5680
1912	5681	to	6738
1913	6739	to	7487
1914	7500	to	7856
1915	None Built		
1916	7857	to	8084
1917	8085	to	8591
1918	8903	to	9177

None Since

* * * * *

"G" 20-40 OILPULL

Year	From		To
1918	10425	to	10750
1919	10751	to	15221
1919	G741	to	G948
1920	G949	to	G1727
1921	G1728	to	G2241
1922	G2242	to	G2689
1923	G2690	to	G3558
1924	G3559	to	G3894

None Since

* * * * *

"H" 16-30 OILPULL

Year	From		To
1917	8627	to	8699
1918	9178	to	10710
1919	10711	to	16284
1919	H3751	to	H4392
1920	H4393	to	H7239
1921	H7240	to	H7395
1922	H7396	to	H8645
1923	H8646	to	H9045
1924	H9046	to	H9645

None Since

* * * * *

"K" 12-20 OILPULL

Year	From		To
1918	12000	to	12100
1919	12101	to	13656
1920	13657	to	15100
1920	16836	to	17639
1921	17640	to	18648
1922	18649	to	19268
1923	19269	to	20510
1924	20511	to	21018

None Since

* * * * *

Advance-Rumely

ADVANCE-RUMELY DO-ALL TRACTOR

Year	From		To
1928	501	to	700
1929	701	to	2115
1930	2116	to	3513
1931	3514	to	3693
	None Since		

RUMELY 6-A TRACTOR

Year	From		To
1930	501	to	502
1931	503	to	1302
	None Since		

"L" 15-25 OILPULL

Year	From		To
1924	1	to	10
1925	11	to	1606
1926	1607	to	4213
1927	4214	to	4855
	None Since		

"M" 20-35 OILPULL

Year	From		To
1924	1 only		
1925	2	to	1013
1926	1014	to	3084
1927	3085	to	3671
	None Since		

"R" 25-45 OILPULL

Year	From		To
1924	1 only		
1925	2	to	138
1926	139	to	647
1927	648	to	761
	None Since		

"S" 30-60 OILPULL

Year	From		To
1924	1	to	4
1925	5	to	34
1926	35	to	234
1927	235	to	434
1928	435	to	514
	None Since		

"W" 20-30 OILPULL

Year	From		To
1928	1	to	2128
1929	2129	to	3733
1930	3734	to	3952
	None Since		

"X" 25-40 OILPULL

Year	From		To
1928	X-1	to	X-1545
1929	1546	to	2259
1930	2260	to	2400
	None Since		

Note: The following "X" tractors were converted from "M", 20-35 OilPull tractors built earlier. 1 — 125, 757, 773, 1616, 1637-37, 1800, 1860, 1870, 1955-57, 1970, 2020-22, 2040, 2046, 2080-82, 2100, 2150, 2200-12, 2220, 2225.

"Y" 30-50 OILPULL

Year	From		To
1929	1	to	245
	None Since		

Note: Tractors Y-1 to Y-100 and Y-221 to Y-245 were converted from "R" tractors to "Y" tractors.

* * * * *

"Z" 40-60 OIL PULL

Year	From		To
1929	1	to	215
	None Since		

Note: Tractors Z-1 to Z-62 were converted from "S" tractors to "Z" tractors, and later converted back again to "S" tractors.

* * * * *

Credits

Compiling a book of this size is certainly not the product on one individual. Without the help of many, many people, this project could never have begun, much less have ultimately been concluded.

At the very top of the list stand C.W. Parker Jr. and Gerald Purser, both of Allis-Chalmers Corporation. Mr. Purser was most helpful and cooperative during our visits to West Allis. An immense amount of insightful history of farm implement developments was gained from Charles Schwab of Allis-Chalmers. Roger Higgins and Don Hurn of the company's Harvey Works provided an immense amount of information, and Mr. Hurn continued to offer additional help, even after Harvey Works had closed. Doug Zoerb at West Allis was also very helpful, as were dozens of others who gave insight, help, and encouragement. To all of you, we extend our thanks.

Our research also included a visit to the company's LaPorte Works and to the LaPorte County (Indiana) Historical Society. In each instance, the help we received was very gratifying, and to all of the LaPorte Works people with whom we worked, plus the many folks at the Historical Society, our gratitude is beyond words.

Also providing immense help and support were a great many individuals. Without these great friends, the job could never have been done, and so we thank Jerry Erickson, Gus Ward, Richard Grimm, Ken Lage, Richard Bockwoldt, Elmer Geigle, and Norm Meinert. Our appreciation is extended to Alan Jones, Lyle Wacker, Wayne Cole, Ron Downer, and Oscar O. Cooke. Additional assistance came from Roy Becker, Dan and Wendell Peterson, Leo Brubaker, Lyle Dumont, and Cass Bowyer. Our thanks to these people. Also our thanks to those who wish to remain anonymous, but to each and all of you, whether named or not, again, our sincere appreciation.

Finally, my personal thanks to George Dammann at Crestline Publishing. His help on this project has run the gamut from shooting countless photographs to evaluating materials and working with this writer in developing the scope and sequence of the entire book.

It's been a long and arduous task, but the kindness and support of each and all of you makes this writer proud to count you as friends!

C.H. Wendel

March 18, 1988

Bibliography

Periodicals

Age of Steel, St. Louis, Missouri
American Machinist, New York, New York
American Thresherman, Madison, Wisconsin
Chilton Tractor Journal, Philadelphia, Pa.
Engineering, London, England
Engineers & Engines Magazine, Joliet, Illinois
Farm Implement News, Chicago, Illinois
Gas Engine Magazine, Lancaster, Pennsylvania
Gas Power Magazine, St. Joseph, Michigan
Implement & Tractor, Kansas City, Missouri
Thresherman's Review, St. Joseph, Michigan
Thresher World, Chicago, Illinois

Books & Manuscripts

Abernathy, R. James, Practical Hints on Mill Building: Privately Published, 1880.

Adams, Edward Dean, Niagara Power: History of the Niagara Falls Power Company, 2 volumes: Privately Printed, 1927.

Allis-Chalmers Mfg. Co., Unpublished manuscripts and other data from company archives.

Benjamin, Park, Appleton's Cyclopedia of Applied Mechanics, 2 volumes, New York: D. Appleton & Co., 1882.

Bjorling, Philip R., Pumps: Historically, Theoretically, and Practically Considered, London: E. & F. N. Spon, 1890.

Crosby, R.C., A Decade of A-C Pioneering, 1930-1940, undated manuscript.

Currie, Barton W., The Tractor and its Influence upon the Agricultural Implement Industry, Philadelphia: Country Gentleman, 1916.

Diesel and Gas Turbine Catalog, Milwaukee: Diesel Engines Inc., 1981.

Ellis, L.W., and Rumely, Edward A., Power and the Plow, New York: Doubleday, Page & Co., 1911.

Ewbank, Thomas, A Descriptive and Historical Account of Hydraulic and Other machines for Raising Water, New York: Derby & Johnson, 1858.

Gibson, Louis H., Gradual Reduction Milling, Minneapolis: C.M. Palmer, Publisher, 1885.

Gueldner, Hugo, Das Entwerfen und Berechnen der Verbrennungskraftmaschinen und Kraftgas-Anlagen, Berlin: Verlag von Julius Springer, 1914.

Horner, Joseph G., Henley's Encyclopedia of Practical Engineering and Allied Trades, 5 volumes, New York: Norman W. Henley, 1906.

Johnson, Alberta Price, Mill Stones to Atom Smashers, 5 typed volumes, unpublished manuscript in A-C Archives.

Loeb, Wilhelm, Encyclopaedie des Gesammten Landwirthschaft, 6 volumes, Leipzig: Verlag von Otto Wiegand, 1850.

Martin, Harold M., Design and Construction of Steam Turbines, New York: Longmans, Green & Co., 1913.

Moody's Manual of Railroad and Corporation's Securities, 2 volumes, New York: Poors Publishing Co., 1924.

Morrison, L.H., American Diesel Engines, New York: McGraw-Hill, 1931.

Diesel Engine Handbook, New York: Diesel Publications, 1943.

Mueller, Wilh., Die Francis-Turbinen. Hannover: Verlag von Gebruder Janecke, 1905.

Peterson, Walter F., An Industrial Heritage, Milwaukee: Milwaukee County Historical Society, 1978.

Richards, Robert H., Textbook of Ore Dressing, New York: McGraw-Hill, 1909.

Shillito, F.W. Jr., Handbook of the Corliss Steam Engine, Bridgeport: American Industrial Publishing Company.

A Note to the Reader

Virtually all of the references cited in this bibiliography are long since out of print. Additional reference to any of the above titles may possibly be secured in selected libraries. Neither the Author nor the Publisher maintains facilities or materials for outside research projects.

The above listings include at least two contemporary magazines. They are: *Gas Engine Magazine and Engineers & Engines Magazine*. Further information may be obtained by contacting these publications directly.

This bibliographical listing contains the major outside reference sources used in this book, with the Allis-Chalmers Archives at West Allis remaining as the primary source. Virtually all the materials used in this book were photographed or photocopied from material in the A-C Archives or with permission of department managers. Considerable information was obtained from the A-C Legal Department.

Illustration, use of, or reference to any Allis-Chalmers trademarks, slogans, or other information which may actually be the property of Allis-Chalmers is purely incidental and is used as a part of a larger overall presentation. The appearance of any such trademarks, slogans, or any other images in connection with A-C products does not in any way imply exclusive ownership to the above, nor does it in any way imply approval of historical context or editorial comment by Allis-Chalmers or its successors. For further information regarding the use of any Allis-Chalmers photographs, logos, or other images, it is suggested that Allis-Chalmers and/or its successors be contacted.

About the author

Completion of this title makes a total of five books on various phases of agricultural history that have been compiled by C.H. Wendel and published by Crestline Publishing Company. In addition, Mr. Wendel has written several small, specialized titles which have been self-published.

Other writing projects include *Reflections*, a regular column in *Gas Engine Magazine*, a monthly journal published at Lancaster, Pennsylvania. The *Reflections* column is comprised mainly of questions by various readers — questions covering virtually all phases of engine and tractor development.

Nearly constant research projects over the past quarter century have resulted in a sizeable research collection which Mr. Wendel has amassed. This data, including a 100-plus year run of the *Patent Office Gazette* has been invaluable for this project as well as those of previous years. Also on hand are hundreds of individual titles relating to steam, gas, water, and electrical power. A virtually complete run of *American Machinist*, plus another complete run of *Engineering* grace the shelves. Recent library acquisitions include an extensive run of *Zeitschrift des Vereines Deutscher Ingenieure* plus a great many issues of *Zeitschrift fur Werkzeugmaschinen und Werkzeuge*. Both of these German publications add immeasurably to the historical and technical background of virtually all phases of mechanical engineering. A major project, and one not yet completed, is to reference pertinent data from these various titles into a computerized data base.

Mr. Wendel has been active for over 30 years in the preservation and restoration of vintage gas engines and tractors. Although there are presently no tractors on hand, the engine collection runs from a small ½ horsepower air-cooled model of the 1920s up to a gigantic Fairbanks-Morse Model 32 diesel engine. The diesel was obtained from a municipal light plant and was saved from a probable trip to the scrap yard. This big engine has cylinders with a 14 x 17 inch bore and stroke, and a rated speed of 257 rpm. Total weight approaches 80,000 pounds!

At the other end of the scale, Mr. Wendel spends sparetime winter hours with the research collection. Due to the poor condition of many titles, Mr. Wendel learned the art of bookbinding in order to restore some of the rather rare books to useable condition. A logical followup was the addition of a century-old printing press, and many, many cases of oldtime lead type.

As might be expected, Mr. Wendel is an inveterate book lover. This has resulted in establishment of his own bookstore, Old Iron Book Co. Located in Amana, Iowa, and in the very heart of Iowa's Amana Colonies, this establishment offers a unique blend of an outstanding bookstore and Old World hospitality.

Continuing research projects demand ever-increasing amounts of material in the form of books, magazines, catalogs, and literature pertaining to the agricultural and engineering industries. If you wish to dispose of any such items, please contact Mr. Wendel directly. Likewise, should you have any comments regarding this book, or discover any errors or ommissions, please write to: C.H. Wendel, R. R. 1, Box 28-A, Atkins, IA 52206.